Fundamental & Applied Aspects of Modern Physics

Lüderitz 2000

Proceedings of the International Conference on
Fundamental and Applied Aspects of Modern Physics

Fundamental & Applied Aspects of Modern Physics

Lüderitz, Namibia   13 – 17 November 2000

Editors

S. H. Connell & R. Tegen
University of the Witwatersrand, Johannesburg
South Africa

World Scientific
New Jersey • London • Singapore • Hong Kong

Published by

World Scientific Publishing Co. Pte. Ltd.
P O Box 128, Farrer Road, Singapore 912805
USA office: Suite 1B, 1060 Main Street, River Edge, NJ 07661
UK office: 57 Shelton Street, Covent Garden, London WC2H 9HE

British Library Cataloguing-in-Publication Data
A catalogue record for this book is available from the British Library.

FUNDAMENTAL AND APPLIED ASPECTS OF MODERN PHYSICS

Copyright © 2001 by World Scientific Publishing Co. Pte. Ltd.

All rights reserved. This book, or parts thereof, may not be reproduced in any form or by any means, electronic or mechanical, including photocopying, recording or any information storage and retrieval system now known or to be invented, without written permission from the Publisher.

For photocopying of material in this volume, please pay a copying fee through the Copyright Clearance Center, Inc., 222 Rosewood Drive, Danvers, MA 01923, USA. In this case permission to photocopy is not required from the publisher.

ISBN 981-02-4589-0

Printed in Singapore by World Scientific Printers

PREFACE

The International Conference on "Fundamental and Applied Aspects of Modern Physics : Lüderitz 2000" took place from the 13th to the 17th of November 2000 in the town of Lüderitz, Namibia. The conference programme both reflected and celebrated the lifelong contribution to Science of one of Southern Africa's most eminent scientists, Professor Jacques Pierre Friederich (Friedel) Sellschop. The scope of the conference, therefore, covered research in atomic, nuclear, elementary particle and astro- physics. Solid state physics featured as well, particularly regarding those aspects where the applications of techniques from the basic disciplines mentioned previously enabled new insights or new technologies. Diamond physics had a special place in this regard. The contributions on science policy marked an additional important theme.

Over one hundred delegates from around the world, including 20 Southern African students participated in Lüderitz 2000. Many of the delegates were very senior and eminent scientists indeed (both from the theoretical and the experimental fraternity), attracted to the conference variously by their scientific and personal relationship with Friedel, by the conference programme, or by a common interest in building science capacity in Africa. Accordingly, the calibre of their contributions has ensured the high standard of these proceedings, and made its compilation a great pleasure. We may whet the reader's appetite for these proceedings by mentioning briefly only some of the highlights :

A keenly anticipated presentation "A confrontation with infinity" was delivered by the 1999 Nobel laureate in Physics, Gerard 't Hooft from Utrecht (The Netherlands). In his paper Gerard 't Hooft reviewed the physics concepts that allowed him to "tame the infinities" that previously plagued theories of the weak interaction. This talk, based on his Nobel lecture, was presented with clarity and insight, and it concluded with a discussion of future directions in our quest for the understanding of basic forces and material particles.

Lüderitz 2000 took place at a most appropriate time for John Ellis to review very recent results from the accelerator facility LEP (i.e. ALEPH and the L3 collaborations at CERN, Geneva) in its dramatic swansong. Tantalising candidate events representing the possible direct observation of the elusive Higgs particle had been seen. John Ellis' overview of the physics of this field was luminary, and set the backdrop for other exciting related talks.

The interface of nuclear and elementary particle physics, probing our understanding of extreme states of matter was richly represented, in overview by Walter Greiner, and then in detail by many other researchers. They showed clearly this is a productive field theoretically, justifying the many new large scale experimental investigations now being mounted at new or soon to be commissioned international facilities.

Coming down in energy, the delegates were treated to Paul Kienle's revelation of the first clear observation of deeply bound pionic states within the nucleus. The editors appreciate his group's decision to use these proceedings as their vehicle for the first announcement of this result.

Interfacing materials and high energy physics, Achim Richter demonstrated the beautiful physics progress that can be achieved with finesse. This demonstrates his clear and original physics intuition (his award of the prestigious Stern-Gerlach medal of the *German Physical Society* was announced to the Conference). His monograph on radiation physics with diamond is most instructive. Others gave talks in a similar vein. For example, Erik Uggerhøj showed how, at very high energies, diamond becomes a laboratory for QED. Next, Mayda Velasco carried this baton further, by expounding on the possibility of a Higgs factory, based on a gamma-gamma collider, most possibly exploiting diamond radiators. Thus, solid state aspects of diamond physics were linked back to high energy physics.

Space constraints prohibit further review of the scientific material of the proceedings. Suffice to say that the diversity of the material nonetheless forms a very coherent whole, in fitting tribute to Friedel. Indeed, these proceedings record most of the 53 oral and 18 poster presentations in a remarkable volume of very excellent material. The interdisciplinary nature of several presentations made it difficult to assign some manuscripts to appropriate chapters. To facilitate the search for a specific paper we have listed the papers under the name of the presenting author (appending *et al.* in the case of multiple authorship). Quite fittingly we have added after Chapter VIII several personal addresses that were made at the conference dinner and during the Closing Ceremony.

A special attempt was launched to harness the manpower development opportunity presented by the exposure of Southern African students to so many experienced and well placed international delegates. This was done via a mentoring scheme, partnering senior physicists to young physicists, to maximize their absorption of the scientific material and development of international linkages. The stark beauty of the Lüderitz environment proved to be the ultimate ice-breaker, and the international delegates took their mentoring responsibilities very seriously. The students made their appreciation known in a moving impromptu speech at the conference banquet. Once again, we have to thank our delegates not only for their scientific contribution, but also for their mentoring role (which we know is continuing in many cases) to the students at the conference.

What role can Science play in nurturing a culture of learning and developing the human resources of Africa? In the light of the pan-African interests of Friedel Sellschop, long before it became popular to do so, the conference appropriately dedicated some sessions to exploring science policy issues relevant to research capacity building and the "African Renaissance". We are indebted to the many high ranking politicians and science policy makers from both South Africa and Namibia, who gave their support to this aspect of the conference and travelled the long distance to the conference venue. In particular, Friedel's close friend, the South African Minister of Arts, Culture, Science and Technology, Dr. Ben Ngubane, and his Director General, Dr. Rob Adam, participated in part of the Conference. Dr Ben Ngubane spoke on the importance of developing a visionary science and technology policy and what the South African government plans to do in conjunction with similar efforts in the SADC region (Namibia, Botswana, Zimbabwe, Zambia, Malawi, Mozambique, Lesotho, Swaziland, and South Africa). The Namibian Deputy Minister on Higher Education, Training and Employment Creation, the Hon. Buddy Wentworth, addressed this issue from the Namibian side. The Regional Councillor for the Karas Region, the Hon. Fluksman Samuehl addressed some of the opportunities and problems of the Karas Region around Lüderitz, and went on to propose that Friedel Sellschop should, with ``the support of the science fraternity'', set up a center to be known as the ``Professor Friedel Sellschop Science Centre'' in Namibia. Under ``his wise guidance, such a center could develop important links with other institutions of higher learning in Southern Africa and elsewhere''. Some of the contributions to the science policy debate are archived within this volume.

While the essence of a conference and of the proceedings is the scientific material presented in talks and subsequent papers, there wouldn't be a Conference or proceedings without extensive sponsorship. The editors record the deep appreciation of the Conference for the funding received from the various organisations displayed on the previous page. Their contributions enabled the Conference to support some of the international delegates, some of the Southern African participants and also the student participation. In particular, their support enabled the Conference to be held at the remote but beautiful venue of Lüderitz. In addition, several national research foundations and individual researchers funded (in some cases) quite a number of researchers and/or graduate students, who could otherwise not have come to Lüderitz.

The idea of a conference in honour of Friedel Sellschop during the year of his 70th birthday had its genesis during a discussion one of us (R.T.) had with Walter Greiner on a rainy and cold day in the November of 1998, in Walter Greiner's Frankfurt office. The sunny desert town of Lüderitz was eventually chosen as the Conference venue as it was indeed the birthplace of Friedel. It also formed a backdrop for the partial focus on diamond physics as well as on development in the Southern African region. Most importantly, it proved to be an outstanding Conference venue, contributing to the scientific deliberations and the new relationships and linkages that were forged, especially in respect of the students. The logistical problems of organizing an international conference at a remote venue with relatively little infrastructure were overcome by careful planning and co-ordination. The organisers were deeply satisfied by the spirit and enthusiasm of the delegates during the conference. The entire town of Lüderitz participated in the success of the conference and deserves our deepest appreciation.

We would like to thank the International Advisory Committee, chaired by Walter Greiner, and the members of the Local Organizing Committee for their contribution to the scholarly standard and organisation of Lüderitz 2000.

Finally the secretarial help by Maddalena Teixeira, of the *Nuclear and Elementary Particle Theory Group,* in getting all contributions in the correct format is highly appreciated. The front and back page design is due to Susan Sellschop.

The editors, Johannesburg, March, 2001

Simon Connell and Rudolph Tegen

ORGANIZING COMMITTEE :

S. H. Connell, Johannesburg (Chair)
R. Tegen, Johannesburg (Vice-Chair)
R. Adam, Johannesburg
K. Bharuth-Ram, Durban
R. Caveney, Johannesburg
N. Comins, Pretoria
M.D. Dlamini, Swaziland
E. Friedland, Pretoria
A.J. Lopes, Mozambique
C.C.P. Madiba, Pretoria
J. Malherbe, Pretoria
M. Mujaji, Zimbabwe
J.S. Nkoma, Botswana

J. Oyedele, Namibia
A. Paterson, Petoria
J. Sellschop, Johannesburg
R. Sellschop, Johannesburg
S. Sellschop, Johannesburg
E. Sideras-Haddad, Johannesburg
S. Sofianos, Pretoria
B. Spoelstra, Zululand
S. Tlali, Lesotho
R. Utui, Mozambique
F. van der Walt, Pretoria
Z. Z. Vililakazi, Cape Town

INTERNATIONAL ADVISORY COMMITTEE :

W. Greiner, Germany (Chair)
J. Als-Nielsen, Denmark
H.H. Andersen, Denmark
J-U. Andersen, Denmark
T. Anthony, South Africa
R. Arndt, South Africa
D.A. Bromley, USA
W. Brown, USA
S.H. Connell, South Africa
M.F. Da Silva, Portugal
J.A. Davies, Canada
G. Dracoulis, Australia
K. Elsener, Switzerland
L. Feldman, USA
A. Freund, France
E. Gadioli, Italy
J. Hamilton, USA
M. Kamo, Japan
P. Kienle, Germany

W.R. Kropp, USA
H. Kanda, Japan
A.E. Litherland, Canada
J. Mayer, USA
W. Mitchell, USA
F. Plasil, USA
A. Richter, Germany
W. Scheid, Germany
F. Seitz, USA
S. Sie, Australia
P. Sigmund, Denmark
T. Suzuki, Japan
R. Tegen, South Africa
C. Toepffer, Germany
E. Uggerhøj, Denmark
E. Vogt, Canada
A. Wolfendale, UK
J. Ziegler, USA

Patronages and Sponsorships

List of Sponsors : The Conference is indebted to the following organisations

De Beers Fund Educational Trust

National Research Foundation (NRF)

Department of Arts Culture Science and Technology

The University of the Witwatersrand - Research Office

The International Atomic Energy Agency

Namdeb Diamond Corporation - Namibia

The National Accelerator Centre

The Alexander Von Humboldt Stiftung

Council for Scientific and Industrial Research

The Schonland Research Centre for Nuclear Sciences

The University of Cape Town - Physics Department

NECSA : the Nuclear Energy Corporation of South Africa (former Atomic Energy Corporation of South Africa)

The University of the Witwatersrand - Physics Department

Namibia Breweries

**This volume is dedicated to
Friedel Sellschop
on the occasion of his seventieth birthday**

Walter Greiner's Dinner Speech: Prologue and Poem

Prologue

Frankfurt's Johann Wolfgang Goethe University,
named after Germany's greatest poetical son,
that's where I teach for more than 1/3 century
and therefore feel obliged to tell a poem; let's hear on

Not like Goethe's is my verse;
remember, I'm a physicist,
which Goethe also tried to be, but worse.
Nevertheless, I couldn't resist.

Friedel Sellschop

At Lüderitz in South-West
We celebrate a Fest:
Friedel's seventieth year
this has brought us here.

Father and mother lived at this bay,
included their children in their pray
shielded and helped them and watched out,
that they became strong; they were so proud.

Born in the midst of diamonds and
around are lots of sand,
he first was small
but soon became tall,

And he took into his hands
his last few Rands.
Friedel moved with his parents away
and chose finally Johannesburg to stay.

At Pretoria the bachellor,
at Stellenbosch the master,
this is what he was out for.
I tell you: it could'nt go faster

His studies were comprehensive; rich
and were completed at Cambridge,
yet, deep in his heart
he felt South African, and he was smart.

At the Witwatersrand
he became Professor young,
built up a nuclear center
and many young could enter.

A school of physics emerged
where various topics were searched:
neutrinos deep down under
was one of the early wonders.

Nuclear structure at the tandem
was a very central item
as were protons and ions
when channelling in crystals like diamonds.

They move straight or bent
and wherever they went
it was for millimeters only,
for Friedel this was most important, holy!

In South Africa at large
Friedel is the founder
of nuclear applied physics and its march
throughout; it could'nt be sounder.

In Aarhus, Frankfurt and Geneva,
In England, France and Hungaria
Friedel had friends and Collaborators everywhere
Even at Yale, at Oak Ridge and Los Alamos: they were there.

International collaboration
this is the way to achieve
not only within the South African nation
progress in science as we all believe.

Friedel's first honoris causa
came from Frankfurt university.
It was there in Goethe's aula
where tribute was given to his activity.

Others followed after years
Stellenbosch and Capetown
they switched to higher gears
and named him doctor of their own

Friedel is a lucky one:
at his side is Sue
who radiates all along
happiness: that is the clue

to his life so rich
giving him strength and dedication
to overcome each
obstacle in health and life's rotation

First two girls, and then two sons
beautiful, handsome and strong
it is a happy family
as we all can see.

Whenever we were in the Sellschop home
we were imbedded warm
and certainly felt never alone,
were overwelmed by Sue's charm.
Liberal in their thinking
patriotic for their land so fine
and occasionally also drinking
an excellent South African wine.

Apropos wine, there's one even finer
it doesn't come from Parl at Cape
no, it's a Friedelsheimer
believe me, this is well made.

We brought it along
all the way so long
as a present for our friend
in that wonderful land.

Friedel, a great scientist
right in our midst.
With imagination and strong will
he could his dreams fulfill.

He's loyal to family and friends
doesn't follow popular trends,
helps wherever he can:
What a man!

> Many happy years
> full of joy and without tears
> Friedel, God bless you
> and also your family and your dear, your Sue.

All the best, my friend!

Laudatio

Friedel Sellschop's scientific achievements are too numerous to be covered in depth here. We mention only a few highlights starting in some detail with the beginning of his distinguished career. In the early 1960s, shortly after Reines and Cowan's detection (14.6.1956) of the first man-made anti-neutrino, Friedel had the foresight to recognize the significance of neutrino research. A collaboration between Reines' and Sellschop's group in Cleveland and Johannesburg, respectively, discovered the first naturally occurring (muon) neutrinos (23.2.1965) in the deepest mine (3200m) at that time with the largest detector at that time (200 000 liters of light-oil scintillator fluid). The first photo* shows the young founding director of the *"Schonland Research Institute for Nuclear Sciences"*, Friedel Sellschop, an engineer from Colorado, John Reid, and the assistant general manager

---

* Photo taken from the article by Johnson and Tegen in S.Afr.J.Sc.**95** (1999) 13-25.

of the *East Rand Proprietary Mine* (ERPM) near Johannesburg, Fred Müller (from left). The plaque commemorating the detection of the first neutrino in nature in 1965, is kept in the main building of the ERPM. The plaque reads *'Detection of the first neutrino in nature on 23rd February 1965 in East Rand Proprietary Mine. This discovery took place in a laboratory situated two miles below the surface of the earth on 76 level of East Rand Proprietary Mine, manned by a group of physicists from the Case Institute of Technology, U.S.A., and the University of the Witwatersrand, Johannesburg. The project was sponsored by:*

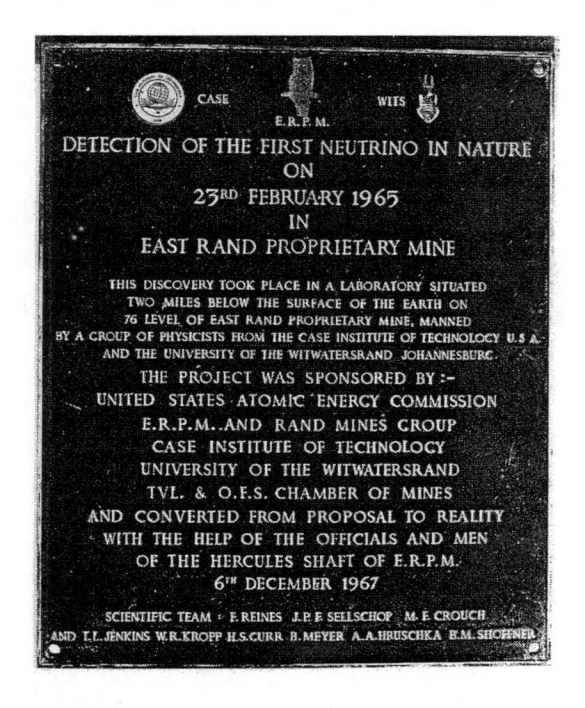

United States Atomic Energy Commission, E.R.P.M. and Rand Mines Group, Case Institute of Technology, University of the Witwatersrand, TVL & O.F.S. Chamber of Mines and converted from proposal to reality with the help of the officials and men of the Hercules shaft of E.R.P.M. 6th December 1967. Scientific team: F.Reines, J.P.F.Sellschop, M.F.Crouch and T.L.Jenkins, W.R.Kropp, H.S.Gurr, B.Meyer, A.A.Hruschka, B.M.Shoffner'

Friedel's career as a gifted experimentalist began with the above pioneering experiment on the "Little Neutral One", the (muon) neutrino. As amply demonstrated on this conference, Friedel (with collaborators) went on to perform many more pioneering experiments, often investigating the "Little Sparkling One", the diamond. On the one hand, he has exploited the unique and most extreme properties of the near perfect diamond lattice to produce and study the highest energy near monochromatic tagged photons ever generated in a laboratory. Diamond is sufficiently perfect a gem of a target that coherent effects are maximized at the expense of the incoherent. In addition, the crystal environment becomes a Lorentz-boosted super-critical equivalent field as viewed by an impinging multi-hundred GeV electron at crystal-aligned incidence. Coherent and Strong Field enhancements in the normal QED processes of bremsstrahlung and pair production have been explored in detail.

On the other hand, material science studies have been pursued in diamond with a view to the scholarly opportunity of this simplest example of a covalent macro-molecule, most easily tractable to microscopic Quantum Chemical calculations. These studies have deployed radio-active ions, stable ion beams, protons, muons, and positrons as probes of the diamond host. They have contributed enormously to the possible deployment of diamond as a 21st century high tech material, which may pervade many aspects of our lives in the future.

Finally, there is Friedel's geological interest, perceiving diamond as a "messenger from the deep". This resilient material is both a chemical and physical "prison" for mantle material, included from a depth of 200km, 2.5 billion years ago, when and where diamond had its genesis. Friedel (and collaborators) unlocked the hidden geochemical secrets from these preserved and priceless inclusions, using again nuclear physics techniques.

Friedel Sellschop is one of South Africa's most eminent and exceptionally honoured scientists. Friedel holds four honorary doctorates from various universities, the first (in 1989) from the University of Frankfurt (Germany). Friedel has received very prestigious international prizes, among others from the *Alexander von Humboldt Stiftung*, the *Max Planck Gesellschaft* (1992 Forschungspreis) in Germany and from other countries all over the world. On the occasion of his 60th birthday an issue of *Zeitschrift für Physik* **A336** (1990) was dedicated to him. During the year 2000 several conferences were dedicated to him. The first one was the 9th Varenna Nuclear Physics conference in Italy and the last one Lüderitz 2000 in Namibia.

In recognition of his life-long dedication to Physics we dedicate this volume to Jacques Pierre Friedrich (Friedel) Sellschop.

Simon Connell and Rudolph Tegen   Johannesburg, March 2001

xx

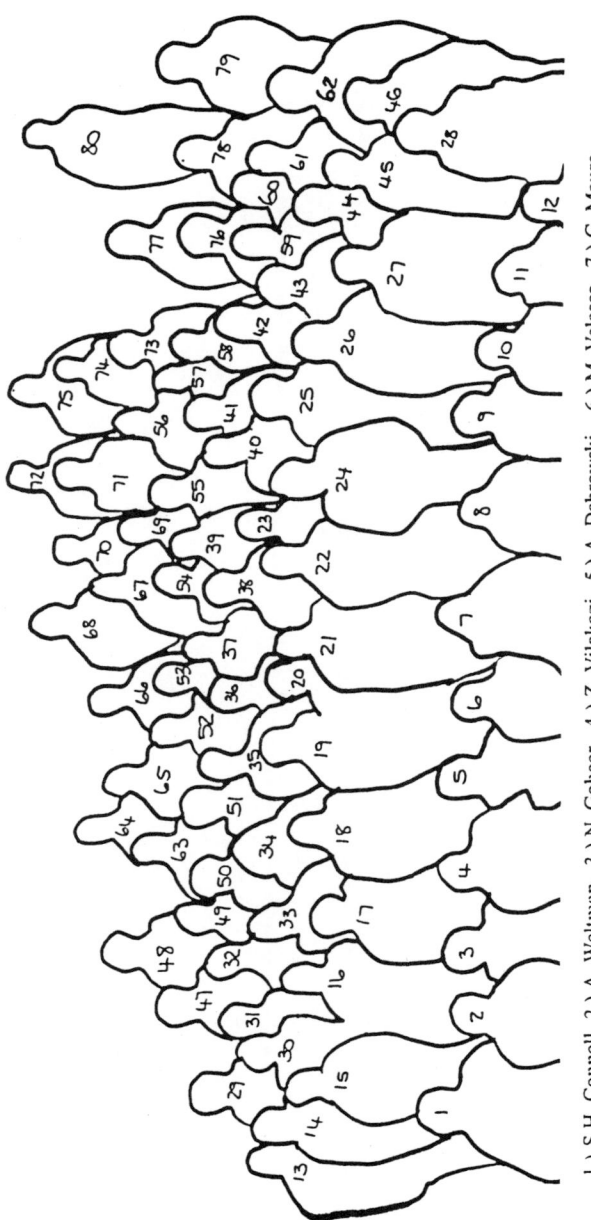

1.) S.H. Connell, 2.) A. Weltman, 3.) N. Goheer, 4.) Z. Vilakazi, 5.) A. Dabrowski, 6.) M. Velasco, 7.) G. Maure, 8.) N. Mhlahlo, 9.) R.Adams, 10.) T.Brandt, 11.) S.Cross, 12.) T. Jili, 13.) H. Appel, 14.) S. Kalbitzer, 15.) E. Friedland, 16.) A. Naran, 17.) G. 'tHooft, 18.) J. Cumbane, 19.) A. Kinyua, 20.) F.R.N. Nabarro, 21.) J. Oyedele, 22.) R. Tegen, 23.) L.M. Lekala, 24.) J.P.F. Sellschop, 25.) P. Kienle, 26.) W. Greiner, 27.) J. Fiase, 28.) E. Gadioli-Erba , 29.) P. Sigmund, 30.) G. Zwicknagel, 31.) A. Da Costa, 32.) B. Doyle, 33.) J. Biersack, 34.) J. Ellis, 35.) U.Rosengard, 36.) A. Zucchiatti, 37.) R. Dutta , 38.) J. Butler, 39.) B. Kämpfer, 40.) H. Machner, 41.) H. Genz, 42.) A. Freund , 43.) J-P. Coffin, 44.) H. Andeweg, 45.) A. Solov'yov, 46.) E. Gadioli, 47.) J. de Wet. 48.) M. Roberts, 49.) J. Cleynans, 50.) H. Backe, 51.) C. Levitte, 52.) P. Rose, 53.) D. Schardt, 54.) D. Ackermann, 55.) S. Krewald, 56.) S. Schramm, 57.) N. Manyala, 58.) C. Toepffer, 59.) B. Becker, 60.) I. Machi, 61.) S. Yacoob, 62.) H. Stöcker, 63.) K. Reed, 64.) D. Rebuli, 65.) C. Fischer, 66.) J. Murugan, 67.) E. Uggerhøj, 68.) J. Sharpey-Schafer, 69.) C. Greiner, 70.) J. Davies, 71.) J.U. Andersen, 72.) J. Hansen, 73.) P. Hodgson, 74.) I. Korir, 75.) M. Rebak, 76.) K. Purser, 77.) R. Groess. 78.) A. Richter, 79.) G. Soff, 80.) R. Uthui

xxi

CONTENTS

Preface v
Committee Members ix
Patronages and Sponsorships x
Friedel Sellschop's Photo xi
Walter Greiner's Dinner Speech: Prologue and Poem xii
Laudatio xvii
Group Photo xx

1. Nuclear Physics and Applied Nuclear Physics

On the need for comprehensive studies of heavy-ion reactions
    E. Gadioli et al. 1

New vistas of fission and neutron rich nuclei
    J. Hamilton et al. 11

The synthesis of superheavy elements - the state of the art
    D. Ackermann 18

Chiral symmetry restoration in nuclei
    P. Kienle 28

Pre-equilibrium reactions
    P. Hodgson 42

Meson production in hadronic reactions
    S. Krewald et al. 54

Meson production in p+d reactions
    H. Machner et al. 62

Semi-empirical effective interactions for inelastic scattering derived
from the Reid potential
    J. O. Fiase et al. 70

2. Atomic Physics and Applied Atomic Physics

Channeling revisited
  J. U. Andersen 78

Radiation physics with diamonds
  A. Richter 87

Channeling of charged particles through periodically bent crystals:
On the possibility of a Gamma Laser
  A. V. Solov'yov et al. 115

Novel interferometer in the X-ray region
  H. Backe et al. 123

Scientific opportunities at third- and fourth-generation X-ray sources
  A. Freund 135

Wave packet molecular dynamics simulations of the equation of state
of hydrogen and deuterium under extreme conditions
  C. Toepffer et al. 160

Heavy-ions stopping in plasmas
  G. Zwicknagel 168

Heavy-ion stopping: Bohr theory revisited
  P. Sigmund et al. 178

Radiation effects microscopy and charge transport simulations
  B. L. Doyle et al. 188

Bombardment-induced topography on semiconductor surfaces
  J. B. Malherbe et al. 201

The activation volume for shear
  F. R. N. Nabarro 211

3. Elementary Particle Physics

Challenges and opportunities in particle physics
  J. Ellis 219

Superstrings: Why Einstein would love spaghetti in fundamental physics
    S. J. Gates 235

Common features of particle multiplicities in heavy-ion collisions
    J. Cleymans et al. 248

The influence of strong crystalline fields on QED-processes investigated using diamond crystals in γ,γ colliders
    E. Uggerhøj 258

Using crystals to solve the nucleon 'spin crisis' TODAY...and looking for Physics beyond the Standard Model TOMORROW
    M. M. Velasco 269

Search for strangeness at new ultra-relativistic heavy-ion colliders
    J. P. Coffin et al. 301

Confinement in the Big Bang and deconfinement in the Little Bangs at CERN-SPS
    B. Kämpfer et al. 309

A confrontation with infinity
    G. 't Hooft 317

Current status of quark-gluon plasma signals
    H. Stöcker et al. 332

Chiral model calculations of nuclear matter and finite nuclei
    S. Schramm et al. 346

Parton showers and multijet events
    G. Soff et al. 354

Signatures of the quark-gluon plasma: a personal overview
    C. Greiner 363

4. Neutrino Physics and Nuclear Astrophysics

Perspectives of Nuclear Physics: From superheavies via hypermatter to antimatter and the structure of a highly correlated vacuum
    W. Greiner 373

H.E.S.S. — an array of stereoscopic imaging atmospheric Cherenkov
telescopes currently under construction in Namibia
R. Steenkamp 403

Cosmic particle acceleration– electron vs. nuclei
O. C. de Jager 411

5. Atomic and Nuclear Physics in the Study of Diamond

Hydrogen mobility in diamond
S. Kalbitzer 422

6. Applications of Pure and Applied Physics in Technology

Tumor therapy with high-energy heavy-ion beams
D. Schardt 433

IBA techniques to study Renaissance pottery techniques
A. Zucchiatti et al. 441

The bias in thickness calibration employing penetrating radiation
J. A. Oyedele 449

The measurement of very old radiocarbon ages by accelerator
mass spectrometry
K. H. Purser et al. 457

7. Science Policy and Anticipations*

Choosing good science in a developing country
R. M. Adam 474

Science partnerships for an African Renaissance: a framework for *Ngumzo*
A. M. Kinyua 482

Policy frameworks in science and technology: Then, Now and Tomorrow
A. Paterson 493

---

* The contents of the papers in this section are the responsibility of the authors.

8. Posters

Localised solutions of the parametrically driven complex Ginzburg-Landau Equation
    S. D. Cross et al. 502

π^0 and η photoproduction off the proton at GRAAL
    A. Zucchiatti et al. 510

Muon(Ium) in nitrogen-rich and ^{13}C diamond
    I. Z. Machi et al. 517

Positrons in diamond
    C. G. Fischer et al. 525

Study of the momentum transfer to target-like residues in heavy-ion reactions by prompt Gamma measurements
    K. A. Korir et al. 535

The Schonland nuclear microprobe – an important tool in Geosciences
    R. K. Dutta et al. 543

Ultra-thin single crystal diamond
    D. B. Rebuli et al. 552

Roasting Speeches at Dinner Party

Richard Sellschop et al. 561

Joseph Hamilton 566

Paul Kienle 568

Ettore Gadioli 569

Helmut Appel 571

Christian Toepffer 573

Closing Ceremony – The Three Devils

For the overseas delegates
    J. A. Davies 577

For Friedel
    A. K. Freund 582

Conference Programme 589

List of Participants 595

Author Index 605

1. Nuclear Physics and Applied Nuclear Physics

ON THE NEED FOR COMPREHENSIVE STUDIES OF HEAVY ION REACTIONS

E. GADIOLI[A,B]

Talk given in honour of Friedel Sellschop on the occasion of his seventieth birthday based on research made in collaboration with

M. CAVINATO[A,B], E. FABRICI[A,B], E. GADIOLI ERBA[A,B],
C. BIRATTARI[A,B], G. F. STEYN[C], S. V. FÖRTSCH[C], J. J. LAWRIE[C],
F. M. NORTIER[C], S. H. CONNELL[D] E. SIDERAS-HADDAD[D],
A. A. COWLEY[E] AND J. P. F. SELLSCHOP[D]

[A] *Dipartimento di Fisica, Universitá di Milano, Italia*
[B] *Istituto Nazionale di Fisica Nucleare, Sezione di Milano, Italia*
[C] *National Accelerator Centre, Faure, South Africa*
[D] *Schonland Research Centre for Nuclear Sciences University of the Witwatersrand, Johannesburg, South Africa*
[E] *Department of Physics, University of Stellenbosch, South Africa*

Even if the study of heavy ion reactions has greatly increased our knowledge of nuclear physics, it often does not provide the quantitative and systematic information which is necessary for the use of these data in applied and trans-disciplinary fields. In order to acquire such knowledge, in addition to review and to supplement what has been done, carefully planned experiments and new theoretical models are required. The Milano-Schonland-NAC-Stellenbosch collaboration's study of the interaction of ^{12}C with nuclei is an example of such research and a few of the results emanating therefrom are discussed.

It is for me a great pleasure to be here to honour Friedel Sellschop on the occasion of his seventieth birthday and I wish to start by congratulating the organizers because the topics covered in this Conference are very representative of the broad range of interests that Friedel showed in his long and distinguished career during which he confronted himself with both the fundamental questions and the use of physics in many applied and trans-disciplinary fields and last but not least with the administration of science and the careful consideration of its impact on society. This kind of approach is especially important in the case of nuclear physics which from the very beginning with the discovery of radioactivity dealt both with fundamental questions and applications which may be at the same time beneficial and harmful to mankind. In many talks at this Conference it was and it will be shown that nuclear physics may still greatly contribute to our understanding of basic questions such as the behaviour of matter under extreme conditions, but at the same time one cannot underestimate the contribution it may provide to other fields of knowledge and to applications useful to mankind. An indication of the growing importance of this branch is the fact that less than one percent of the world's particle accelerators are used for basic research in nuclear and particle physics [1]. One must also take into account that a rather considerable part of the beam time of major accelerators such as the GSI SIS or here in Africa, the NAC, is used for cancer therapy and production of radio-isotopes for medical diagnosis and therapy. New large high energy accelerators to be used only for medical or industrial applications will soon come into operation. The importance of this trend should not be underestimated and one must realize that it is beneficial for the future of nuclear physics which otherwise could be negatively considered by the public. In this context we have to ask ourselves if our present knowledge is adequate for the use of nuclear physics in related and trans-disciplinary fields and industrial and medical applications. This use requires a very quantitative and systematic knowledge. For instance, the use of hadron beams (protons and ^{12}C) for the therapy of deep tumours requires a knowledge of the cross-sections and of the spectra of the particles and γ rays which are produced in the interaction of these particles with nuclei of both the biological tissue and the materials used for beam collimation and degradation. An example of the required know-how is given by the recent ICRU Report on Nuclear Data for Neutron and Proton Radiotherapy and for Radiation Protection [2]. No less extended and detailed studies are needed for the interaction of hadron beams with structural materials also in, e.g., fusion reactors, aircraft and satellites [3,4].

The creation of data bases of experimental and evaluated cross-sections and other relevant information and of evaluation codes and reference input parameters for neutron and proton reactions (which at present are of more immediate utility) is actively pursued. Much less satisfactory is the collection of similar data for heavy ion reactions in spite of the enormous number of experiments on heavy ion reactions which have greatly increased our knowledge. In fact even basic information is often lacking. Just to give an example: systematic measurements of reaction cross-sections are quite rare and little is known about the elastic scattering of most ions so that even in the case of a basic reaction model such as the Optical Model one does not know much about the best parameters to use. Another point that should be stressed is that, contrary to what has been the case in light particle induced reactions, one often does not look for complete (inclusive) knowledge preferring to study exclusive processes which, however important, provide only partial information. It would not be correct to say that information of this type is absolutely lacking, however it is not systematic, nor is it easy to find, making it highly advisable to undertake a systematic search of what has been published in heavy ion dynamics summarizing this information in a easy to access form. The same should be done for the theoretical models which have been proposed, selecting the best fitting parameters to be used and comparing their predictions when many of them can be used to describe a given reaction. It must also be remarked that many analyses of published data use very simplistic models or unproven assumptions which often lead to unjustified conclusions [5].

Obviously heavy ion reactions are much more complex than those induced by light ions because two heavy ions may interact in many different ways which depend on their structure, their relative energy and angular momentum. One may ask if it is really possible to provide a comprehensive description of all the processes which may occur. While a formal comprehensive theory seems, at the moment, to be beyond our possibilities, it seems that semi-classical approaches may have some success. I cannot give here a balanced account of the many approaches which have been proposed to reach such an aim (which include the Vlasov-Uehling-Uhlenbeck (VUU) and the Boltzmann-Nordheim-Vlasov (BNV) equations [6-12], Quantum Molecular Dynamics (QMD) [13-17] and Antisymmetrized Molecular Dynamics (AMD) [18,19]). I will limit myself to discuss, as an example of a specific research program, some results of the Milano - Schonland - NAC - Stellenbosch collaboration which aims to obtain comprehensive information on the reactions induced by ^{12}C and ^{16}O with nuclei and to provide a comprehensive phenomenological description of these. The data, at incident energies varying from 5 to about 45 MeV/amu, include a large number of excitation functions for residue formation [20,21], residue angular and forward recoil range distributions [22], spectra of fragments produced in the projectile's break-up [23,24], spectra of protons and neutrons emitted in complete fusion reactions [25], and more recently Doppler shifted and broadened prompt γ lines emitted by the residues [26]. The experimental results which we have collected so far, for nuclei with A$\geq$60, are reproduced quite satisfactorily in a comprehensive and consistent way (i.e. by means of a single calculation) using a model which is essentially based on the hypothesis that only a few interaction mechanisms contribute incoherently to the reaction cross section. These are the complete fusion of ^{12}C with the target, its break-up into α-particles (two of which loosely bound to form a ^{8}Be) followed in most cases by the fusion of one fragment with the target, the transfer of nucleons from ^{12}C to the target and the projectile inelastic scattering. In each of these *primary* interaction modes an excited nucleus is created in a state far from statistical equilibrium to which it proceeds by means of a cascade of nucleon-nucleon interactions which is described by the Boltzmann Master Equation theory [25].

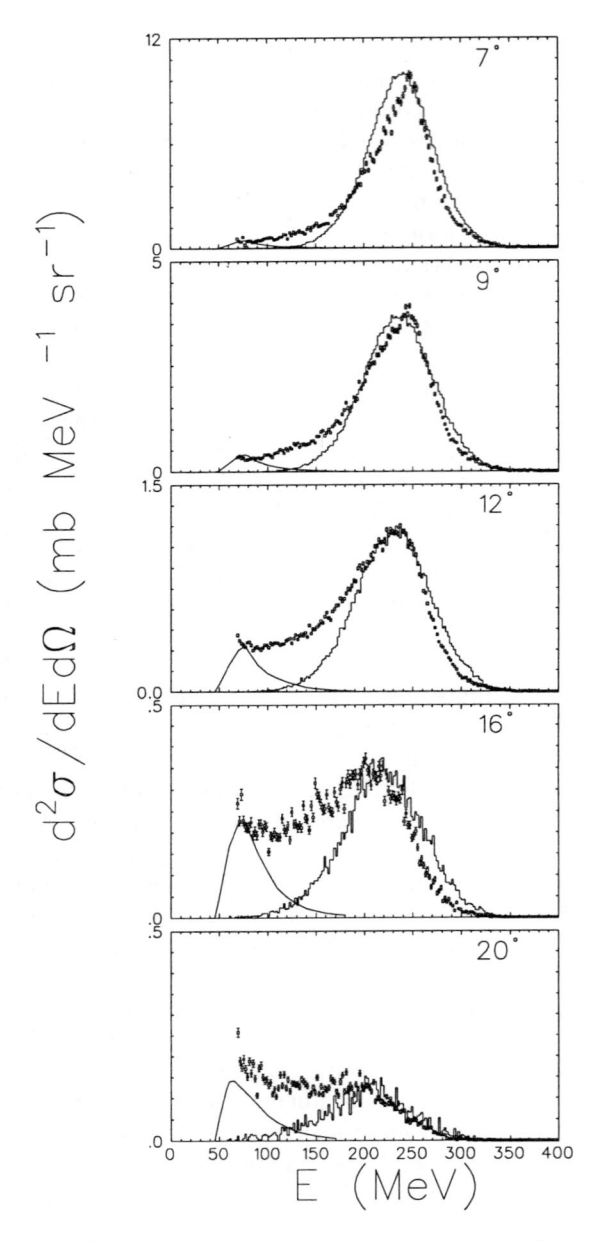

Figure 1. Spectra of ^{8}Be produced in the interaction of 400 MeV ^{12}C ions with ^{93}Nb.

In order to give more quantitative information, and to show the comprehensiveness of our results, let me briefly show a few representative results. Fig. 1 shows the spectra of the ^{8}Be fragments emitted at forward angles (between 7° and 20 °) in the interaction of 400 MeV ^{12}C ions with ^{93}Nb. They provide significant information on the mean field interaction between the projectile and the target. The open squares represent the experimental values and the full line histograms the contribution of the fragments produced in the binary fragmentation of ^{12}C. The average ^{8}Be energy is considerably less than the one expected from *pure* break-up process and this suggests that before breaking up ^{12}C may suffer a considerable energy loss [24,27]. The full line gives the contribution of ^{8}Be produced by nucleon coalescence in the course of the two-body interaction cascade by means of which the composite nucleus created in the complete fusion of carbon with niobium thermalizes [25,28]. Fig. 2 shows the spectra of the α particles emitted at angles varying from 10° to 120° in the interaction of 400 MeV ^{12}C ions with ^{93}Nb. The experimental data are given by the open squares. The theoretical spectra, given by the full line histograms, are the incoherent sum of the spectra of: (a) the *spectator* α particles from ^{12}C break-up, (b) the break-up α particles which fused with the target nucleus and were re-emitted with most of their energy, (c) the α particles produced by nucleon coalescence mainly during the thermalization of the composite nuclei produced in a complete fusion, (d) the α particles evaporated by the equilibrated nuclei which are eventually produced after the fast stage of the de-excitation process. Thus these spectra reflect the full complexity of the initial interaction and the subsequent composite nucleus de-excitation [23]. Fig. 3 shows the forward recoil range distributions of near target residues which are produced in the interaction of 400 MeV ^{12}C ions with ^{103}Rh with high cross-sections and a very low linear momentum. In these reactions most of the projectile energy is given to fast fragments produced both in binary fragmentation and nucleon transfer reactions and to high energy pre-equilibrium particles emitted in two-body de-excitation cascades [22]. Fig. 4 shows, as a function of the residues' mass, the ratio of the experimental and the theoretical cross-sections for residues' production in the interaction of 400 MeV ^{12}C ions with ^{103}Rh [22]. The data are reproduced with quite remarkable accuracy which compares favourably with those obtained in the analysis of nucleon induced reactions. Finally, Fig. 5 shows, to the left, the experimental (black dots) and theoretical velocity spectra of the evaporation residues emitted at an angle of 7.5°±1.2° in the interaction of 300 MeV ^{12}C ions with ^{165}Ho and, to the right, the experimental and theoretical spectra of the neutrons emitted in coincidence with these residues in the complete fusion of the two ions [25,29].

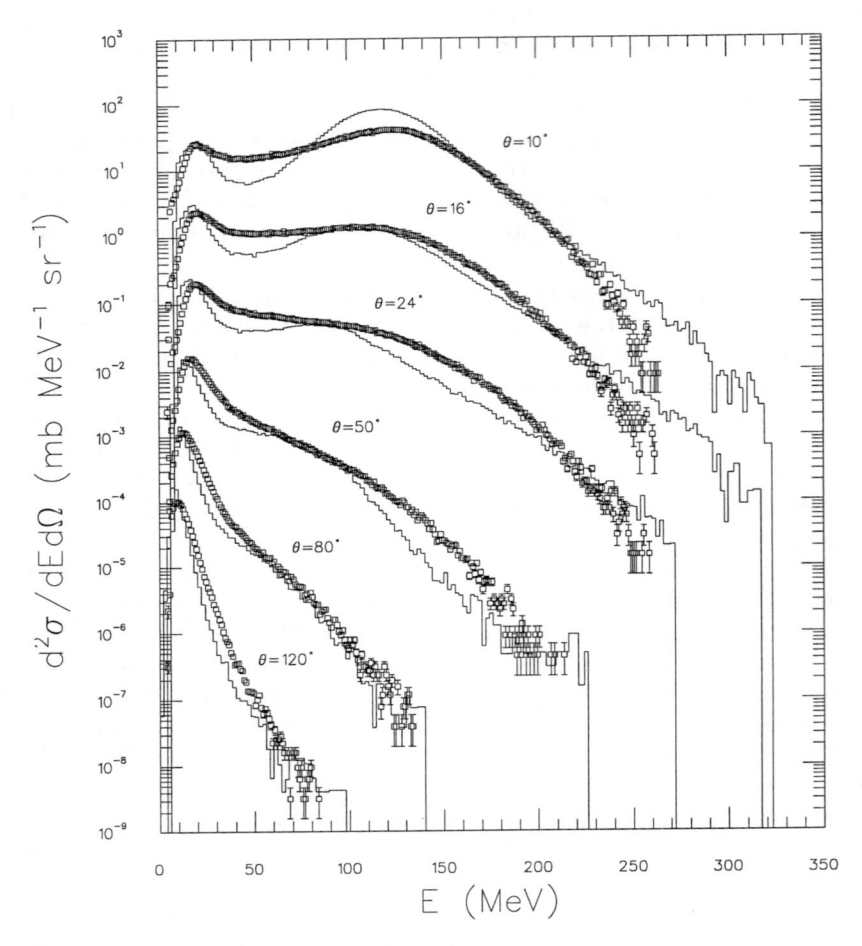

Figure 2. Spectra of the α particles produced in the interaction of 400 MeV ^{12}C ions with ^{93}Nb.

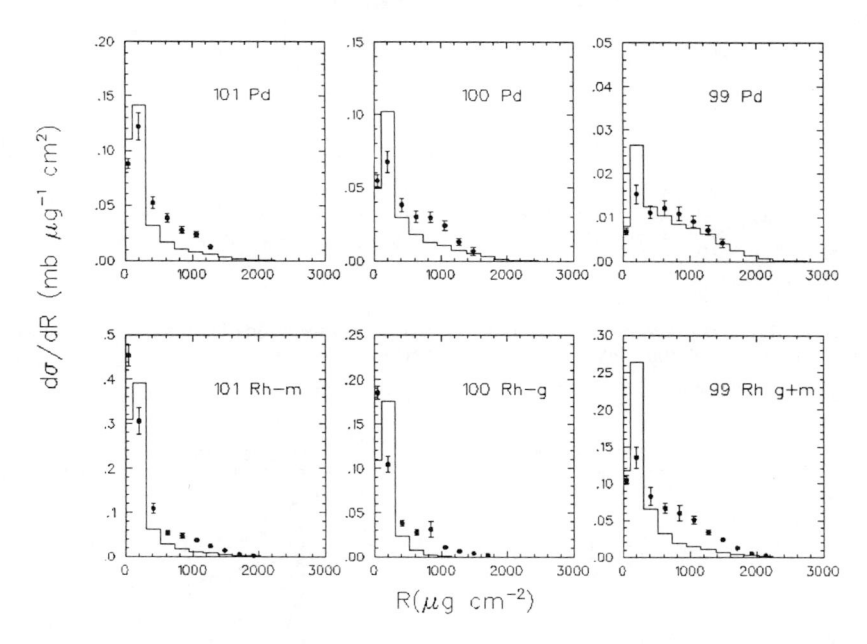

Figure 3. Forward recoil range distributions of near target residues produced in the interaction of 400 MeV ^{12}C ions with ^{103}Rh.

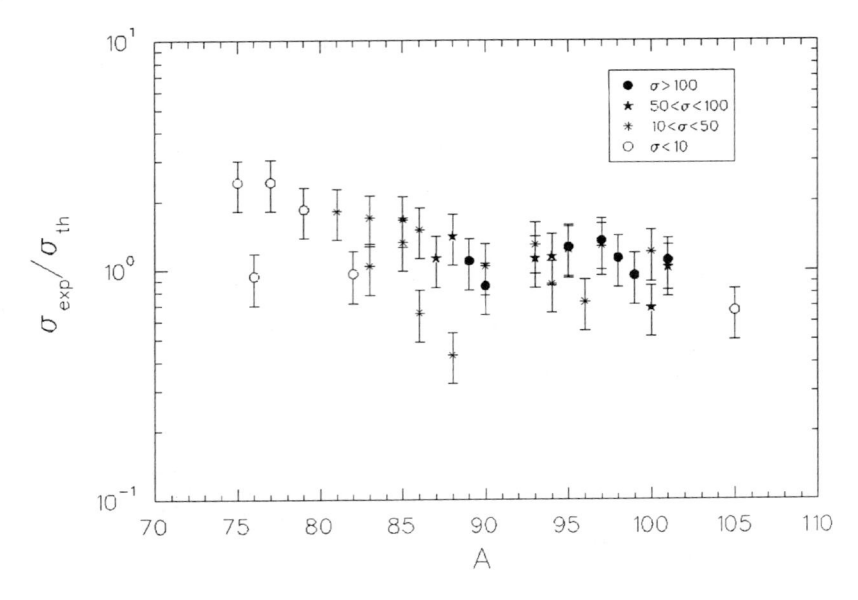

Figure 4. Ratio of the experimental and theoretical cross-sections for residues' production in the interaction of 400 MeV ^{12}C ions with ^{103}Rh.

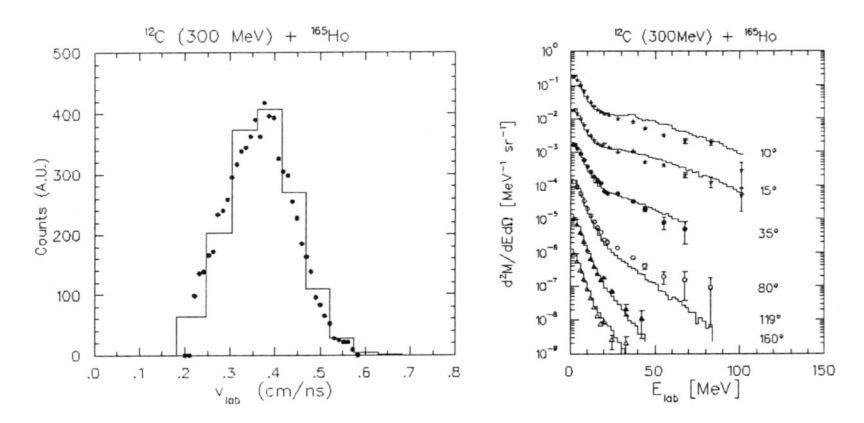

Figure 5. Left part: velocity spectra of residues emitted at 7.5°±1.2° in the interaction of 300 MeV ^{12}C ions with ^{165}Ho. Right part: spectra of neutrons emitted in coincidence with these residues in the complete fusion of the two ions [29].

These comparisons suggest that presumably it is possible to describe with fair accuracy most of the reactions which are induced by a light and rather simple nucleus such as ^{12}C at incident energies up to about 45 MeV/amu. A considerable effort is required to obtain similar results for a sufficiently representative number of projectile - target combinations and larger energies and provide a more firm theoretical basis to calculations of this type. However we suggest that such an effort should be made if one wishes to transform our knowledge of heavy ion reactions from essentially a qualitative to a quantitative one.

References

1. U. Amaldi, *Nucl. Phys.* A **654**, 375c (1999).
2. ICRU 63, Report on Nuclear Data for Neutron and Proton Radiotherapy and for Radiation Protection (2000).
3. IAEA-TECDOC- 1034, Handbook for calculations of nuclear reaction data, Reference input parameter library (1998).
4. INDC(NDS)-416, Nuclear model parameter testing for nuclear data evaluation (Reference input parameter library: phase II) (2000).
5. E. Gadioli et al., Report INDC(NDS)-41, 63 - 72 (2000). and Proceedings of the 9th International Conference on Nuclear Reaction Mechanisms, Varenna, 5-9 June, 2000, Ricerca Scientifica ed Educazione Permanente, Suppl. **115**, 527 - 535 (2000).
6. J. Aichelin and G. Bertsch, *Phys. Rev.* C **31**, 730 (1985).
7. G. Bertsch, S. Das Gupta and H. Kruse, *Phys. Rev.* C **29**, 673 (1984).
8. H. Stöcker and W. Greiner, *Phys. Rep.* **187**, 277 (1986).
9. C Gregoire et al., *Nucl. Phys.* A **465**, 317 (1987).
10. D. R. Bowman et al., *Phys. Rev.* C **46**, 1834 (1992).
11. K. Hagel et al., *Phys. Rev. Lett.* **68**, 2141 (1992).
12. B. Borderie et al., *Z. Phys.* A **338**, 369 (1991).
13. J. Aichelin and H. Stöcker, *Phys. Lett.* B **176**, 14 (1986).
14. J. Aichelin et al., *Phys. Rev.* C **37**, 2451 (1988).
15. J. Aichelin, *Phys. Rep.* **202**, 233 (1991).
16. T. Maruyama, K. Niita and A. Iwamoto, *Phys. Rev.* C **53**, 297 (1996).
17. R. Neubauer et al., *Nucl. Phys.* A **658**, 67 (1999).
18. A. Ono and H. Horiuchi, *Phys. Rev.* C **53**, 845, 2341 and 2958 (1996).
19. A. Onishi and J. Randrup, *Nucl. Phys.* A **565**, 474 (1993).
20. C. Birattari et al., *Phys. Rev.* C **54**, 3051 (1996).
21. E. Gadioli et al., *Phys. Lett.* B **394**, 29 (1997).
22. E. Gadioli et al., *Nucl. Phys.* A **641**, 271 (1998).

23. E. Gadioli et al., *Nucl. Phys.* A **654**, 523 (1999).
24. E. Gadioli et al.,*Eur. Phys. J.* A**8**, 373 (2000).
25. M. Cavinato, *Nuc. Phys. A , in course of publication*
26. Milano-Wits-NAC-Stellenbosch collaboration, *to be published.*
27. E. Gadioli et al., Proceedings of the 9th International Conference on Nuclear Reaction Mechanisms, Varenna, 5-9 June, 2000, *Ricerca Scientifica ed Educazione Permanente, Suppl.* **115**, 487-497 (2000).
28. M.Cavinato et al., *Z. Phys.* A **347**, 237 (1994).
29. E. Holub et al., *Phys. Rev.* C **33**, 143 (1986).

NEW VISTAS OF FISSION AND NEUTRON RICH NUCLEI

J.H. HAMILTON[1], A.V. RAMAYYA[1], J.K. HWANG[1], G. M. TER-AKOPOPIAN[2,3],
A.V. DANIEL[2,3], J.O. RASMUSSEN[4], S.-C. WU[4], T.N. GINTER[1,4], R. DONANGELO[4,5],
S.J. ZHU[1,3,6], E.F. JONES[1], P. M. GORE[1], C.J. BEYER[1], J. KORMICKI[1], X.Q. ZHANG[1],
W. GREINER[1,3,7], D. POENARIU[1,3,8], I. Y. LEE[4]. A.M. RODIN[2], A.S. FORMICHEV[2],
J. KLIMAN[2,9], L. KRUPA[2,9], M. JANDEL[2,9], YU. TS. OGANESSIAN[2], G.
CHUBARIAN[10], D. SEWERYNIAK[11], R.V.F. JANSSENS[11], W.C. MA[12], R.B.
PIERCEY[12], J.D. COLE[13] AND M. DRIGERT[13]

[1] *Department of Physics, Vanderbilt University, Nashville, Tennessee 37235, USA*
[2] *Flerov Laboratory for Nuclear Reactions, JINR, Dubna, Russia*
[3] *Joint Institute for Heavy Ion Research, Oak Ridge, Tennessee 37831, USA*
[4] *Lawrence Berkeley National Laboratory, Berkeley, California 94720, USA*
[5] *Instituto de Fisca, Univ. Federal do Rio de Janeiro, 21945-970 Brazil*
[6] *Physics Department, Tsinghua University, Beijing 10084, PRC*
[7] *Institut for Theoretische Physik, Frankfurt, Germany*
[8] *National Institute of Physics and Nuclear Engineering, Bucharest, Romania*
[9] *Institute of Physics, Bratislava, Slovakia*
[10] *Cylotron Institute, Texas A & M University, Texas 77843*
[11] *Argonne National Laboratory, Argonne, Illinois 60439*
[12] *Dearment of Physics, Mississippi State University, Mississippi 39762, USA*
[13] *Idaho National Engineering Laboratory, Idaho Falls, Idaho 83415, USA*

Binary and ternary spontaneous fission of ^{252}Cf and the structure of neutron rich nuclei have been studied via γ-γ-γ coincidences and γ-γ-light charged particle coincidences with Gammasphere. New nuclear structure effects observed in neutron rich nuclei include octupole deformation, the coexistence of symmetric and asymmetric shapes and a new phenomena of shifted identical bands with identical moments of inertia in neighboring nuclei when Eγ of one nucleus are shifted by the same constant for every spin state. Remeasured yields of correlated Mo-Ba pairs in binary fission confirm the previous hot fission mode with 8-10 neutron emission but with lower intensity. By gating on the light charged particles detected in ΔE-E detectors and γ-γ coincidences, the relative yields of correlated pairs in alpha ternary fission with zero to 6n emission are observed for the first time. A new γ-γ-γ data set (August, 2000) support the non-Doppler broadened but shifted energies of peaks assigned to the 2-0 transitions in ^{10}Be ternary fission. The data support but still need improved statistics before one can definitely establish long lived nuclear molecules in ^{10}Be ternary spontaneous fission.

Introduction

Studies of prompt γ-rays emitted in spontaneous fission (SF) with large detector arrays have given new insight into the fission process [1-4] and the structure of neutron rich nuclei [5]. A few selected examples [6,7] of new nuclear structure phenomena in the region of octupole deformation brought on by reinforcing shell gaps [8] for protons (Z=56) and neutrons (N=88) for the same $\beta_3 \sim 0.15$ and the discovery of shifted identical bands in neighboring nuclei are briefly presented. A redetermination of the Ba-Mo yields supports the ultra hot fission mode with the Ba nuclei hyperdeformed but with lower intensity [9]. From the light charged particle -γ-γ data, we extracted for the first time α ternary fission yields accompanied by 0 to 6n emission. Our new SF data (August, 2000) with 2.3 times the statistics of our previous data yield high energy peaks non-Doppler broadened but shifted in energy by 6 to 26 keV as seen earlier in ^{10}Be ternary fission. The data support but cannot definitely establish the existence of extremely long-lived nuclear molecules ($\tau \sim 10^{-12}$s) in ^{10}Be ternary fission.

New Nuclear Structure Vistas

Our new level scheme for ^{145}Ba is shown in Fig. 1 [6]. The new bands 1 and 5 and the intertwined enhanced E1 transitions to bands 2 and 4 now provide evidence for the long standing theoretical prediction [10] of stable octupole deformation in ^{145}Ba. The

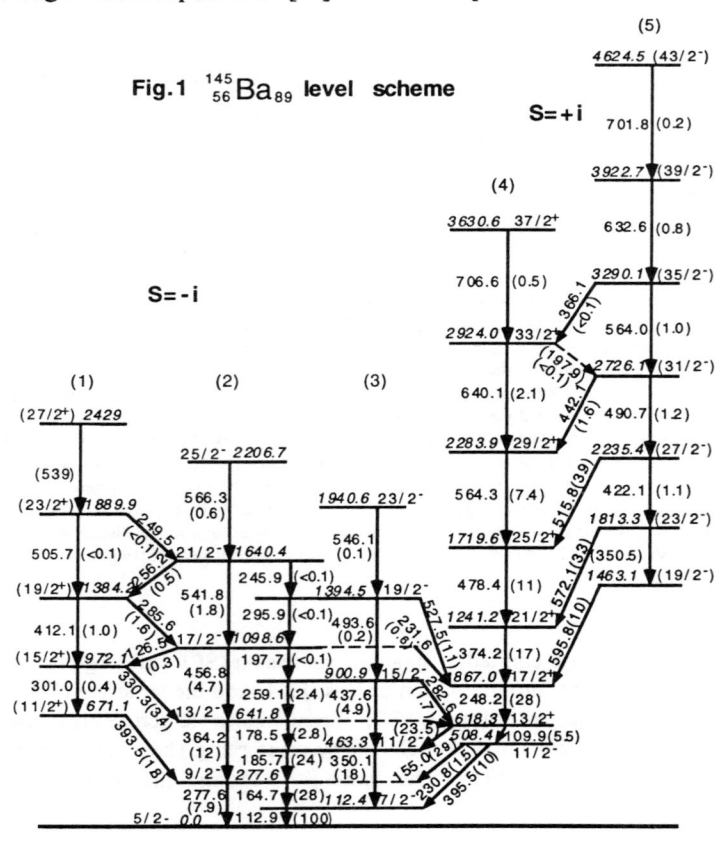

Fig. 1 $^{145}_{56}Ba_{89}$ level scheme

prediction was based on shell gaps for both N=88 and Z=56 for $\beta_3 \approx 0.15$, another example of the importance of reinforcing shell gaps [8]. The ground state band is built on a symmetric deformed shape. Nuclear rotation then enhances octupole deformation as predicted [11] and there is a change to an asymmetric shape around spin 19/2. Similar symmetric-asymmetric shape coexistence and rotation enhancement of octupole deformation is also seen in ^{145}La [12]. These are the first examples of such shape coexistence. Band 5 has an unusual structure, it has $J_1 \approx J_2 \approx$ constant ($\Delta E_\gamma \approx$ constant) as a function of spin as expected for a rigid rotor and as found for superdeformed nuclei. This may be the first superdeformed band in neutron rich nuclei.

Our identification of levels in ^{160}Sm and ^{162}Gd lead to the discovery of a new phenomenon we call shifted identical bands [7]. In a shifted identical band, the energies in two neighboring nuclei [a,b] separated by 2-8 nucleons have identical E_γ, J_1 and J_2 when E_γ in one nucleus are shifted by the same constant amount, κ, for every spin state from 2^+ up to 16^+, $E_{\gamma a} = (1+\kappa) E_{\gamma b}$, where the spread in κ is required to be $\leq \pm 1\%$ to form a SIB. For example, $E_\gamma (^{158}\text{Sm}) = [1.034 \, (^5_3)] E_\gamma (^{160}\text{Sm})$ and $E_\gamma (^{158}\text{Sm}) = [0.968 \, (^1_2)] E_\gamma (^{160}\text{Gd})$ ($\kappa = 3.2 \, ^{+0.1}_{-0.2}$ for this 2p separation, note the ± are not statistical errors but the spread of the maximum to minimum values of κ here −3.1% to −3.4%) and ^{156}Nd = 0.894 $(^4_2)$ ^{160}Gd ($\kappa = -10.6 \, ^{+0.4}_{-0.2}$ % for this 4p separation). Note the change in sign adding 2p and 2n to ^{158}Sm and the change in magnitude for adding 4p. In over 700 comparisons of neighboring nuclei separated by 2n, 2p, 4n, 4p, α, and others, we found 55 cases of shifted identical levels (SIB). The percentage spreads in J_1 and J_2, $\Delta J_1/J_1$ and $\Delta J_2/J_2$ are in general smaller than those of the "most spectacular" identical bands like SD1 and SD3 in ^{192}Hg and ^{194}Hg [13]. So after the shifts, SIB's are more identical than the "most spectacular" identical bands. These ground state SIB are found in stable to neutron rich nuclei with none in proton rich nuclei. Their occurrence is not correlated with size of deformation, E_4/E_2 ratios, N_pN_n, or the interaction strengths of the crossing ground-S bands. This new phenomena provides new challenges for microscopic models.

The Ba-Mo yield matrix for ^{252}Cf

We carried out pioneering work on the quantitative determination of yield matrices, using γ-γ and γ-γ-γ coincidence data [2]. Of particular note was the discovery of a new type of biomodel fission with the second mode having an unusually low average total kinetic energy [2]. In that work [2], about 7% of the ^{252}Cf Ba-Mo goes via a "hot fission" mode, with up to 10 neutrons emitted. Some skepticism has arisen since the hot fission mode was reported only in the Ba-Mo pairs in ^{252}Cf and not in ^{248}Cm spontaneous fission. Because of the importance of this mode, a new analysis was carried out with uncompressed triple coincidence spectra [9] with special attention to the degeneracy of several γ-rays in the 8-10 neutron emission yields for Mo-Ba. Fig. 2 shows semi-log plots of the summed Ba-Mo fission yields vs. neutron-emission number found in our new analysis (9) and in our previous work [2]. One sees that the hot fission mode is still present but its intensity is reduced by about a factor of 3 from the 7% reported earlier [2]. Since our work was completed, Biswas et al. [14] also reported analogous data that show a similar small irregularity around 8 neutrons lost. At scission, the Ba nuclei associated with the 8-10ν provide the first example of hyperdeformation; 3:1 long to short axis.

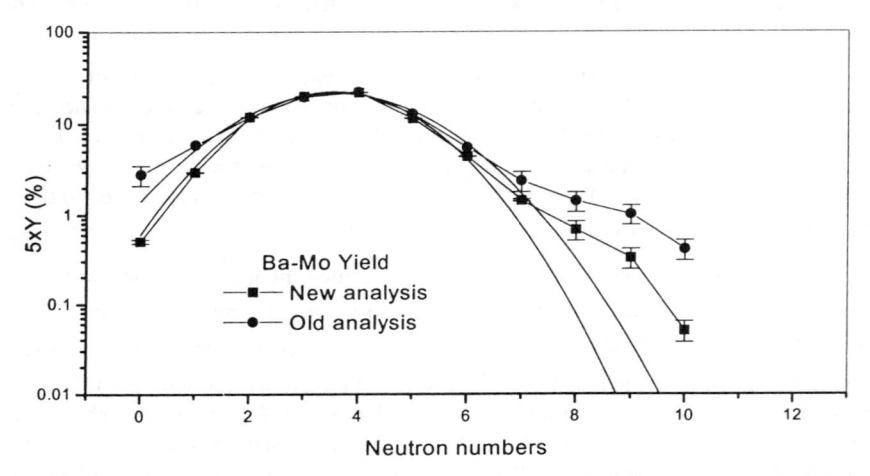

Fig. 2. Ba-Mo yields from previous and new analysis vs. neutron multiplicity

Light Charged Particle Ternary Fission

Ternary fission is very rare process that occurs roughly only once in every 500 spontaneous fission (SF) dominated by α ternary fission. Roughly, a ^{10}Be particle is emitted once per 10^5 spontaneous fissions. The maximum yield in the binary spontaneous fission is located around 3 to 4 neutrons. More recently, we performed an experiment incorporating charged particle detectors to detect light ternary particles in coincidence with γ-rays in Gammasphere. The energy spectrum of charged particles emitted in the spontaneous fission of ^{252}Cf was measured by using two ΔE-E Si detector telescopes installed at the center of Gammasphere. The ΔE-E telescopes provided unambiguous Z and A identification for the light charged particles of interest. The α-gated γ-spectrum in coincidence with the $2^+ \rightarrow 0^+$ transition in ^{142}Xe, is shown in Fig. 3 where various α, xn fission channels are marked. From the analysis of the γ-

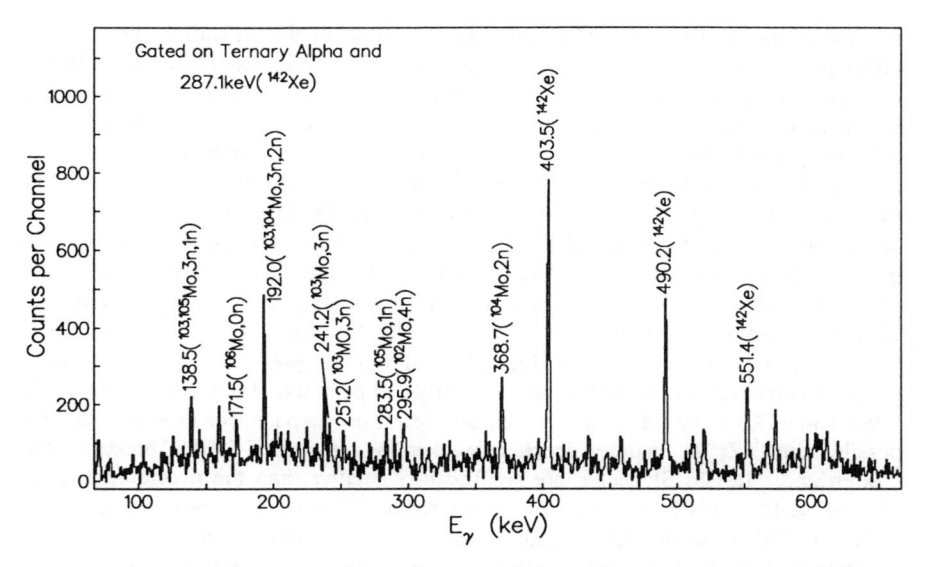

Fig. 3. Spectrum gated on α and 287.1 keV transition in ^{142}Xe

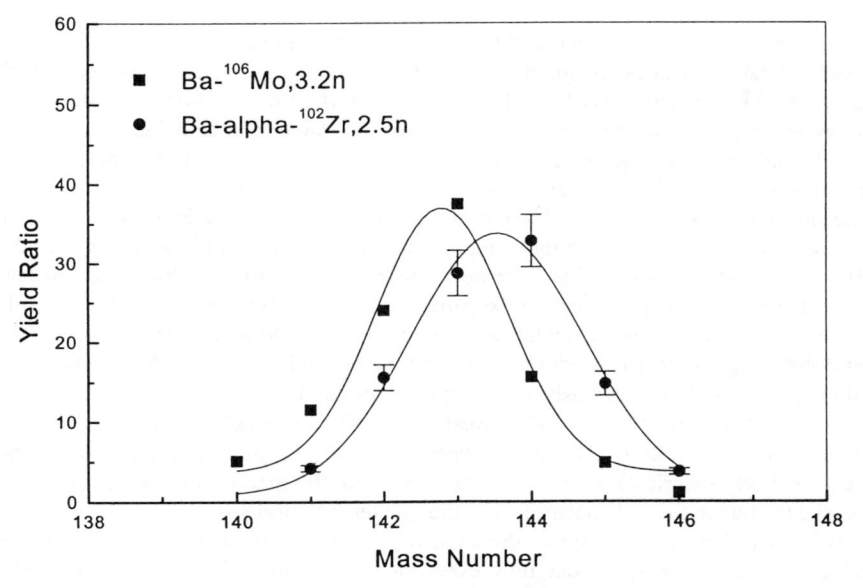

Fig 4. Relative 0-6n yields for channels shown

ray intensities in these types of spectra, one can calculate the yield distributions. The yields for binary fission and ternary α fission from 0 to 6n emission are shown in Fig. 4 for two particular channels. These are the first relative 0-6n yields for any ternary α SF. Note the average neutron emission in this α ternary SF channel is shifted down by 0.7n compared to the binary channel. This is the first observation of such a shift.

The first case of neutronless ^{10}Be ternary (SF) in ^{252}Cf was the pair ^{96}Sr and ^{146}Ba [4]. In our LCP-γ-γ data, the cover foils allowed only the high energy tail of the ^{10}Be energy spectrum to be observed in the particle detectors. For example, in the ^{10}Be gated γ-γ matrix, we set a gate on the 212.6 keV energy in ^{100}Zr and saw the 352.0 ($4^+ \to 2^+$) and 497.0 ($6^+ \to 4^+$) keV transitions in ^{100}Zr. New correlated pairs associated with ^{10}Be ternary SF including ^{100,102,104}Zr – ^{142,140,138}Xe, $^{104\ (or\ 108),106}$Mo – $^{138\ (or\ 134),136}$Te and ^{110}Ru – ^{132}Sn were identified. From the γ-γ-γ cube, we could easily establish coincidences between these new correlated pairs too. By double gating on the 376.7 and 457.3 γ-rays in ^{140}Xe, we can see clearly the zero neutron channel ^{102}Zr and possibly the ^{100}Zr 2n channel which is weaker by a factor of 5-10 if present. The ^{10}Be+n and ^{10}Be+2n SF yields are significantly smaller than the neutronless (cold) ^{10}Be SF yield. This is in contrast to the α ternary fission which peaks around 2.5n emission. This difference in the α and ^{10}Be yields as a function of neutron multiplicity is a unique discovery in the study of the cold (zero neutron) fission processes.

Long Lived Nuclear Molecules

In our first report of evidence for long lived nuclear molecules, we found a non-Doppler broadened peak assigned as the 2-0 transition in ^{10}Be in coincidence with transitions in ^{96}Sr and ^{146}Ba [4]. The peak was shifted 6 keV from the known 2-0 energy in ^{10}Be. A non-Doppler broadened peak at 3368 keV in coincidence with ^{10}Be particles has also been seen in work with NaI detectors [15]. There, small energy shifts could not have been observed. The stopping time of ^{10}Be in the cover foils in our experiment was the order of 10^{-12}s and the lifetime of the 2^+ state is 0.1 x 10^{-12}s. We suggested that what could be happening is that part of the time in ^{10}Be ternary SF, the ^{10}Be is captured in a potential well between the fragments forming a long lived nuclear molecule then the 2-0 transition can be emitted while the ^{10}Be is at rest in the molecule before break up and is not Doppler broadened. This would mean the molecule lived somewhat longer than 10^{-13}s which is the lifetime of the level. This lifetime would be truly remarkable since previously nuclear molecules lived $\leq 10^{-20}$s.

Next we found for ^{134}Te – ^{108}Mo a weak peak shifted to 3342 keV and in other ^{10}Be cases, peaks at 3352 keV. The energy shifts correlate with nuclear deformation. The more compact (spherical) the nuclei in the molecule the greater the nuclear potential pocket can change the ^{10}Be potential and the greater the energy shift.

To seek a definitive answer to the existence of such extremely long-lived nuclear molecules, we are carrying out new experiments at Gammasphere, now with 110 detectors (compared to 72 in our previous run), and with a stronger ^{252}Cf source.

Acknowledgement

Work at Vanderbilt U. and MiSU, LBNL, ANL and INEEL is supported in part by the U.S. Department of Energy under Grants and Contract No. DE-FG05-88ER40407, DE-FG05-95ER40939, DE-AC03-76SF00098, DE-AC07-76ID01570. and W-31-109-ENG-38.

References

[1] J. H. Hamilton et al., J. Phys. G **20**, L85 (1994).
[2] G. M. Ter-Akopian et al., Phy. Rev. Lett. 77, 32 (1996) and Phys. Rev. C **55**, 1146 (1997).
[3] A. V. Ramayya et al., *Heavy elements and related new phenomena*, Volume I, ed. R. K. Gupta and W. Greiner, (World Scientific, Singapore 1999) p. 477.
[4] A. V. Ramayya et al., Phys. Rev. Lett **81**, 947 (1998).
[5] J. H. Hamilton et al., Prog. In Part. And Nucl. Phys. **35**, 635 (1995) and J. H. Hamilton et al., ibid **38**, 273 (1997).
[6] S. J. Zhu et al., Phys. Rev. C **60**, 051304(R), 1999.
[7] E. F. Jones et al., Phys. Rev. Lett. (submitted).
[8] J. H. Hamilton et al., J. Phys. G: Nucl. Phys. Lett **10**, L25 (1984).
J. H. Hamilton, Int. Conf. On Nuclear Phys: Shells – 50, ed. Yu. Ts. Oganessian and R. Kalpakchieva, (World Scientific, Singapore 2000) p. 88.
[9] S.-C. Wu et al., Phys. Rev. C **62**, 041601(R) (2000).
[10] G. A. Leander et al., Phys. Lett. **152B**, 284 (1985).
[11] W. Nazarewicz and S. Tabor, Phys. Rev. C **45**, 222 (1992).
[12] S. J. Zhu, et al., Phys. Rev. C **59**, 1316 (1999).
[13] C. Baktash, et al., Am. Rev. Nucl. Part. Sc. **45**, 485 (1995).
[14] D. C. Biswas et al., Eur. Phys. J. **A7**, 189 (2000).
[15] P. Singer, et al., 3rd Int. Conf. Dynamical Aspects of Nucl. Fission, ed. J. Kliman and B. I. Pastylnik (Dubna Press, Dubna 1996) p. 250.

THE SYNTHESIS OF SUPER HEAVY ELEMENTS
- THE STATE OF THE ART -

D. ACKERMANN

University of Mainz/Gesellschaft für Schwerionenforschung GSI, Planckstr. 1,
D-64291 Darmstadt, Germany
E-Mail:d.ackermann@gsi.de

Throughout the passed two decades isotopes of the elements with atomic numbers 107-112 have been synthesized and unambiguously identified at the velocity filter SHIP at GSI. In a recent experiments at SHIP the results for element 112 and 111 have been confirmed and a third decay chain of the isotope ^{277}112 and three additional chains for ^{272}111 have been observed. Cold fusion reactions using Pb and Bi targets and evaporation residue(ER)-α-α correlations together with an efficient separation and detection system are the major ingredients for the success of these experiments. The sensitivity limit of the set-up at GSI has reached the 1pb level. For a systematic investigation in this region of the chart of nuclei and to synthesize heavier nuclei this limit has to be pushed to even lower values. An extensive development program is pursued at SHIP in order to reach at least an order of magnitude lower cross sections. Systematic investigations, the construction of decay chain networks and mass measurements are some of the possible approaches to study the decay chains attributed to isotopes of the elements 114, 116 and 118 at Dubna and Berkeley, which are, in contrast to those observed at GSI, not connected to decays of known isotopes. For the Berkeley results, in particular, several trials of confirmation have been undertaken at various laboratories including GSI.

1 Introduction

The search for superheavy elements, predicted close to the double magic nucleus ^{298}114 [1] - more recent theoretical results are found in [2,3] - was a substantial motivation for the construction of the UNILAC and the velocity filter SHIP [4] at GSI in Darmstadt. To reach the "island of superheavy elements" in the beginning of the experimental work at SHIP in 1976 only one method seemed possible: to jump across the "sea of instability". Although this method was tempting, it contained severe uncertainties. Decay properties of nuclei in the intended region, such as decay modes, decay energies and half-lives, were not known and could only be estimated on the basis of predicted mass excesses, shell effects, fission barriers etc., and were therefore extremely uncertain. The same held for the prediction of production cross sections using fusion-evaporation codes optimized to reproduce data in the region of known elements. Experiments, performed at SHIP, to produce superheavy elements in bombardments of ^{170}Er with ^{136}Xe or ^{238}U with ^{65}Cu[5], as well as by the reaction ^{48}Ca + ^{248}Cm [6] did not show positive results. It turned out to be more successful to approach the heavier elements step by step. Following the concept of "cold" fusion of lead or bismuth targets with medium heavy projectiles like ^{40}Ar or ^{50}Ti, first applied successfully by Oganessian et al. [7], the SHIP group succeeded to produce and identify about 25 new isotopes with atomic numbers from Z=98 up to Z=112. Mutual interaction of experimental results and theoretical calculations led to a better understanding of their stability, while measured excitation functions allowed for a reliable empirical extrapolation of optimum bombarding energies and cross sections for 1n deexcitation channels. Continuous technical development pushed the sensitivity of the set-up down to a cross section value of about 1 pb. To proceed towards higher Z an extensive development program is being followed at present. Recently the synthesis of isotopes of the elements 114, 116 and 118 has been reported at Dubna and Berkeley. The unambiguous assignment of those events, however, is not yet possible. An attempt to confirm the Berkeley results for element 118 at SHIP did not yield a positive result. A recent review on the discovery of the heaviest elements [8] gives a complete overview over the recent achievements in the field. There also a detailed description of the experimental set-up at GSI can be found.

2 Excitation Functions

Complete fusion reactions appear as most successful method for the production of transactinide nuclei. The formation cross section of a specific nuclide in a given reaction, however, is strongly dependent on the excitation energy E^* of the compound nucleus, according to the relation $E^* = E_{cm} + Q$ (where E_{cm} denotes the energy in the center-of-mass system and Q the Q-value of the reaction), and thus on the bombarding energy $E_{lab} = (mp + mt)/mt \times E_{cm}$. Since maximum production cross sections are decreasing rapidly with increasing atomic numbers, the choice of the optimum E_{lab} is crucial for the production of the heaviest nuclei. Measured excitation functions for reactions of ^{208}Pb, ^{209}Bi targets with various projectiles producing heavy nuclei in the range Z=104 to 112 are presented in fig. 1. Excitation energies were calculated using experimental mass excesses published by Audi and Wapstra [9] and values predicted by Myers and Swiatecki [10]. They were calculated for the center of the target using energy losses of the projectiles according to [11]. In all shown cases the cross section maxima are approximately centered between zero and the interaction barrier according to the Bass model [12].

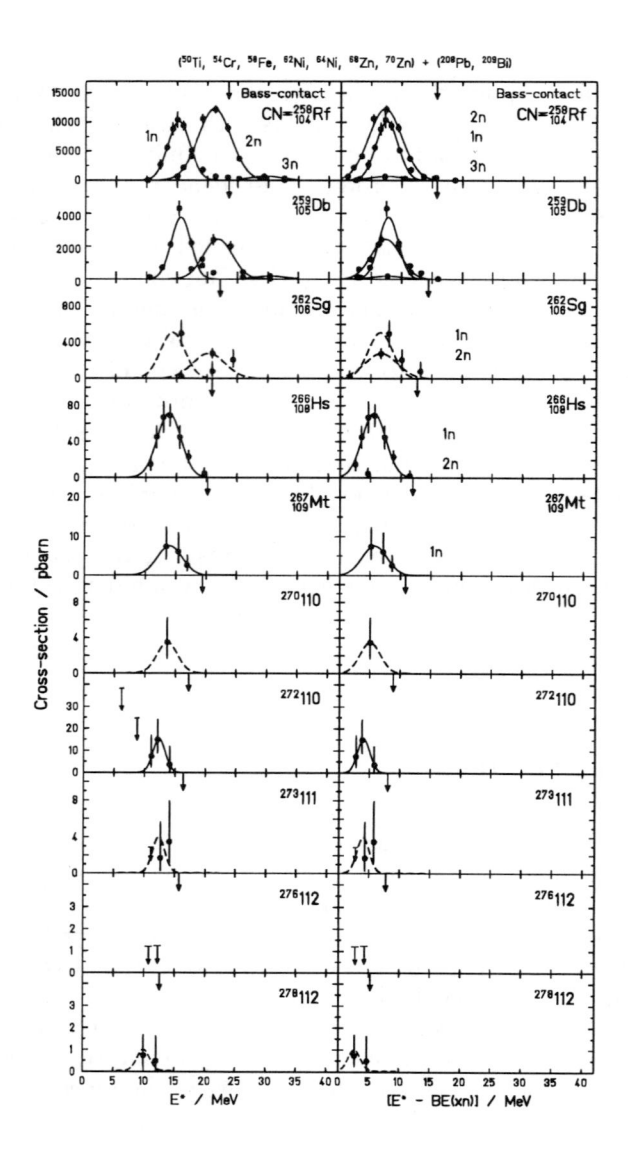

Figure 1. Measured excitation functions for Z=104 to 112. Cross sections are plotted as a function of the excitation energy (left panel) and the excitation energy lowered by the neutron binding energy according Myers and Swiatecki[10] for the various evaporation channels (right panel). The continuous curves are Gaussian fits through the data points, the dashed curves are interpolations. The arrows in both mark the interaction barriers of the reaction according to the Bass model[12].

3 Recent Results on the Synthesis of Heavy Elements with Z=110-112 at GSI

The elements with Z=107-112 have been synthesized and unambiguously identified at SHIP. The elements 107-109 have already been named and have been entered as Bohrium (Bh, Z=107), Hassium (Hs, Z=108) and Meitnerium (Mt, Z=109) in the periodic table of elements. The properties found for the elements 110, 111 and 112 are presented in this section.

A linear extrapolation of the optimum excitation energies for the production of ^{257}Rf and ^{265}Hs (see fig. 1) resulted in an 'optimum' value of E* = 12.3 MeV for the production of ^{269}110 via the reaction ^{62}Ni + ^{208}Pb. In an experiment in November 1994, where a total projectile dose of 2.2×10^{18} was collected, four α-decay chains were observed, which were attributed to the isotope with the mass number 269 of the new element 110 [13]. The assignment was based on the observation that the α-decays directly preceded the well established α-decay chain of ^{265}Hs and, therefore, have to origin from the α-mother ^{269}110. From the measured decay data an average decay energy of E = (11.112±0.020) MeV and a half-life of $T_{1/2} = (170^{+160}_{-60})$ μs was obtained. The production cross section was $\sigma = (3.5^{+2.7}_{-1.8})$ pb.

Since it is well established in the region of transfermium nuclei that more neutron rich projectiles lead to higher formation cross sections, one could expect for the combination ^{64}Ni + ^{208}Pb a still higher ER cross section than for ^{62}Ni + ^{208}Pb. In a directly following experiment in November/December 1994 the ER production by the reaction ^{64}Ni +^{208}Pb was investigated at E* = (8-13) MeV. Nine α-decay chains observed in this experiment could be attributed to ^{271}110. A maximum cross section of $\sigma = 15 \; (^{+9}_{6})$pb was measured at E* = 12.1 MeV. Details of the decay chains can be found in ref. [16].

In an experiment in October 2000 we observed in the reaction ^{64}Ni+^{207}Pb eight decay chains correlated ER-α-fission events which we attribute to the decay of the new isotope ^{270}110. Also the daughter and grand daughter products ^{266}Hs and ^{262}Sg have not been observed before. The data are presently still being analyzed.

On the basis of these encouraging results for the synthesis of element 110 in the reactions ^{62,64}Ni + ^{208}Pb the production of an isotope of element 111 by the reaction ^{64}Ni + ^{209}Bi was undertaken in an experimental run in December 1994. Three bombarding energies at 10.0 MeV, 11.6 MeV, and 13.0 MeV were chosen using the predicted mass excess of [10] for the compound nucleus ^{273}111 excitation energies. Projectile doses of 1.0×10^{18} at E* = 10.0 MeV, 1.1×10^{18} at E* = 11.6 MeV and 1.1×10^{18} at E* = 13.0 MeV were collected. While no decay chain that could be attributed to ^{272}111 was registered at E* = 10.0 MeV, one event was observed at E* = 11.6 MeV, and two events at E* = 13.0 MeV [14], referring to a cross section of $\sigma = 3.5^{+4.6}_{-2.3}$ pb. In the series of experiments performed in October 2000 we also confirmed the synthesis of ^{272}111 observing additional three decay chains of this isotope.

In early 1996 the search for element 112 was undertaken using the projectile target combination ^{70}Zn + ^{208}Pb. A total projectile dose of 3.4×10^{18} was collected. Following the systematics on optimum excitation energies a bombarding energy according to E* = 10.1 MeV was chosen. Two decay chains which could be attributed to ^{277}112 were observed, the resulting production cross section was $\sigma = 1.0$ pb [15]. The most striking result, however, was the significant difference in the decay energies and lifetimes of the daughter isotope ^{273}110 of E = 9.73 MeV, Δt = 170 ms (chain 1) and E_α = 11.08 MeV, Δt = 110μs (chain 2). Due to the large differences in lifetime the two transitions must be assigned to different levels in ^{273}110. In a recent experiment in May 2000 a third decay chain of ^{277}112 has been recorded. It is shown together with the first two chains in fig. 2. This latter chain has been observed at an excitation energy of about 2 MeV higher at E* = 12 MeV. During an irradiation time of 19 days a total of 3.5×10^{18} projectiles were sent onto the target. The resulting cross section at this energy is $(0.5^{+1.1}_{-0.4})$ pb. This value fits well

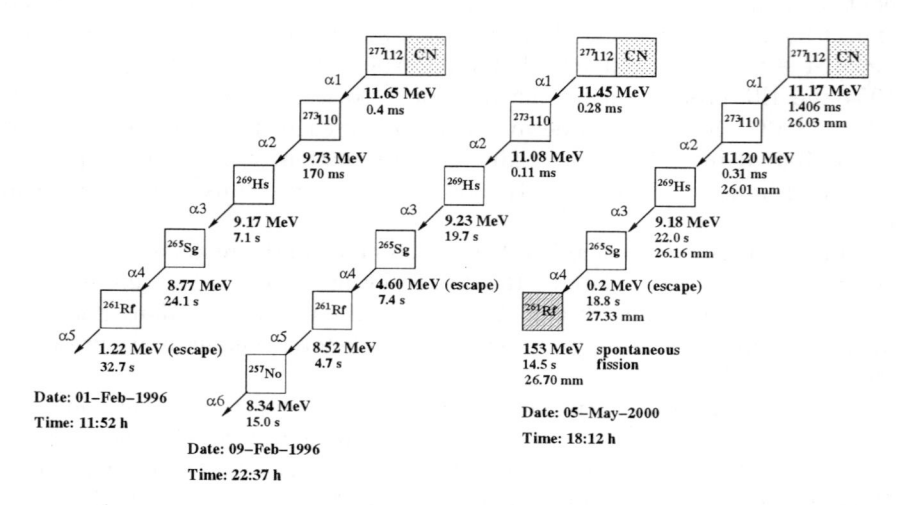

Figure 2. The three decay-chains observed for the isotope ^{277}112, including the chain observed in the confirmation run in May 2000. For this chain also the position in vertical direction on the 5 mm wide detector strip, where this event was observed, is given in mm.

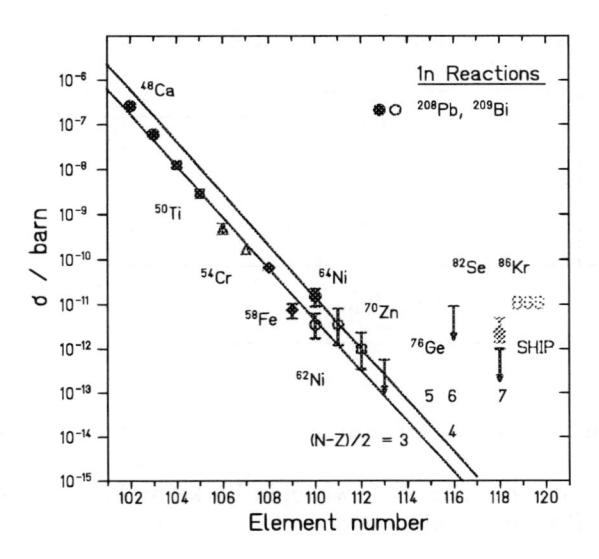

Figure 3. Maximum cross sections and cross section limits for heavy elements in fusion reactions with Pb and Bi targets for various projectiles at SHIP and the recent result from the BGS at the LBNL (see text).

into the cross section systematics shown in fig. 1. The first two α decays have energies of 11.17 and 11.20 MeV, respectively. They are succeeded by an α of only 9.18 MeV, an energy step of 2 MeV. Correspondingly, the lifetime increases by about five orders of magnitude between the second and third α decay. This decay pattern is in agreement with the one observed for the second chain in the first experiment and supports the explanation of a local minimum of the shell correction energy at neutron number N = 162, which is crossed by the α-decay of ^{273}110. The α energy of 9.18 MeV for ^{269}Hs is within the detector resolution identical to the one observed in the first chain. A new result is the occurrence of fission ending the new chain at ^{261}Rf, for which fission was not observed so far, but is likely to occur taking into account the high fission probabilities of the neighboring isotopes. For more details see ref. [8].

Table 1. Main parameters for the experiments investigating the reaction ^{86}Kr+^{208}Pb at the LBNL and GSI.

	BGS	SHIP
Time	11.0 d	24 d
Current	300 pnA	224 pnA
Dose	2.3×10^{18}	2.9×10^{18}
Efficiency	75% d	50%
E*	13 MeV	13 MeV
Events	3	0
σ	$2.2^{+2.6}_{-0.8}$ pb	< 1.0 ($0.5^{+1.1}_{-0.4}$)

4 Hints for the Synthesis of Elements with Z=114,116, and 118

In spring 1999 the group running the newly built Berkeley Gas-filled Separator (BGS) [17] at the LBNL in Berkley reported the observation of three long α decay chains which were tentatively attributed to the decay of the isotope ^{293}118 [18]. The observed ER-α-α correlations indicated the possible synthesis of ^{293}118 and ^{289}116 in the reaction ^{86}Kr+^{208}Pb. The energy of the ^{86}Kr beam was 459 MeV. Before impinging on the 300-450 μg/cm^2 ^{208}Pb-target the projectiles had to pass through an entrance window into the gas-filled chamber of the separator of 0.1 mg/cm^2 carbon and the target backing of 40 μg/cm^2 carbon. The beam energy in the middle of the target thickness corresponded to a calculated excitation energy of about 13 MeV. The detector system was similar to the one used at SHIP. In a first five day experiment from April 8 to April 12, 1999 with a total beam dose of 0.7×10^{18} projectiles two chains were observed. The experiment was repeated in Berkeley from April 30 to May 5, 1999. In this experiment one additional chain was observed at a total beam dose of 1.6×10^{18} projectiles. The cross section resulting from these two irradiations was given by Ninov et al. [18] as $(2.2^{+2.6}_{-0.8})$pb. In order to examine this result the same reaction was investigated at SHIP in summer 1999 [8]. No decay chain was observed which could have confirmed the LBNL result. With a slightly higher beam dose a cross section limit of 1pb was established on the basis of $0.5^{+1.1}_{-0.4}$ pb for one event at the given beam dose of 2.9×10^{18} collected projectiles. The parameters of both experiments are compared in table 1. In fig. 3 the values for the cross section measured at Berkeley and the limit obtained at SHIP are shown together with maximum cross sections for reactions with Pb and Bi targets leading to ER with Z$\geq$102.

A major problem, however, of the decay chains observed here is that they are not connected to α decays of known isotopes, as it is the case for the chains observed at SHIP for the elements 107-112. The same problem exists for the decay chains found at the FLNR in Dubna for reactions with ^{48}Ca projectiles on ^{242,244}Pu and ^{248}Cm. Here chains were seen which tentatively have been assigned to isotopes of the elements 114 and 116 [20]. Here additional information is needed to identify mass and atomic charge of the detected ER. For direct mass measurements using time of flight techniques detector development is needed. Another possibility to attack the problem of the non-connected decay chains is the systematic investigation of decay chain networks. Redundant information for the same isotopes can be provided using different reactions and decay paths producing the same isotopes. A given chain can be entered at different entry points using different reactions, as e.g. in the case of the Berkley chain for 118 one could exchange the projectile of ^{86}Kr by ^{82}Se and produce via the 1n channel the daughter of ^{293}118: ^{289}116. Similarly, as what has been reported at this conference by Oganessian [20], the exchange of the ^{244}Pu with ^{248}Cm leads via the 4n channel to the mother ^{292}116 of ^{288}114 which is the 4n ER produced in the reaction ^{48}Ca+^{244}Pu. Varying excitation energy and isotope one can in principle synthesize the same nucleus via different evaporation channels. A limitation for this kind of investigations is certainly the low cross sections which are to expect. Therefore, the state-of-the-art sensitivity of the experimental equipment, which presently limits the accessible cross section regime to 1 pb, has to be lowered by orders of magnitude. This sensitivity improvement is required for a successful continuation of the Pb and Bi based reactions as well as for those using actinide targets.

5 Technical Development at SHIP

The three areas presently under technical investigation at SHIP are:

- beam development
- target development
- background reduction.

To access a region of lower cross section the number of interactions and, therefore, the number of projectiles has to be increased. The UNILAC at GSI delivers the beam with a duty cicle of about 28%. Apart from raising the beam current, the use of an accelerator with 100% duty cycle (DC) would already provide a factor of 3.5 in higher beam intensity. The increased beam current, together with a higher Z of the projectiles in some cases, asks for measures to protect the Pb and Bi targets, both having a low melting point. A first step is to spread the beam as homogeneous as possible over a maximum area. With the target wheel presently in use we have already reached the limit for the presently available beam intensities. The planned introduction of ion optical elements like octupole magnets in the UNILAC beamline will help to approach the desired optimum of a rectangular beam profile illuminating the target as uniformly as possible. Besides those "passive" measures also an "active" target cooling is now under development. A set-up providing a gas jet blown onto the spot where the beam hits the target is currently being developed. Chemical compounds of Pb or Bi with higher melting temperatures are also under investigation. The higher projectile rate required for a successful investigation of reactions with lower cross section will have as a consequence an increase of background per time unit. To improve the background suppression we test the use of foils to stop scattered beam particles which pass SHIP with low kinetic energy. With all those measures and an increase of the beam intensity from presently 3×10^{12} particles s^{-1} to 3×10^{13} particles s^{-1} a cross section regime of one order of magnitude lower than the present limit could be reached.

6 Acknowledgments

The recent experiments were performed together with P. Armbruster, H.-G. Burkhard, H. Folger, F.P. Heßberger, S. Hofmann, B. Kindler, B. Lommel, V. Ninov, S. Reshitko, H.-J. Schött, C. Stodel (GSI Darmstadt), A.N. Andreyev, A.Yu. Lavrentev, A.G. Popeko, A.V. Yeremin (FLNR-JINR Dubna), S. Antalic, P. Cagarda, R. Janik, Š. Šaro (Uiversity of Bratislava), and M. Leino (University of Jyväskylä).

References

1. H. Meldner, *Arkiv f. fysik* **36**, 593-598 (1967).
2. P. Möller, J.R. Nix, *J. Phys. G. Part. Phys* **20**, 1681-1747 (1994).
3. S. Cwiok, A. Sobiczewski, *Z. Phys. A* **342**, 203-213 (1992).
4. G. Münzenberg, W. Faust, S. Hofmann, P. Armbruster, K. Güttner, H. Ewald, *Nucl. Instrum. Meth.* **161** 65-82 (1979).
5. G. Münzenberg, P. Armbruster, W. Faust, S. Hofmann, W. Reisdorf, K.-H. Schmidt, K. Valli, H. Ewald, K. Güttner, H.G. Clerc, W. Lang, *GSI Jahresbericht 1977* **GSI-J-1-78**, 75 (1978).
6. Y.K. Agarwal, P. Armbruster, S. Hofmann, F.P. Heßberger,

G. Münzenberg, K. Poppensieker, W. Reisdorf, K.-H. Schmidt, J.R.H. Schneider, W.F.W. Schneider, G. Vermeulen, A. Ghiorso, M. Leino, K.J. Moody, *GSI Scientific Report 1983* **GSI-84-1**, 79 (1984).
7. Yu.Ts. Oganessian, ; editors Harney, H.L. Braun-Munzinger, P. Gelbke, C.K., *Lecture Notes in Physics Vol. 33*, Berlin, Heidelberg, New York: Springer, 1974, pp. 221-252.
8. S. Hofmann, and G. Münzenberg, *Rev. Mod. Phys.* **72**, 733 (2000).
9. G. Audi, and A.H. Wapstra, *Nucl. Phys. A* **565**, 409-480 (1993).
10. W.D. Myers, and W.J. Swiatecki, *Nucl. Phys. A* **601**, 141-167 (1996).
11. F. Hubert, R. Bimbot, H. Gauvin, *Atomic Data and Nuclear Data Tables* **46**, 1 (1990).
12. R. Bass, *Nucl. Phys. A* **231**, 45 (1974).
13. S. Hofmann, V. Ninov, F.P. Heßberger, P. Armbruster, H. Folger, G. Münzenberg, G. H.-J. Schött, A.G. Popeko, A.V. Yeremin, A.N. Andreyev, SS. aro, R. Janik, M. Leino, *Z. Phys. A* **350**, 277-280 (1995).
14. S. Hofmann, NV. inov, F.P. Heßberger, P. Armbruster, H. Folger, G. Münzenberg, H.-J. Schött, A.G. Popeko, A.V. Yeremin, A.N. Andreyev, S. Saro, R. Janik, M. Leino, *Z. Phys. A* **350**, 281-282 (1995)
15. S. Hofmann, V. Ninov, F.P. Heßberger, P. Armbruster, H. Folger, G. Münzenberg, H.-J. Schött, A.G. Popeko, A.N. Yeremin, S. Saro, R. Janik, M. Leino, *Z. Phys. A* **354**, 229-230 (1996)
16. F.P. Heßberger, S. Hofmann, V. Ninov, P. Armbruster, H. Folger, A. Lavrentev, M.E. Leino, G. Münzenberg, A.G. Popeko, S. Saro, SCh. todel, A.N. Yeremin, *Proceedings of the Tours Symposium on Nuclear Physics III, Tours, France, 2.-5. September 1997*, AIP Conference Proceedings 425, Woodbury, New York, 1998, pp. 3-17.
17. V. Ninov, K.E. Gregorich, C.A. MacGrath, *Proceedings of the 2nd International Conference on Exotic Nuclei and Atomic Masses, ENAM-98*, AIP Conference Proceedings No.455, Bellaire, Michigan, June 23-27 1998, edited by B.M. Sherril, D.J. Morissey, and C.N. Davids, Woodbury, New York, 1998, p. 704.
18. V. Ninov, et. al., *Phys. Rev. Lett.* **83**, 1104 (1999).
19. Yu.Ts. Oganessian, et. al., *Nature* **400**, 242-245 (1999).
20. Yu.Ts. Oganessian, et. al., *contribution to this proceedings.*

CHIRAL SYMMETRY RESTORATION IN NUCLEI

P. KIENLE

Physik-Department, Technische Universität München
E-mail: Paul.Kienle@physik.tu-muenchen.de

The recent discovery of deeply bound pionic states in heavy nuclei, using pion transfer reactions of the type ^{206}Pb (d,^{3}He) π^- ⊗ ^{205}Pb is reviewed. Binding energies and widths of pionic 1s- and 2p-states are determined for ^{207}Pb and recently also ^{205}Pb. The characteristic narrowness of these states, generated by a strong repulsive s-wave pion-nucleus interaction is naturally explained by a reduced and density dependent pion decay constant, which in turn reflects the change of the QCD vacuum structure in dense matter.

1 Introduction

The current masses of u- and d-quarks [m_u = (5.1±1.5) MeV, m_d = (8.4±2.4) MeV] are two orders of magnitude smaller than typical hadron masses of about 1 GeV. This large mass gap between the QCD groundstate and the hadrons is assumed to originate from spontaneous chiral symmetry breaking [1]. It produces a quark condensate as QCD vacuum with a finite expectation value of $<0|\bar{q}q|0>$.

For u- and d-quarks $<0|\bar{q}q|0>$ was derived by QCD sumrules to have a value [2] of

$$<0|\bar{q}q|0> \equiv <\bar{q}|q> = -(225 \pm 40)\text{MeV}^3 \cong -1.5\text{fm}^{-3} \qquad (1)$$

Each spontaneously broken symmetry leads to massless Goldstone-Bosons, with the quantum number of the vacuum. In the SU(2) representation of the light quarks one expects 3 Goldstone-Bosons, which are identified as the pion triplet with $J^{\pi}=0^-$ and with unusual small masses (135, 140 MeV) compared with typical hadron masses of 1 GeV. The pions are excitations of the vacuum and thus can annihilate into the vacuum, characterized by a matrix element of the axial current $A_a^{\mu}(x)$, which defines the so called pion decay constant f_{π} in the following way:

$$<0|A_a^{\mu}(x)|\pi_b(q)> = iq^{\mu}f_{\pi}\delta_{ab}e^{iqx} \qquad (2)$$

The pion decay constant f_{π}, can be extracted from the weak leptonic decay of a pion ($\pi^+ \rightarrow \mu^+\nu_{\mu}$) and has a value [3] of

$$f_{\pi} = (92.4 \pm 0.3)MeV \qquad (3)$$

It is used as an order parameter of spontaneous chiral symmetry breaking. The pion decay constant f_{π} and the expectation value of the quark condensate are related by the Gell-Mann, Oakes, Renner relation [4] which connects hadron properties such as the pion mass m_{π}, the pion decay constant f_{π} with QCD properties, the average current quarkmass m_q = 1/2 (m_u+m_d), and the quark condensate $<0|\bar{u}u|+\bar{d}d|0>$

$$m_{\pi}^2 f_{\pi}^2 = -m_q <0|\bar{u}u+\bar{d}d|0> \qquad (4)$$

Our goal is to study the change of the quark condensate, respectively f_{π} of the vacuum, at the presence of a nuclear medium with density ρ.
The density dependence of the quark condensate can be described with the pion-nucleon σ-term, σ_N = (45±8) MeV in the following form [5]

$$\frac{<\bar{u}u+\bar{d}d>(\rho)}{<\bar{u}u+\bar{d}d>(0)} = 1 - \frac{\sigma_N}{f_\pi^2 \cdot m_\pi^2} \cdot \rho \qquad (5)$$

This can be used to formulate a density dependant pion decay constant $f_\pi^*(\rho)$ in leading order of ρ.

$$f_\pi^{*2}(\rho) = f_\pi^2 \left(1 - \frac{\sigma_N}{f_\pi^2 m_\pi^2}\rho\right) \qquad (6)$$

A nuclear medium with density ρ reduces f_π and the chiral condensate $<\bar{q}|q>$ correspondingly. Our experimental goal is to determine the renormalized value of f_π^* in a heavy nucleus, such as ^{207}Pb or ^{205}Pb.

2 Pion-Nucleon and Pion-Nucleus s-wave Interaction.

In the low energy theorem of Tomazawa and Weinberg [6] the pion-nucleon T-matrix is in leading order determined by the following relations

$$T^{(+)} = \frac{1}{2}[T_{\pi^- p} + T_{\pi^- n}] = 0 = 4\pi b_0 \qquad (7)$$

$$T^{(-)} = \frac{1}{2}[T_{\pi^- p} - T_{\pi^- n}] = \omega/2f_\pi^2 = -4\pi b_1 \qquad (8)$$

The isoscalar T matrix $T^{(+)}$ is in leading order vanishing and so the isoscalar scattering length b_0, while the isovector part of the pion nucleon interaction, $T^{(-)}$ and b_1, is determined by the pion decay constant f_π with the T-matrix evaluated at $\omega = m_\pi$. Note also that the π^-p-interaction is attractive, while the π^-n-interaction is repulsive.

Recently the pion-nucleon s-wave scattering lengths were accurately determined [7] from a precise measurement of 3p-1s x-ray transitions in pionic hydrogen and deuterium. The values for the isoscalar and isovector scattering length derived from these measurements are

$$b_0 = (1.6 \pm 1.3)10^{-3} m_\pi^{-1}$$

$$b_1 = (-86.8 \pm 1.4)10^{-3} m_\pi^{-1} \quad (9)$$

This result is in good agreement with the expectations of the Tomazawa-Weinberg low energy theorem.

If we put a pion in a nuclear medium with density ρ, its wave equation is modified by a self energy term $\Pi(\omega, \vec{q}, \rho)$ which is related to an s-wave local potential, U, by the relation $\Pi = 2\omega U$. The wave equation for a pion with energy ω, momentum $\vec{q}$ in a medium with density ρ, is given by the expression

$$[\omega^2 - \vec{q}^2 - m_\pi^2 - \Pi(\omega, \vec{q}; \rho)]\phi = 0 \qquad (10)$$

The interaction leads to an effective mass of the pion in the nuclear medium given by the expression

$$m_\pi^{*2}(\rho) \equiv \omega^2(\vec{q}=0; \rho) = m_\pi^2 + \text{Re}\,\Pi(\omega = m_\pi, \vec{q}=0; \rho) \qquad (11)$$

The self energy Π of a negative pion in a nucleus with proton-density ρ_p and neutron-density ρ_n can be expressed in T-matrices in the following way

$$\Pi = 2\omega U = -T(\pi^- p)\rho_p - T(\pi^- n)\cdot\rho_n \qquad (12)$$
$$= -T^{(+)}(\rho_p + \rho_n) - T^{(-)}(\rho_p - \rho_n)$$

Using the Tomazawa Weinberg Theorem of eq. 7 and 8 and taking into account that the pion decay constant is modified in a nuclear according to equation (6), the following expression can be derived

$$\Pi = \frac{-\omega}{2f_\pi^*}(\rho_p - \rho_n) \qquad (13)$$

from which follows a s-wave-pion optical potential

$$U = -\frac{\rho_p - \rho_n}{4f_\pi^{*2}} = \Delta m(\pi^-) \qquad (14)$$

For heavy nuclei ρ_n is larger than ρ_p from which one concludes that the pion nucleus potential is positive, which means repulsive. As a consequence of relation (14) one expects in pionic atoms of heavy, neutron rich nuclei due to a balance of the attractive Coulomb potential and the repulsive optical potential, a potential pocket at the edge of the nucleus in which the pion density is located. This leads to a halo like distribution of the deeply bound pions especially in the 1s- and 2p states. This feature was first pointed out by Friedman and Soff [8] and elaborated by Toki and Yamazaki [9].

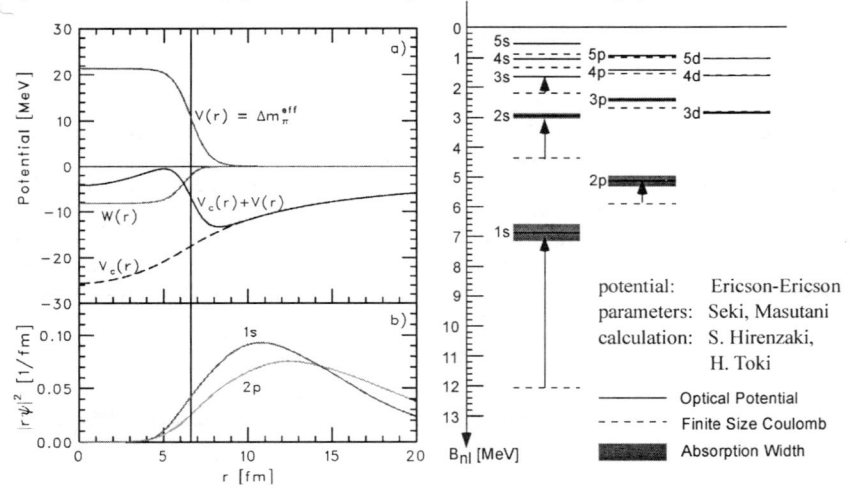

Figure 1 Various potential contributions for a π^- bound to a ^{208}Pb-nucleus and its wavefunctions in the 1s- and 2p-state (left side). Level scheme of pionic ^{208}Pb (right side).

Fig. 1 shows schematically the various contributions to the real and imaginary parts of the potential, a negative pion, bound to a ^{208}Pb, is exposed. The repulsive nuclear potential pushes the 1s- and 2p-wavefunctions out of the nucleus to form a halo like density distribution. This has several consequences as shown by the level scheme on the right hand side of fig. 1 calculated with an Ericson-Ericson potential [10] using the Seki-Matsutani parameterization [11]. Note that the 1s- and 2p levels are considerably shifted towards lower binding energies due to the nuclear repulsion, compared with the pure Coulombforce, whereas the states with quantum numbers accessible by the pionic x-ray cascades are only slightly affected. In addition the widths of the 1s- and 2p states are decreased to values below 1 MeV, so no overlap of the states are expected. Yet the absorption in the deep lying states is so large that they can not by studied by x-ray spectroscopy of pionic atoms.

So a new method, based on reaction spectroscopy in which a π^--meson is created and transfered into deeply bound states has to be developed. The pion has to be created with small momentum transfer, because the momentum distribution of the bound pions is rather narrow, due to the small binding energies. In order to meet the last requirement neutron pick up reactions, in which in addition a negative pion is produced have been proposed [12]. Its elementary process can be seen as $n(d^3He)\pi^-$, which when performed with a ^{208}Pb-nucleus as target can be written as

$$^{208}Pb(d,^3He)\pi^- \otimes ^{207}Pb \tag{15}$$

with $\pi^- \otimes {}^{207}$Pb denoting a pion bound to a ^{207}Pb-nucleus. The herewith produced states are bound π^--neutron hole configurations in ^{207}Pb of the type $(nl)_{\pi^-}(n'l'j')_n^{-1}$. For each of these configurations, the excitation energy E_x with respect to the groundstate of ^{207}Pb is related to the pion binding energy B_{nl} by the expression

$$E_x = [M_x - M(^{207}Pb)]c^2 = m_{\pi^-}c^2 - B_{ne} + E(n'l'j') \tag{16}$$

M_x is the mass of the pionic lead atom, m_{π^-} denotes the pion mass and $E(n',l',j')$ the excitation energy of the neutron hole states in ^{207}Pb. An accurate measurement of E_x will give us the binding energy B_{nl} of the pion, but also Γ_{nl} the width of the pionic states.

In order to achieve recoilless conditions, with small momentum and angular momentum transfer in the (d^3, He) reaction one must choose a specific bombarding energy, as indicate in

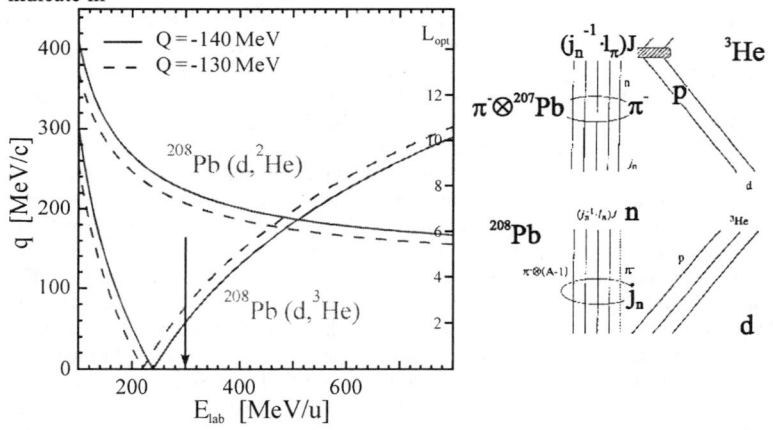

Figure 2 Momentum transfer q as function of the bombarding energy E_{lab} for the ^{208}Pb $(d,^3He)$ $\pi^- \otimes ^{207}$Pb-reaction, which is sketched on the right hand side of the figure.

fig. 2, which shows the momentum transfer q as function of the bombarding energy E_{lab}. We have chosen 300 MeV/u to meet the recoilless conditions approximately (about 60 MeV/c momentum transfer). The cross-section of the (d^3He) pion transfer reaction is strongly forward focused ($\Theta_{1/2} \sim 1.3°$) and is predicted to be about 50µb/sr for the dominant $(2p)_\pi \otimes (p_{1/2}, p_{3/2})_n^{-1}$ configuration [12].

3 Experimental Procedures

Deuterons were accelerated in the heavy ion synchrotron SIS of GSI, Darmstadt to energies of 600 MeV, slowly extracted with spill times of 1s and a repetition period of 2.8 s. The extracted beams had the desired small momentum width of $\Delta p/p < 0.5$ MeV FWHW, which is a prerequisite for the high resolution reaction spectroscopy.

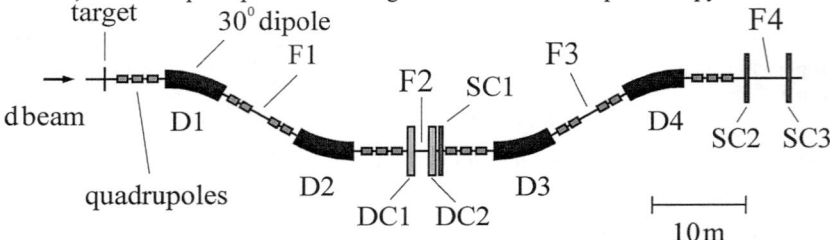

Figure 3 Experimental set up using the fragment separator as a high resolution forward spectrometer. D1-D4 30° dipolemagnets. F1-F4 focal planes, DC1 and DC2 minidrift-chambers, SC1-SC3 plastic scintillators.

Fig. 3 shows schematically the experimental set up for the observation of the (d ^{3}He)-reaction. The ^{3}He ejectiles emitted at 0° with respect to the deuteron beam, were identified and momentum analysed with a high resolution forward spectrometer, FRS [13], which also served for effective background suppression. It consists of four 30° dipole magnets $D_1 \rightarrow D_4$ with focusing quadrupole triplets at the entrance and exit of each of the two sections and two quadrupole doublets between the dipole magnets.

The first section with the dipole magnets D_1 and D_2 serves as a high resolution momentum analyser in a dispersion mode. The acceptance of the FRS is about 0.15 msr with a dispersion at the focal plane F2 of about 6 cm per 1 % momentum acceptance. The total momentum acceptance at a fixed bending field is about 6 %. High rate mini drift chambers (DC1 and DC2) were inserted around the focal plane F2 for an accurate measurement of the position and the direction of the ^{3}He-particles in the dispersion plane. The direction information was used for small optical corrections of the spectrometer. Behind DC1 and DC2 a 5 mm thick plastic scintillator detector SC1 is placed. It serves two purposes, first it gives a timing signal for the drift chambers and the time of flight measurement through the second magnet section (D3 and D4) and it acts as an absorber for an energy loss measurement by D3 and D4 for a very clear ^{3}He-identification. The scintillators SC2 and SC3 were placed around the focal plane F4 of the second magnet sector. They give stop signals for the time of flight in the second section and are also used to measure the position and direction of the ^{3}He-particles at F4. The pulse hight measured in all scintillation counters and the time of flight between SC1 and SC2 were used for an additional identification of ^{3}He-particles beside the selective magnetic transport through D3 and D4 following the energy loss in the detectors at F2. Together with the time signal from SC3, this redundant procedure allowed a completely background free identification of the ^{3}He-particles. It should be noted that there is a potentially high proton background from deuteron break up in the target. Especially the detectors around F2 are exposed to a high flux of break up protons (10^5-10^6 protons/spill) because under recoil free conditions the magnetic rigidity of the protons is the same as that of the ^{3}He-ions. In order to reduce the proton background at F2, we have introduced a small momentum transfer in the reaction. This cuts down the number of break up protons in the selected momentum bite for ^{3}He-ions appreciably.

We have carried out up till now two experiments the first one [14] with a ^{208}Pb- and the second one with a ^{206}Pb-target [15], both enriched in ^{208}Pb and ^{206}Pb respectively. The targets were used in form of metallic strips 2 mm (^{208}Pb) and 1.5 mm (^{206}Pb) wide and thickness of 45.2 mg/cm^2 (^{208}Pb) and 25 mg/cm^2 (^{206}Pb). The beam intensities were typically (1-1.5) x 10^{11} deuteron per spill (1s) and 2.8 s cycling time. The measured energy spread of the beam including the FRS was less than 0.5 MeV FWHM. The total resolution including the energy spread introduced by the target was about 0.65 MeV (FWHM) for the ^{208}Pb-experiment and could be reduced to 0.31 MeV FWHM for the ^{206}Pb-experiment. A major problem is the need of an absolute calibration of the ^{3}He-momentum in the S2 focal plane of the FRS. For this we made use of the monoenergetic ^{3}He-line from the p (d, ^{3}He) π°-reaction with a (CH$_2$)$_n$-target bombarded with an energy calibrated deuteron beam using a Schottky noise frequency measurement of the deuteron revolution frequence in the synchrotron. The calibration runs were frequently (1-2 hr) repeated and used to correct the energy calibration of the individual measurements. A detailed description of the experimental procedures including the data analysis is given by Gilg et al [16] and Itahashi et al. [17].

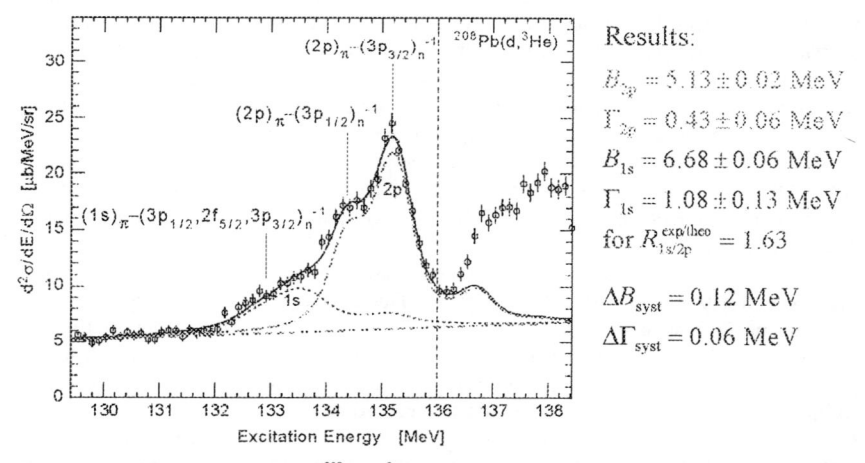

Results:

$B_{2p} = 5.13 \pm 0.02$ MeV

$\Gamma_{2p} = 0.43 \pm 0.06$ MeV

$B_{1s} = 6.68 \pm 0.06$ MeV

$\Gamma_{1s} = 1.08 \pm 0.13$ MeV

for $R_{1s/2p}^{exp/theo} = 1.63$

$\Delta B_{syst} = 0.12$ MeV

$\Delta \Gamma_{syst} = 0.06$ MeV

Figure 4 Excitation energy spectra of the ^{208}Pb (d,^{3}He) reaction (upper panel). Analysis and results for the binding energies and widths of the pionic states in ^{207}Pb (lower panel) [17].

Fig. 4 shows the double differential cross section d^{2}σ/dEdΩ of the ^{208}Pb (d ^{3}He) reaction as function of the excitation energy in the ^{207}Pb-nucleus in the high energy excitation range between 120 and 150 MeV which led to the discovery of deeply bound pionic states in ^{207}Pb [14]. The spectrum contains as an insert the calibration line from the p(d, ^{3}He) π° production marked as "π°-peak".

The discussion of the excitation energy spectrum can be devided in three qualitative different spectral regions. At excitation energies above the π-emission threshold (vertical broken line) E_x = 139.57 MeV, a continuum with rising cross section for quasi free pion production is seen. In the excitation energy region below 130 MeV down to 120 MeV, a rather flat continuum with nuclear excitation without pion production and a cross section of about 5μb/sr MeV shows up. The central region between 130 MeV and 140 MeV excitation energy is most interesting. It shows a clear, but complicated line structure, which can be quantitatively assigned to deeply bound pions coupled to well known neutron hole states in ^{207}Pb, such as the $(3p_{1/2}, 2f_{5/2}, 3p_{3/2})_n$ neutron hole states with excitation energies of 570 and 898 keV respectively. The largest line at around 135 MeV is mainly the expected doublet of a substitutional configuration with the pion bound in the 2p-state and the neutron hole in the $(3p_{1/2})_n^{-1}$-groundstate or with larger strength in the $(3p_{3/2})_n^{-1}$ 898 keV excited state. This configuration is expected to show the strongest excitation, because it requires angular momentum transfer l=0, and its respective population should be proportional by its statistical weight, namely 1 : 2, well born out by the spectrum. Thus we observe with high cross section the "recoil-free" production of a pion in the 2p-state coupled to two neutron hole states in ^{207}Pb as expected theoretically. A closer look at the spectral tail at excitation energies below 134 MeV reveals a shoulder which is energetically close to states predicted for a pion in the 1s-state coupled to $p_{1/2}$, $f_{5/2}$ and $p_{3/2}$ neutron hole states. An analysis of the boundstate region of the spectrum is shown in the lower part of Fig. 4. In order to determine the binding energies and widths of the pionic 1s- and 2p-states, B_{1s}, B_{2p}, Γ_{1s} and Γ_{2p} we decomposed the region of interest in a 2p and 1s component coupled to the neutron hole states as indicated. The best fit to the data were achieved with a ratio of experimental to theoretically intensities of 1s- to 2p-states of 1.63. The results for the binding energies and widths of the 2p- and 1s-states are summarized also in fig. 4 with the statistical errors quoted. The systematical errors for the binding energies are ΔB_{syst} = 0.12 MeV, originating from the absolute energy calibration and for the widths Γ_{syst} = 0.06 MeV.

In order to get a better chance for a clear observation of the pionic 1s-state, we carried out a second experiment with ^{206}Pb as a target. Due to the more favourable neutron hole structure of ^{205}Pb with less $3p_{1/2}$-hole contribution a spectrum without double lines is expected. Thus we hoped to separate the 1s- and 2p-pionic configuration much better [18]. In addition we were able to improve our resolution to estimated 0.31 MeV full width half maximum by several optimalisation procedures.

Fig. 5 shows an excitation energy spectrum of the ^{206}Pb(d, ^{3}He)-reaction from a preliminary analysis [19]. The central part from 130 to 140 MeV shows now clearly separated line. The strongest one at E_x = 135 corresponds to the (2p)-pionic states coupled with the three neutron hole states $(f_{5/2}, p_{1/2}, p_{3/2})_n^{-1}$. The well separated line at Ex ~ 133 MeV is assigned to the 1s-pionic states, also coupled to three neutron hole states. The line structure with 136 MeV $\leq E_x \leq$ 138 is due to the formation of bound pionic states in the 3p, 3d, ... orbits. The small line near the threshold of π^- production is due to the p(d, ^{3}He) π° reaction caused by a hydrogen contaminant of the target.

In order to determine the binding energies and widths of the 1s and 2p-states a similar decomposition into the relevant pion-neutron hole contributions like with the ^{207}Pb-data

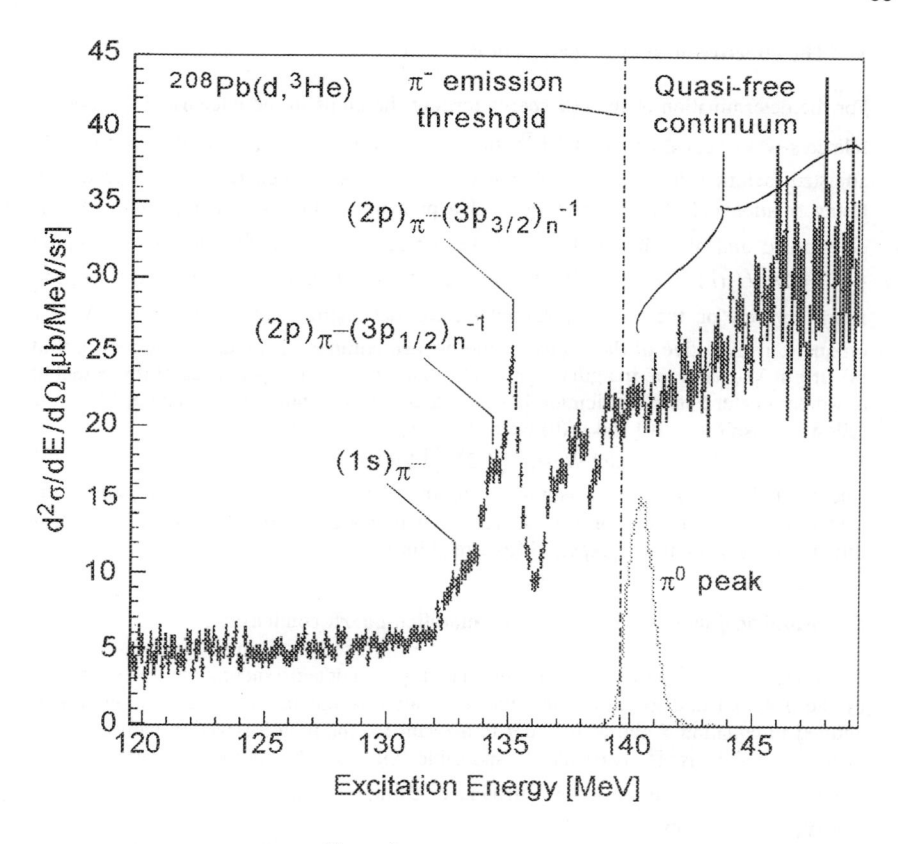

Figure 5 Excitation energy spectra of ^{206}Pb (d,^{3}He) reaction (upper panel). Analysis and results for the binding energies and widths of the pionic states in ^{205}Pb [19].

were performed and are shown in the lower part of Fig. 5. Each pionic $(nl)_\pi$-state is coupled with ten neutron hole states using experimental value for their excitation energies and calculated strengths. The fitting parameters are the binding energies (B_{nl}), widths (Γ_{nl}) and intensities (I_{nl}) of both $(1s)_\pi$ and $(2p)_\pi$. The result of this fit is very good in the central regime but it produces a $(2p)_\pi (f_{7/2})_n^{-1}$ line outside the fitting region which is stronger than the experimental data. A 40 % reduction of the $f_{7/2}$ and $f_{5/2}$ neutron hole strength improves the overall fit and yields B_{1s}- and B_{2p}-values slightly smaller than with using the theoretical f-neutron hole strength. The solid curve in Fig. 5 (lower part) shows this fit function, with the broken and dotted curves representing the 1s- and 2p-contributions respectively. The results for B_{2p}, Γ_{2p}, B_{1s} and Γ_{1s} are tabulated also in fig 5, with statistical errors quoted only. The systematic errors of the binding energies are estimated to about +0.03 and -0.05 MeV. So in summary the results for ^{205}Pb are more accurate and free of systematical errors. So in the following discussion we will mainly use the ^{205}Pb-results.

4 The effective mass of π^--mesons in Pb-nuclei.

For the determination of the self energy term of the pions in the nucleus $\Pi(\vec{r})$, and its related s-wave optical potential $U(\vec{r})$, the Klein-Gordon equation was solved with $\Pi(\vec{r})$ adjusted such that the experimental values of the pion binding energies were reproduced. Using relation (11) the effective mass of a pion in the nuclear medium $m_\pi^*(\vec{r})$ was determined and also the local part of the optical potential $U_s(\vec{r})$ using the relation $\Pi(\vec{r}) = 2\omega \cdot U_s(\vec{r}) = 2m_\pi \cdot U_s(\vec{r})$. Since the $(1s)_\pi$-binding energy and width depend nearly entirely on the s-wave potential we can determine $U_s(\vec{r}) = V(\vec{r}) + i\, W(\vec{r})$ uniquely, irrespective of the choice of the p-wave parameters, by using the real part at the origin V(0) and the imaginary part W(0) only. For the r-dependence of the potential a two parameter Fermi distribution has been used. For the more precise data of ^{205}Pb, the following result for V(0) and W(0) were obtained

$V(0) = (26.1 \pm 2.5)$ MeV and $W(0) = \left(13.2^{+4.5}_{-5.1}\right)$ MeV.

The result V(0) = 26 MeV is equivalent to an effective mass of a negative pion in the center of a ^{205}Pb-nucleus of 166 MeV, which indicates a strong repulsive π^--nucleus interaction as qualitatively expected from relation 14.

5 Empirical deduction of a medium modified quark condensate.

The s-wave part of the Ericson-Ericson optical pion-nucleus potential can be expressed by the isoscalar and isovector pion-nucleon scattering length, b_0 and b_1 respectively as defined in equation 7 and 8. In order to account for the pion interaction with a pair of nucleon, which is in particular responsible for the absorption, a complex term proportional to B_0 and $\rho^2(\vec{r})$ is introduced. Using b_0, b_1 and B_0 one can express V(r) and W(r) in the following way [17]

$$V(r) = -\frac{2\pi}{\mu}[\varepsilon_1\{b_0\rho(r)+b_1[\rho_n(r)-\rho_p(r)]\}+\varepsilon_2\,\mathrm{Re}\,B_0\rho^2(r)] \qquad (17)$$

$$W(r) = -\frac{2\pi}{\mu}\varepsilon_2\,\mathrm{Im}\,B_0\rho^2(r) \qquad (18)$$

with $\varepsilon_1 = 1+\mu/M = 1.147$ and $\varepsilon_2 = 1+\mu/2M = 1.073$ and M, μ denoting the nucleon respectively the pion mass. Using the conventional form for the nuclear densities $\rho_n(r)$ and $\rho_p(r)$ and $\rho(r) = \rho_n(r) + \rho_p(r)$ we can derive simple relations for V(0) and W(0) [17]

$$V(0) = -455\left[b_0 + \frac{N-Z}{A}b_1\right] - 192 R_e B_0 \text{ in MeV} \qquad (19)$$

$$W(0) = -197\,\mathrm{Im}\,B_0 \text{ in MeV} \qquad (20)$$

with b_0 and b_1 in units of m_π^{-1} and b_0 in units of m_π^{-3}.

If one assumes Re $B_0 = 0$, which is supported by a detailed analysis of the deuteron scattering length [20], one arrives at simple relation between the experimentally determined value of V_0 (in (MeV) and the isoscalar and isovector scattering lengths, b_0 and b_1 respectively

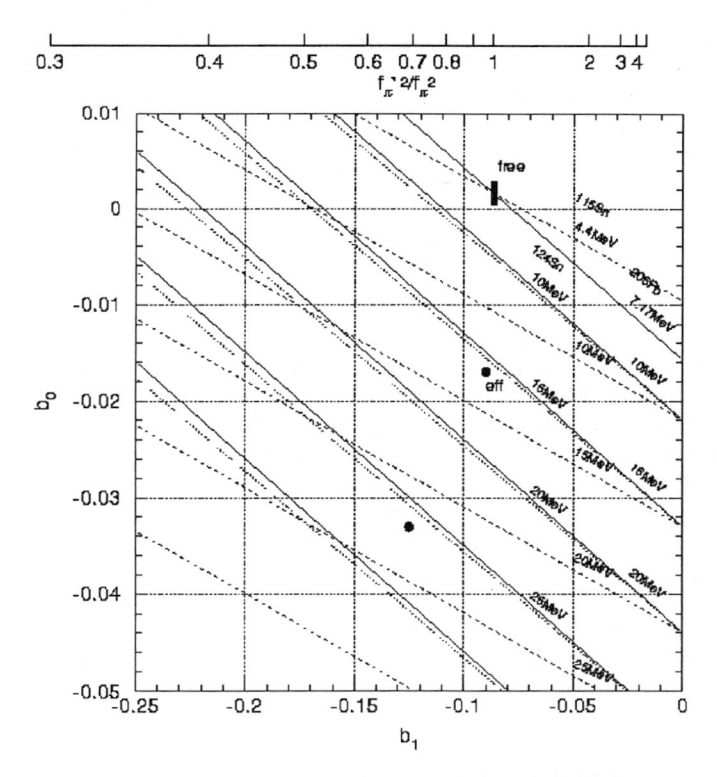

Figure 6 Plot of the isoscalar, b_0, versus the isovector b_1, scattering length of negative pions in various nuclei with the local potential depth V(0) as a parameter.

$$b_0 + \frac{N-Z}{A}b_1 = -\frac{V(0)}{455} \quad (21)$$

so in principal one can determine b_0 and b_1 separately if one extracts V(0) of the pion interaction with nuclei with different values of (N-Z)/A. Recently we have proposed [21] a procedure to compare the pion 1s- binding energies of two nuclei which have different (N-Z)/A values such as ^{205}Pb and ^{115}Sn. A more direct way is to deduce b_1 from an observation of the isotope shift of the 1s pion energy in nuclei, such as ^{115}Sn and ^{123}Sn.

Fig. 6 shows in a b_0-b_1 plane straight lines for experimentally determined V(0)–values and nuclei with different values of (N-Z)/A, such as ^{115}Sn ((N-Z)/A = 0.130), ^{205}Pb ((N-Z/A = 0.200) and ^{123}Sn ((N-Z)/A = 0.187). One notes that one can determine from measured values of V(0) for nuclei with sufficient different values of (N-Z)/A from the crossing of the corresponding straight lines in medium values of b_0^* and b_1^* empirically. We can use for such a procedure three typical cases, π^-–^{205}Pb, ^{115}Sn and ^{123}Sn. For ^{205}Pb we have already obtained an accurate value V(0) = 26.1 MeV for the s-wave potential at the nuclear origin. Experiments on Sn-isotopes are in the planning stage. Note that the

b_0 versus b_1 line for ^{205}Pb wit V(0) is far away from the free values b_0^{free} and b_1^{free} also shown in Fig. 6.

As a first step to derive in medium values of b_0^* and b_1^* we have to consider in medium corrections on the scattering lengths arising from double π-N scattering and higher terms. Following a recent detailed study by Weise et al. [22] we use the following expressions for the effective b-values, including in medium modified b- values, b_0^* and b_1^*, and higher order corrections

$$b_0^{eff} = b_0^* - [(b_0^*)^2 + 2(b_1^*)^2]\left\langle\frac{1}{r}\right\rangle \approx -2b_1^{*2}\left\langle\frac{1}{r}\right\rangle \quad (22)$$

$$b_1^{eff} = b_1^* \quad (23)$$

Note that only the isoscalar scattering length is modified appreciably by the double scattering term which is essentially determined by $(b_1^*)^2$ and the <1/r> value, the expectation value of the inverse relative distance between two nucleons. For a Fermi-distribution with Fermi momentum p_F, $<1/r> = 3p_F/2\pi \cong 0.91 m_\pi$.

Insertion of b_0^{eff} and b_1^{eff} in relation 22 gives the following approximate expression for a determination of the in medium modified value of b_1^*:

$$-2b_1^{*2} <1/r> + \frac{N-Z}{A}b_1^* = \frac{V_0}{455} \quad (24)$$

For the case of ^{205}Pb, with $\frac{N-Z}{A}$ = 0.20 and <1/r> = 0.91 m_π, one can deduce the in medium scattering lengths b_1^* = -0.127 and b_0^* = -0.029, which we include in Fig. 6. Using the relation

$$b_1^*/b_1^{free} = \left(\frac{f_\pi^*}{f_\pi}\right)^{-2} \quad (25)$$

we get for f_π^*/f_π the relative change of the pion decay constant in the medium a value of about

$$f_\pi^*/f_\pi = 0.82 = \frac{<\bar{q}|q>^*}{<\bar{q}|q>} \quad (26)$$

The conclusion is that our measurement of the 1s-binding energy of a negative pion in ^{205}Pb indicates that the pion decay constant and thus the quark condensate is reduced in a Pb-nucleus by a factor of 0.82.

6 Deeply bound pionic states in chiral perturbation theory

Very recently Leisibach and Weise (20,23] calculated the binding energies and widths of pionic 1s- and 2p-states in Pb- and Sn-isotopes using chiral perturbation theory. The main goal was to study systematically the sensitivity of the energies and widths of theses states on the density dependence of the pion decay constant, which determines the chiral s-wave potential. Using relation (6) with σ_N = (45 ± 8) MeV and f_π in the isovector $T^{(-)}$ matrix replaced by $f_\pi^*(\rho_0) \cong 0.82 f_\pi$, leads to a roughly twice as large repulsive pion-nucleon potential and therefore a reduction of both the binding energies as well as the widths of the 1s- and 2p-states. The results for ^{207}Pb are shown in Fig. 7 in which the widths of 2p- and 1s-states is plotted versus its binding energies. The points denoted with f_π are obtained with the vacuum value of the pion decay constant in $T^{(-)}$ as it enters in the s-wave optical potential. The dark ellipses, denoted with $f_\pi^*(\rho)$ are obtained with $f_\pi^* = 0.82\ f_\pi$ and using $\text{Re} B_0 = 0$ and $\text{Im } B_0 = 0.06 m_\pi^{-4}$. The neutron radius is taken 3% larger than the proton radius in the local neutron and proton density distributions $\rho_n(r)$ and $\rho_p(r)$ respectively. Clearly the "vacuum" f_π value based predictions fail to reproduce the experimental binding energies and especially the narrow widths of the states indicated by light shaded ellipses. On the other hand replacing the vacuum decay constant in $T^{(-)}$ by $f_\pi^*(\rho)$ as given by equation 6, with ρ treated as local density distribution the missing repulsion in the s-wave potential is adequately supplied. The theoretical predictions indicated by dark ellipses reproduces the experimental binding energies and the especially narrow widths of both the 2p- and 1s-states rather well.

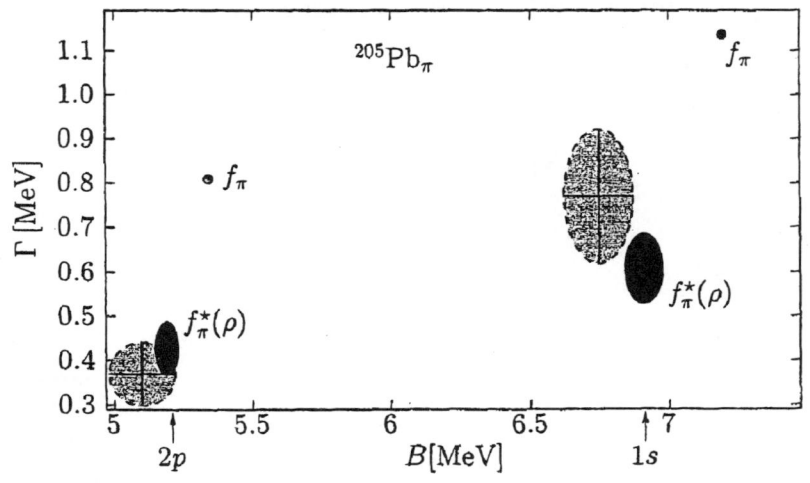

Figure 7 Binding energy B and width Γ of 1s- and 2p-pionic states in ^{205}Pb. Points, marked "f_π" are obtained using the chiral s-wave optical potential with the vacuum pion decay constant ($f_\pi = 92.4$ MeV) and Re $B_0 = 0$. Dark ellipses ($f_\pi^*(\rho)$) are results when f_π was replaced by $f_\pi^*(\rho)$ with $\sigma_N = (45\pm 8)$ MeV. Light shaded ellipses represent the experimental results for ^{205}Pb [23].

Empirically the missing s-wave repulsion can be generated by simply adjusting Re B_0. This would require a large negative value, Re $B_0 \cong -0.07 m_\pi^{-4}$, which is not expected by many body calculations and which is at odds with the π^--deuteron scattering length.

7 Summary and perspective

We have observed for the first time deeply bound pionic states in Pb-nuclei using recoilfree pion transfer reactions. The measured binding energies and narrow widths of the 1s- and 2p-states indicate a strong nuclear repulsion of the negative pions, which leads together with the attractive Coulombforce to potential pockets at the edge of the nuclei in which halo like pionic states with narrow decay width are formed.

It could be shown empirically as well as in the frame work of a chiral perturbation theory that the strong s-wave nuclear repulsion of the negative pions is naturally explained by a reduced density dependent pion decay constant which gives a direct evidence of a density dependent reduction of the quark condensate in the QCD vacuum. In the future systematic studies of pionic 1s-states in S_n-nuclei and isotones with $N \cong 82$ should help to establish a large and precise enough data set to extract the density dependent pion decay constant uniquely.

The author would like to thank W. Weise and T. Yamazaki for many fruitful discussions, which led to the presented understanding of the complex problem. The work is supported by the Gesellschaft für Schwerionenforschung, the Japanese Ministry of Education, Science, Culture and Sports. The author owes thanks to the Alexander von Humboldt Stiftung for a special support of this work.

References

1. Nambu Y. and Jona-Lasinio G., Phys. Rev. **122** (1961) pp. 345.
2. Reinders L, Rubinstein H., Yazaki S., Phys. Rep. **127** (1985) pp. 2.
3. Carrasco R.C., Phys. Rev. **C48** (1993) pp. 2333.
4. Gell-Mann M., Oakes R.J., Renner B., Phys. Rev. **175** (1968) pp. 2195.
5. Drukarev E.G. and Levin E.M., Nucl. Phys. **A511** (1990) pp. 679.
   Lutz M., Klimt S., Weise W., Nucl. Phys. **A542** (1992) pp. 52
   Cohen T.D., Furnstahl R.J., Griegel D.K., Phys. Rev. **C45** (1992) pp. 1881.
6. Tomozawa Y.; Nuovo Cimento **A46** (1966) pp. 707.
   Weinberg S., Phys. Rev. Lett. **17** (1966) pp. 616.
7. Simon L., Phys. Lett. **B469** (1999) pp. 25.
   Ericson T.E.O., Loiseau B., Thomas A.W., Nucl. Phys. **A663-664** (2000) pp. 541c.
8. Friedmann E. and Soff G., J. Phys. G. Nucl. Phys. **11** (1985) pp. L37.
9. Toki H. and Yamazaki, T., Phys. Lett **213B** (1988) pp. 129.
10. Ericson M. and Ericson T.E.O., Ann. Phys. **36** (1966) pp. 323.
11. Seki R. and Matsutani , Phys. Rev. **C27** (1983) pp. 2799.
12. Toki H. et al., Nucl. Phys. **A501** (1989) pp. 653.
    Toki H., Hirenzaki S., Yamazaki T., Nucl. Phys. **A530** (1991) pp. 679.
    Hirenzaki S., Toki H., Yamazaki T., Phys. Rev. **C44** (1991) pp. 2472.
13. Geissel H. et al., Nucl. Instr. Meth. **B70** (1992) pp. 286.
14. Yamazaki T. et al., Z. Phys. **A355** (1996) pp. 219.
15. Geissel H. et al., to be published.
16. Gilg H. et al., Phys. Rev. **C62** (2000) pp. 025201.
17. Itahashi K. et al., Phys. Rev. **C62** (2000) pp. 025202.
18. Hirenzaki S. and Toki H., Phys. Rev. **C55** (1997) pp. 2719.
19. Suzuki K., Yoneyama T., unpublished.
20. Leisibach R. and Weise W., in preparation.
21. Kienle P. and Yamazaki T., in preparation.
22. Weise W. et al., in preparation.
23. Weise W., Acta Physica Polonica **B31** (2000) pp. 2715.

PRE-EQUILIBRIUM REACTIONS

P.E. HODGSON
Nuclear Physics Laboratory, Oxford, UK

Analyses of (p, α) and $(p,{}^3\text{He})$ reactions to the continuum are reviewed, and the importance of analysing powers for determining the reaction mechanism is discussed. Recent analyses using the Feshbach-Kerman-Koonin multistep direct theory of the (p, α) reactions using the alpha-particle knock out model and of the $(p,{}^3\text{He})$ reaction using the deuteron pickup model are described. All this work shows how spin-dependent phenomena provide a powerful tool for the study of the nuclear reaction mechanisms.

1. Introduction

Many analyses of differential cross-sections and analysing powers of (p, α) reactions to discrete states have been made (Gadioli and Hodgson, 1989) using either the triton pickup or the alpha-particle knock-out models. Both models give rather similar results (Gadioli et al, 1984), but nuclear structure arguments favour the pick-up model. For a long time the calculations gave absolute magnitudes for the cross-section that were far too low (Brunner et al, 1983; Hoyler et al, 1985) but detailed analyses (Walz et al, 1988) have largely removed the discrepancy, although some difficulties still remain (Kajihara, 1992). Comparativley few analyses have been made of (p, α) reactions at higher incident energies that proceed to unresolved continuum states. At these energies the contributions of two, three and higher step processes become more important, and so it is appropriate to use multistep reaction theories. Quantum-mechanical theories of multistep reactions have been developed by Feshbach, Kerman and Koonin (1980) (FKK), by Tamura et al (1977ab, 1981, 1982) and by Nishioka, Weidenmüller and Yoshida (1988, 1990). These theories make somewhat different statistical assumptions, and they have been compared in detail by Koning and Akkermans (1991, 1993). All these theories give the same result for the first step of the reaction, but differ for the second and higher steps. The FKK theory gives expressions for the cross-section that have a simple convolution structure which greatly facilitates the calculation of the contributions of the higher steps, and mainly for this reason it has been used more often than the other theories to analyse experimental data.

2. The Reaction Mechanism

A purely classical argument can be used to show that the triton pick-up mechanism is more likely than alpha knockout to populate low-lying final states. The kinematics of the proton-alpha collision do not allow all the energy of the incident proton to be transferred to the alpha-particle; it retains some energy and therefore tends to go to final states of higher energy, mainly in the continuum. The choice of the pick-up model is supported by consideration of the states in the final nucleus populated by the reaction (Gadioli et al, 1984). Calculations of the cross-sections of the reaction ^{58}Ni $(p, \alpha)^{55}$Co to the ground state by Bonetti et al (1989) at 22 and 72 MeV show that the pickup model gives approximately the correct absolute magnitude, whereas the knockout model gives abolute cross-sections that are too low by a factor of a thousand. Most calculations of reactions to resolved final states, especially those used for nuclear spectroscopy, have therefore used the pickup model. Several other results, however, indicated that the reaction may also proceed by the knockout process, in particular the qualitative evidence that unpaired valence target nucleons can act as spectators (Gadioli et al, 1981, 1982, 1983).

Initially, calculations of reactions to continuum states also used the pickup model. Subsequently, some analyses were made of the knockout cross- section using the exciton model and assuming that the reaction can be treated as a quasi-free proton-alpha interaction (Scobel et al, 1977), using the free nucleon-alpha-particle cross-section and describing the distortions of the incident proton and outgoing alpha-particle wave functions classically as refraction at the nuclear boundary (Ferrero et al, 1979). This theory predicts that the energy and angular distributions of the alpha- particles should be nearly independent of the target nucleus, and this is indeed the case. As in the case of reactions to discrete states, the calculations using the pick-up and the knock-out models give equally good angular distributions, but the work of Bonetti et al, (1989) indicates that the analysing powers are successfully predicted by the knock-out model but not by the pick-up model, and this is supported by the phase-space arguments. The usefulness of analysing powers to determine the reaction mechanism is further discussed in section 4.

3. Analyses of (p, α) reactions using the Feshbach-Kerman- Koonin theory

The first analysis of (p, α) reactions to continuum states using the FKK theory was made by Bonetti et al, (1989) using data of Lewandowski et al (1982). Due to the difficulties of evaluating the magnitude of the continuum cross-section they calculated only the analysing power. This work is discussed in Section 4. A detailed analysis of double-differential cross-sections was made by Olaniyi *et al* (1995), using the data of Ferrero *et al* (1979). These data comprised

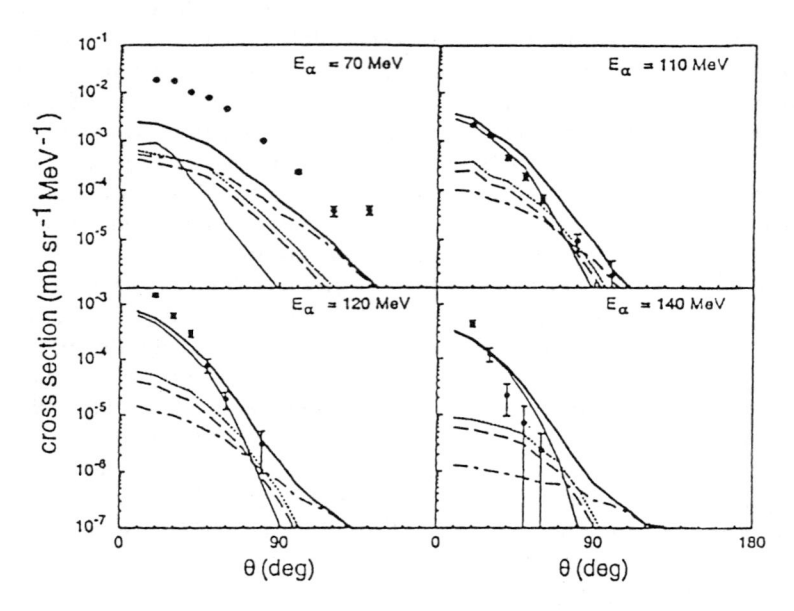

Figure 1. Double-differential cross-sections for the ^{59}Co (p, α) reaction at an incident energy of 160 MeV compared with Feshbach-Kerman-Koonin calculations for one-step knockout (thin solid curves), the two-step (p, p', α) (dotted curves) and (p, n, α) (dashed curves) processes and the sum of the three-step contributions (dot-dashed curves). The sum of all these processes is given by the thick solid curves (Cowley *et al*, 1996).

double differential cross-sections on several nuclei for incident energies of 30 and 44 MeV. The knock-out model was used, following Bonetti *et al* (1989). At these energies two-step processes are unlikely and the incident proton is captured into a bound state of the residual nucleus. In these experiments, the $(p, ^3\text{He})$ cross-section was included in the (p, α) cross-section since the ^{3}He and alpha-particles were not resolved. However it is estimated that the $(p, ^3\text{He})$ cross-section is less than 10% of the (p, α) cross-section, so this does not appreciably affect the analysis. At these incident energies there is a large contribution from the compound nucleus process, so this was removed using the subtraction method (Demetriou *et al*, 1993). The remaining cross-section was analysed using the multistep direct FKK theory, which proved able to give a good fit to the data. The compound nucleus cross-section was calculated using the theory of Hauser and Feshbach (1952) and, when added to the multistep direct cross-section, gave a good overall fit to the angle-integrated energy spectra of the outgoing alpha-particles.

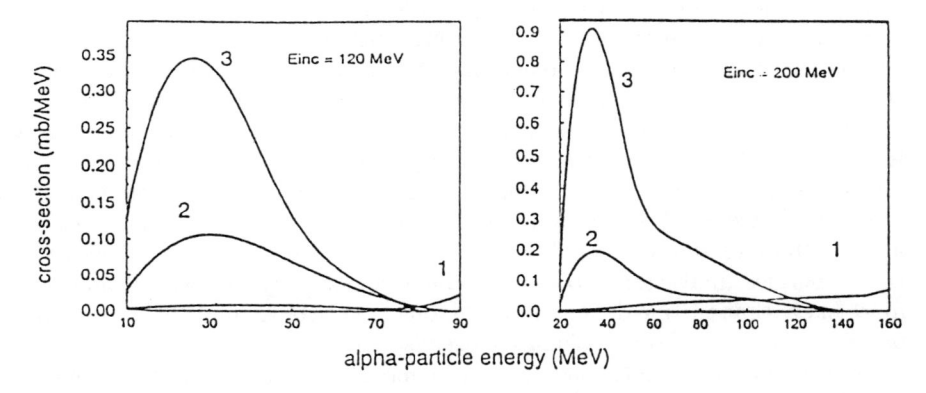

Figure 2. The cross-sections of the first, second and third steps of the ^{59}Co (p, α) reactions for two incident energies as a function of the outgoing alpha-particle energy, (Hodgson et al, 1997).

These calculations were made using a zero-range effective interaction and the direct form of the transition matrix element, which is justified because then the direct and exchange forms of the matrix element are identical. Calculations with a finite-range interaction require an exchange matrix element, and so far these have not been made. Calculations with finite range Gaussian and Yukawa interactions gave very similar results to those obtained with the zero-range reaction, and this may indicate that it is sufficiently accurate to use the zero-range interaction. In all these calculations, the cross-sections were normalised to the experimental data by a factor that depends on the spectroscopic amplitudes and on the alpha-particle preformation probability. Pairing and spectator effects in (p, α) reactions were studied by Guazzoni et al (1996). A more detailed study of the energy dependence of the (p, α) cross-section was made by Demetriou and Hodgson (1996). As in the case of nucleon-induced reactions, the only energy dependence in the FKK theory is that of the effective interaction, and the folding model indicates that this is the same as that of the corresponding optical potential. In the case of the (p, α) reaction, this is the alpha-particle optical potential, and the effective interaction was indeed found to have the same, rather weak, energy dependence, within the experimental uncertainties. As in previous analyses, it was found that the calculated cross-sections are relatively insensitive to the proton optical potential, but are very sensitive to that of the alpha-particle, even when they are in accord with the corresponding elastic scattering cross-sections. This is a serious difficulty in all analyses using optical potentials obtained from analyses of elastic scattering cross-sections. The reason for this sensitivity is to be found in the different contributions of the S-matrix elements to the elastic and non-elastic-reactions, so that a set determined from elastic scattering does not necessarily give the best physically-correct values for non-elastic reactions. The only way to tackle this problem phenomenologically is to choose the sets of proton and alpha-particle optical potentials that also give the best fit to the reaction data, and this procedure was adopted. These calculations were subsequently extended to analyse the (p, α) cross-sections at 120, 160 and 200 MeV on ^{27}Al, ^{59}Co and ^{197}Au (Cowley et al, 1996). At these energies, multistep processes become increasingly important, and it is more likely that the incident proton remains in the continuum. The multistep processes can be readily calculated because of the convolution structure of the multistep direct **FKK** theory, but it is more difficult to include the effects of unbound protons in the final state. In these calculations the device was adopted of deepening the proton potential to make these protons just bound. This may be a reasonable approximation at energies just above the theshold, but it is likely that it becomes increasingly unsatisfactory at higher energies. In these calculations, the two-step reactions (p, p', α) and (p, n, α) were included, and also the three-step (p, N, N', α) reactions, where N stands for a neutron or a proton. The multi-step reactions require the nucleon-nucleon effective interaction, and its energy dependence was included in the calculations. The calculations were normalised to the experimental data at the highest outgoing energies, which are dominated by the one-step process. This normalisation factor includes the pre-formation factor of the alpha-particle in the target nucleus and other uncertainties in the calculation. The results of some of these calculations are shown in Fig. 1 and the contributions of the first, second and third steps to the ^{59}Co (p, α) reaction at 120, 160 and 200 MeV are shown in Fig. 2.

It is apparent that the multistep processes become increasingly important at the lower outgoing energies, in accord with physical expectations.

4. Analysing Powers in (p, α) Reactions

If the incident proton beam is polarised, the asymmetry in the angular distribution of the emitted alpha-particles enables the analysing power to be determined. Measurements of the analysing power of (p, α) reactions to the continuum have been made in this way by Sakai *et al* (1980) and by Lewandowski *et al* (1980).

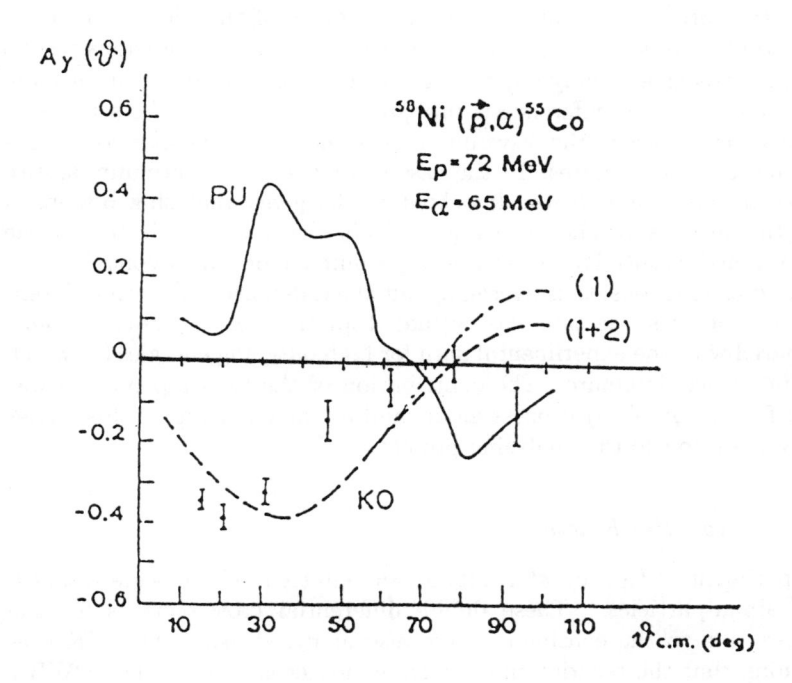

Figure 3. Analysing powers for the ^{58}Ni $(p, \alpha)^{55}$Co reaction at 72 MeV to continuum states compared with Feshbach-Kerman-Koonin calculations of the one and two-step contributions using the pickup and knockout theories (Bonetti *et al*, 1989).

The first analyses of analysing powers in (p, α) reactions were made by Tamura *et al* (1982) using the triton pickup theory. The fits to the one-step reaction, and to the two-step (p, α, α') and (p, p', α) reactions are quite good for the two lower outgoing energies, but there are serious differences for the highest outgoing energy. Tamura *et al* suggest that these may be removed by a more detailed consideration of the individual nuclear states. The calculations for all outgoing energies show a broad maximum from 70° to 160°, and this is due mainly to the (p, α, α') reaction. The analysing powers of the (p, α) reaction at 22 and 72 MeV on ^{58}Ni to both discrete and continuum states have been calculated by Bonetti et al (1989) using both the triton pickup and the alpha- particle knockout reaction mechanisms. They find that the pickup reaction to discrete states gives qualitative agreement with the analysing powers, whereas the knockout reaction fails to give the observed structure. This confirms the applicability of the pickup model to reactions to discrete states, but may appear contrary to the previous results that both models give equally good fits to both the angular distributions and analysing powers. This difference is probably attributable to the sensitivity of the analysing powers to the wavefunctions of the few states that contribute to reactions to discrete states. In the case of reactions to continuum states, Bonetti et al find that only the knockout model gives analysing powers in accord with the data, as shown in Fig. 3. This result is not affected by the above-mentioned sensitivity because a large number of states contribute to continuum cross-sections, thus averaging out the differences between different transitions. The cross- section was evaluated for the one-step reaction only, and compared with the experimental data for high outgoing energies for which one-step processes dominate. The contribution of the two-step process was estimated for the (p, p', α) process alone, and as shown in Fig. 3 this makes a small contribution to the analysing power.

5. Analysis of $(p, {}^3He)$ Reactions

In the experiment of Cowley et al (1996), the emitted helions were resolved from the alpha-particles. These double-differential cross-sections for the $(p, {}^3He)$ reaction to the continuum were also analysed using the FKK theory, assuming that the reaction mechanism is deuteron pickup. The DWBA cross-section is given as a sum over all possible neutron- proton configurations and isospin transfers, with the appropriate Clebsch-Gordon coefficients. The form factor of the deuteron was obtained using the well-depth procedure with geometrical parameters adjusted so that the microscopic and macroscopic form factors are almost identical.

The ^{3}He optical potential was chosen to give the best overall fit to both elastic scattering and reaction data. The cross-sections of the two-step $(p, p', {}^3He)$ and three-step $(p, p', p'', {}^3He)$ reactions were also calculated. The cross-sections were normalised to the data at the highest outgoing energy. Some typical results for the cross-sections are shown in Fig. 4, and for the individual contributions of the one, two and three-step processes in Fig. 5. It is notable that as the energy transfer increases the two-step process becomes increasingly more important and becomes comparable with the one-step pro-

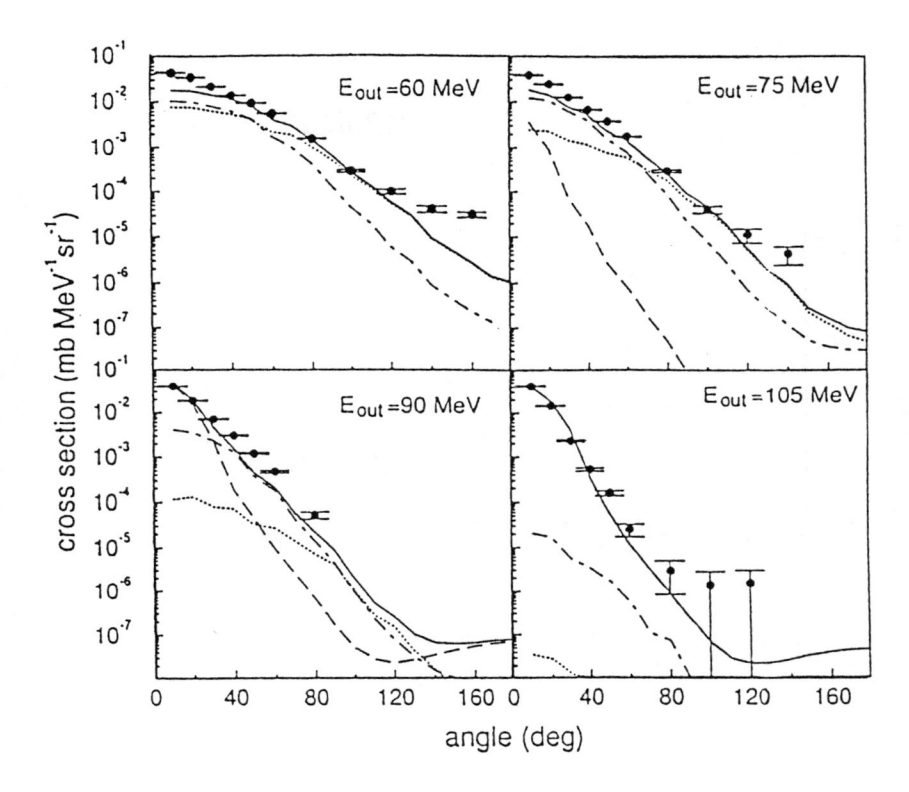

Figure 4. Double-differential cross-sections for the ^{59}Co $(p, ^{3}\text{He})^{57}$Fe reaction at an incident energy of 120 MeV compared with Fesbach-Kerman-Koonin calculations for one step (long-dashed curves), two-step (dot-dashed curves) and three-step (dotted curves) processes. The sum of these contributions is shown by the solid curves (Cowley et al, 1997).

cess for energy differences between incident and outgoing energies around 30 MeV. Thereafter the one-step cross-section decreases rapidly with decreasing outgoing energy, finally becoming negligible for energy differences around 50 MeV where two- and three-step processes dominate the cross-section.

The analysing powers provide further information on the reaction mechanism. We therefore studied the cross-sections and analysing powers of $(p, ^{3}\text{He})$ reactions at 72 MeV measured by Lewandowski et al (1982) for five nuclei. It is notable that the analysing powers for the ^{58}Ni $(p, ^{3}\text{He})$ reaction at 72 MeV are very close to zero for all except the highest outgoing energy of 50

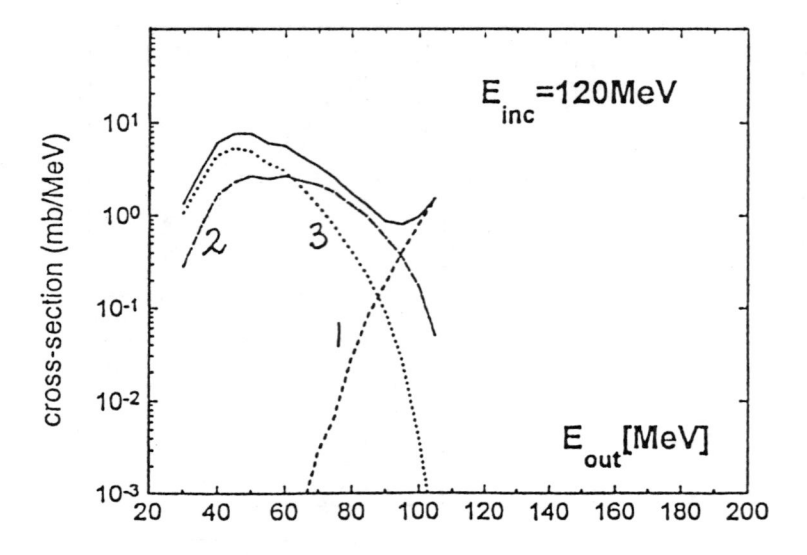

Figure 5. The cross-sections of the first, second and third steps of the ^{197}Au $(p, {}^3\text{He})$ reaction at 120 MeV (Hodgson et al, 1997).

MeV. As shown in Fig. 6, the one-step process dominates the reaction for small energy losses (high outgoing energies) and the two and three-step processes rapidly become more important as the energy loss increases, while the one-step process rapidly becomes insignificant.

It is thus reasonable to conjecture that the analysing powers of the second and higher steps of this reaction are negligible, and that the values observed for the outgoing energy of 50 MeV is due to the first step, diluted by the presence of contributions from the higher steps. These results already show the usefulness of analysing powers for studying nuclear reaction machanisms, in this case the contributions of multistep processes. While the physics of the reaction is quite well understood, there remain two outstanding problems. The first is the absolute magnitude of the two-step process. The shortfall found in the present calculations may be due to the omission of the successive pickup reaction mechanism, whereby the two nucleons are picked up one by one (Hagiwara and Tanifuji, 1957). The second problem is the sensitivity to the helion optical potential. Further studies are in progress.

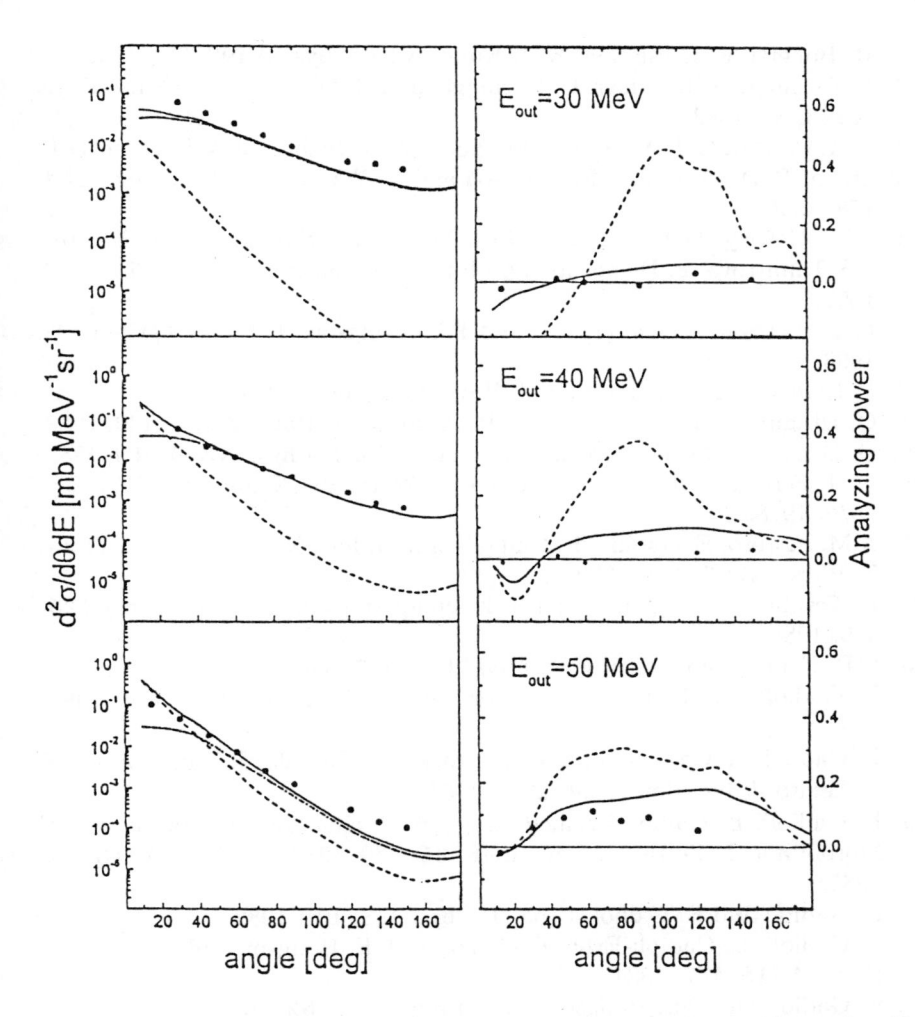

Figure 6. Double - differential cross sections and analyzing powers for the ^{58}Ni $(p, {}^3\text{He}){}^{56}$Co reaction at an incident energy of 72 MeV and three outgoing energies, compared with Feshbach-Kerman-Koonin calculations for one-step (dashed curves) and two-step (dot-dashed curves) processes. The sum of the two contributions is given by the solid curves.

References

1. F. E. Bertrand and R. W. Peelle, Oak Ridge National Laboratory Report, QUIT-4469, 1970.

2. R. Bonetti, F. Crespi and K.-I. Kubo, Nucl. Phys. **A499**, 381, (1989).
3. F. Brunner, H.H. Müller, C. Dorninger and H. Oberhummer, Nucl. Phys. **A398**, 84, 1983.
4. A.A. Cowley, G.J. Arendse, J.W. Koen, W.A. Richter, J.A. Stander, G.F. Steyn, P. Demetriou, P.E. Hodgson and Y. Watanabe, Phys. Rev. **C54**, 778, 1996.
5. A.A. Cowley, G.J. Arendse, G.F. Steyn, J.A. Stander, W.A. Richter, S.S. Dimitrova, P. Demetriou and P.E. Hodgson, Phys. Rev. **C55**, 1843, 1997.
6. P. Demetriou, P. Kanjanarat and P.E. Hodgson, J. Phys. **G19**, L193, 1993.
7. P. Demetriou, P.E. Hodgson, J. Phys. **G22**, 1813, 1996.
8. O. Dragun, A.A. Ferrero and A. Pacheco, Nucl. Phys. **A369**, 149, 1981.
9. O. Dragun, A.M. Ferrero and A. Pacheco, Nucl. Phys. **A369**, 169, 1981.
10. A.M. Ferrero, I. Iori, N. Molho and L. Zetta, INFN Report, INFN/BE, 78/6, 1978.
11. A.M. Ferrero, E. Gadioli, E. Gadioli-Erba, I. Iori, N. Molho and L. Zetta, Z. Phys. **A293**, 123, 1979.
12. H. Feshbach, A. Kerman and S. Koonin, Ann. Phys. (New York) **125**, 429, 1980.
13. C.B. Fulmer and J.C. Hafele, Phys. Rev. **C7**, 631, 1973.
14. E. Gadioli, E. Gadioli-Erba, I. Iori and L. Zetta, J. Phys. **G6**, 1391, 1980.
15. E. Gadioli, E. Gadioli-Erba, L. Glowacka, M. Jaskola, J. Turkiewicz and L. Zemto, Phys. Rev. **C24**, 2331, 1981.
16. E. Gadioli, E. Gadioli-Erba, R. Gaggini, P. Guazzoni, P. Machelato, P. Moroni and L. Zetta, N. Cim. Lett. **35**, 460, 1982; Z. Phys. **A310**, 43, 1983.
17. E. Gadioli and E. Gadioli-Erba, J. Phys. **G8**, 101, 1982.
18. E. Gadioli, E. Gadioli-Erba, P. Guazzoni, P.E. Hodgson and L. Zetta, Z. Phys. **A318**, 147, 1984.
19. E. Gadioli and P.E. Hodgson, Rep. Prog. Phys. **52**, 301, 1989.
20. P. Guazzoni, L. Zetta, P. Demetriou and P.E. Hodgson, Z. Phys. **A354**, 53, 1996.
21. P. Guazzoni, I. Iori, S. Micheletti, N. Molho, M. Pagnanelli and G. Semenescu, Phys. Rev. **C4**, 1092, 1971.
22. P. Guazzoni, M. Pignanelli, E. Colombo and F. Crescenti, Phys. Rev. **C13**, 1424, 1976.
23. H. Hagiwara and M. Tanifuji, Prog. Theor. Phys. **18**, 322, 1957.
24. W. Hauser and H. Feshbach, Phys. Rev. **87**, 366, 1952.

25. P.E. Hodgson, Contemp. Phys. **31**, 99, 1990.
26. P.E. Hodgson, P. Demetriou and S.S. Dmitrova. Il Nuovo Cimento **110A**, 967, 1997.
27. F. Hoyler, H. Oberhummer, T. Rohwer, G. Standt and H.V. Klapdor, Phys. Rev. **C31**, 17, 1985.
28. T. Kajihara, M.Sc. Thesis, Tokyo Metropolitan University, 1992.
29. A.J. Koning and J.M. Ackkermans, Ann. Phys. **208**, 216, 1991; Phys. Rev. **C47**, 724, 1993.
30. I. Kumabe, M. Matoba and K. Fukuda, Proc. 1980 RCNP Int. Symp. on Highly Excited States in Nuclear Reactions. Ed. H. Ikegami and M. Muraoka (Osaka: RCNP), 61, 1980.
31. Z. Lewandowski, E. Loeffler, R. Wagner, H. H. Mueller, W. Reichart and P. Schober, E. Gadioli and E. Gadioli-Erba, N. Cim. Lett. **28**, 15, 1980.
32. Z. Lewandowski, E. Loeffler, R. Wagner, H.H. Mueller, W. Reichart and P. Schober, Nucl. Phys. **A389**, 249, 1982.
33. H. Nishioka, H.A. Weidenmuller and S. Yoshida, Ann. Phys. **183**, 166, 1988 ; Z. Phys. **A336**, 197, 1990.
34. H.B. Olaniyi, P. Demetriou and P.E. Hodgson, J. Phys. **G21**, 361, 1995.
35. C.M. Perey and F.G. Perey, At. Nuc. Data Tables **17**, 1, 1976.
36. B. W. Ridley, Bull. Am. Phys. Soc. **13**, 117, 1968.
37. H. Sakai, K. Hosono, N. Matsuoka, S. Nagamachi, K. Okada, K. Maeda and H. Shimizu, Nucl. Phys. **A344**, 41, 1980.
38. W. Scobel, M. Blann aand A. Mignerey, Nucl. Phys. **A287**, 301, 1977.
39. J.S. Sens and R.J. de Meijer, Nucl. Phys. **A407**, 45, 1983.
40. T. Tamura and T. Udagawa, Phys. Lett. **B71**, 273, 1977a.
41. T. Tamura, T. Udagawa, D.H. Feng and K.K. Kim, Phys. Lett. **B66**, 109,1977b.
42. T. Tamura, H. Lenske and T. Udagawa, Phys. Rev. **C23**, 2769, 1981.
43. T. Tamura, T. Udagawa and H. Lenske, Phys. Rev. **C26**, 379, 1982.
44. T. Udagawa, T. Tamura and K.S. Low. Phys. Rev. Lett. **34**, 30, 1975.
45. M. Walz, R. Neu, G. Standt, H. Oberhummer and H. Cech, J. Phys. **G14**, L91, 1988.

MESON PRODUCTION IN HADRONIC REACTIONS

S.KREWALD, O.KREHL, Z. S. WANG, AND J.SPETH

Institut für Kernphysik, Forschungszentrum, D-5242 Jülich, Germany
E-mail: s.krewald@fz-juelich.de

A meson-theoretical model for two-pion production in the pion-nucleon reaction has been developped for ultrarelativistic energies of the incoming pion. It can explain the $f_0(980)$ production anomaly seen in the $\pi^- p \to \pi^0 \pi^0 n$ reaction by both the GAMS and the BNL experimental collaborations. The new data are compatible with a non-$q\bar{q}$ structure of the f_0. The Roper resonance $N^*(1440)$, in the Jülich meson-exchange model, likewise shows a strong deviation from a three valence quark structure.

1 The $f_0(980)$ production puzzle in the pion-nucleon reaction

A large amount of our knowledge about the pion-pion interaction has been obtained by the reaction $\pi N \to \pi\pi N$. For pion beam momenta above approximately 10 GeV/c and for small values of the square $t = (p_N - p_{N'})^2$ of the momentum transferred between the incoming and outgoing nucleon, the reaction is a peripheral one. In the language of heavy ion physics, the incoming pion performs a "grazing collision" with the nucleon and therefore touches only the pion cloud which dresses the bare nucleon. Assuming this simple reaction mechanism (see Fig.1), one has derived the known pion-pion phase shifts.

During the last decade, there has been a significant experimental progress due to new detector developments which allow high statistics studies of two-pion production [1,2]. In the charge exchange reaction $\pi^- p \to \pi^0 \pi^0 n$, which eliminates the odd angular momenta from the partial wave analysis of the two-pion system and is selective to the isospin $I = 0$ and $I = 2$, a production anomaly has been observed recently in the vicinity of the $f_0(980)$ meson. The GAMS collaboration has employed a π^- beam momentum of $38 GeV/c$ [1], while the BNL-E852 collaboration uses an incident beam momentum of $18.3 GeV/c$ [2]. The experimental results can be summarized as follows. For momentum transfers $-t < 0.2 (GeV/c)^2$, the squares of the S-wave amplitudes show a broad enhancement above the threshold with a sharp dip near the invariant two-pion mass $m_{\pi\pi} = 980 MeV$. This dip corresponds to the excitation of the $f_0(980)$. A similar dip has been seen in the reaction $\bar{p}p \to 3\pi^0$ by the Crystal Barrel collaboration [3]. For momentum transfers above $-t < 0.4 (GeV/c)^2$, however, a puzzle emerges: at $m_{\pi\pi} = 1 GeV$ a distinct peak is seen. In the

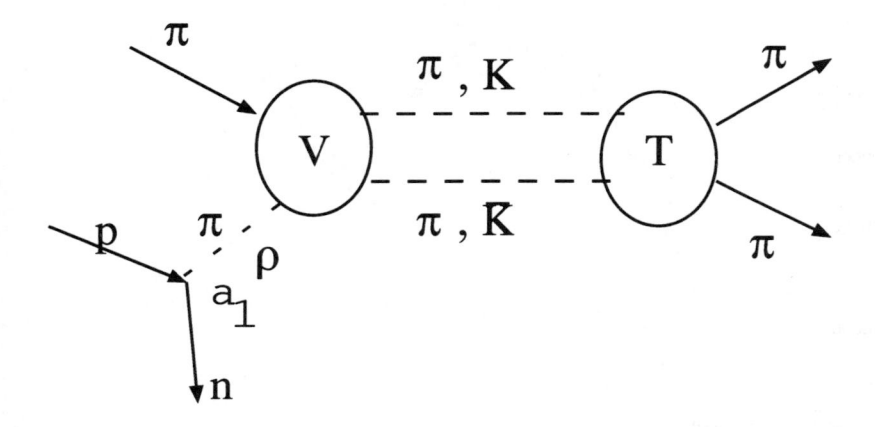

Figure 1. Meson-exchange model for the peripheral reaction $\pi N \to \pi\pi N$.

GAMS data, the peak is taller than in the corresponding BNL-data. In Fig.2, we show the most recent BNL-data.

Despite a large body of experimental and theoretical work, the structure of the $f_0(980)$ remains a controversial issue. Since at least three scalar-isoscalar mesons have been established by now, i.e. the $f_0(980)$, $f_0(1370)$, and the $f_0(1500)$, and the low-lying scalar-isoscalar strength might be summarized as a meson $f_0(400-1200)$, there is no obvious single candidate for the scalar member of the $q\bar{q}$ nonet. The $f_0(980)$ has been interpreted as a multi-quark state, a $K\bar{K}$ molecule, or as a unitarized $q\bar{q}$ state. The issue could not be decided by the analyses of the $\gamma\gamma \to \pi\pi$ reaction.

In the meson-exchange model developped in Jülich, the pions interact by the exchange of mesons[4]. A coupled-channel scattering equation $T = V + VGT$ has to be solved. Although ideally one would like to solve the full Bethe-Salpeter equation, in pratical calculations one relies on a three-dimensional reduction, such as the Blankenbecler-Sugar reduction. Close to the reaction threshold, the $\rho-$exchange in the t-channel is dominant and provides a large attraction in the scalar-isoscalar partial wave which generates a strongly rising phase shift. The attraction is not strong enough, however, to generate a quasibound two-pion state. This is different in the $K\bar{K}$-channel. Here, there is a coherent attraction provided by the exchange of the $\rho-, \omega$, and $\Phi-$ meson-exchange in the t-channel, which generates a quasibound $K\bar{K}$-state.

Recently, the Jülich meson-exchange model has been extended and applied to the $\gamma\gamma \to \pi\pi$ reaction[5]. The square of the S-wave partial wave ampli-

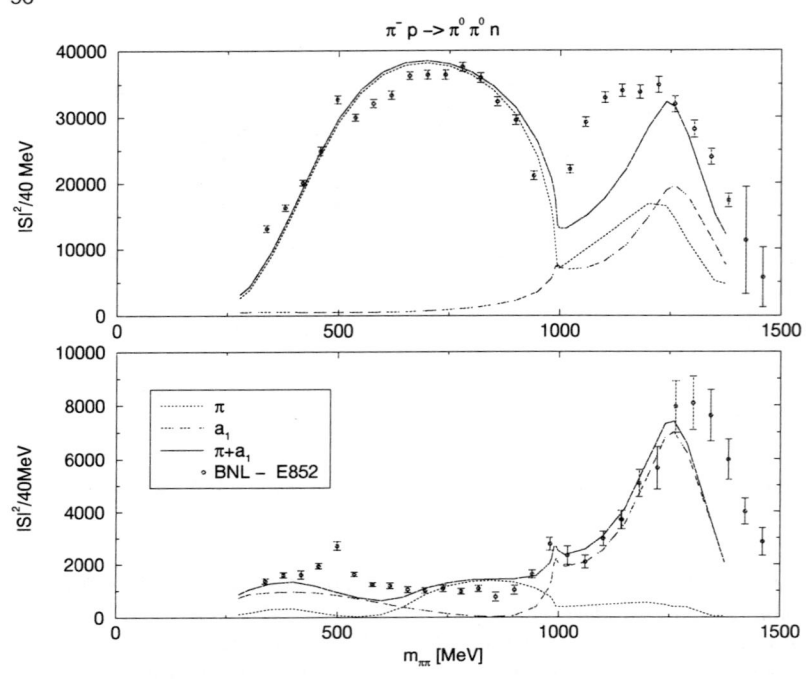

Figure 2. The squares of the absolute values of the S-wave partial wave amplitude is shown as a function of the invariant two-pion mass $m_{\pi\pi}$ for events in the region $0.01 < -t < 0.10(GeV/c)^2$ (upper panel) and $0.40 < -t < 1.40(GeV/c)^2$ (lower panel). The present calculation is given by the solid line. Assuming the exchange of a virtual pion only in the reaction mechanism shown in Fig.1, one obtains the dotted line. The contribution due to virtual a_1 exchange only is given by the dashed line.

tude is shown in Fig.2 as a function of the invariant two-pion mass. For values of the momentum transfer summed over $0.01 < -t < 0.1(GeV/c)^2$ (upper panel), the data show a broad strength distribution above threshold which is interrupted by a sharp dip in the vicinity of $m_{\pi\pi} = 1 GeV$. For these small momentum transfers, the reaction mechanism is dominated by one-pion exchange. The a_1-exchange becomes noticeable only above 1 GeV. The appearance of a dip can be understood microscopically in meson-exchange models: because of the strong attraction between the two pions in the scalar-isoscalar partial wave, the pion-pion phase shift rises rapidly to 90^0, and therefore interferes destructively with the amplitude generated by the opening of the $K\bar{K}$ channel. For large momentum transfers, the contribution of the virtual pion in the reaction mechanism (see Fig.1) becomes negligible. The broad strength distribution seen in the upper panel for invariant pion masses below 1 GeV, which often is interpreted as the "sigma"-meson, has disappeared. A small bump remains near 500 Mev, but this has nothing to do with the sigma meson, but can be traced to a contamination of the pion beam with K^- mesons. This produces a two-pion background via the production of the K_S^0. Now the virtual a_1-exchange dominates the reaction mechanism. Therefore the $f_0(980)$ resonance appears as a clear bump.

In the Jülich model, only those f_0-mesons above 1 GeV are considered which are listed in the Particle Data Group. Now we note that the present calculation misses some S-wave strength above $m_{\pi\pi} = 1 GeV$. A straightforward possibility to reproduce the BNL-E852 data would be to follow Anisovich and Sarantsev and postulate that the $f_0(1370)$ resonance summarizes two scalar resonances, namely a $f_0(1300)$ and a broad $f_0(1550)$[6]. In summary, we find that the $K\bar{K}$ molecule structure of the $f_0(980)$ resonance, as obtained in the Jülich model, is fully compatible with the new high-energy data.

2   The structure of the Roper resonance

Pion-nucleon scatttering is an important experimental tool to obtain information about the masses, widths and decays of baryon resonances. Quark models of the baryon resonances show impressive successes, but also raise many new questions. A recent summary can be found in ref.[7]. One of the "big" problems concerns the missing-resonances problem, i.e. the models predict more states than known experimentally. A "smaller" problem addressed here concerns the Roper resonance $N^*(1440)$, the first excited state of the nucleon with the quantum numbers of the nucleon. Non-relativistic quark potential models obtain a too large mass for this state which requires quark excitations over two harmonic oscillator shells.

A remarkable difference between the $N^*(1520)$ and the $N^*(1440)$ is seen in the partial wave amplitudes in Fig.3. The $N^*(1520)$ causes a change in the phase shift of the partial wave D_{13} up to $180°$ and crosses $90°$ at ≈ 1520 MeV. This is also the position of the maximum in the inelasticity. After passing the resonant phase of $90°$, the amplitude goes back to being almost elastic. The situation is completely different for the $N^*(1440)$. Here the phase shift in the P_{11} increases slowly, which corresponds to a very broad resonance, but the inelasticity opens very rapidly (almost as fast as in the D_{13}) and remains inelastic over a very large energy range. Furthermore, the suggested resonance position of $m_R = 1440$ MeV does not correspond to $\delta = 90°$. The shape of the P_{11} partial wave amplitude in the region of the Roper resonance also looks

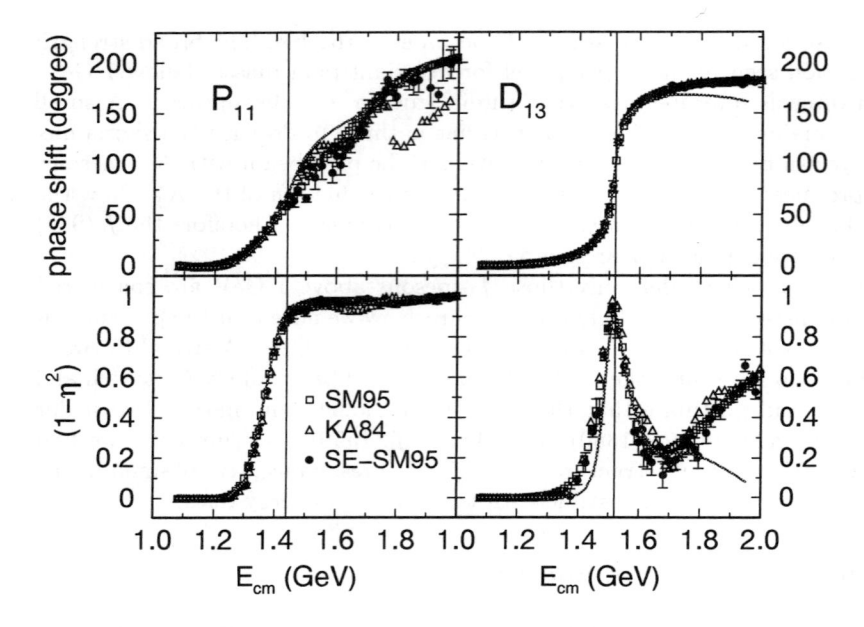

Figure 3. Experimental P_{11} and D_{13} phase shifts and inelasticities compared with our theoretical model.

very different from a typical Breit-Wigner resonance.

Given these experimental facts, a detailed study of the reaction mechanism, in our opinion, is prerequisite to a detailed understanding of the Roper resonance. The dominant decay modes of the nucleon resonances below 2 GeV are the πN and the $\pi\pi N$ reaction channels, as well as the ηN channnel for the $N^*(1535)$. Since a three-body calculation is much too complicated for realistic potentials, we must reduce the $\pi\pi N$ channel into effective two-body channels. In doing this we are guided by studying strong interactions between two-body clusters of the three-body $\pi\pi N$ state. The dominant clusters are the Δ in the πN interaction, the ρ in the vector isovector $\pi\pi$ interaction and the strong correlation in the scalar-isoscalar $\pi\pi$ interaction, which we call σ. Therefore – besides the πN and ηN channels, which are needed for a complete description of the $N^*(1535)(S_{11})$ – our model includes the reaction channels $\pi\Delta$, σN and ρN.

As in the case of meson-meson scattering, one has to solve a three-dimensional reduction of the Bethe-Salpeter equation. The pseudopotential

V is constructed from the effective Wess-Zumino Lagrangian, which we have supplemented with additional terms for including the Δ isobar, the ω, η, a_0, f_0 meson and the σ. We also have included terms that characterize the coupling of the resonances $N^*(1535)$, $N^*(1520)$ and $N^*(1650)$ to various reaction channels. Details can be found in ref.[8].

Our model is able to describe πN data very well up to energies of about 1.9 GeV. Only in the partial wave S_{31}, the model deviates from the data, and that is because we have not yet included the resonance $\Delta(1620)$. Our model does not give significant contributions to the inelasticity in this partial wave. The model is then a good starting point for an investigation of the Roper resonance.

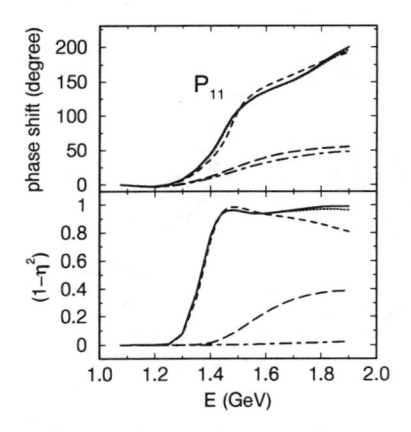

Figure 4. Phase shift and inelasticity in the partial wave P_{11}. The curves are calculated using the full model (solid line), the channels $\pi N/\sigma N/\pi\Delta$ (dotted line), $\pi N/\pi\Delta$ (long dashed line), $\pi N/\sigma N$ (short dashed line), and the elastic model (dot-dashed line). The common parameters are the same in all five cases.

As can be seen in Fig.3, our model results in a very good description of the P_{11}, *without including a genuine Roper resonance*. The resonance is created entirely by meson-dynamics. The rise of the phase shift and the opening of the inelasticity is generated by the coupling to the inelastic channels. In Fig.4 we show how the different reaction channels contribute to the P_{11}. The potential of the elastic model (i.e., where πN is the only reaction channel) is attractive due to the ρ exchange, and leads to a rising phase shift without generating a resonant behavior. Including the $\pi\Delta$ channel hardly improves the situation for the phase shift but leads to some inelasticity, which starts at about 1.4 GeV. As soon as we couple to the σN channel, a resonant shape of the phase shift is generated. The inelasticity opens at 1.3 GeV and reproduces the rapid rise of the experimental data. Since the reaction channels ρN and ηN scarcely contribute to the P_{11}, decoupling the $\pi\Delta$ channel from the full model leaves us basically with a $\pi N/\sigma N$ model, which does not differ much from the full result. Only at higher energies does the $\pi\Delta$ channel contribute to the inelasticity.

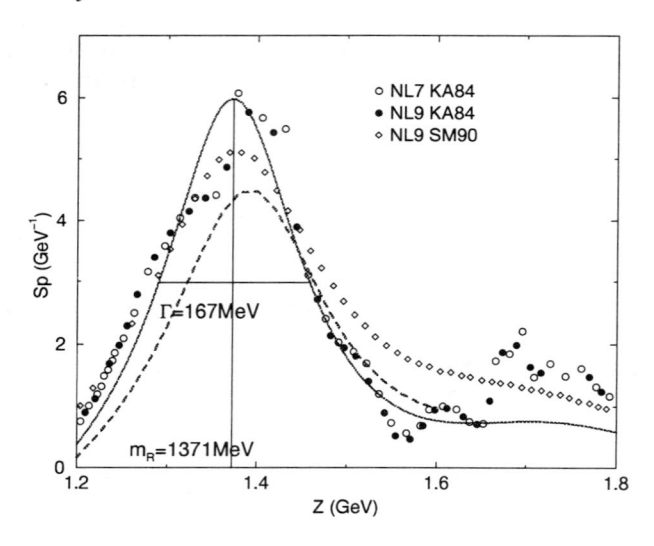

Figure 5. Speed in the partial wave P_{11}. The calculation performed with the full model is given by the solid line. The result obtained after the removal of the $\pi\Delta$ channel is represented by the dashed line.

In order to define the mass and the width of the Roper resonance, we calculate the speed, which is defined by

$$Sp^{IJLS}(E) = \left|\frac{d\tau^{IJLS}}{dE}\right|, \quad (1)$$

and gives some information about the time delay in the reaction. A resonance causes a large time delay and will, therefore, form a peak in a diagram in which the speed is plotted against the energy E. From the speed plot, one can extract the mass of the Roper resonance to be $m_R = 1371 MeV$ and the width $\Gamma = 167 MeV$. By switching off the $\pi\Delta$ channel, one can investigate the sensitivity of the resonance to the presence of this reaction channel. We are led to the conclusion that the Roper resonance, in our model, is dominated by the σN channel. This suggests that the Roper decay should be dominated by two-pion final states, not by one-pion final states.

3 Conclusions

We have shown that the new high-energy data on the $\pi^- p \to \pi^0 \pi^0 n$ reaction are compatible with a non-$q\bar{q}$ structure of the $f_0(980)$ meson. The Roperresonance $N^*(1440)$ is found to be a resonance made of two pions and a bare nucleon.

Acknowledgments

Discussions with T.Barnes, E.Klempt and H.P. Morsch are gratefully acknowledged. Financial support by the Graduiertenkolleg "Die Erforschung der subnuklearen Strukturen der Materie" of the university of Bonn is acknowledged.

References

1. Alde, D. et al., Z. Phys. C **66**, 375 (1995).
2. Gunter, J. et al., hep-ex/0001038.
3. Amsler, C., Rev. Mod. Phys. 7012931998
4. D.Lohse, J. W. Durso, K. Holinde, J.Speth, Phys. Rev. C49, 2671(1990).
5. Z.Wang, S. Krewald, J.Speth, to be publ.
6. V.V. Anisovich and A.V. Sarantsev, Phys. Lett. B **413**, 137 (1997).
7. S. Capstick and W. Roberts, nucl-th/0008028.
8. O.Krehl, C. Hanhart, S. Krewald, J.Speth, Phys. Rev. C62, 25207(2000).

MESON PRODUCTION IN P+D REACTIONS

H. MACHNER

Institut für Kernphysik, Forschungszentrum Jülich, Jülich, Germany
representing the GEM Collaboration

M. BETIGERI [I], J. BOJOWALD [A], A. BUDZANOWSKI [D], A.
CHATTERJEE [I], J. ERNST [G], L. FREINDL [D], D. FREKERS [H], W.
GARSKE [H], K. GREWER [H], A. HAMACHER [A], J. ILIEVA [A,E], L.
JARCZYK [C], K. KILIAN [A], S. KLICZEWSKI [D], W. KLIMALA [A,C], D.
KOLEV [F], T. KUTSAROVA [E], J. LIEB [J], H. MACHNER [A], A. MAGIERA [C],
H. NANN [A], L. PENTCHEV [E], H. S. PLENDL [K], D. PROTIĆ [A], B. RAZEN [A],
P. VON ROSSEN [A], B. J. ROY [I], R. SIUDAK [D], A. STRZAŁKOWSKI [C], R.
TSENOV [F], K. ZWOLL [B]

A. *Institut für Kernphysik, Forschungszentrum Jülich, Jülich, Germany*
B. *Zentrallabor für Elektronik, Forschungszentrum Jülich, Jülich, Germany*
C. *Institute of Physics, Jagellonian University, Krakow, Poland*
D. *Institute of Nuclear Physics, Krakow, Poland*
E. *Institute of Nuclear Physics and Nuclear Energy, Sofia, Bulgaria*
F. *Physics Faculty, University of Sofia, Sofia, Bulgaria*
G. *Institut für Strahlen- und Kernphysik der Universität Bonn, Bonn, Germany*
H. *Institut für Kernphysik, Universität Münster, Münster, Germany*
I. *Nuclear Physics Division, BARC, Bombay, India*
J. *Physics Department, George Mason University, Fairfax, Virginia, USA*
K. *Physics Department, Florida State University, Tallahassee, Florida, USA*

Total and differential cross sections for the reactions $p + d \to {}^3He + m^0$ with $m = \pi, \eta$ and $p + d \to {}^3H + \pi^+$ were measured with the GEM detector at COSY for beam momenta between threshold and the maximum of the corresponding baryon resonance. For both reactions a strong forward–backward asymmetry was found. The data were compared with model calculations.

1 Introduction

The deuteron is a loosely bound system with large distance between the two nucleons. It seems therefore a good testing ground for the impulse approximation. In addition, its wave function as well as those of the produced light nuclei are believed to be well known and one can hope that a theoretical treatment in the three nucleon sector might be possible.

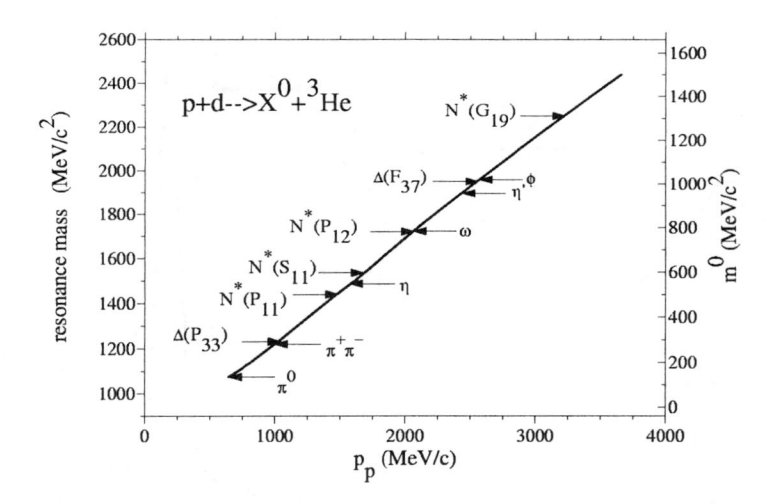

Figure 1. Excitation of baryon resonances as function of the proton beam momentum for the indicated meson. Also shown are the thresholds for neutral meson production. Some selected resonances and mesons are indicated.

2 Kinematical Aspects

Meson production mechanism in the $p + d \rightarrow {}^3He + meson$ reaction can have several contributions. There can be a direct production, i.e. radiation of a meson off a nucleon. However, nuclear resonances can be involved as intermediate state. These are different N^*- and Δ-resonances, depending on the quantum numbers of the meson. Fig. 1 shows the dependence of mass of the baryon resonance as function of the beam momentum. The range corresponds approximately of the momentum range of the COSY accelerator in Jülich. Also shown are some selected resonances. The right hand y-axis shows the mass of the produced neutral meson, or the total mass of the electrical neutral multi-meson system. Also indicated are the thresholds for neutral meson production. Here we have concentrated on narrow states. Also the threshold for two charged pion production is shown.

Almost all resonances above the two pion thresholds decay into a nucleon plus two pions, sometimes with a pion and a Δ as intermediate states. Two resonances are different: The P_{33} $\Delta(1232)$ can decay only into pion and nucleon and the S_{11} $N^*(1535)$ couples strongly to the η-nucleon channel. We,

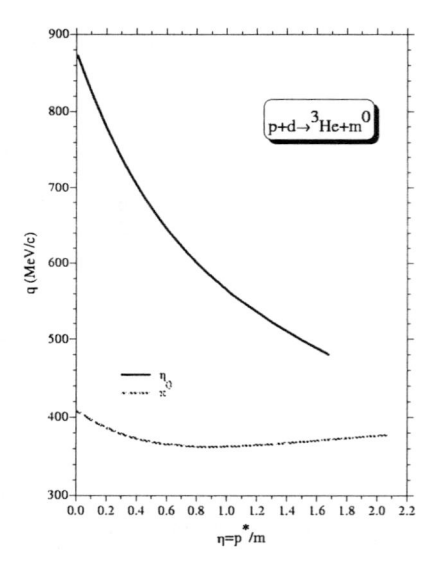

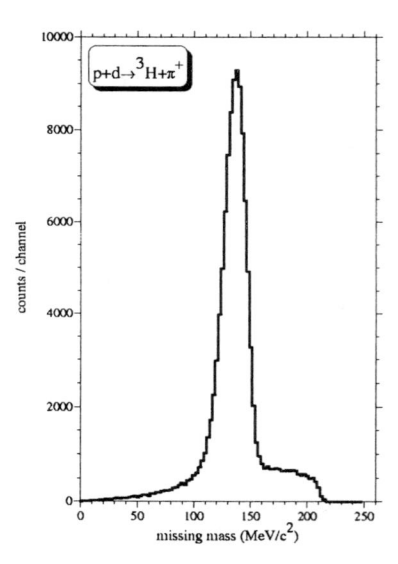

Figure 2. The momentum transfer q as function of the meson relative centre of mass momentum η.

Figure 3. Deduced missing mass for the charged pion from the kinematical complete measurement of the triton. The small background was subtracted for each angular bin.

therefore, have studied the reaction mechanism in the range of these two resonances.

In the reactions of interest one is dealing with large momentum transfers. This is shown in Fig. 2 for the reactions with the neutral mesons. While the momentum transfer for the $p + d \to {}^3He + \eta$ decreases strongly with increasing beam momentum is the dependence of q in the case of $p + d \to {}^3He + \pi^0$ reaction rather weak. From the numbers involved one can estimate that pion production can occur on one target nucleon, whereas the large momentum transfer in the case of η- production makes such a mechanism unlikely. Both target nucleons have to participate making two step processes very likely.

The theoretical attempts in $p + d \to (A = 3) + meson$ have not been particular successful although a lot of effort was devoted to this subject over the years [1].

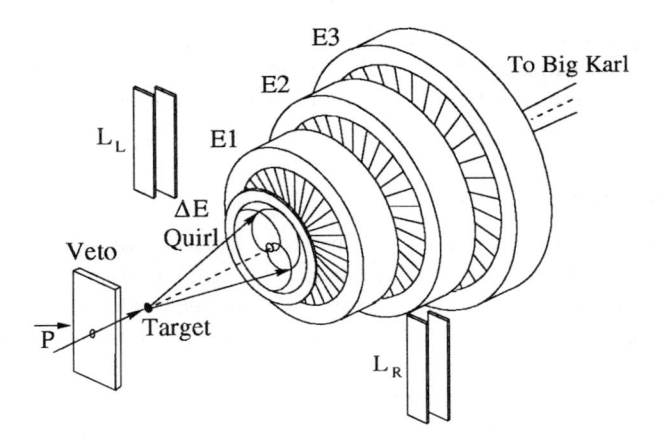

Figure 4. The Germanium wall. The response of the Quirl detector to a two hit event is shown. To the left and right there are two luminosity monitors, consisting of two scintillator paddles each. Each pair acts in coincidence and is denoted by L_R and L_L, respectively.

3 Experiments

The proton beams were extracted from the cooler– synchrotron COSY in Jülich. Although the beam was not cooled it had typically emittance of 2.5π mm mrad in all directions. The detector used is a stack of diodes made from high purity Germanium. The diodes have structures on the front side as well as on the rear side allowing track reconstruction, energy measurement and particle identification. With this detector called Germanium Wall Ref. [2] we measured the heavy $A = 3$ recoils. It allows to measure the particle type, the energy of the recoil and its direction. Thus the four momentum vector of the reaction product can be constructed. Then, by making use of conservation laws the four momentum vectors of the unobserved mesons could be extracted. Recoils being emitted under zero degree in the laboratory were measured with the magnetic spectrograph BIG KARL. Both detector elements together form the GEM detector.

4 Results

In Fig. 3 the missing mass spectrum for charged pions from the $pd \rightarrow {}^3H\pi^+$ reaction at 850 MeV/c is shown. Two things are worth mentioning: the high statistics in the experiment and the small background. In case of the η production the background is larger due to multi–pion production. The background was subtracted by fitting a smooth curve to it. The resulting counting rate was converted to cross sections and the emission angle in the laboratory into those in the centre of mass system.

4.1 π Production

For the π-production very close to threshold isotropic angular distributions are observed. For higher beam momenta the angular distribution becomes backward peaked for the heavy recoil. For a beam of 750 MeV/c it has an almost exponential slope. For higher momenta an isotropic component shows up with increasing importance with increasing beam momentum. As examples the angular distributions for charged pion production at 850 MeV/c and neutral pion production at 1050 MeV/c are shown in Fig.'s 5 and 6, respectively. The data up to 850 MeV/c are published in Ref. [3], the other data are preliminary. One point measured at $\cos(\theta) = -1$ which corresponds to zero degree in the laboratory was measured with the magnetic spectrograph BIG KARL.

As examples the data are compared with several model calculations also shown in Fig.'s 5 and 6. Here, we restrict ourselves to comparisons with published calculations for beam momenta close to the present ones or perform such calculations in the very transparent Locher–Weber model [4]. The differential cross section in this model is given by

$$\frac{d\sigma}{d\Omega} = S\,K\,|F_D(q) - F_E(q)|^2\,\frac{d\sigma}{d\Omega}(pp \to d\pi^+) \qquad (1)$$

with S, a spin factor, K a kinematical factor, and F_D, the direct form factor and F_E, the exchange form factor, i.e. an elastic πd scattering after pion emission from the incident proton. The graph corresponding to $F_D(q)$ is treated by most calculations. The form factors were evaluated with emphasis on the short range–components of the deuteron and the triton. This is achieved by fitting the free parameters in a Hulthén function to the deuteron–charge form factor and similarly for tritium, the parameters for the Eckart function and the 3-pole function to the tritium–charge form factor. Also other functional dependencies taken from the work of Fearing [5] were tried applying the same normalization as in the work of Locher and Weber [4]. All calculations need normalization factors when compared to the data. Best results were obtained for the Eckart wave function and an exponential (see Fig.'s 5 and 6). We also compare the data with calculations from Ueda [6]. The normalization factor is 0.01. For details we refer to the original work. The local minimum at $\cos(\theta) \approx 0$ for 850 MeV/c as well as the strong rise at forward angles for

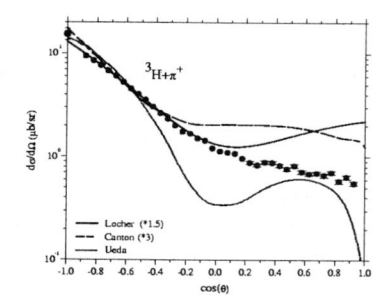

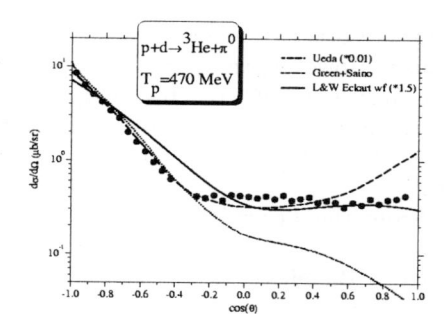

Figure 5. Angular distribution for pion emission for the reaction $pd \to {}^3H\pi^+$ at a beam momentum of 850 MeV/c. The curves are calculations from Ref.'s [1,4,6].

Figure 6. Same, but for $pd \to {}^3He\pi^0$ emission at a beam momentum of 1050 MeV/c. These data are preliminary. The calculations are according to Ref.'s [4,6,7].

1050 MeV/c is not supported by the data. This range is rarely reproduced by a calculation. Green and Sainio [7] even claim their calculations to be unreliable in this range. The large and almost exponential shaped component may be attributed to the direct production graph. Therefore, new physics may be hidden in the small more isotropic component.

4.2 η Production

For the η-production we obtained an angular distribution [8] which is dominated by p-wave (see Fig. 7) in contrast to near threshold, where the angular distributions are s-wave dominated. A Legendre polynomial of second order was found to account for the data.

Kingler [11] has extended a model originally developed for pion production [12] to include higher-resonances. The original model was limited to only pion exchange, Δ resonance excitation as well as non resonant contributions. The extension treats other meson exchanges as well as higher baryon resonances than the Δ. The vertex function for the different baryon-baryon-meson couplings were calculated in a simple quark model and are momentum dependent. For the present reaction the largest contribution to the cross section comes from the $NN\rho$ and $NN\omega$ interactions while the contribution due to $NN^*(1535)\pi$ interaction is one order of magnitude smaller. The contributions of other resonances like $N^*(1440)$, $N^*(1650)$, and $N^*(1710)$ are even smaller. However, the form factor is calculated only for harmonic oscillator wave functions with the frequency being a free parameter varied to fit the experimental data. The model predictions are also shown in Fig. 7 as dashed curve. Obviously, the calculation shows structures not observed in the present data.

From the Legendre polynomial fit a total cross section of $\sigma = 573 \pm 83$ (*stat.*) ± 69 (*syst.*) nb was deduced. This result is close to the one obtained by Banaigs et al. [9] at a slightly higher energy. All cross section from threshold up to $T_p \approx 1500$ MeV could be nicely accounted by a simple calculation. This energy region corresponds to the centre of the N^* S_{11} resonance ($\Gamma \sim 200$ MeV) known to couple strongly to the η–N channel [13]. One may therefore attempt to describe the cross section by an intermediate $N^*(1535)$ resonance excitation:

$$\sigma(E) = \frac{p_\eta}{p_p} |M(E)|^2 \qquad (2)$$

with E the excitation energy and M the matrix element which is calculated as in photoproduction on the proton [14] as Breit–Wigner form with an energy dependent width. All parameters were taken from Ref.'s [14] and [13]. The only free parameter is the strength fitted to the present data point. The trend of the data is reproduced, which may be taken as an indication that production of the $N^*(1535)$ resonance is the dominant reaction mechanism and that the product of kinematics and form factor changes only very little over the present energy range. Deviations occur near threshold which might be an indication of strong final state interactions.

References

1. L. Canton, G. Cattapan, G. Pisent, W. Schadow, J. P. Svenne, Phys. Rev. **C 57** (1998) 1588; L. Canton and W. Schadow, Phys. Rev: **C 56** (1997) 1231
2. M. Betigeri *et al.*, Nucl. Instrum. Methods Phys. Res. **A421**, 447 (1999).
3. M. Betigeri *et al.*, Nucl. Phys. **A** (in press).
4. M. P. Locher and H. J. Weber, Nucl. Phys. **B76**, 400 (1974).
5. H. W. Fearing, Phys. Rev. **C 16** (1977) 313.
6. T. Ueda, Nucl. Phys. **A505**, 610 (1974).
7. A. M. Green and M. E. Sainio, Nucl. Phys. **A 329** (1979) 477.
8. M. Betigeri *et al.*, Phys. Lett. **B472**, 267 (2000).

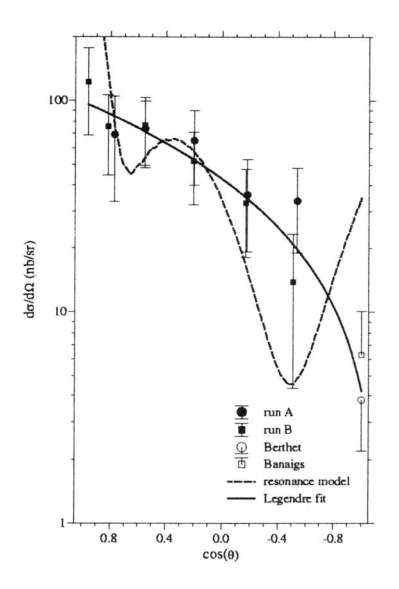

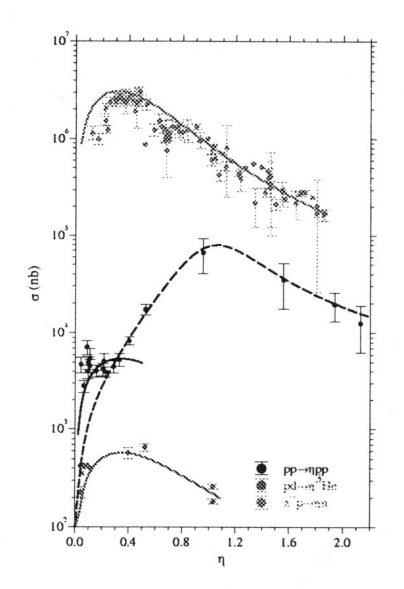

Figure 7. Angular distribution of η production in the cm system. Two points from the indicated references are added (Ref. [9,10]). The solid curve indicates a Legendre polynomial fit. The dashed curve is from a model calculation [11].

Figure 8. Excitation functions of η production in different reactions as function of $\eta = p_\eta^*/m_\eta$. The upper band is the data from $\pi^- p \to n\eta$ reactions, the middle band from $pp \to pp\eta$ reactions and the lowest one from $pd \to {}^3He\eta$ reactions. The solid curves are predictions assuming the matrix element from photoproduction of η mesons off the proton assuming only resonance production of the $N^*(1535)$. For the $pp \to pp\eta$ reaction an additional resonance at 1740 MeV/c was assumed (dashed curve).

9. J. Banaigs et al., Phys. lett. **B45**, 394 (1973).
10. P. Berthet et al., Nucl. Phys. **C443**, 589 (1985).
11. J. Kingler, Ph. D. Thesis, Univ. Bonn 1995 and private communications to H. M.
12. M. Harzheim et al.., Z. Physik 340 (1991) 399.
13. Particle Data Group: C. Caso et al.., The European Phys. J. 3 (1999) 1.
14. B. Krusche, Acta Phys. Polonica 27B (1996) 3147.

SEMI-EMPIRICAL EFFECTIVE INTERACTIONS FOR INELASTIC SCATTERING DERIVED FROM THE REID POTENTIAL

J. O. FIASE, L.K. SHARMA AND D. P. WINKOUN
Department of Physics, University of Botswana Private Bag 0022 Gaborone, Botswana

A. HOSAKA
Numazu College of Technology 3600 Ooka, Numazu 410 Japan

An effective local interaction suitable for inelastic scattering is constructed from the Reid soft - core potential. We proceed in two stages: We first calculated a set of relative two - body matrix elements in a variational approach using the Reid soft-core potential folded with two-body correlation functions. In the second stage we constructed a potential for inelastic scattering by fitting the matrix elements to a sum of Yukawa central, tensor and spin-orbit terms to the set of relative two - body matrix elements obtained in the first stage by a least squares fitting procedure. The ranges of the new potential were selected to ensure the OPEP tails in the relevant channels as well as the short - range part of the interaction. It is found that the results of our variational techniques are very similar to the G - matrix calculations of Bertsch and co - workers in the singlet - even, triplet - even, tensor - even and spin-orbit odd channels thus putting our calculations of two - body matrix elements of nuclear forces in these channels on a sound footing. However, there exist major differences in the singlet - odd, triplet - odd, tensor - odd and spin - orbit even channels which casts some doubt on our understanding of nuclear forces in these channels.

1 Introduction

For sometime now the usual approach to the derivation of semi-empirical potentials for inelastic scattering has been the use of the G-matrix techniques (Bertsch et al., [1], Hosaka and Toki [2]). The prescription usually used consists of two stages: in the first stage a set of G- matrix elements are calculated from realistic forces in an oscillator basis using suitable approximations. In the second stage the set of relative matrix elements are fitted by those of a sum of Yukawa, spin - orbit and tensor terms to produce the issuing potential. This approach has been very popular over the years and has produced quite a number of very successful potentials; of particular note being the M3Y interaction (Anantaraman et al. [3]), which is very popular among researchers (see e.g Satchler and co-workers [4], Ismail et al. [5]).
In this article a similarly motivated interaction is presented but here we use the method of correlated basis functions. Using such method, Fiase et al.[6]

produced an effective interaction for shell-model calculations by folding together a Hamiltonian for the rest-frame of the nucleus based on the Reid [7] soft- core potential. This interaction gave results that were in excellent agreement with the fitted interactions of Wilthenthal and co-workers [8].

Motivated by the success of that interaction we now construct an effective interaction for inelastic scattering based on the prescription of that earlier work. We first calculate a set of two-body relative matrix elements in a variational approach using the Reid soft-core potential folded with two-body correlation functions. In the second stage we fit the relative matrix elements by those of a sum of Yukawa, spin-orbit and tensor components. In doing so the OPEP tail and the short -range part of the interaction is maintained.

This interaction appears via a different calculational procedure from the G-matrix calculations but as will be shown, it is extraordinary similar to the original M3Y interaction described by Bertsch et al., [1] in the singlet-even (SE), triplet-even (TE), tensor-even (TNE), spin-orbit odd (LSO) channels but differ in the singlet-odd (SO), triplet - odd (TO), tensor - odd(TNO) and spin - orbit even(LSE) channels. Moreover, as will be shown the matrix elements in the channels in which the results of the two methods differ poorly fit the interaction . In fact in the original M3Y interaction, the SO, TO, TNO and the LSE relative two - body matrix elements were replaced by a set of Elliot [9] matrix elements in these channels because they poorly fit the interaction (Anantaraman et al., [3]). Since potential models are built on phase-shift data, one can view these two-body matrix elements as direct link to phase-shift data. That is to say, in the odd and LSE channels where these differences mainly appear, it is possible that either our calculational procedures described here are deficient or the odd forces and LSE forces are not yet fully understood.

2 Derivation of Matrix Elements

In this section we briefly describe the method discussed by Fiase et al. [6] for the derivation of effective interactions. In this approach we define an effective two - body Hamiltonian for the rest frame of the nucleus as (Irvine [10]):

$$H_{eff}^{(2)} = \sum_{i>j}(f_2(ij))(P_{ij}^2/M + V_{ij})f_2(ij) \qquad (1)$$

where $\vec{P}_{ij} = \frac{1}{\sqrt{2}}(\vec{p}_i - \vec{p}_j)$ is the relative momentum of the two particle - system; $M = m_N A$ is the total mass of the nucleus and V_{ij} is taken to be the Reid [7] soft - core potential and $f_2(ij)$ are the two-body correlation operators. In their lowest order constrained variational approach (Irvine, [10]), the two - body correlation functions were allowed to reflect the freedom of the chosen Hamiltonian which was taken to be the Reid [7] soft - core potential for the N - N sector and the Green - Niskanen -Sanio [11] interaction for the $N - \Delta(1232)$ sector. Three features were found which included the 'wound' induced in the two - body wave function by the repulsive core of the N - N interaction, the tensor correlations especially in the $^3S_1$ - $^3D_1$ channel and the meson exchange correction. It was found that the most important feature of these was the tensor correlations and Irvine et al. [12] parameterised the two - body correlation function in the form:

$$f(r_{ij}) = 0, r_{ij} < r_c$$
$$f(r_{ij}) = (1 - e^{-\beta(r_{ij}-r_c)^2})(1 + \alpha S_{ij}), r_{ij} \geq r_c \qquad (2)$$

where $r_c = 0.25 fm$ and $\beta = 25 fm^{-2}$. The parameter, α represents the strength of the tensor correlation and is non-zero only in the $^3S_1$ - $^3D_1$ channel. The cluster expansion converges if

$$\frac{3}{r_0^3} \int < (1 - f_2(ij))^2 > r_{ij}^2 dr_{ij} \ll 1 \qquad (3)$$

Using the above values of r_c and β Irvine [10] found the value of this quantity to be less than 0.1 and hence the cluster expansion rapidly converged giving a possible justification of the use of $H_{eff}^{(2)}$ in eq. (1). The matrix elements of $H_{eff}^{(2)}$ were calculated in a harmonic oscillator basis with the oscillator size parameter and the strength of the tensor correlation as the only adjustable free parameters. In our calculation we have separated out the relative two-body matrix elements into the various channels: The $^1S_0$ and $^1P_1$ gave the singlet even (SE) and singlet odd (SO) channels while the coupled $^3S_1$ - $^3D_1$ channels give the triplet-even (TE) and the tensor-even (TNE) components. The triplet-odd (TO), tensor-odd (TNO) and the two components of the spin-orbit force were obtained by ignoring the quadratic term according to the following equation (Anantaraman et al., [3])

$$V(TO) = V(3P0) + 2V(LSO) + 4V(TNO)$$
$$V(TNO) = -5/72[2V(3P0) - 3V(3P1) + V(3P2)]$$
$$V(LSO) = -1/12[2V(3P0) + 3V(3P1) - 5V(3P1)]$$
$$V(LSE) = 1/3[V(TE) - 2V(TNE) - V(3D1)]. \qquad (4)$$

3 Construction of the N - N Effective Interaction

We now define an effective interaction which is consistent with effective N- N interactions suitable for inelastic scattering. These take the form (Anantaraman et al. [3])

$$V = \Sigma_i A_i Y(r_{12}/R_i), central$$
$$V = \Sigma_i A_i Y(r_{12}/R_i)\vec{L}.\vec{S}, spin - orbit$$
$$V = \Sigma_i A_i r_{12}^2 Y(r_{12}/R_i) S_{12}, tensor \tag{5}$$

where $Y(x) = e^{-x}/x$ and $i \leq 4$. The coefficients A_i are the strength of the interaction which are determined by a least squares fitting procedure to the present relative matrix elements. The choice of the ranges $i \leq 4$ are $R_i = 0.25, 0.4, 0.7$ and 1.414 fm are theoretically motivated such that they ensure the OPEP tails in the relevant channels as well as the short-range part which account for the ρ- and ω- meson exchange (Bertsch et al. [1]).

4 Results

In Table 1 we present the relative two - body matrix elements of the effective interaction obtained by using $\hbar\omega = 14 MeV$ and $\alpha = 0.075$ for the SE, SO, TE, TO, TNE, TNO, LSO and LSE channels. The matrix elements were calculated in a harmonic oscillator basis. If we consider the conservation of energy and angular momentum for the two - particle system with the quantum numbers (n_1, l_1) for particle 1 and (n_2, l_2) for particle 2 with their relative and centre of mass counterparts (n, l) and (N, L) respectively, then

$$2n_1 + l_1 + 2n_2 + l_2 = 2n + l + 2N + L$$
$$\mid \vec{l} - \vec{L} \mid \leq \vec{l} + \vec{L}, \vec{l_a} + \vec{l_b} = \vec{\lambda} = \vec{l} + \vec{L} \tag{6}$$

For the sd-shell model calculation in a harmonic oscillator basis i.e., $0d_{5/2}, 1s_{1/2}, 0d_{3/2}$ the maximum value n or n' can take is 2 whereas in the G-matrix calculation, the actual wave functions used were the Wood-Saxon wave functions which were expanded in an oscillator basis. Due to this expansion there are non-zero expansion coefficients up to indefinetely large n values. The expansion was truncated at n = 3 (Anantaraman, [13]). Our comparison with the G-matrix result is therefore limited to the $n, n' = 0, 1$ and 2 values. From the table we immediately see the striking similarities between the relative matrix elements of the present calculation and those for the G-matrix calculations for the SE, TE, TNE,and LSO channels. For example, the TNE relative matrix elements presently calculated and the G-matrix elements calculated with both the Paris or the Reid interactions do not differ from each other by more than 5 per cent. This similarity is noteworthy considering the different calculational procedures and the various approximations used. For the SO, TO, TNO and LSE channels we encounter major differences. For example, The SO matrix elements for $n, n' = 2$ are roughly twice as large as those for the G-matrix elements calculated with the Paris potential.

We next performed a least squares fitting of our calculated two - body matrix elements in the various channels to a sum of Yukawa potentials for the central, tensor and spin-orbit forms as given in eq. (5). These forms are consistent with potentials for inelastic scattering as earlier discussed.

As can be seen from our calculations, the SE, TE and TNE strengths of our calculations agree very well with the G-matrix calculations both for the Reid [7] and the Paris [14] potentials but major differences exist in the SO, TO, TNO and LSE channels.

5 Conclusions

The two calculational produres presented here show excellent agreement in the SE, TE, TNE and LSO channels while they differ significantly in the SO, TO, TNO and LSE channels. These means that the two-body nuclear forces in the even angular momentum channels are fairly well understood as can be seen from these similarities. However we conjecture that our knowledge of two-body nuclear forces in the odd nuclear forces require some further scrutiny. Indeed the problem of odd nuclear forces is not new in nuclear structure calculations for example as in the so-called analysing power puzzle where there is an intricate interplay of two- and three-body p-wave forces on calculations of analysing power. Thus, apart from the necessary inclusion of density dependence in our effective two-body interactions, one should not ignore perhaps the effects of three-body forces and the mass dependence.

Acknowledgments

One of us (JOF) would like to thank Professor J. M. Irvine who introduced him to the theory of effective interactions. Financial assistance from the Conference Organisers and Commonwealth Science Council travel grant is gratefully acknowledge.

Table 1. Calculated relative matrix elements for $\hbar\omega = 14 MeV$, $\alpha = 0.075$. The first entry for the column is the result of the present calculation. The second entry in bracket is the G- matrix calculation with the Paris [14] potential while the third entry is the G- matrix calculation with the Reid [7] potential.

S/S singlet even	n = 0	1	2
$n' = 0$	-7.81	-6.94	-5.59
	(-6.73)	(-5.27)	(-3.73)
	-6.64	-5.41	-3.80
1		-6.70	-5.61
		(-4.84)	(-3.47)
		-4.73	-3.30
2			-4.92
			(-2.61)
			-2.23

S/S triplet-even	n = 0	1	2
$n' = 0$	-10.45	-7.91	-5.17
	(-9.63)	(-8.90)	(-7.55)
	-9.93	-8.84	-6.81
1		-6.98	-4.77
		(-8.43)	(-6.82)
		-7.84	-5.71
2			-3.49
			(-5.83)
			-3.76

S/D tensor-even	n = 0	1	2
$n' = 0$	-5.76	-8.12	-9.71
	(-5.54)	(-7.62)	(-8.66)
	-5.26	-7.49	-8.95
1	-2.87	-5.47	-7.50
	(-2.61)	(-5.09)	(-6.90)
	-2.81	-5.52	
2	-1.69	-3.47	-5.42
	(-1.35)	(-2.95)	(-4.72)
	-1.52	-3.23	-5.30

P/P singlet-odd	n = 0	1	2
$n' = 0$	2.32	3.49	4.43
	(2.52)	(2.37)	(2.11)
1		5.31	6.49
		(3.03)	(3.00)
2			7.96
			(3.39)

P/P	n = 0	1	2
LSO			
$n' = 0$	-0.79	-1.19	-1.46
	(-0.57)	(-0.90)	(-0.96)
	-0.60	-0.89	-1.06
1		-1.74	-2.13
		(-1.28)	(-1.40)
		-1.26	-1.52
2			-2.62
			(-1.71)
			-1.84

P/P	n = 0	1	2
TO			
$n' = 0$	0.15	0.26	0.37
	(-.08)	(-.23)	(-.09)
1		0.46	0.63
		(-.24)	(-.05)
2			0.86
			(0.04)

D/D	n = 0	1	2
LSE			
$n' = 0$	0.26	0.284	0.276
	(-0.14)	(-0.13)	(-0.12)
	-0.60	-0.89	-1.06
1		0.39	0.43
		(-0.18)	(-0.21)
		-1.26	-1.52
2			0.51
			(-0.26)
			-1.84

P/P	n = 0	1	2
TNO			
$n' = 0$	0.66	0.59	0.45
	(0.78)	(0.75)	(0.63)
1		0.67	0.55
		(0.92)	(0.87)
2		.	0.50
			(0.95)

Table 2. Best-fit interaction strength (MeV).

	Channel	$R_1 = 0.25$ fm	$R_2 = 0.40$ fm	$R_3 = 0.7$ fm	$R_4 = 1.414$ fm
1	SE	10279	-3393		-10.463
		(11466)	(-3556)		(-10.463)
		12455	-3835		-10.463
2	TE	21857	-6908		-10.463
		(13967)	(-4594)		(-10.463)
		21227	-6622		-10.463
3	SO	59323	-5913		31.389
		(-1418)	(950)		(31.389)
		29580	-3464		31.389
4	TO	2273	-103		3.488
		(11345)	(-1900)		(3.488)
		12052	-1990		3.488
5	TNE		-1369	-10.15	
			(-1096)	(-30.9)	
			-1260	-28.4	
6	TNO		-19.71	27.06	
			(244)	15.6	
			263	13.8	
7	LSE	-35973	4295		
		(-5101)	(-337)		
		-4382	-352		
8	LSO	-61.9	-1342		
		(-1897)	(632)		
		-2918	-483		

References

[1] G. Bertsch, J. Borysowicz, H. McManus and W.G. Love, Nucl. Phys. **A284**, 399 (1977).
[2] A. Hosaka, K. I. Kubo and H. Toki, Nucl. Phys. **A444**, 76 (1985).
[3] N. Anantaraman, H. Toki and G.F Bertsch, Nucl. Phys. **A398**, 269 (1983).
[4] Dao T. Khoa and G. R. Satchler, Nucl. Phys. **A668**, 3 (2000).
[5] M. Ismail and Kh. A. Ramadan, J. Phys. G. Nucl. Part. Phys. **26**, 1621 (2000).
[6] J. Fiase, A. Hamoudi, J.M. Irvine and F. Yazici, J. Phys. G. Nucl. Phys. **14**, 27 (1988).
[7] R.V. Reid, Ann. Phys. (N.Y) **50**, 411 (1968).
[8] B.H. Wildenthal, Prog. in Part. and Nucl. Phys. **11**, 5 (1984)
[9] J.P. Elliot et al., Nucl. Phys. **A121**, 241 (1968).
[10] J.M. Irvine, Prog. in Part. and Nucl. Phys. **5**, 1 (1980).
[11] A.M. Green, J.A. Niskanen and M. Sanio, J. Phys. **G4**,1055 (1978).
[12] J.M. Irvine, G.S. Mani, V.F.E. Pucknell, M. Vallieres and F. Yazici, Ann. Phys. (N.Y.) **102**, 129 (1976).
[13] N. Anantaraman, Private Cmmunication (1991).
[14] M. Lacombe, B. Loiseau, J.M. Richard, R. Vinh Mau, J.Cote, P. Pires and R. de Tourreil, Phys. Rev **C21**, 861 (1980).

2. Atomic Physics and Applied Atomic Physics

CHANNELING REVISITED

J.U. ANDERSEN

Institute of Physics and Astronomy, University of Aarhus
DK – 8000 Aarhus C, Denmark
E-mail: jua@ifa.au.dk

The relation between classical and quantum mechanics was a central issue in Lindhard's theory of channeling. While ion channeling can be described by classical mechanics, there are strong quantal features for positrons and electrons. At high energy the description becomes classical also for these particles, and the transition provides an illustration of the correspondence between quantum and classical mechanics in the limit of large quantum numbers. With the discovery of channeling radiation in the late 70'ies, electron channeling gained new interest. Applications to the observation of solid-state properties were demonstrated, and electron channeling may provide interesting sources of hard photons.

1 Introduction

The title of this paper was suggested to me for a talk at the conference in Lüderitz, honouring J.P.F. Sellschop on his 70'th birthday. The phenomenon of channeling of charged particles in crystals has played an important role in the scientific life of many of the participants in the conference, including y, whic Sellschop and his coworkers at Wits University. I took the invitation as an opportunity to look back at highlights of my own involvement in channeling.

When I became a graduate student in Aarhus, Jens Lindhard had just completed his classical paper on the theory of channeling,[1] and his lectures were a great inspiration. Building on Niels Bohr's analysis of binary scattering of atomic particles, [2] Lindhard showed that the classical theory of channeling applies at all velocities for particles which are heavy compared with the electron.[1,3] However, for electrons and positrons there can be strong quantal features, as is well known from electron microscopy which is based on wave interference. Clear classical features of electron channeling were first demonstrated in an experiment by my fellow student, Erik Uggerhøj.[4,5] The classical interpretation was strongly criticised by proponents of the wave theory of electrons,[6] and the heated debate made for an exciting time until the issue of correspondence was clarified.[7]

In the quantum theory of channeling, particle trajectories are replaced by energy eigenstates of channeled particles, just as in the old Bohr theory of hydrogen. The crucial step towards the discovery of channeling radiation was

Kumakhov's derivation of the correct relativistic transformation of photon energies,[8] which showed that for channeling of MeV electrons and positrons, transitions between states separated by a few eV could lead to emission in the forward direction of photons in the x-ray region. Both planar channeling radiation from positrons[9] and electrons,[10] and axial channeling radiation from electrons [11] was soon observed. This development also led to new insight into other coherent radiation phenomena in crystals, such as coherent bremsstrahlung.[12,13,14]

2 String Effect

The basic phenomenon in channeling, the governing of particle trajectories by correlated collisions with crystal atoms, is illustrated in Fig. 1. A positively charged particle is approaching a string of atoms with spacing d at an angle of incidence ψ. A number of small-angle deflections by string atoms reflect the

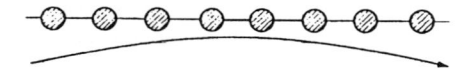

Figure 1. Illustration of the deflection of an ion by a string of atoms.

particle, which leaves the string at an exit angle equal to ψ. The trajectory of the particle can to a good approximation be described as motion in a continuum string potential,

$$U(\mathbf{r}) = \frac{1}{d} \int_{-d/2}^{d/2} V(x,y,z) dz, \qquad (1)$$

where V is the ion-crystal potential and $\mathbf{r} = (x,y)$ is the distance from the string, which is chosen as the z axis. When vibrations of crystal atoms are included, the continuum potential becomes the thermally averaged potential $U_T(\mathbf{r})$.

In this continuum model, there is a separation between the longitudinal motion with velocity $v = \beta c$ and the transverse motion governed by a Hamiltonian,

$$H(\mathbf{r}, \mathbf{p}_\perp) = \frac{p_\perp^2}{2M_1\gamma} + U_T(\mathbf{r}), \qquad (2)$$

where $\mathbf{p}_\perp = (p_x, p_y)$ is the transverse momentum and M_1 is the projectile rest mass.[1,3] The factor $\gamma^{-1} = (1-\beta^2)^{1/2}$ can be considered a constant, independent of the transverse motion.[a]

At angle ψ to the string, the magnitude of the transverse momentum is $p\sin\psi \simeq p\psi$ and the transverse kinetic energy in Eq. (2) therefore equals $1/2 pv\psi^2$. Since the maximum of the string potential is of order $2Z_1 Z_2 e^2/d$, where $Z_1 e$ and $Z_2 e$ are the nuclear charges of the projectile and the target atoms, a projectile approaching a string at an angle less than a critical angle, given by

$$\psi_1 = \left(\frac{4Z_1 Z_2 e^2}{pvd}\right)^{1/2}, \qquad (3)$$

cannot penetrate to the centre of strings. This gives rise to a strong reduction in yield of nuclear reactions or large-angle scattering for $\psi < \psi_1$,[16] and this effect has become the basis for most applications of ion channeling.[17,18,19] Also negative particles (e.g. electrons) can be channeled for small incidence angles. The potential in Eq. (1) then has the opposite sign and for $\psi < \psi_1$ the particles can be trapped into localized motion near one string.

3 Electrons and Positrons

From several arguments Lindhard found that a classical description of channeling is always justified for particles heavy compared to the electron.[1,3] This was not an obvious conclusion since Bohr's analysis of binary Coulomb scattering showed that a classical description fails at high velocities, where instead a quantal perturbation treatment may be applied.[2] A simplified argument for Lindhard's result is the following: The definition of a wave packet following the trajectory of a classical particle requires that the reduced wavelength in the transverse motion, $\lambda_\perp = \hbar/p\psi$, be small compared to a characteristic distance in the string potential. This distance we may take to be the Thomas–Fermi screening length in atoms, $a \simeq a_0 Z_2^{-1/3}$, where a_0 is the Bohr radius of hydrogen. If we choose a transverse momentum corresponding to the characteristic angle ψ_1, we obtain the condition

$$2\gamma^{1/2}\left(\frac{M_1}{m_0}\right)^{1/2}\left(\frac{a_0}{d}\right)^{1/2}|Z_1|^{1/2}Z_2^{1/6} > 1, \qquad (4)$$

---

[a]This is always a good approximation for determination of the transverse motion but the variation of the longitudinal velocity is important for the emission of channeling radiation at very high energies.[15]

where m_0 is the electron mass. This relation is indeed always fulfilled for particles heavy compared to the electron, but for low–energy electrons and positrons, with $|Z_1| = 1$ and $\gamma \simeq 1$, the left–hand side is of order unity, only.

Investigation of electron and positron channeling was therefore the key to a clarification of the relation between classical and quantum mechanics in channeling. Uggerhøj's experiment is illustrated in Fig. 2.[4,5] The radioactive isotope Cu^{64} was implanted in a small spot on a copper single crystal. This isotope emits both positrons and electrons, and hence the influence of the crystal lattice on emission yields could be studied simultaneously for positive and negative particles. The observation of a strong axial dip in yield for positrons and a peak for electrons, as illustrated in Fig. 3, confirmed qualitatively the predictions from the classical channeling theory. After emission from an atom on a string, a positron is repelled by the string potential while an electron is attracted. This leads to a blocking dip for positrons and a focusing in the string direction for electrons.

The classical interpretation of these experiments immediately came under attack from physicists familiar with the wave theory of electron diffraction, as applied in electron microscopy [6]. An exciting time followed where we students of Lindhard felt as knights defending our king. The spirit was reflected in a letter from Lindhard after I had written to him about my new calculations using the many–beam formulation of diffraction theory: "*Et tu, Brute,...*" it began! Of course it was all in jest, and Lindhard gave constant support and encouragement. According to the correspondence principle, quantum mechanics approaches classical mechanics in the limit of large quantum numbers. As seen in Eq. (2), the effective mass for the transverse motion of channeled

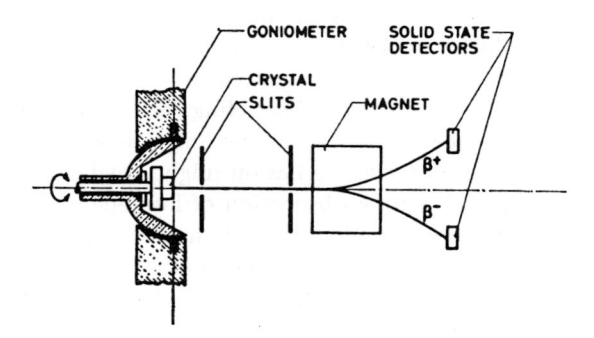

Figure 2. Experimental assembly used for the emission experiments in Refs.4 and 5.

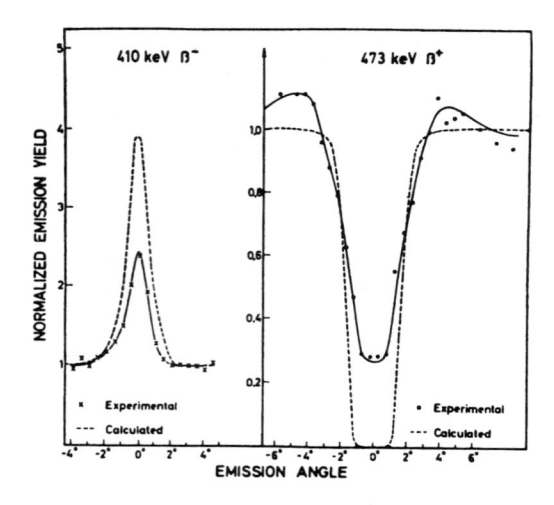

Figure 3. Results of the experiment in Ref. 5. The intensities of emitted particles as functions of the angle to a ⟨110⟩ direction in the Cu crystal are compared with the prediction from a classical description of electron and positron channeling.

particles is the relativistic mass $M_1\gamma$, and hence the density of states increases with γ (linearly in two dimensions). The resulting approach to classical results was illustrated by explicit calculations in Ref. 7.

4 Channeling Radiation

In Bohr's development of a model of the hydrogen atom, the connection to spectroscopy through the formula $\Delta E = h\nu$ was very important, and it would seem a natural idea to look for a similar connection of jumps between channeling states to emission of characteristic channeling radiation. However, it was not clear that the coherence lengths would be long enough for definition of discrete lines and, furthermore, such radiation might not be observable since it is emitted in a solid with strong absorption of low–energy radiation. For this latter question, Kumakhov's derivation of the correct relativistic transformation of the photon energy was very important.[8]

Consider the motion of an axially channeled particle in a reference frame following the particle motion parallel to the axis.[15] If γ is not too large, i.e. for $\gamma\psi_1 < 1$, the motion is non–relativistic in this frame. The Hamiltonian governing the motion is obtained simply through multiplication of the expression in Eq. (2) by γ : The transverse momentum is invariant under the transformation and the first term is the non–relativistic kinetic energy; the increase of the potential energy is caused by Lorentz contraction of the strings, reducing the spacing d in Eq. (1) by the factor γ. In this 'rest frame', the Bohr formula connects the photon frequency to the change, $\gamma\Delta\varepsilon$, in energy of the particle. The photon energy in the laboratory frame is obtained from the Doppler transformation, and at an angle θ to the axis we obtain

$$h\nu = \frac{\gamma\Delta\varepsilon}{\gamma(1-\beta\cos\theta)} \simeq \frac{2\gamma^2\Delta\varepsilon}{1+\gamma^2\theta^2}. \qquad (5)$$

The factor $2\gamma^2$ gives a strong boost of the photon energy in the forward direction, even for electron energies of a few MeV, only. These transformations are also very useful for a qualitative description of coherent bremsstrahlung,[14] which is associated with the periodic perturbation of a particle trajectory crossing crystal planes.[12,13] With the Lorentz boost, the channeling radiation from MeV electrons is in the x–ray region where the absorption is weak. There remained the problem of incoherent scattering limiting the coherence length. This scattering should be very strong for channeled electrons moving in the vicinity of crystal atoms on a string or plane, and the first observation of a line spectrum from planar channeled electrons came as a surprise.[10] For axially channeled electrons, the lines are only resolved for fairly low electron

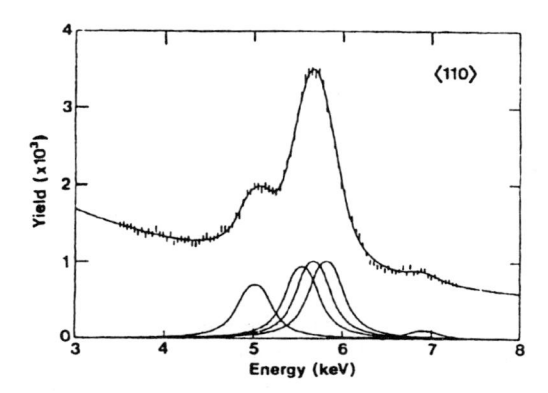

Figure 4. Measured $2p1s$ line of channeling radiation from 4–MeV electrons along a $\langle 110 \rangle$ direction in diamond.[22] The line is decomposed into transitions from the four $2p$ levels indicated in Fig. 5. The channeling radiation is superposed on a sloping background of incoherent bremsstrahlung.

energies.[11] The general analysis of incoherent scattering and line broadening turned out to be an interesting and challenging task.[20,21]

Diamond is an ideal crystal for observation of channeling radiation because of the low atomic number and high Debye temperature, which lead to weak thermal scattering. In a collaboration with Friedel Sellschop, we carried out a study of axial channeling radiation from 4–MeV electrons with a beautiful thin diamond crystal (3.5 μm thick, type 2A with low defect concentration).[22] The result for the $\langle 110 \rangle$ direction is shown in Fig. 4. In the diamond lattice the $\langle 110 \rangle$ strings are pairwise very close to each other (separation 0.89 Å in diamond) and the channeling potential U_T is illustrated in Fig. 5. The lowest bound states, the 1s states, are well localized on one

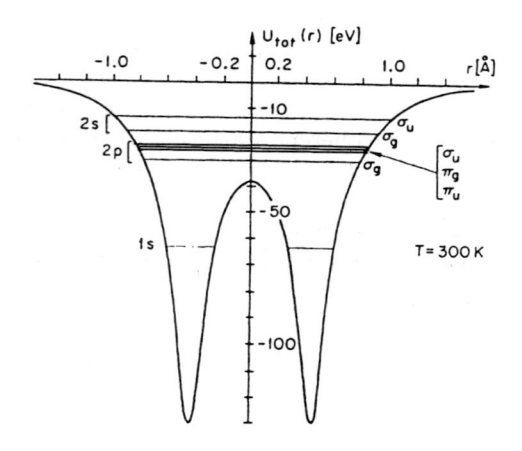

Figure 5. Potential and energy levels for 4–MeV electrons channeled along a $\langle 110 \rangle$ direction in diamond.[22] The 2p levels are classified according to the symmetry of 'molecular' states, shared between two neighbouring atomic strings.

string, but the 2p states overlap and the eigenstates have a molecular character. They are labelled according to their symmetry, under reflection in the midpoint between the two strings, gerade and ungerade, and under reflection in the line connecting the strings, σ and π. The $\sigma_g 2p$ state, which is analogous to a bonding orbital in a molecule, is much lower in energy than the others, and the transition from this state to the ground state is seen as a shoulder on the observed 2p1s line in Fig. 4. The splitting depends on the magnitude of the potential in the region between the strings, which in turn is sensitive to the amount of charge accumulated in the tetrahedral bonds between carbon atoms in diamond. The measurement therefore gives detailed information on the electron density in diamond.

As illustrated by this example, the spectroscopy of channeling radiation can be a useful tool for solid state observations.[15] The potential of channeling radiation as an x-ray source has been investigated in detail by the Darmstadt group.[23,24] Comprehensive investigations of channeling radiation at very high energies have been carried out by Uggerhøj and his group at CERN,[25] and, as discussed in Refs. 25 and 26, channeled electrons in diamond may provide an interesting source of gamma rays.

5 Final Remarks

I have given a brief review of highlights of the development of channeling from a personal perspective. As a justification for my focus on quantum aspects of channeling I may remind you that in the year 2000 we celebrate not only Friedel Sellschop's 70th birthday but also the 100th anniversary of Planck's discovery of h, the quantum of action! Looking back, I find the struggle with the difficulties in the quantum description my most exciting experience with channeling. In the words of the Danish physicist and poet, Piet Hein: "Problems worthy of attack prove their worth by hitting back." The discussion of channeling radiation has given me an opportunity to recall an enjoyable collaboration with Friedel Sellschop, on an experiment with electrons penetrating his favourite target, a diamond crystal. Channeling radiation is still an active field: At the conference in Lüderitz, there were several reports on new experimental and theoretical investigations, and exciting new applications were suggested, including production of the Higgs boson, which had perhaps just been seen at CERN for the first time.

Acknowledgment

It is a pleasure to acknowledge support from the Danish National Research Foundation through the Aarhus Center for Atomic Physics.

References

1. J. Lindhard, *Mat. Fys. Medd. Dan. Vid. Selsk.* **34**, No. 14 (1965).
2. N. Bohr, *Mat. Fys. Medd. Dan. Vid. Selsk.* **18**, No. 8 (1948).
3. Ph. Lervig, J. Lindhard, and V. Nielsen, *Nucl. Phys.* A **96**, 481 (1967).
4. E. Uggerhøj, *Phys. Lett.* **22**, 382 (1966).

5. E. Uggerhøj and J.U. Andersen, *Can. J. Phys.* **46**, 543 (1968).
6. R.E. De Wames, W.F. Hall, and L.T. Chadderton, *Proc. of Int. Conf. on Solid State Physics with Accelerators*, ed. A.N. Goland (Brookhaven Natl. Lab., 1968) p. 3. A. Howie, ibid p. 15.
7. J.U. Andersen, S.K. Andersen, and W.M. Augustyniak, *Mat. Fys. Medd. Dan. Vid. Selsk.* **39**, No. 10 (1977).
8. M.A. Kumakhov, *Phys. Lett.* A **57**, 17 (1976).
9. M.J. Alguard, R.L. Swent, R.L. Pantell, B.L. Berman, S.D. Bloom, S. Datz, *Phys. Rev. Lett.* **42**, 1148 (1979).
10. R.L. Swent, R.H. Pantell, M.J. Alguard, B.L. Berman, S.D. Bloom, S. Datz, *Phys. Rev. Lett.* **43**, 1723 (1979).
11. J.U. Andersen and E. Laegsgaard, *Phys. Rev. Lett.* **44**, 1079 (1980).
12. M.L. Ter-Mikaelian, *High-energy Electromagnetic Processes in Condensed Media*, (Wiley & Sons, New York, 1972).
13. H. Überall, *Phys. Rev.* **103**, 1055 (1956).
14. J.U. Andersen, *Nucl. Instrum. Methods* **170**, 1 (1980).
15. J.U. Andersen, E. Bonderup, and R.H. Pantell, *Ann. Rev. Nucl. Part. Sci.* **33**, 453 (1983).
16. J. Lindhard, *Phys. Lett.* **12**, 126 (1964). E. Bøgh, J.A. Davies, and K.O. Nielsen, ibid p. 129.
17. J.W. Mayer, L. Eriksson, and J.A. Davies, *Ion Implantation into Semiconductors.* (Academic Press, New York 1970).
18. L.C. Feldman, J.W. Mayer, and S.T. Picraux, *Materials Analysis by Channeling.* (Academic Press, New York, 1982).
19. W.M. Gibson, *Ann. Rev. Nucl. Sci.* **25**, 465 (1975).
20. J.U. Andersen, E. Bonderup, E. Laegsgaard, and A.H. Sørensen, *Phys. Scripta* **28**, 308 (1983).
21. L. Vestergaard Hau and J.U. Andersen, *Phys. Rev.* A **47**, 4007 (1993).
22. J.U. Andersen, S. Datz, E. Laegsgaard, J.P.F. Sellschop, A.H. Sørensen, *Phys. Rev. Lett.* **49**, 215 (1982).
23. A. Richter, these proceedings.
24. H. Genz, J. Freudenberger, S. Fritzler, A. Richter, A. Zilges, R.A. Carrigan, J.P. Carneiro, P.L. Colestock, W.H. Hartung, K.P. Koepke, M.J. Fitch, and J.P.F. Sellschop, these proceedings.
25. E. Uggerhøj, these proceedings.
26. M.M. Velasco, these proceedings.

RADIATION PHYSICS WITH DIAMONDS [*]

A. RICHTER

*Institut für Kernphysik, Technische Universität Darmstadt,
Schlossgartenstrasse 9, D-64289 Darmstadt*

Some salient features and examples of the numerous results obtained during the last decade in close collaboration with Friedel Sellschop in a Bonn/Darmstadt/Erlangen/Kharkov/ Munich/Rossendorf/Wits Collaboration at the superconducting Darmstadt electron linear accelerator (S-DALINAC) on the interaction of relativistic electrons with diamonds are presented. These studies started with investigations of channeling radiation (CR) and were later on extended into the field of parametric X radiation (PXR), with the main aim to possibly develop a tunable radiation source with a small bandwidth. But also basic properties of these types of radiations, like e.g. coherence and occupation lengths, linewidths, polarization and interference with coherent bremsstrahlung have been investigated as well as special properties of diamonds. An example for such a measurement is the susceptibility of diamond through PXR. Lately CR has also been explored in the context of a new acceleration mechanism.

1 Introduction

This talk is dedicated to Friedel Sellschop, real scholar, gentle teacher, strong collaborator and true friend since a quarter of a century, who unlike anyone else has put in a remarkable way all of his efforts and energy for decades into developing in particular South African and in general African Science. I am thus very grateful to be able to be here at Luederitz where Friedel was born seventy years ago to praise him and his scientific work together with close colleagues and friends at this international conference.

The subject of this talk – radiation physics with diamonds – is one much to the heart of Friedel who infact has introduced me to it by presenting me in June of 1989 in Johannesburg a 1979 edition of John Field's book *The Properties of Diamond* with a personal dedication (see Fig. 1).

Since then I have built up a research group around the superconducting Darmstadt linear electron accelerator S-DALINAC which has excellent electron beam properties [1] for performing radiation physics experiments with diamonds provided by Friedel. We were soon joined by collaborators from outside Darmstadt and within a Bonn/Darmstadt/Erlangen/Kharkov/ Munich/Rossendorf/Wits Collaboration we have worked during the last decade

---

[*]WORK SUPPORTED BY THE BMFT UNDER GRANT NUMBER 06DA915I, BY A MAX-PLANCK-FORSCHUNGSPREIS AND BY THE DEUTSCHE FORSCHUNGSGEMEINSCHAFT THROUGH A TRAVEL GRANT.

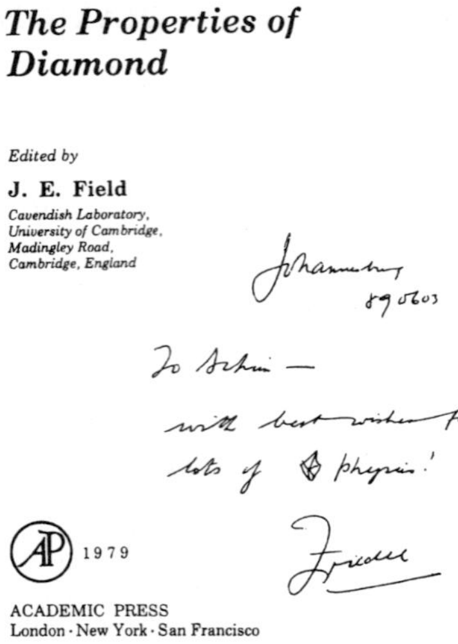

The Properties of Diamond

Edited by

J. E. Field

Cavendish Laboratory, University of Cambridge, Madingley Road, Cambridge, England

Johannesburg 890603

To Achim — with best wishes for lots of ◊ physics!

Friedel

1979

ACADEMIC PRESS
London · New York · San Francisco

Figure 1. Copy of the front page of the 1979 edition of John Field's book *The Properties of Diamonds* with an insertion from June 1989 *To Achim – with best wishes for lots of diamond physics! Friedel.*

successfully on topics like axial-channeling radiation from MeV electrons in diamond and silicon[2], high intensity electron channeling and perspectives for a bright tunable X-ray source[3], intensity and occupation lengths in diamond crystals [4], parametric X-ray radiation observed in diamond at low energies[5], measurements of the linear polarization of channeling radiation in silicon and diamond [6], coherence and occupation lengths in diamonds [7], experimental determination of the linewidth, lineshape and spectral density of parametric X-ray radiation at low electron energy in diamond[8,9], the boost of intensity of parametric X-rays under Bragg-condition[10], and the theoretical description and experimental detection of the interference between parametric X-radiation and coherent bremsstrahlung [11].

There is neither time nor space in this presentation to cover all of the listed research topics, and in fact the types of radiation that can be observed from the interaction of relativistic electrons with single crystals are still more numerous than just channeling radiation (CR) and parametric X-radiation (PXR) which form the basis of this talk. As Fig. 2 shows schematically, besides CR and PXR one might observe bremsstrahlung (BS), coherent bremsstrahlung (CBS), transition radiation (TR), polarization radiation (PR), characteristic X-radiation (CX), self reflected transition radiation (SDTR) and self deflected bremsstrahlung (SDBS).

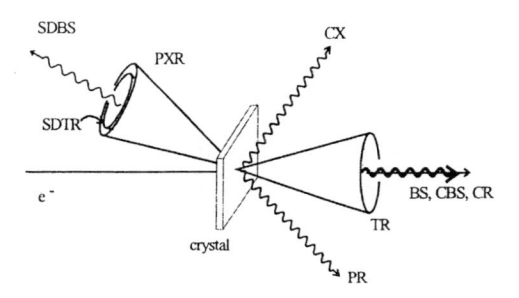

Figure 2. Various types of radiation from the interaction of relativistic electrons with single crystals. The symbols stand for: bremsstrahlung (BS), coherent bremsstrahlung (CBS), channeling radiation (CR), polarization radiation (PR), characteristic X-radiation (CX), transition radiation (TR), parametric X-radiation (PXR), self deflected transition radiation (SDTR) and self deflected bremsstrahlung (SDBS).

Let us next ask the question: Why study radiation physics with diamond? One primary goal of our earlier experiments in this field has been to investigate if the claim often made in the past (see e.g. [12]) is justified that CR and PXR can be generated with photon fluxes comparable to those produced by synchrotrons with and without wigglers. This meant – expressed more modestly – to prove if an efficient production of beams of about 10^{12} photons/s within a bandwidth of about 10% and a tunability between photon energies of about 10 to 40 keV would become possible. In this case the use of only MeV instead of GeV electrons for the production of intensive and tunable X-rays would be a real alternative [13]. Thus during the last years systematic studies have been performed at the S-DALINAC with respect to both mechanisms, CR and PXR. We have in particular investigated such properties like the energy, the intensity, the linewidth and the polarization of both types of radiation and have thereby also been able to learn something about basic crystal properties like form factors and the dielectric susceptibility. In all of these studies it has been clearly very advantageous to use diamond which, because of its high Debye temperature, is able to withstand large electron beam currents without showing signs of radiation damage or mechanical destruction. This latter aspect might become very relevant for an experimental proof of acceleration in a dense plasma produced and maintained for some time in solids. It will be shown towards the end of this talk that first results of a Darmstadt/Fermilab/Wits collaboration show indeed that channeling still is observed at about 10^{10} electrons per beam bunch [14].

2 Experimental matters

The experiments on CR and PXR were performed at the S-DALINAC, a new generation continuous wave (cw) electron accelerator [1] which provides beams with currents up to 40 μA in the energy range between 3 and 130 MeV with a small energy spread between 10^{-3} and 10^{-4}. The normalized beam emittance is around 2πmm mrad only and the beams are thus of a very high quality. Therefore intensity reducing (and background producing) apertures in front of the crystal are not needed any longer to reach conditions for channeling.

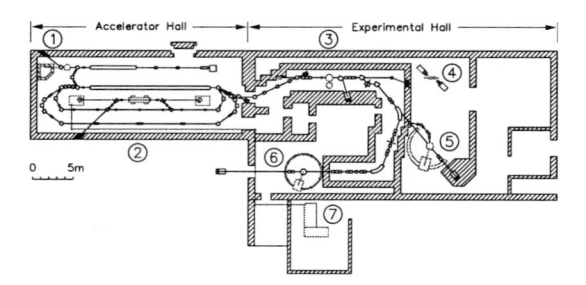

Figure 3. Schematic layout of the accelerator hall and the experimental hall and of various experimental stations for nuclear physics and radiation physics at the S-DALINAC. The experiments discussed have been performed at positions 1, 3 and 4.

Figure 3 shows a schematic layout of the accelerator hall and the experimental hall attached to it. Furthermore, the experimental stations where the various experiments performed at the S-DALINAC in nuclear physics (photon and radiation physics CR and PXR at Electron Laser – the first FEL built and operated successfully in Germany – at positions 2 and

The results from CR and PXR at low electron energies (3-9 MeV) presented here originated from experiments performed behind the injector of the S-DALINAC. A schematic layout of the experimental set-up (described in detail e.g. in [15]) is shown in Fig. 4. Measurements of CR and PXR at high electron energies (some results at 62 and 87 MeV, respectively, will in Fig. 3.

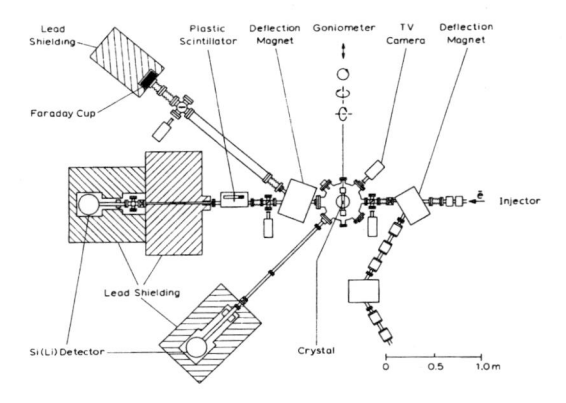

Figure 4. Schematic layout of the experimental set-up behind the 10 MeV injector of the S-DALINAC to observe CR at 0° and PXR at 44° with respect to the incoming electron beam from the crystal mounted on the goniometer within the target chamber (from [15]).

3 Channeling radiation

The basic processes underlying CR are now well understood. It is emitted when a relativistic electron penetrates a single crystal along one of its major axes or planes. Axial CR is created when the electron orbits around a string of crystal atoms and planar CR when it is wiggling along a crystal plane. The radiation is emitted into a cone with a polar angle $\theta \sim \gamma^{-1}$ in the direction of the electron path, where the Lorentz factor $\gamma = (E_0 + m_0c^2)/m_0c^2$ with E_0 being the energy of the incoming electron and m_0c^2 its rest mass.

3.1 Spectra, energy dependence and tunability

Spectra of CR are dominated by lines as is shown in Fig. 5 for a diamond crystal aligned along the $< 110 >$ axis. The lines that diamond emits at the particular electron energies chosen can be identified as being due to various transitions between bound states in the molecular type of crystal potential (Fig. 6) proposed [16] originally in 1982 within a single-string approximation approach [17]. The lines observed in Fig. 5 are predominantly due to 2p-1s and 3p-1s transitions at the lowest electron energy $E_0 = 4.8$ MeV, 3d-2p, 2p-1s and 3p-1s transitions at $E_0 = 4.8$ MeV and 2p-2s, 3d-2p and 2p-1s transitions in the crystal potential of Fig. 6. The broad shoulder underneath the line spectra results from transitions from the continuum into bound states of the potential.

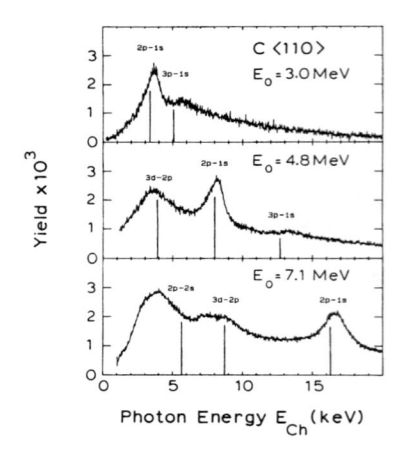

Figure 5. Axial CR spectra (background subtracted) of diamond at three different electron energies. The vertical lines with the indicated transitions are calculated in the single-string approximation (from [13]).

The energy dependence and tunability of CR can also be inferred from Fig. 5. Let us for simplicity only consider the dominant 2p-1s transition which lies at the lowest electron energy of $E_0 = 3$ MeV already in the region of the self-absorption of the diamond crystal, i.e. CR produced in the first atomic layers of the crystal is absorbed and reduced in the subsequent layers. The character of a "single-line source" with a photon energy of 3.5 keV at $E_0 = 3$ MeV is essentially preserved also at $E_0 = 4.8$ MeV and 7.1 MeV where the corresponding photon energies have increased to 8 and 16 keV, respectively. Many detailed investigations have shown that the photon energy in CR changes to a very good approximation with electron energy as $E_{ch} \sim \gamma^{3/2}$. Likewise, it can be stated that a CR source is tunable in energy with $\gamma^{3/2}$.

We have also searched for CR lines in planar channeling of electrons in diamond [7]. As a typical spectrum (Fig.7) taken at an energy $E_0 = 9$ MeV with a 14 μm thick diamond crystal shows, the radiation from the (110) plane exhibits one very pronounced isolated line at about 8 keV. Different planes yield lines at different energies and with different intensities, but since those lines are obviously much less suited for an intense CR source our interest focussed primarily on the 1-0 transition (in a potential model formalism for planar CR similar to the one discussed for axial CR in Fig. 6).

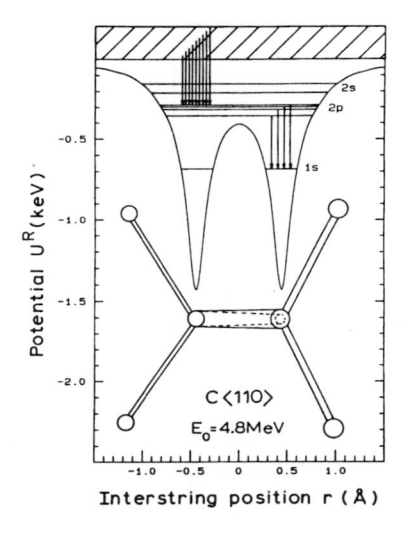

Figure 6. Crystal potential of diamond [16] calculated for an electron energy of $E_0 = 4.8$ MeV with a computer code [18] in the so-called many-beam formalism. The various 2p-1s transitions, that cause the broad 2p-1s peak in the spectra of Fig. 5 are indicated schematically. Also some continuum-to-bound state transitions causing the broad shoulder underneath the peaks in Fig. 5 are shown (from [13]).

3.2 Intensity and its dependence on energy and crystal thickness

As we have just seen, the energy of the spectral lines in CR can be tuned easily by varying the electron impact energy or by selecting different crystal axes or planes. The intensity or photon production rate (number of photons per electron at unit solid angle) also depends upon the electron energy. Furthermore, the angle between the electron beam and the axis or plane affects the initial population of the individual states and thus the intensity of CR. Also the scattering of the electron beam inside the crystal influences the photon production rate as well as the linewidth. We have investigated all those effects carefully [4,7]. First of all, for the 1-0 transition of the (110) plane we confirm the theoretically expected $\gamma^{5/2}$ dependence of the intensity (Fig. 8) by combining our data with previously measured ones at higher electron energies [19]. Second, the photon yields as
a function of crystal thickness are shown in Fig. 9. The 1-0 transition of the (110) plane is by far the strongest transition with 0.08 photons/e^-sr measured with the thickest crystal (55 μm) at $E_0 = 9$ MeV. Third, for the

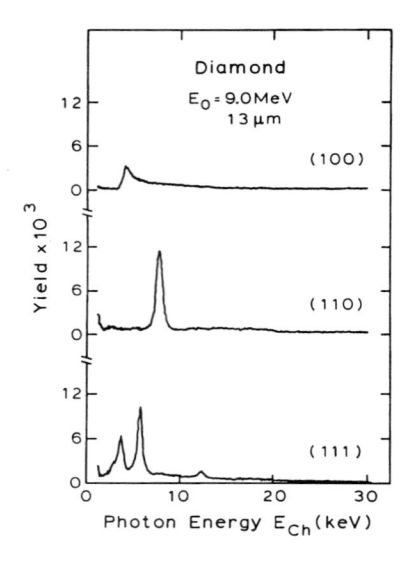

Figure 7. Planar CR spectra of a 13 µm thick diamond crystal bombarded with electrons of energy $E_0 = 9$ MeV. The background determined with the crystal randomly oriented has been subtracted. The spectra of the different planes show lines at different energies and intensities. The most pronounced CR line is located at a photon energy of about 8 keV in the spectrum of the (110) plane (from [7]).

experiment at 9 MeV an increase of intensity with crystal thickness is observed for both planar CR transitions displayed in Fig. 9. The behavior at the lower electron energy $E_0 = 5.2$ MeV is somewhat different. In this case the transition energy is only a few keV, which causes major self-absorption of the radiation in the crystal. Fourth, with increasing electron energy the intensity changes dramatically according to the expected $\gamma^{5/2}$ dependence, in the present case by a factor of 20 from one energy to the other.

Compared to the data in Fig. 9 are results from a full quantum mechanical treatment of planar channeling by Toepffer and Weber formulated in detail in [7]. This description incorporates fully the interaction of the electron with thermal vibrations and crystal electrons and thus is able to model the rather complex population dynamics of the CR process realistically. While in case of the 9 MeV measurements the agreement between experiment and theory is on the 10% level or better, there is at 5.2 MeV about a factor of two more intensity predicted than measured. The origin of this discrepancy is not quite clear at present, but it might possibly be traced to a greater uncertainty in

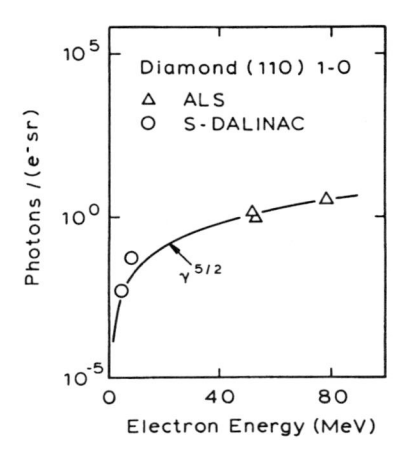

Figure 8. Axial CR intensity of the 1-0 transition in diamond as a function of the electron impact energy. The data points denoted by ALS are from [19] and are combined with data points from the S-DALINAC (from [4]).

the prediction of scattering probabilities and a greater importance of lattice defects at lower electron energies.

We have recently performed additional measurements at $E_0 = 9$ and 10 MeV with two natural diamond crystals of the type IaA and a thickness of 42 and 204 µm, respectively, in order to understand CR in thick crystals better and to determine the optimum crystal thickness for maximum radiation yield. The latter is obviously very much influenced by a delicate balance between multiple scattering and consequently by the population dynamics determining the CR process and by self-absorption in the crystal [20]. After normalization of the measured photon yields by a factor of 3.45 (platelet inclusions in the type IaA diamonds account for that) to the previously obtained data from diamonds of type Ia (Fig. 9) the dependence of the 1-0 transition of the (110) plane shown in Fig. 10 is obtained. It is interesting to note that the photon yield scales with the square root of the crystal thickness, in agreement with one possible solution of a set of differential equations describing the population dynamics [21,4]. Furthermore the maximum intensity for CR at electron energies E_0 between 9 and 10 MeV is expected for diamond crystals with a thickness between 500 and 1000 µm.

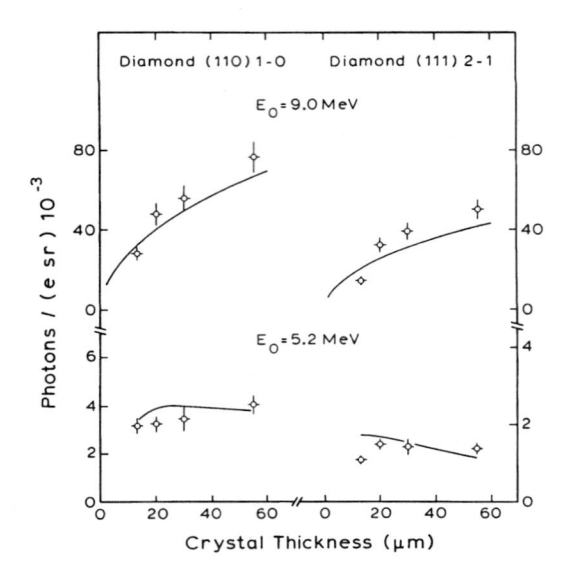

Figure 9. Axial CR photon yields as a function of crystal thickness for two different planes and energies. The solid curves are the results of a full quantum mechanical treatment of the CR process including the population dynamics of the levels. In case of 5.2 MeV the calculated results had to be multiplied by a factor of 0.55 and 0.38 for the (110) and (111) planes, respectively. At 9.0 MeV no normalization factor was needed (from [7]).

3.3 Linewidths

Great care was taken in the analysis of the numerous measured CR spectra by taking into account background and bremsstrahlung subtraction, corrections for the detector efficiency, the self absorption of photons inside the crystal, energy calibration and charge normalization. The spectra were subsequently deconvoluted by fitting a so-called Voigt function, i.e. a convolution of a Gaussian and a Lorentzian as parameterized in [22], to the bound-to-bound transitions, a Gaussian to describe the free-to-free and free-to-bound transitions, and with a $1/E$-varying function to account for the background. The result of such a fitting procedure is illustrated for an arbitrary spectrum shown in Fig. 11.

It is obvious that by this procedure also experimental linewidths for the observed transitions are obtained which then could be compared to theoretical predictions. Rather than comparing width by width this way, however a comparison has been made for the whole spectrum. Since the full quantum

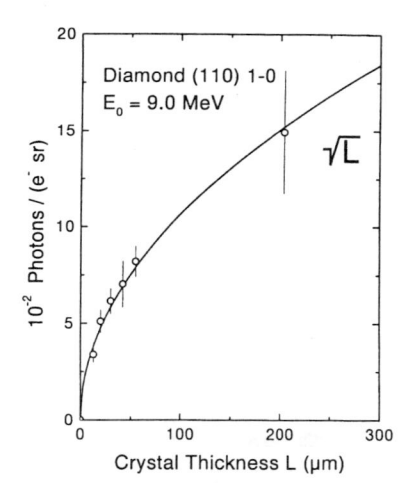

Figure 10. Axial CR photon yield as function of crystal thickness showing a square root dependence on L (from [20]).

mechanical treatment of [7] allows to calculate the entire CR spectrum consisting of the dominant bound-to-bound transitions as well as the most important free-to-bound and free-to-free transitions this procedure is meaningful. An example for such a comparison is shown in Fig. 12. The theoretical calculations take into account thermal and electronic scattering, both for the occupation dynamics and the intrinsic linewidth and, furthermore, Bloch and Doppler broadening of states. The initial population at the finite beam divergence has been taken into account as well as the fact, that the radiation intensity is modified by self absorption inside the crystal. The agreement with the experimental spectrum is very satisfactory, although the calculated width of the peak is to small with increasing crystal thickness. This may be explained by an underestimation of the Doppler broadening, since an additional beam divergence growth could be caused by scattering at crystal impurities which was not taken into account in the calculations.

Finally, from the fair overall agreement between the theoretical and experimental linewidth, the intrinsic linewidth that entered into the calculation can be used to estimate the coherence length through the expression $\Gamma_{int} = 2\gamma^2 \hbar c / l_{coh}$ which turns out to be 0.7 μm for axial channeling in diamond, and thus about a factor of two larger than observed before in Silicon (for more details see [7]).

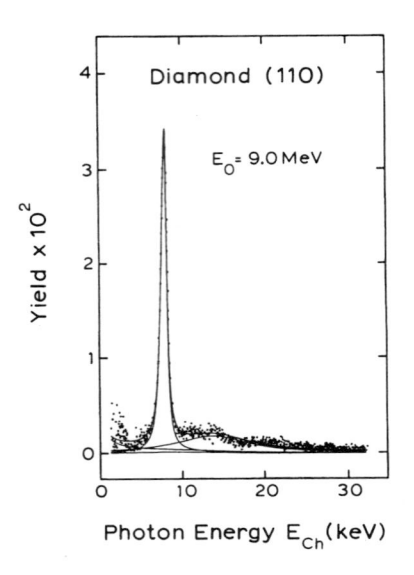

Figure 11. Result from the fitting procedure to deconvolute a CR spectrum of the (110) plane with the procedure described in the main text (from [7]).

3.4 Polarization

We have also studied the linear polarization of planar and axial CR in the photon energy range from 50 to about 400 keV produced by $E_0 = 62$ MeV electrons from the S-DALINAC incident on thin silicon and diamond crystals [6]. In those experiments, Compton scattering under 90° has been employed as the analyzing process. As is demonstrated in Fig. 13, planar CR due to transitions between transverse bound as well as unbound states is completely linearly polarized perpendicular to the channeling plane. This is in agreement with expectations for pure dipole radiation in spontaneous transitions between energy levels of the one-dimensional continuum potential used in the many-beam formalism [23] – mentioned already above in connection with Fig. 6 – to approximate the three-dimensional lattice potential. The measured linear polarization thus renders additional support to the continuum approximation. Furthermore our polarization measurements establish planar CR as a useful tunable narrow band source. Axial CR does not show linear polarization. In passing it should be mentioned that polarization measurements of CR and PXR from crystals at the S-DALINAC are performed nowadays by a

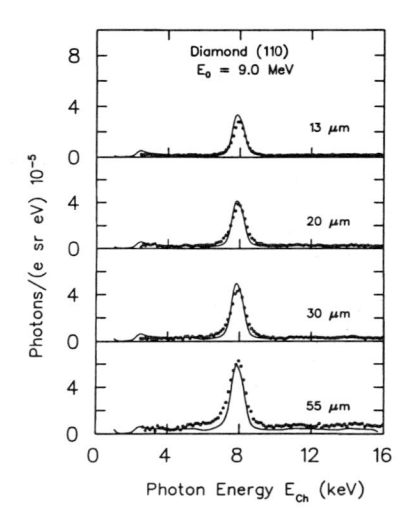

Figure 12. Experimental points and theoretical predictions (solid line) of the (110) planar CR spectra of diamonds of various thickness. The theoretical description includes bound-to-bound, free-to-bound and free-to-free transitions. For better representation only every fourth experimental data point has been displayed (from [7]).

novel method exploiting directional information of the photoelectric effect in a charged coupled device consisting of 6.8 μm × 6.8 μm pixels [24].

4 Parametric X-radiation

Like in the case of CR the basic processes underlying PXR are now also well understood. In contrast to CR, where the radiation is caused by a transition of electrons between states in the crystal potential, for PXR the crystal itself is the source of the radiation. A relativistic electron moving through a crystal polarizes the crystal atoms located in the vicinity of its trajectory. Due to constructive interference of radiation emitted as a result of this polarization intense X radiation occurs. This process might thus be viewed as Bragg diffraction of the virtual photons of the electromagnetic field of the electron at the crystal planes [25,26], or alternatively as de-excitation of crystal atoms which were virtually polarized and excited by the relativistic electrons [27].

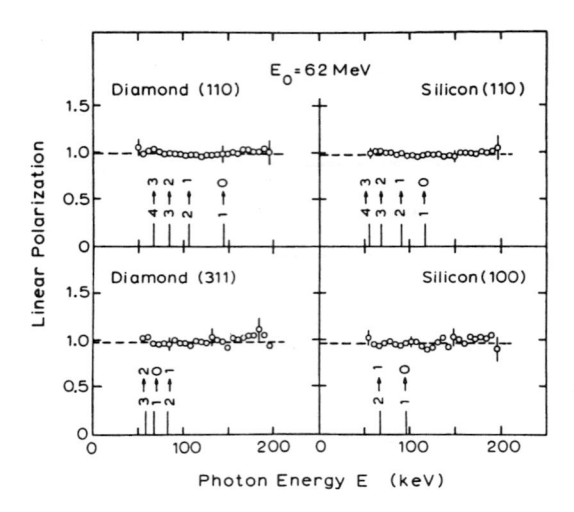

Figure 13. Energy dependence of the degree of linear polarization of planar CR for the (110) plane of both crystals (top row) and the (311) plane of diamond and the (110) plane of Silicon (bottom row). The energy E is the photon energy after 90° Compton scattering. The vertical lines indicate the energies corresponding to transitions between the lowest lying bound states. The error bars give representative 1σ errors composed of statistical and systematic contributions added in quadrature. The horizontal dashed line corresponds to the average polarization over the entire range (from [6]).

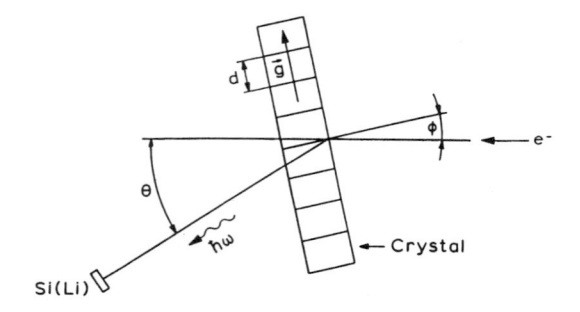

Figure 14. Definition of the geometry and the angles in the PXR experiments. The reciprocal lattice vector $\vec{g}$ and the lattice constant d are also indicated schematically. The PXR photons of energy $\hbar\omega$ are detected in a Si(Li) semiconductor detector placed at the angle θ.

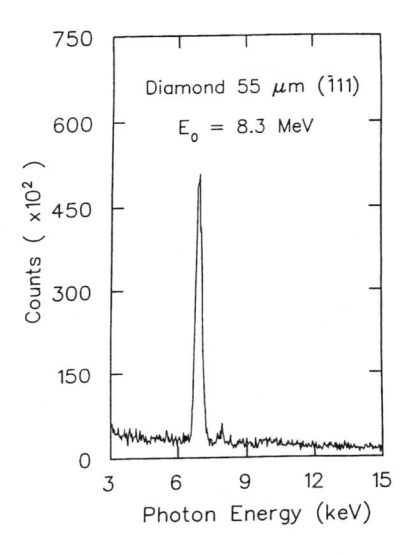

Figure 15. Typical PXR spectrum obtained with 8.3 MeV electrons recorded by a Si(Li) detector placed under $\theta = 44°$ with respect to the electron beam axis (from [5]).

4.1 Spectra, energy dependence and tunability

As Fig. 14 shows schematically, the incoming relativistic electron interacts with the crystal which is characterized by its lattice constant d and the reciprocal lattice vector $\vec{g}$ of its planes reflecting the virtual photons observed under the angle θ with a Si(Li) detector. A typical spectrum from our earlier work [5] at $\theta = 44°$, i.e. twice the Bragg angle, for the scattering at the $(\bar{1}11)$ plane of diamond is displayed in Fig. 15. Low energy electrons of energy $E_0 = 8.3$ MeV, i.e. $\gamma \approx 17$, produce a very sharp and isolated PXR line at 7 keV above a low background level.

Since the photon energy dependence of PXR is given by $\hbar\omega = \hbar v g \sin\phi/(1 - v\cos\theta)$, where v is the velocity of the electron and ϕ the angle between the normal to the crystal surface and the electron beam (Fig 14), it is immediately evident that the photon energy is nearly independent of the electron energy since for relativistic electrons $v/c \approx 1$. However, the energy of the PXR photons can be very effectively varied as a function of the tilt angle ϕ as is shown in Fig. 16. For a change of ϕ between 17° and 32° $\hbar\omega$ lies between 4 and 12 keV. This easily achievable variation in energy makes PXR also a potential candidate for a tunable X ray photon source.

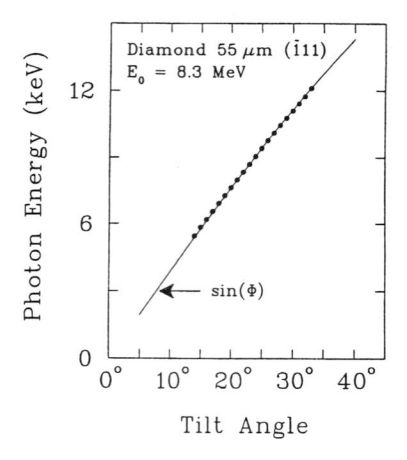

Figure 16. Variation of the PXR photon line energy as a function of the crystal tilt angle ϕ. The solid line represents a fit of a fucntion $\hbar\omega = a\sin\phi$ to the data points (from [5]).

4.2 Intensity and its dependence on tilt angle and electron energy

The spectral intensity of PXR has been obtained by fitting a Gaussian to the line and a polynomial to the spectra like the one displayed in Fig. 15. The results for the spectra measured at different tilt angles ϕ of the crystal for an electron energy $E_0 = 8.3$ MeV are shown in Fig 17. Note that the intensity is given in absolute units. The angular distribution exhibits a distinct minimum near the Bragg angle and two maxima of different intensity at $\phi = 20°$ and $25°$, i.e. below and above the Bragg angle $\phi_B = 22°$. The distance $\Delta\phi$ between the two maxima is roughly $1/\gamma$. The solid line compared to the data represents the results from a theoretical calculation starting from "first principles of PXR" incorporating multiple scattering of the electrons in the crystal (for details see [5]). Note, however, that the calculated angular distribution had to be multiplied by a factor of 1.12 in order to describe the experimental one, i.e. we have – a still not understood – discrepancy between theory and experiment within 12%. As pointed out in [5], the small deviations at large tilt angles of the crystal come presumably from the particular distribution of core and valence electrons in a diamond lattice. For PXR wavelengths smaller than the spatial extension of the valence electron (sp^3) distribution, the PXR radiation can originate from various points. Therefore the angular distribution may especially be affected at the high energy tail (large tilt angles) pulling the theoretical curve down to the experimental values.

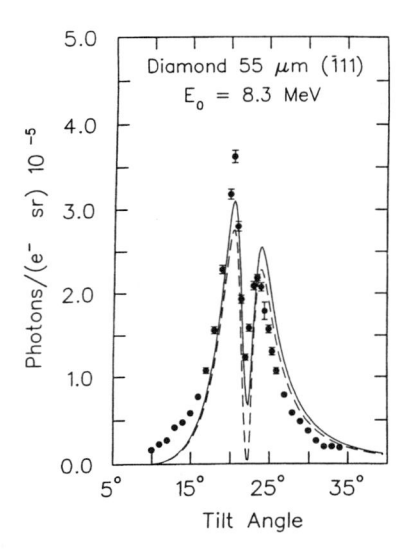

Figure 17. The intensity of the PXR line of Fig.15 as a function of the crystal tilt angle ϕ. The experimental points have been obtained by rotating the diamond crystal stepwise in the vicinity of the Bragg angle. The solid line represents the result from the theoretical calculation explained in detail in [5].

The dependence of PXR intensity as a function of electron energy is shown in Fig 18 where the number of photons of the maximum below the Bragg angle is plotted for five electron energies. The number of photons at the tilt angle ϕ varies proportional to $\gamma(\theta\gamma - 1) \approx \gamma^2$, as is indicated by the solid line in the figure, in very good agreement with the experiment. Note, however, that the PXR lines in diamond have an intensity about three orders of magnitude smaller to compared CR lines in diamond [4].

4.3 Linewidths

The lineshape and linewidth in the spectra are mainly determined by multiple scattering of the electrons in the crystal [8,9] and by geometrical effects [28]. By applying an absorber technique we were able to measure the shape and width of a PXR line at 9 keV produced by bombarding a diamond crystal of 55 μm thickness with electrons of 6.8 MeV [9]. The variance of the spectral line distribution was found to depend on the tilt angle of the crystal and to have a magnitude of $\sigma = 51$ eV. Simulations based on a Monte Carlo method exhibit

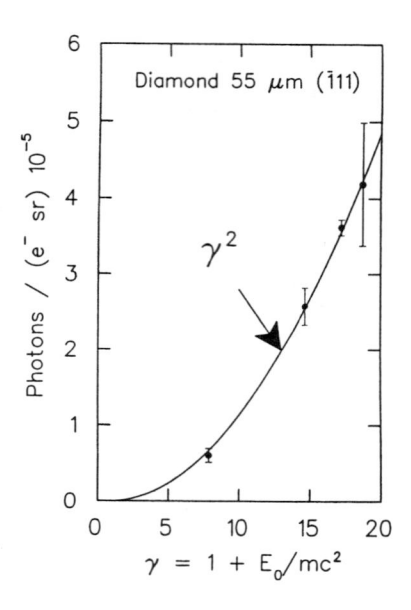

Figure 18. Intensity of the first maximum of the PXR line of Fig. 17 as a function of different electron energies.

indeed that the observed variance is mainly influenced by multiple scattering of electrons passing through the crystal ($\approx$ 43 eV) and the finite detector opening ($\approx$ 18 eV, leaving for the intrinsic linewidth a value of the order of 1 eV. This value has been substantiated by a precise experiment at the Mainz Microtron (MAMI) electron accelerator at a much higher bombarding energy where multiple scattering effects are much smaller that at low energies [28].

4.4 Polarization

The polarization character of PXR so far is still largely unexplored. There exist only two older measurements at high electron energies in which the linear polarization of some PXR reflex in a silicon crystal has been investigated [29,30]. Since the results from those measurements were of conflicting nature with the theory we have recently determined the polarization degree and direction of PXR for experimental conditions similar as in [29,30] but at an energy of $E_0 = 80.5$ MeV. With the results of the experiment also the appropriate expressions for the observables describing the polarization properties of PXR derived for the first time were tested. Details can be found in [24,31]. Polarization measurements of PXR from diamond have so far not been made.

4.5 PXR and crystal properties

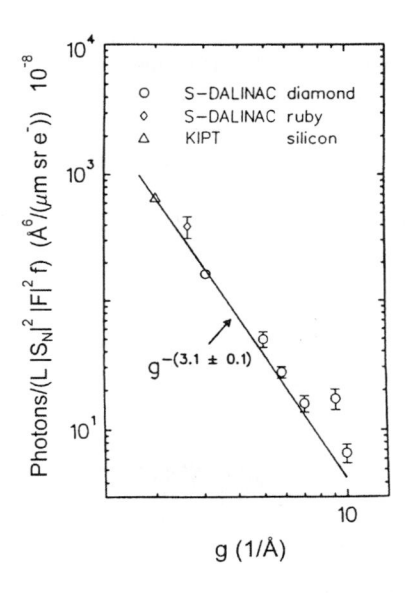

Figure 19. Dependence of the number of PXR photons, divided by the crystal thickness, the squares of the structure and atomic form factors, respectively, and the Debye-Waller factor, on the magnitude of the reciprocal lattice vector. The expected g^{-3} scaling behaviour is evident (from [32]).

For possible applications it is of interest to compare PXR from different crystals. It is known from the theory of PXR (see e.g. [25,27]) that at low electron energies the number of PXR photons depends upon crystal properties according to $N_{PXR} \sim [|S_N|^2 |F|^2 f / g^3] \cdot L \cdot P(\phi, \theta, \gamma)$. The first term in square brackets in this expression contains the structure factor S_N normalized to the volume of one elementary crystal cell, the atomic form factor F which is the Fourier transform of the electron density distribution of the crystal atoms, the Debye-Waller factor containing the temperature dependent transverse vibrational amplitude of the crystal atoms and g, the magnitude of the reciprocal lattice vector already introduced above. The quantity L denotes the crystal thickness which for the validity of the simple proportionality relation for N_{PXR} to hold is assumed to be $L < L_a$, with L_a being the absorption length of the crystalline medium. Finally, P is a known function depending on the angles ϕ and θ defined in Fig. 14 above and on the electron energy γ, and is for a set of those experimental parameters a constant. Dividing the number of PXR photons by $|S_N|^2|F|^2 \cdot f \cdot L \cdot P$ one should find a proportionality to g^{-3}. Recently this relation has been investigated [32] for all PXR experiments with electron energies $E_0 < 30$ MeV in which absolute photon yields were determined. The result is shown in Fig. 19. Besides our data points from the S-DALINAC on diamond and ruby one point taken on silicon in [33] has been included in the plot. Fitting a function of the form ag^{-b} to the data indeed confirms the expectation of a g^{-3} scaling of the PXR photon intensity.

4.6 PXR at backward angles: spectra, atomic form factor and dielectric susceptibility

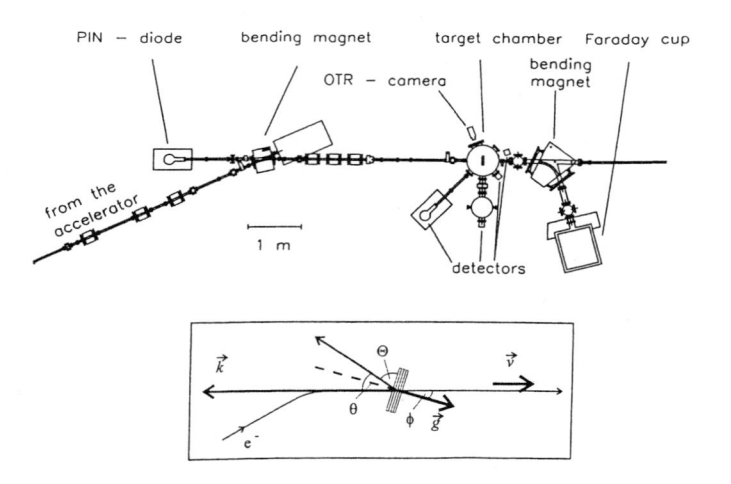

Figure 20. Experimental set up for the detection of high energy CR and PXR at the S-DALINAC at position 3 in Fig. 3 above. In the insert underneath the geometry of crystal induced PXR at $\theta = 180°$ is plotted (from [34]).

So far in this talk I have put a strong emphasis on PXR at low electron energies and its suitability for a tunable X ray source. We have, however, recently also studied PXR for the first time at $\theta = 180°$, i.e. under kinematic conditions which simplify the theoretical approaches to the interpretation of the clean experimental data enormously and allow a straightforward deduction of interesting basic crystal properties [34]. Two of those, the atomic form factor and the dielectric susceptibility, are briefly discussed. Let us first look at the experimental set up (Fig. 20). The electron beam from the accelerator is bent magnetically into a straight line path with a target chamber housing the crystal mounted on a three-axes goniometer and is then bent into a Faraday cup [35] for beam normalization. An opening at the back of the first bending magnet allows the emitted PXR to be detected with a pin-diode at $\theta = 180°$. Underneath the experimental set up a schematic representation of the Bragg-geometry (the only geometry possible at extreme backward angles) is plotted. The quantities $\vec{v}$ and $\vec{k}$ are the velocity vector of the relativistic electron and the wave vector of the emitted PXR, respectively. All other variables shown are self-explanatory and have already been used above.

Figure 21 shows a PXR spectrum of 86.9 MeV electrons interacting through their virtual photon field with diamond at various crystal orientations. The spectrum is extremely clean and almost background free. The linewidth is entirely dominated by the photon detector resolution.

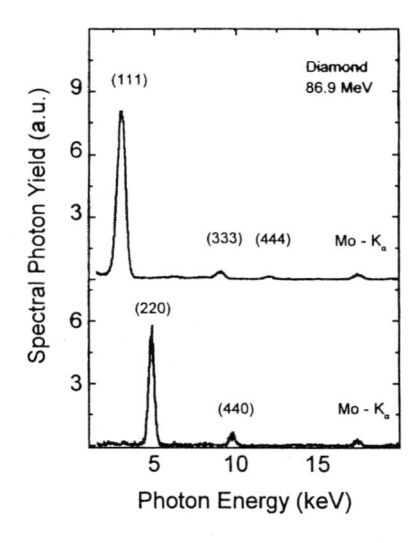

Figure 21. Typical PXR spectrum taken at $\theta = 180°$ and $E_0 = 86.9$ MeV. In the first orientation at which the (111) lattice vector is almost parallel to the direction of the electron beam, one observes a (111) reflex. The (222) reflex is forbidden by the structure factor and only higher order reflexes – (333), (444) and (555) – appear. Another geometrical orientation of diamond allowed a showing of the (220) reflex and its higher orders. Seen is also always a molybdenum K_α-line resulting from the interaction of electrons from a slight beam halo with the crystal holder (from [34]).

These spectra allow a precise determination of the PXR intensity as a function of crystal orientation and hence a measurement of atomic form factors. In Fig. 22 for diamond such measured form factor points from PXR are plotted and compared to values from the literature [36] obtained from the diffraction of real photons and to a standard atomic form factor description [37]. The agreement between the two experiments – one using virtual and the other one real photons – and with the phenomenological form factor is very satisfying.

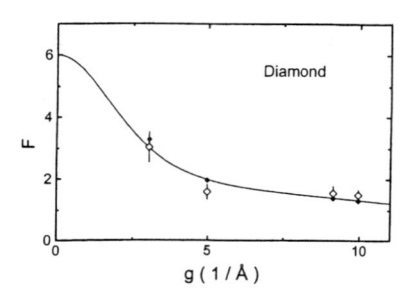

Figure 22. Measurement of the atomic form factor of diamond from PXR (open symbols) and real photons[36] (full symbols) compared to a standard phenomenological description [37]. For details, see[34].

Another benchmark observation at 180° is the PXR yield and its particular dependence on the crystal orientation controlled by the tilt angle ϕ defined in Fig. 20. For an angle of observation $\theta = 180°$ the resulting angular distribution is rotational symmetric about the direction of the electron beam. It thus suffices to make an appropriate cut in order to obtain the photon intensity as a function of crystal orientation. If this intensity is plotted as a function of the angle ϕ that determines the direction between the reciprocal lattice vector $\vec{g}$ and the electron velocity $\vec{v}$ one gets the data shown in Fig. 23 for the PXR (220) line of diamond observed at $E_0 = 86.9$ MeV. Several points are worth noticing. First, at $\phi = 0°$ the intensity has a minimum. Second, it rises to a maximum at an angle $\Delta\phi \approx 0.29°$ which by the theory of PXR (see e.g.[35]) is connected to the dielectric susceptibility χ of the crystal by the relation $\Delta\phi \approx \sqrt{1/\gamma^2 - \chi}/2$. Third, the solid line in Fig. 23 represents the result of the PXR photon yield in standard PXR theory. However, agreement between experiment and theory is only achieved by multiplying the latter by a factor of two. This most recent result of our collaboration [10] is still not understood. The dashed line has been obtained by taking into account multiple electron scattering inside the crystal, finite electron beam size and divergence, mosaic structure of the crystal detector acceptance and photon absorption inside the crystal. It has again to be multiplied by a factor of two to agree with the data. Fourthly, the dotted curve and the dash-dotted curve represents contributions of self deflected transition radiation (SDTR) and self deflected bremsstrahlung (SDBS), respectively, defined in Fig. 2 above (from [34]).

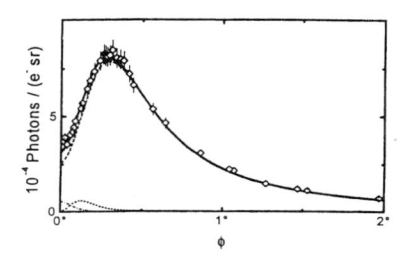

Figure 23. Intensity of the PXR (220) line in diamond measured at $E_0 = 86.9$ MeV as a function of ϕ, the angle between the direction of the reciprocal lattice vector and the electron velocity (compare Fig. 20). The open symbols represent the experimental points. For the meaning of the theoretical curves (full, dashed, dotted, and dash-dotted lines, respectively), see the main text (from [34]).

Finally, the detected dielectric susceptibility – the authors of [38] were the first ones to propose the use of PXR for that – of diamond from the tilt angle difference $\Delta\phi$ of the PXR line observed under $\theta = 180°$ and different electron and consequently photon energies is given in Fig. 24. The experimental points are described perfectly well by the so-called high energy approximation $\chi(\omega) \xrightarrow{\omega > \omega_k} -\omega_p^2/\omega^2$, where ω_k denotes the position of the K absorption edge and ω_p the plasma frequency [39].

5 Summary and outlook

In this talk I have reported briefly on some of the many results we have been able to accumulate during the last decade on the study of radiation sources in the X ray region. An absolute prerequisite to this has been the S-DALINAC with its excellent continuous wave electron beams. Both CR and PXR from diamond (and other crystals) are intensive, tunable and of a small bandwidth. However it took great time and experimental effort to disprove the claim that has been in the literature [12] since 1985 that under certain conditions CR and PXR might reach photon fluxes similar to monochromatic X rays provided by (now third generation) synchrotrons and wigglers. If we look at the so-called brilliance defined as the number of photons per (s · mrad^2 mm^2 · 0.1% bandwidth) then CR and PXR at $E_0 = 9$ MeV studied

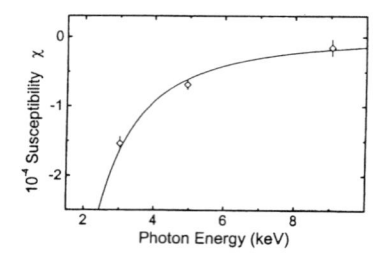

Figure 24. Dielectric susceptibility of diamond. The data points are perfectly well described by the so called high energy approximation $\chi(\omega) \xrightarrow{\omega > \omega_k} -\omega_p^2/\omega^2$ (from [34]).

at the S-DALINAC using diamond crystals yield a brilliance of at most 10^6 compared to 10^{14} for transition radiation investigated at MAMI with electron beams of $E_0 = 856$ MeV [40] and of third generation synchrotrons with numbers between 10^{14} and 10^{16} (in the not too distant future to be raised by several orders of magnitude by X-ray Free Electron Lasers). This limits severely the applicability of CR and PXR sources. But as discussed by way of examples along the struggle for more intense sources a number of basic properties of CR and PXR as well as of diamond could be studied.

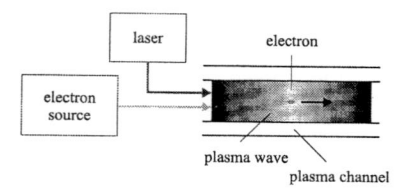

Figure 25. The principle of plasma acceleration is based on a longitudinal wave propagating through the medium and pulling thereby the electron to higher energies.

In closing this talk, however, I would like to risk a look into the future. In a recent very interesting article by one of the worlds most renown accelerator physicist Maury Tigner on the future of accelerator-based particle physics, which has just appeared [41] when I finished writing up this manuscript, remarked on "significant studies toward lowering cost per unit energy" of future accelerators the following: *"One possible approach would be to implement some substantially new acceleration schemes, such as lasers with plasma mode conversion* [42], *to give the transverse electromagnetic field of the laser beam a longitudinal accelerating component."* The scheme of plasma acceleration is displayed in Fig. 25. The plasma can be produced either by the highly intense electron beam itself or by a powerful external laser. Using a crystal instead of a gas firstly high accelerating fields (~ 100 GV/cm) might be achieved due to the higher plasma density and furthermore confinement of the accelerator beam is expected because of the channeling. For this however, one first has to prove experimentally if channeling still occurs at high electron beam densities, i.e. at much higher values hitherto observed [3,43]. First results of a Darmstadt/Fermilab/Wits collaboration [14] show indeed (Fig. 26) that channeling is still observed at about 10^{10} electrons per beam bunch. At present this result (for which novel detection techniques had to be developed) is still orders away from the plasma acceleration regime but very encouraging.

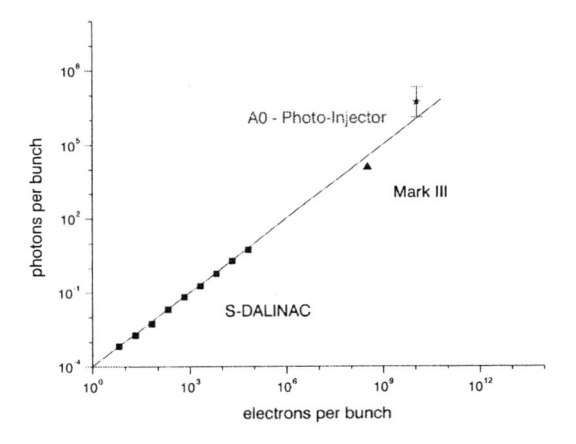

Figure 26. Observation of CR as a function of electrons per bunch. The data taken at the S-DALINAC are from [3], the ones at Mark III from [43] and the latest data point measured at the superconducting A0-photo-injector at Fermilab has been evaluated in [44].

Acknowledgments

I am very much indebted to all members of the Bonn/Darmstadt/Erlangen/Kharkov/ Munich/Rossendorf/Wits and of the Darmstadt/Fermilab/Wits collaborations – all names are in the list of references – for sharing their insight with me into the various topics presented in this talk. I mention in particular only my two last graduate students in the field of radiation physics with diamonds, Jörg Freudenberger and Victor Morokhovskii, and Harald Genz who now for years with untiring efforts has coordinated the work of the Darmstadt group and the collaborations and has filled it with inspiring physics. Without the technical help of Christian Dembowski with this manuscript I truly would have been lost. Many thanks to him. Finally, all of the work described in this talk would not have been possible without Friedel Sellschop and "his diamonds", to whom I dedicate this article with deep admiration for his seminal work in physics.

References

1. A. Richter, Proc. 5th European Particle Accelerator Conference, eds. S. Meyer *et al.* (IOP Publishing, 1996), p. 110.
2. W. Lotz, H. Genz, P. Hoffmann, U. Nething, A. Richter, A. Weickenmeier, H. Kohl, W. Knüpfer, and J.P.F. Sellschop, Nucl. Instr. Meth. **B48** (1990) 256.
3. H. Genz, H.-D. Gräf, P. Hoffmann, W. Lotz, U. Nething, A. Richter, H. Kohl, A. Weickenmeier, W. Knüpfer, and J.P.F. Sellschop, Appl. Phys. Letters **57** (1990) 2956.
4. U. Nething, M. Galemann, H. Genz, M. Höfer, P. Hoffmann-Stascheck, J. Hormes, A. Richter, and J.P.F. Sellschop, Phys. Rev. Lett. **72** (1994) 2411.
5. J. Freudenberger, V.B. Gawrikow, M. Galemann, H. Genz, L. Groening, V.L. Morokhovskii, V.V. Morokhovskii, U. Nething, A. Richter, J.P.F. Sellschop and N.F. Shul'ga, Phys. Rev. Lett. **74** (1995) 2487.
6. M. Rzepka, G. Buschhorn, E. Diedrich, R. Kotthaus, W. Kufner, W. Rößl, K.H. Schmidt, P. Hoffmann-Stascheck, H. Genz, U. Nething, A. Richter, and J.P.F. Sellschop, Phys. Rev. B **52** (1995) 771.
7. H. Genz, L. Groening, M. Höfer, P. Hoffmann-Stascheck, J. Hormes, U. Nething, A. Richter, J.P.F. Sellschop, C. Toepffer, and M. Weber, Phys. Rev. B **53** (1996) 8922.
8. J. Freudenberger, M. Galemann, H. Genz, L. Groening, P. Hoffmann-Stascheck, V.L. Morokhovskii, V.V. Morokhovskii, U. Nething, H. Prade, A. Richter, J.P.F. Sellschop, and R. Zahn, Nucl. Instr. Meth. B **115** (1996) 408.
9. J. Freudenberger, H. Genz, L. Groening, V.V. Morokhovskii, A. Richter, V.L. Morokhovskii, U. Nething, R. Zahn, and J.P.F. Sellschop, Appl. Phys. Lett. **70** (1997) 267.
10. J. Freudenberger, H. Genz, V.V. Morokhovskii, A. Richter, and J.P.F. Sellschop, Phys. Rev. Lett. **84** (2000) 270.
11. V.V. Morokhovskii, J. Freudenberger, H. Genz, V.L. Morokhovskii, A. Richter, and J.P.F. Sellschop, Phys. Rev. B **61** (2000) 3347.

12. V.G. Baryshevsky and I.D. Feranchuk, Nucl. Instr. Meth. **228** (1985) 490.
13. A. Richter, Mat. Sci. Eng. B **11** (1992) 139.
14. J. Freudenberger, S. Fritzler, H. Genz, A. Richter, A. Zilges, R.A. Carrigan, Jr., J.-P. Carneiro, P.L. Colestock, H.T. Edwards, W.H. Hartung, K.P. Koepke, M.J. Fitch, N. Barov, and J.P.F. Sellschop, Contribution to the Int. Workshop on 2nd Generation Laser and Plasma Accelerators, Kardamyli, Greece, June 27 - July 2, 1999.
15. J. Freudenberger, H. Genz, L. Groening, P. Hoffmann-Stascheck, W. Knüpfer, V.L. Morokhovskii, V.V. Morokhovskii, U. Nething, A. Richter, J.P.F. Sellschop, Nucl. Instr. Meth. B **119** (1996) 123.
16. J. U. Andersen, S. Datz, E. Laegsgaard, J.P.F. Sellschop, and A.H. Sørensen, Phys. Rev. Lett. **49** (1982) 215.
17. A.W. Saenz and H. Überall (eds.), Coherent Radiation Sources (Springer, Berlin, 1985) p. 162.
18. G. Buschhorn, E. Diedrich, W. Kufner, and D. Pollmann, Nucl. Instr. Meth. B **30** (1988) 29.
19. M. Gouanere, D. Sillou, M. Spighel, N. Cue, M.J. Gaillard, R.G. Kirsch, J.-C. Poizat, J. Remillieux, B.L. Berman, P. Catillon, L. Roussel, and G.M. Temmer, Nucl. Instr. Meth. **194** (1982) 225 and Phys. Rev. **338** (1988) 4352.
20. I. Reitz, J. Freudenberger, H. Genz, V.V. Morokhovskii, A. Richter, and J.P.F. Sellschop, to be published.
21. J. U. Andersen, E. Bonderup, E. Laegsgaard, and A.H. Sørensen, Phys. Scr. **28** (1983) 308.
22. A.K. Hui, B.H. Armstrong, and A.A. Wray, J. Spectros. Radiat. Transfer **19** (1978) 509.
23. J.U. Andersen, E. Bonderup, E. Laegsgaard, and A.H. Sørensen, Nucl. Instr. Meth. **194** (1982) 209.
24. V.V. Morokhovskii, K.H. Schmidt, G. Buschhorn, J. Freudenberger, H. Genz, R. Kotthaus, A. Richter, M. Rzepka, and P.M. Weinmann, Phys. Rev. Lett. **79** (1997) 4389.
25. V.L. Morokhovskii, Coherent X-rays of Relativistic Electron in Crystals, CSRI atominform **39** (Moscow, 1989). An English translation is available from the author upon request.
26. M.L. Ter-Mikaelian, High Energy Processes in Condensed Media (Wiley, New York, 1972).
27. H. Nitta, Nucl. Instr. Meth. B **115** (1996) 401.
28. K.-H. Brenzinger, B. Limburg, H. Backe, S. Dambach, H. Euteneuer, F. Hagenbuck, C. Herberg, K.H. Kaiser, O. Kettig, G. Kube, W. Lauth, H.

Schöpe, and Th. Walcher, Phys. Rev. Lett. **79** (1997) 2462.
29. Yu. Adishchev, V.A. Verzilov, S.A. Vorobiev, A.P. Potylitsyn, and S.R. Uglov, Pis'ma Zk. Eksp. Teor. Fiz. **48** (1988) 311; JETP Lett. **48** (1988) 342.
30. Yu. Adishchev, V.A. Verzilov, A.P. Potylitsyn, S.R. Uglov, and S.A. Vorobyev, Nucl. Instr. Meth. B **44** (1989) 130.
31. K.H. Schmidt, G. Buschhorn, R. Kotthaus, M. Rzepka, P.M. Weinmann, V.V. Morokhovskii, J. Freudenberger, H. Genz, and A. Richter, Nucl. Instr. Meth. B **145** (1998) 8.
32. J. Freudenberger, in X-Ray and Inner-shell Processes, eds. R.L. Johnson et al., AIP Conf. Proc. **389** (Woodburry, N.Y., 1996) p. 73.
33. A.V. Shchagin, V.I. Pristupa, and N.A. Khizhnyak, in Proc. Int. Symp. on Radiation of Relativistic Electrons in Periodical Structures, Tomsk, Russia (1993) 62.
34. J. Freudenberger, Dissertation D17, Darmstadt University of Technology, 1999, and to be published.
35. V.V. Morokhovskii, Dissertation D17, Darmstadt University of Technology, 1998, and to be published.
36. Z.W. Lu, A. Zunger, and M. Deutsch, Phys. Rev B **47** (1993) 9385.
37. J.A. Ibers and W.C. Hamilton (eds.), International Tables of X-ray Crystallography, Vol. IV (The Kynoch Press, Birmingham, 1974).
38. R.B. Fiorito, D.W. Rule, M.A. Piestrup, X.U. Maruyama, R.M. Silzer, D.M. Skopik, and A.V. Shchagin, Phys. Rev. E **51** (1995) R2795.
39. B.L. Henke, E.M. Gullikson, and J.C. Davis, Atom. Nucl. Data Tabl. **54** (1993) 181.
40. H. Backe, S. Gampert, A. Grendel, H.J. Hartmann, W. Lauth, C. Weinheimer, R. Zahn, F.R. Buskirk, H. Euteneuer, K.H. Kaiser, G. Stephan, and Th. Walcher, Z. Phys. A **349** (1994) 87.
41. M. Tigner, Physics Today, January 2001,p. 36.
42. E. Esarey, in Handbook of Accelerator Physics and Engineering, eds. A. Chao and M. Tigner (World Scientific, Singapore, 1998) p. 543.
43. C.K. Gary, A.S. Fisher, R.H. Pantell, J. Harris, and M.A. Piestrup, Phys. Rev. B **42** (1990) 7.
44. S. Fritzler, Diploma thesis, Darmstadt University of Technology, 2000, and to be published by the Darmstadt/ Fermilab/Wits collaboration listed above

CHANNELING OF CHARGED PARTICLES THROUGH PERIODICALLY BENT CRYSTALS: ON THE POSSIBILITY OF A GAMMA LASER

A. V. KOROL, W. KRAUSE, A. V. SOLOV'YOV AND W. GREINER

Institut für Theoretische Physik der Johann Wolfgang Goethe-Universität, 60054 Frankfurt am Main, Germany
E-mail: solovyov@th.physik.uni-frankfurt.de

We discuss radiation generated by positrons channeling in a crystalline undulator. The undulator is produced by periodically bending a single crystal with an amplitude much larger than the interplanar spacing. Different approaches for bending the crystal are described and the restrictions on the parameters of the bending are established. We present the results of numeric calculations of the spectral distributions of the spontaneous emitted radiation and estimate the conditions for stimulated emission. Our investigations show that the proposed mechanism provides an efficient source for high energy photons, which is worth to be studied experimentally.

1 Introduction

We discuss a new mechanism, initially proposed in [1,2], for the generation of high energy photons by means of the planar channeling of ultra-relativistic positrons through a periodically bent crystal. In this system there appears, in addition to the well-known channeling radiation, an undulator type radiation due to the periodic motion of the channeling positrons which follow the bending of the crystallographic planes. The intensity and the characteristic frequencies of this undulator radiation can be easily varied by changing the positrons energy and the parameters of the crystal bending.

The mechanism of the photon emission by means of the crystalline undulator is illustrated in Figure 1. It is important to stress that we consider the case when the amplitude a of the bending is much larger than the interplanar spacing d ($\sim 10^{-8}$ cm) of the crystal ($a \sim 10\, d$), and, simultaneously, is much less than the period λ of the bending ($a \sim 10^{-5} \ldots 10^{-4}\, \lambda$).

In addition to the spontaneous photon emission by the crystalline undulator, the scheme we propose leads to a possibility to generate stimulated emission. This is due to the fact, that photons emitted at the points of the maximum curvature of the trajectory travel almost parallel to the beam and thus, stimulate the photon generation in the vicinity of all successive maxima and minima of the trajectory.

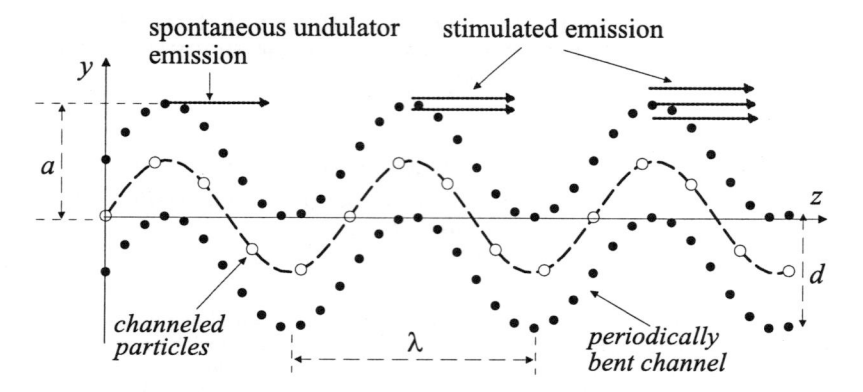

Figure 1. Schematic representation of spontaneous and stimulated radiation due to positrons channeling in a periodically bent crystal. The y- and z-scales are incompatible!

2 The bent crystal

The bending of the crystal can be achieved either dynamically or statically. In [1,2] it was proposed to use a transverse acoustic wave to dynamically bend the crystal. The important feature of this scheme is that the time period of the acoustic wave is much larger than the time of flight τ of a bunch of positrons through the crystal. Then, on the time scale of τ, the shape of the crystal bending doesn't change, so that all particles of the bunch channel inside the same undulator. One possibility to couple the acoustic waves to the crystal is to place a piezo sample atop the crystal and to use radio frequency to excite oscillations.

The usage of a statically and periodically bent crystal was discussed initially in [1,2] and later in [3]. In the latter work the idea to construct a crystalline undulator based on graded composition strained layers was suggested.

Let us now consider the conditions to be fulfiled in the stable channeling regime. The channeling process in a periodically bent crystal takes place if the maximum centrifugal force in the channel, $F_{\rm cf} \approx m\gamma c^2/R_{\min}$ ($R_{\min}$ being the minimum curvature radius of the bent channel), is less than the maximal force due to the interplanar field, $F_{\rm int}$ which is equal to the maximum gradient of the interplanar field (see [2]). More specifically, the ratio $C = F_{\rm cf}/F_{\rm int}$ is better to keep smaller than 0.1, because otherwise the phase volume of channeling trajectories becomes significantly reduced (see also [4]). The inequality $C < 0.1$ connects the energy of the particle, $\varepsilon = m\gamma c^2$, the parameters of the bending (these enter through the quantity $R_{\min}$), and the characteristics of

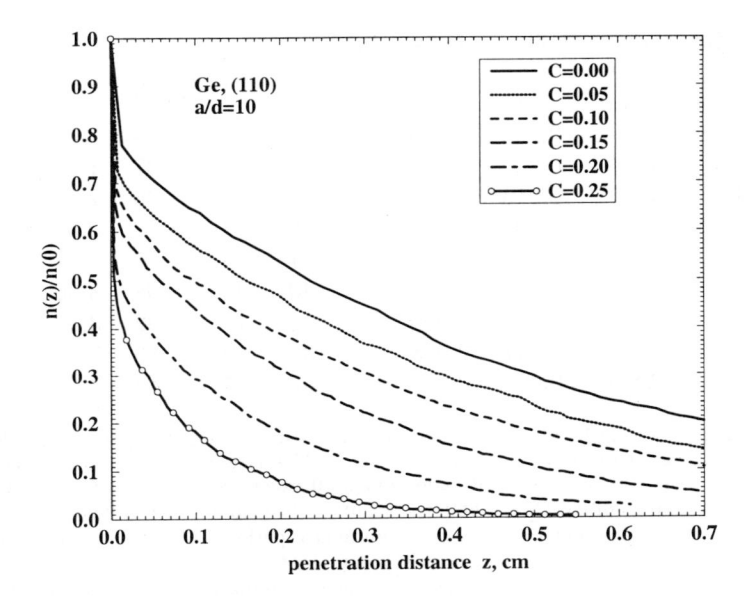

Figure 2. The calculated beam dencity dependences $n(z)/n(0)$ versus penetration distance z for 5 GeV positrons channeling along the (110) in Ge crystal for various values of the parameter C [8]. The data correspond to the shape function of the channel: $S(z) = a\sin(2\pi z/\lambda)$. The a/d ratio equals 10. For each indicated C the corresponding values of λ, and the calculated magnitudes of the dechanneling lengths L_d^c and the number of undulator periods $N_d^c = L_d^c/\lambda$ are presented in Table 1.

the crystallographic plane.

A particle channeling in a crystal (straight or bent) undergoes scattering by electrons and nuclei of the crystal. These random collisions lead to a gradual increase of the particle energy associated with the transverse oscillations in the channel. As a result, the transverse energy at some distance L_d from the entrance point exceeds the depth of the interplanar potential well, and the particle leaves the channel. The quantity L_d is called the dechanneling length [5]. To estimate L_d one may follow the method described in [6,7,8]. Thus, to consider the undulator radiation formed in a crystalline undulator, it is meaningful to assume that the crystal length does not exceed L_d. The detailed numerical analysis of the dechanneling phenomena in periodically bent

C	R_{min} cm	λ μm	L_d^e cm	L_d^c cm	N_d^e	N_d^c	ω_1 MeV	p
Crystal: Ge, $d = 2.00$Å, $R_c = 0.42$ cm								
0.00	∞	-	0.263	0.513	-	-	-	-
0.05	8.465	81.8	0.237	0.450	29	55	1.37	1.50
0.10	4.232	57.8	0.213	0.364	36	63	1.26	2.13
0.15	2.822	47.2	0.190	0.269	40	57	1.15	2.61
0.20	2.116	40.9	0.168	0.176	41	43	1.05	3.01
0.25	1.693	36.6	0.148	0.095	40	26	0.98	3.36
0.30	1.411	33.4	0.129	0.060	38	18	0.92	3.68
0.35	1.209	30.9	0.111	0.028	35	9	0.86	3.98
0.40	1.058	28.9	0.095	0.012	32	4	0.82	4.25

Table 1. Dechanneling lengths for 5 GeV positron channeling along the (110) planes for the Ge crystal and various values of the parameter C [8]. The data correspond to the shape function $S(z) = a \sin(2\pi z/\lambda)$. The a/d ratio equals 10 except for the case $C = 0$ (the straight channel). The quantity L_d^c presents the accurately calculated dechanneling length [8], $N_d^c = L_d^c/\lambda$ is the corresponding number of the undulator periods, L_d^e is the dechanneling length derived from a simple estimate (see [6,7,8]), $N_d^e = L_d^e/\lambda$. Other parameters are: d is the interplanar spacing, $R_c = \varepsilon/U'_{max}$ is the critical (minimal) radius consistent with the condition $C \ll 1$, ω_1 is the energy of the first harmonic of the crystalline undulator radiation for the forward emission, p is the undulator parameter. R_{min} is the minimum curvature radius of the bent channel centerline.

crystals and its influence on the spectral characteristics of the undulator radiation has been performed in [8]. An example of these calculations for 5 GeV positrons channeling along the (110) in Ge crystal is presented in figure 2 and in 1. In [8], the similar calculations have been also performed for the Si and W crystals.

Let us demonstrate how one can estimate, for a given crystal and energy ε, the ranges of the parameters a and λ which are subject to the conditions formulated above. For doing this we assume that the shape of the centerline of the periodically bent crystal is $a \sin(2\pi z/\lambda)$. Figure 3 illustrates the above mentioned restrictions in the case of $\varepsilon = 0.5$ GeV positrons planar channeling in Si along the (110) crystallographic planes. The diagonal straight lines correspond to various values (as indicated) of the parameter C. The curved lines correspond to various values (as indicated) of the number of undulator periods N related to the dechanneling length L_d through $N = L_d/\lambda$. The horizontal lines mark the values of the amplitude equal to d (with $d = 1.92 \cdot 10^{-8}$ cm being the (110) interplanar distance in Si) and to $10\,d$. The vertical line marks the value $\lambda = 2.335 \cdot 10^{-3}$ cm, for which the spectra (see section

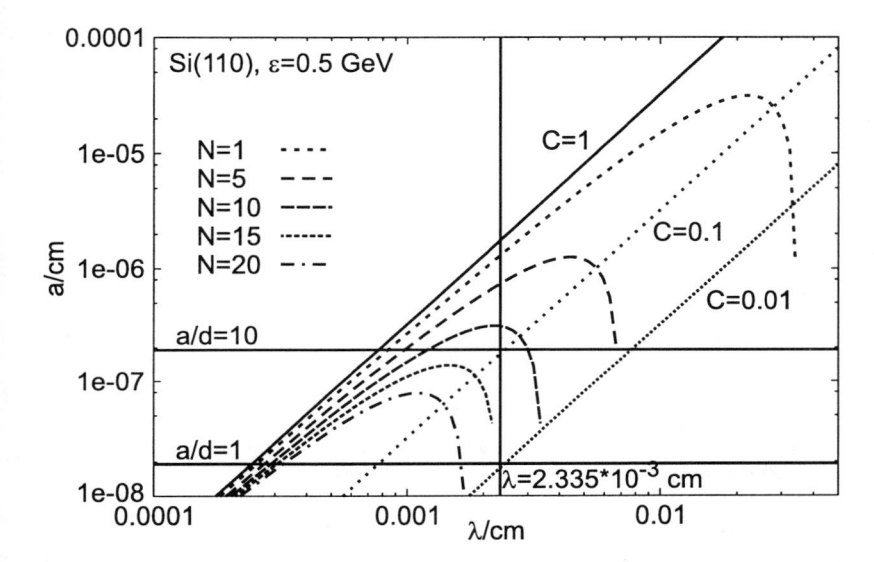

Figure 3. The range of parameters a and λ for a bent Si(110) crystal at $\varepsilon = 500$ MeV.

3) were calculated.

3 Spectra of the spontaneous emitted radiation

Let us consider the spectra of the spontaneous emitted radiation calculated in [7] using the quasiclassical method [9]. In [7] the trajectories of the particles were calculated numerically and then the spectra were evaluated. The latter include both radiation mechanisms, the undulator and the channeling radiation.

The spectral distributions of the total radiation emitted in forward direction for $\varepsilon = 500$ MeV positrons channeling in Si along the (110) crystallographic planes are plotted in figure 4. The wavelength is fixed at $\lambda = 2.335 \cdot 10^{-3}$ cm, while the ratio a/d is changed from 0 to 10. The length of the crystal is $L_d = 3.5 \cdot 10^{-2}$ cm and corresponds to $N = 15$ undulator periods. The first graph in figure 4 corresponds to the case of the straight channel ($a/d = 0$) and, hence, presents the spectral dependence of the ordinary channeling radiation only. The spectrum starts at $\hbar\omega \approx 960$ keV, reaches its maximum value at 1190 keV, and steeply cuts off at 1200 keV. This peak corresponds to the radiation into the first harmonic of the ordinary channeling radiation, and there is almost no radiation into higher harmonics. The latter

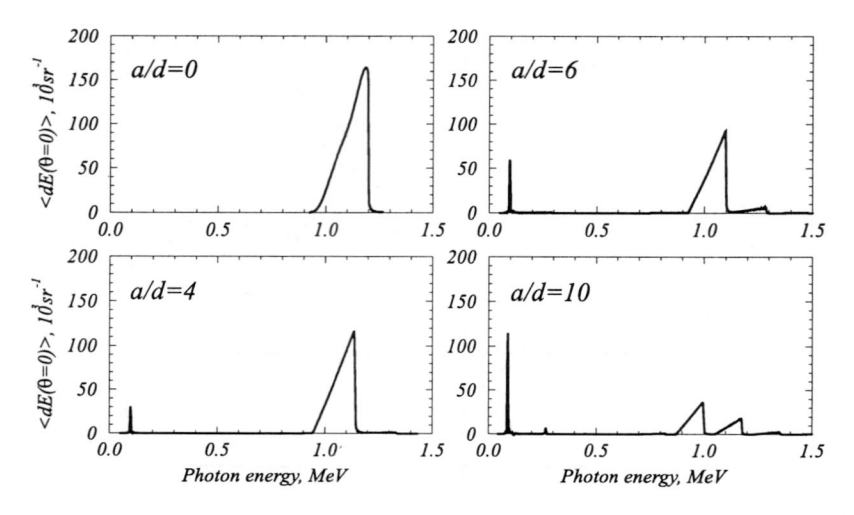

Figure 4. Spectral distributions of the total radiation emitted in forward direction for $\varepsilon = 500$ MeV positrons channeling in Si along the (110) crystallographic planes for different a/d ratios.

fact is consistent with general theory of dipole radiation by ultra-relativistic particles undergoing quasiperiodic motion. The dipole approximation is valid provided the corresponding undulator parameter $p_c = 2\pi\gamma(a_c/\lambda_c)$ is much less than 1. In this relation a_c and λ_c stand for the characteristic scales of, correspondingly, the amplitude and the wavelength of the quasiperiodic trajectory. In the case of 0.5 GeV positrons channeled along the (110) planes in Si one has $p_c \approx 0.2 \ll 1$ and all the channeling radiation is concentrated within some interval in the vicinity of the energy of the first harmonic.

Increasing the a/d ratio leads to modifications in the spectrum of radiation. The changes which occur manifest themselves via three main features, (i) the lowering of the ordinary channeling radiation peak, (ii) the gradual increase of the intensity of undulator radiation due to the crystal bending, (iii) the appearing of additional structure (the sub-peaks) in the vicinity of the first harmonic of the ordinary channeling radiation. A more detailed analysis of these spectra can be found in [7,10].

4 Discussion of stimulated photon emission

The scheme illustrated by figure 1 allows to discuss the possibility to generate stimulated emission of high energy photons by means of a bunch of ultra-relativistic positrons moving in a periodically bent channel. Indeed the photons emitted in the nearly forward direction at some maximum or minimum point of the trajectory by a group of particles of the bunch stimulate the emission of photons with the same energy by another (succeeding) group of particles of the same bunch when it reaches the next maximum/minimum.

In [2] estimates for the gain factor for the spontaneous emission in crystalline undulators were obtained. It was demonstrated that to achieve a total gain equal to 1 on the scale of the crystal length (equal to the dechanneling length), one has to consider volume densities n of the channeling positrons on the level of $10^{20} \ldots 10^{21}$ cm^{-3} for positron energies within the range $0.5 \ldots 5$ GeV. These magnitudes are high enough to be questioned whether they can be really reached.

Let us estimate the volume density n of a positron bunch which can be achieved in modern colliders. To do this we use the data presented in [11] (see p.142) for a beam of 50 GeV positrons available at SLC (SLAC, 1989). The bunch length is 0.1 cm and the beam radius is 1.5 μm (H) and 0.5 μm (V), resulting in the volume of one bunch $V = 2.4 \cdot 10^{-9}$ cm^3. The number of particles per bunch is given as $4.0 \cdot 10^{10}$. Therefore, one obtains $n = 1.7 \cdot 10^{19}$ cm^{-3}. This value, although being lower by an order of magnitude than the estimates obtained in [2], shows that the necessary densities should be reachable in the future with accelerators optimized for high particle densities.

Finally let us discuss the required transverse emittance of the beam. For doing this we need to consider the angle between the particle's trajectory and the crystal plane. If this angle is larger than the Lindhard angle Ψ_P, the particle will not be captured into the channeling mode and leaves the channel immediately [12]. For 5 GeV positrons channeling along the (110) plane of silicon, we have $\Psi_P = 72$ μrad and for 50 GeV positrons it equals to $\Psi_P = 23$ μrad.

One can compare these values with the divergence of the SLC beam, which transverse emittance in vertical direction is given as 0.05π rad nm [11]. With a vertical beam radius of 0.5 μm we get for the vertical beam divergence $\Psi = 100$ μrad. Thus the divergence of the beam is about four times higher than the acceptance of the channel and so only a quarter of all particles will participate in the channeling process. Evidently it is necessary to reduce the divergence of the beam, for example by increasing the beam radius. But then it is also necessary to reduce the bunch length, to keep the particle density high enough. Fortunately, like for the particle density, the values achievable today differ only about one order of magnitude from the values estimated above for the stimulated emission.

5 Conclusion

To conclude we point out that the crystalline undulators discussed in this work can serve as a new efficient source for high energy photons. As we have shown above, the present technology is nearly sufficient to achieve the necessary conditions to constract not only crystalline undulator, but also achieve the stimulated photon emission regime. The parameters of the crystalline undulator and the radiation generated with its use differ substantially from what is possible to achive with the undulators based on magnetic fields.

In our opinion the effects described above is worth experimental study. As a first step, one may concentrate on measurements of the spontaneous undulator radiation spectra.

The related problems not discussed in this paper, but which are under consideration, include the investigation of the crystal damage due to the acoustic wave, photon flux and beam propagations.

References

1. A.V. Korol, A.V. Solov'yov and W. Greiner, J. Phys. G **24**, L45 (1998).
2. A.V. Korol, A.V. Solov'yov and W. Greiner, Int. J. Mod. Phys. E **8**, 49 (1999).
3. U.Mikkelsen and E.Uggerhøj, Nucl. Inst. and Meth. B **160**, 435 (2000).
4. A.V. Korol, A.V. Solov'yov and W. Greiner, Int.J. Mod.Phys. E **9**, 77 (2000).
5. D.S. Gemmel, Rev. Mod. Phys. **46**, 129 (1974).
6. V.M. Biruykov, Y.A. Chesnokov and V.I. Kotov, *Crystal Channeling and its Application at High-Energy Accelerators* (Springer, Berlin, 1996).
7. W.Krause, A.V. Korol, A.V. Solov'yov and W. Greiner, J. Phys. G **26**, L87 (2000).
8. A.V. Korol, A.V. Solov'yov and W. Greiner, J. Phys. G **27**, 95 (2001).
9. V.N. Baier, V.M. Katkov and V.M. Strakhovenko, *High Energy Electromagnetic Processes in Oriented Single Crystals* (World Scientific, Singapore, 1998).
10. A.V. Korol, W.Krause, A.V. Solov'yov and W.Greiner, to be submitted to Phys. Rev. E , (2001).
11. C.Caso *et al*, *Review of Particle Physics*, Eur. Phys. J. C **3**, (1998).
12. J. Lindhard, K. Dan., Viddensk. Selsk. Mat. Phys. Medd. **34**, 14 (1965)

NOVEL INTERFEROMETER IN THE X–RAY REGION

H. BACKE, N. CLAWITER, S. DAMBACH, H. EUTENEUER,
F. HAGENBUCK, K.-H. KAISER, O. KETTIG, G. KUBE, W. LAUTH,
TH. WALCHER

Institut für Kernphysik, Johannes Gutenberg–Universität, D–55099 Mainz, Germany

Novel interferometers have been developed with which the complex index of refraction of thin self–supporting foils can be measured. The principle is based on the coherence of the x–rays emitted in forward direction by ultra relativistic electrons in two consecutive x–ray emitting devices. For the vacuum ultraviolet and soft x–ray region the interferometer consists of two collinear undulators, between which a foil can be placed, and a grating spectrometer. For the hard x–ray region it consists of two foils, at which the electron beam produces transition radiation, and a single crystal spectrometer in Bragg geometry. Taking advantage of the low emittance 855 MeV electron beam at the Mainz Microtron MAMI distinct intensity oscillations have been observed as a function of the distance between the undulators or foils. The complex index of refraction has been investigated at the K–absorption edge of carbon and nickel at 284 eV and 8333 eV, respectively.

1 Introduction

Resonant anomalous x–ray scattering plays an increasingly important role in many disciplines of physics, biology, and material sciences. Using the brilliant and tuneable x–ray beams from modern synchrotron radiation sources it is now possible to fully exploit the information in the strong energy and polarisation dependence of the atomic scattering amplitude $f(\omega, \vec{q}) = f_0(\vec{q}) + f'(\omega) + if''(\omega)$ near absorption edges[1]. This is the ratio of the scattering amplitude of an atom to that of a free (Thomson) electron. With these quantities the complex index of refraction $n(\omega) = 1 - \delta(\omega) - i\beta(\omega)$ can be determined using the relations for the dispersion $\delta(\omega) = (1/2)(\omega_p/\omega)^2 (f_0(0) + f'(\omega))/Z$ and the absorption $\beta(\omega) = (1/2)(\omega_p/\omega)^2 f''(\omega)/Z$. In these expressions Z is the atomic number, ω_p the plasma frequency with $\omega_p^2 = 4\pi r_0 c^2 n_a Z$, r_0 the classical electron radius, n_a the number of atoms per volume, and $f_0(0) = Z$ neglecting relativistic corrections.

The imaginary scattering factor f'' can be directly determined from the total photon cross section $\sigma(\omega)$ by employing the optical theorem: $f''(\omega) = \omega\,\sigma(\omega)/(4\pi r_0 c)$. The total cross section is well approximated by the absorption cross section which can be measured by a transmission experiment. The real part f' can be calculated from f'' by means of Kramers–Kronig dispersion relations. However, this method is suited for a relative comparison

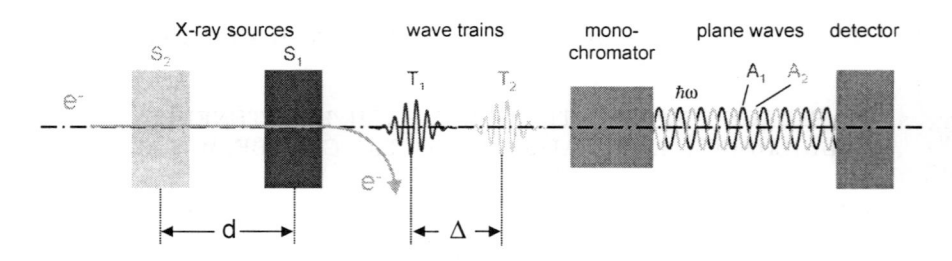

Figure 1. Interferometer with two spatially separated, phase correlated x–ray sources.

only, since it requires precise absorption data for all frequencies from zero to infinity[2]. If precise absolute values are needed, a direct measurement of $f'(\omega)$ is required. Direct measurements are based on x–ray interferometry[3], refraction through a prism[4,5], diffraction from perfect crystals and pendellösung fringes[6,7], determination of the angle of total reflection[2,8], and Fresnel bi–mirror interferometry[9,10]. Whereas most of these methods are based on splitting of either wave amplitudes or wave fronts the novel type of interferometer which is described here uses two spatially separated coherent x–ray emitters.

The basic idea of the interferometer will be explained by means of the schematic experimental setup shown in Fig. 1. Relativistic electrons create two wave trains in source 1 and source 2, the relative distance Δ of which is given in leading order by $\Delta(\theta, d) = \frac{1}{2}(\gamma^{-2} + \theta^2)d$. Here d is the distance of the sources, γ the Lorentz factor of the electron and θ the observation angle with respect to the electron beam direction. The monochromator serves as a Fourier analyser of the wave trains. The two resulting plane waves with a phase difference $\Phi = \frac{\omega}{c}\Delta(\theta, d)$, c is the velocity of light, interfere in the detector, resulting in oscillations of the intensity $I(d)$, if the distance d is varied. A sample foil placed between the two sources produces an additional phase shift and attenuation of wave 2. Consequently, both quantities, i.e. dispersion and absorption, can be extracted from the measured interference oscillations $I(d)$ with and without the foil between the sources. This holds independently of the nature of the emission process provided that the produced x–rays remain coherent.

2 The Soft X–Ray Interferometer

For photon energies in the range of about 100 eV and 2 keV we use two identical undulators with period length L_U=12 mm, number of periods 10, undulator parameter K=1.1 and a variable line spacing grating spectrometer.

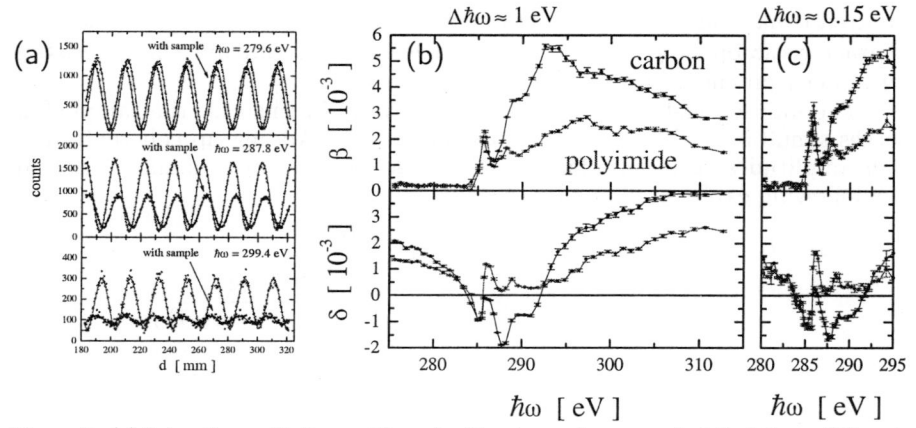

Figure 2. (a) Intensity oscillations with and without a carbon sample foil at three different energies. Note the change of sign of the phase shift. (b) Extracted optical constants of a carbon sample foil (65 µg/cm^2) and a polyimide sample (35 µg/cm^2). (c) Same as (b) with improved energy resolution of 0.15 eV.

The recorded intensity with a foil between the undulators is given by

$$I(d) = |A_1|^2 + |A_2|^2 e^{-2\frac{\omega}{c}\beta(\omega)t_0} + 2|A_1||A_2|e^{-\frac{\omega}{c}\beta(\omega)t_0}$$
$$\times \cos\left\{\frac{\omega}{c}\left[\Delta(\theta,d) + \delta(\omega)t_0 + \frac{K^2}{4\gamma^2}L_U\right]\right\} \quad , \tag{1}$$

with A_2 being the amplitude of the upstream, A_1 that of the downstream undulator and t_0 the thickness of the foil. All other quantities have been defined above.

The interferometer has been developed at the Mainz Microtron MAMI and its performance was demonstrated with measurements at the K-absorption edge of carbon at 284 eV, see Fig. 2. The visibility (coherence), defined by $V = (I_{max} - I_{min})/(I_{max} + I_{min})$ without sample foil, is close to its maximum value $C = 1$. No loss of coherence was observed over the scanning distance of 15 cm. Therefore, the optical constants δ and β could be extracted by a fit with simple cosine functions. Details of this experiment can be found elsewhere[11].

3 The Hard X–Ray Interferometer

3.1 Basics

High photon energies can be produced at MAMI if the 855 MeV electron beam traverses the interface between two media of different polarisabilities. The emitted transition radiation (TR) is sharply peaked into a forward cone with characteristic opening angle of $2/\gamma$ and features broadband characteristics with a cut–off energy at $\hbar\omega_p \approx 40$ keV.

The interferometer consists of two foils only. In this case the sample foil serves simultaneously as the downstream emitter. Adding up coherently the four amplitudes of the interfaces results after some algebraic manipulation in the following expression for the intensity:

$$I(\theta, d) = |A_1|^2 + |A_2|^2 e^{-\sigma_1} + 2|A_1||A_2|e^{-\sigma_1/2}$$
$$\times \cos\left\{\frac{\omega}{v}\Delta'(\theta, d) + \phi_1 + \arctan\left(\frac{e^{-\sigma_1/2}\sin(\phi_1)}{1 - e^{-\sigma_1/2}\cos(\phi_1)}\right)\right.$$
$$\left. - \arctan\left(\frac{e^{-\sigma_2/2}\sin(\phi_2)}{1 - e^{-\sigma_2/2}\cos(\phi_2)}\right)\right\}, \quad (2)$$

with $A_i = \frac{\sqrt{\alpha}}{\pi}\theta\left[\frac{1}{\gamma^{-2} + \theta^2 + 2\delta_i} - \frac{1}{\gamma^{-2} + \theta^2}\right](1 + e^{-\sigma_i} - 2e^{-\sigma_i/2}\cos\phi_i)^{\frac{1}{2}}, (3)$

and $\Delta'(\theta, d) = 1/2(\gamma^{-2} + \theta^2)(d - t_2)$, $\phi_i = \frac{\omega t_i}{2c}(\gamma^{-2} + \theta^2 + 2\delta_i)$, and $\sigma_i = 2\frac{\omega}{c}\beta_i t_i$. The quantities A_i, t_i, δ_i and β_i are the TR amplitude[12], thickness, dispersion and absorption, respectively, of foils $i = 1, 2$. The parameters of the upstream foil t_2, δ_2 and β_2 must be known. They can either be obtained from an independent measurement or be taken from literature[13]. The quantities of interest are the dispersion δ_1 and the absorption β_1 of the downstream sample foil. In principle both can be extracted from the interference oscillation either as function of the observation angle θ with respect to the electron beam direction or the distance d between the foils, provided the foil thickness t_1 is known from a separate measurement. In the experiment described in the following a rather thin sample foil was choosen with a small absorption. As a consequence, δ_1 and β_1 correlated strongly in the analysis procedure and could not be determined independently with the envisaged high accuracy. Therefore, also the absorption β_1 had to be kept fixed at values taken from literature[13,14,15].

3.2 Experimental

The experimental setup is shown in Fig. 3. The two foils with an area of 10×10 mm^2 each are glued onto frames with apertures of 6 mm diameter. The upstream beryllium and the nickel foil had thicknesses of (10.04 ± 0.09) μm and (2.109 ± 0.010) μm, respectively. The monochromator is simply a flat silicon single crystal, cut with the [111] parallel to the surface, which acts as energy dispersive mirror. It was positioned in Bragg geometry at a distance of 5.5 m

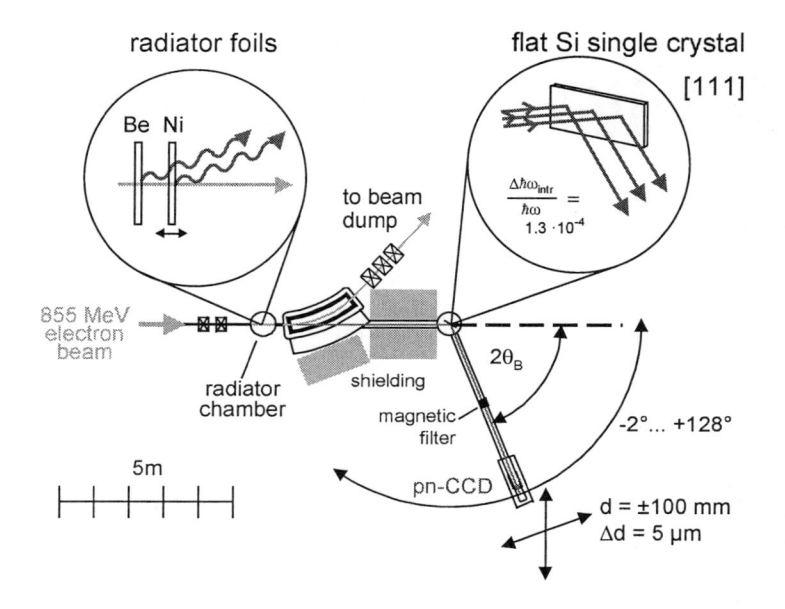

Figure 3. Experimental setup of the transition radiation interferometer.

from the foils. As detector a silicon pn–CCD has been used with 10×30 mm^2 active area, 150×150 μm^2 pixel size and 0.27 mm active thickness[16]. The pn–CCD detector is located at a distance of 5.5 m from the monochromator crystal. The two dimensional resolving pn–CCD detector allows a simultaneous recording of the TR angular distribution in the limits $\Delta\theta_x = \pm 1.4$ mrad horizontally, and $\Delta\theta_y = \pm 0.45$ mrad vertically. Typical examples of measured TR interference patterns are shown in Fig. 4. Note that the energy decreases in the horizontal coordinate (abszissa) by about 100 eV from column 1 (left) to column 200 (right). This fact allows the simultaneous measurement of the dispersion δ in the mentioned energy band as will be described below in more detail.

Measurements have been performed around the K–absorption edge at 8333 eV as well as around 9929 eV, well above the K–absorption edge, where extended diffraction anomalous fine structures (EDAFS) in the dispersion spectra are negliable. The latter are the equivalents to the well known EXAFS structures in the absorption spectra. In these measurements the foil distance was varied with a positioning accuracy of 0.1 μm in the limits between 75 μm and 11,000 μm.

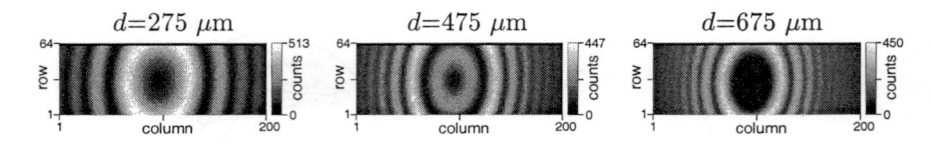

Figure 4. Angular Distribution of TR interference patterns in gray scale representation as measured with a 10×30 mm^2 pn–CCD detector for a beryllium and a nickel foil combination at the quoted distances d. The photon energy at column 100 amounts to 9929 eV. From the left column 1 to the right column 200 an energy band of 121 eV is covered.

3.3 Data Analysis

There are two possibilities to extract the dispersion δ_1 of the downstream nickel foil from the TR interference patterns shown in Fig. 4. In the first one the information is obtained from the interference oscillations as function of the observation angle θ. Since the pn–CCD detector was arranged horizontally the interference oscillations could be observed only along the energy dispersive angular coordinate θ_x. This method yields only reliable results, if the dispersion $\delta_1(\omega)$ can be approximated with reasonable accuracy by a linear expansion as function of the photon energy $\hbar\omega$. This is possible for the example shown in Fig. 4. But generally such an approximation is not valid, particulary not close to an absorption edge. In this case, as a second possibility, the dispersion $\delta_1(\omega)$ can be extracted from the interference oscillations observed as function of foil distance d for the selected photon energy $\hbar\omega$ and at constant observation angle θ. Both methods will be described in the following in more detail.

3.3.1 Interference Oscillations as Function of the Observation Angle

The determination of the dispersion δ_1 of the downstream nickel foil is further hampered by the fact that only a *relative* distance d_r between the foils can be measured with the goniometric stage. To obtain the true distance $d = d_r + d_0$ an unknown offset d_0 must be added. This offset can, in principle be determined by an independent measurement. However, we felt that, on the one hand, rather involved developments are necessary to achieve an accuracy in the order of 0.1 μm, and that, on the other hand, such an accuracy might be obtainable also from the analysis of the angular distributions as shown in Fig. 4. It follows from Eqn. (2), that an offset d_0 causes a phase change proportional to $\theta^2 d_0$ which very sensitively affects the interference pattern as function of observation angle θ while this is not the case for ϕ_1 as long as $\theta^2 \ll 2\delta_1(\omega)$ holds. The latter condition is indeed very well fulfilled at $\hbar\omega_0 = 10$ keV for $\theta \leq 1.5$ mrad.

Unfortunately, Eqn. (2) can not directly be fitted to the experimental interference pattern because both, the angular spread of the electron beam and the small angle scattering of the electrons in the foil material, deteriorate and, at large angles θ or distances d, finally spoil the visibility of the interference. Our approach to take these effects into account is based on a Monte Carlo simulation in which the electron beam is described by a two dimensional Gaussian with the standard deviation Σ', and the small angle scattering in the foil material according to refs.[17,18]. It turned out[19] that for the experiment described in this paper the individual directional changes of the electron inside the foils due to plural and multiple scattering in the foil material are unimportant. This means that amplitudes and phases can be calculated with sufficient accuracy from the actual electron direction at the foil surfaces. The angular distribution function were calculated on the basis of 1000 electron trajectories and fitted to the measurements with the distance offset d_0, the angular divergence Σ', a common amplitude factor A, and the dispersion $\delta_1(\omega_0)$ as free parameters where $\hbar\omega_0 = 9929$ eV denotes the photon energy in column 100, at the center of the pn–CCD. The linear variation of $\partial\delta/\partial\hbar\omega = -3.110 \cdot 10^{-9}$ eV^{-1} was taken from ref.[13] and was kept fixed in the fit. In addition, a misalignement of the pn–CCD as well as a residual curvature of the "flat" analysing crystal were taken into account in the fit. The residual curvature brings about a horizontal squeezing of the angular distribution by a factor $f = (0.745 \pm 0.002)$ which can easily be recognized in Fig. 4.

The accuracy with which the mentioned fit parameters can be determined depends on the choosen foil distance. Since it is a priori not known at which distance which parameter can be determined with best accuracy, totally nine angular distributions have been analyzed as taken at different relative distances 0 μm $\leq d_r \leq$ 1000 μm. Two examples are shown in Fig. 5. The high quality of the fit is expressed best in panels (c) which show the fit residues, and (d) which represent the data from a horizontal cuts through the center of the angular distributions. As weighted averages over all the nine meaurements the following results have been obtained: distance offset $d_0 = (75.0 \pm 0.2)$ μm, standard deviation of the electron beam divergence $\Sigma' = (29.4 \pm 1.2)$ μrad, common amplitude factor $A = 1.088 \pm 0.007$, and dispersion of the nickel foil $\delta_1 = (1.77 \pm 0.03) \cdot 10^{-5}$ at $\hbar\omega_0 = 9929$ eV. The distance offset d_0 can be determined most precisely from angular distributions taken at small foil distances while for the angular divergence Σ' of the electron beam the these distances are less sensitive. Together with the measured horizontal and vertical beam spot sizes in the focus at the TR foils of $\sigma_x = (250 \pm 12)$ μm and $\sigma_y = (138 \pm 4)$ μm a beam emittance of $\varepsilon_x = 4.7$ π nm rad and $\varepsilon_y = 3.7$ π nm rad can be esti-

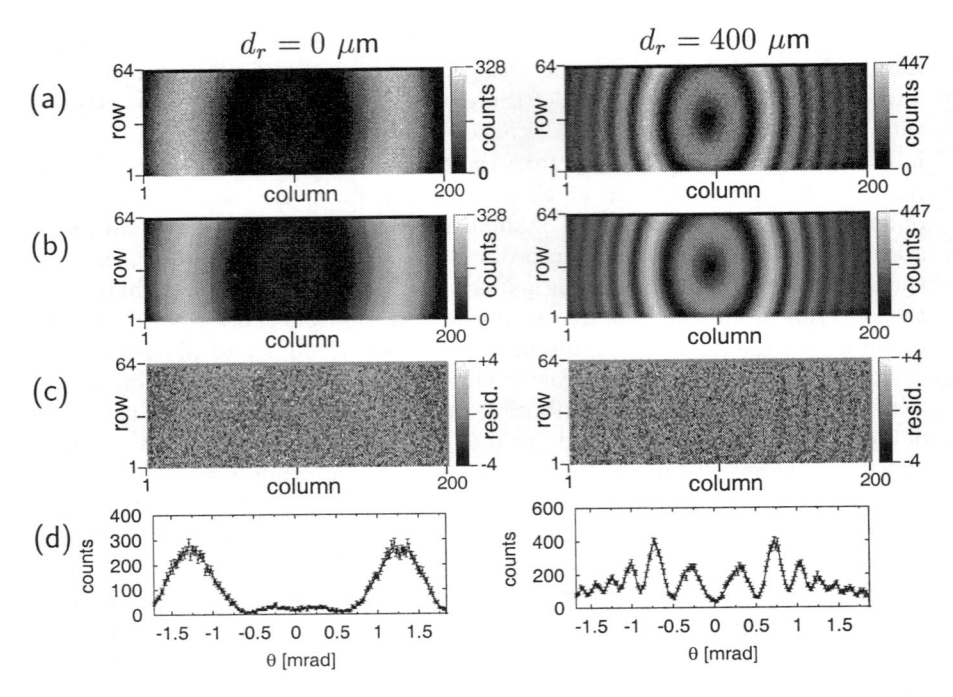

Figure 5. (a) Measured and (b) simulated interference pattern at fixed foil distance. Left panels for the smallest relative distance $d_r = 0$ µm, right panels for an intermediate distance $d_r = 400$ µm. Central photon energy $\hbar\omega = 9929$ eV. Gray scales indicate the intensities. Graphs (c) depict the fit residues, i.e. the difference between measurement and fit, normalized to the statistical errors, and (d) a horizontal cut through the angular distributions at row 30. Error bars indicate measurements, full lines simulation calculations.

mated, respectively. These values deviate somewhat from the known values $\varepsilon_x = 7\ \pi$ nm rad and $\varepsilon_y = 1\ \pi$ nm rad at MAMI[20], but judged against the background that the angular divergence Σ' is a mean value of Σ'_x and Σ'_y, with the horizontal Σ'_x stronger weighted than the vertical Σ'_y, this is a fully satisfactory result. The common amplitude factor A deviates only 8.8% from the expected value unity. This good agreement indicates that beam current, and solid angle measurements, etc. have been got fully under control in this experiment. The experimental value of the dispersion δ_1 is in good agreement with the tabulated value[13] $\delta_1^{\text{calc}} = 1.74 \cdot 10^{-5}$.

In conclusion of this analysis, the measured angular distributions of the two foil TR interference pattern are described very well by Monte Carlo simulations in which the angular divergence of the electron beam, and the small

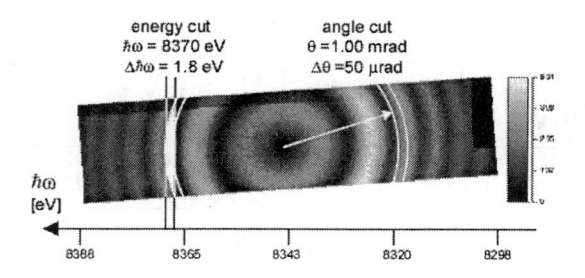

Figure 6. Selection of photon energy $\hbar\omega$ and observation angle θ. The tilted installation of the CCD detector by an angle of 4.16° against the horizontal plane and a residual curvature of the "flat" crystal have been corrected for. The example, $\hbar\omega = 8370$ eV and $\theta = 1.00$ mrad, demonstrates the simultaneous energy and angle selection in the overlap of the 1.8 eV wide energy stripe with the 50 μrad wide angular cone.

angle scattering of the electrons in the foils are taken into account. The fit results are consistent with what is expected from the known experimental parameters and the tabulated dispersion $\delta_1(\omega_0)$ of the nickel foil. Therefore, we are confident that the fit value for the distance offset d_0 is fully reliable as well. In the analysis procedure of the dispersion $\delta_1(\omega)$ with high energy resolution, as described in the following, $d_0 = 75.0$ μm was kept fixed, as well as the angular divergence $\Sigma' = 29.4$ μrad.

3.3.2 Interference Oscillations as Function of the Foil Distance

For the determination of the dispersion $\delta_1(\omega)$ of the nickel foil with the instrumental energy resolution of (2.4 ± 0.2) eV at 8333 eV, advantage has been taken from the fact that the horizontal coordinate along the pn–CCD detector is energy dispersive. The energy can be selected by a vertical stripe with a width of 600 μm which corresponds to $\Delta\hbar\omega = 1.8$ eV, see Fig. 6. Interference oscillations have been generated as function of the distance d between the foils under the constraint of a well defined observation angle θ with constant width of $\delta\theta = 50$ μrad, see Fig. 7. The best fit of the simulation calculation to the data, with only the dispersion $\delta_1(\omega)$ and an amplitude factor A as free parameters, precisely describe the measurement in all cases. The rapid damping of the visibility after about three full oscillations originates mainly from the small angle scattering of the electrons in the upstream beryllium foil.

The results of the dispersion measurements are shown in Fig. 8. Dispersion data are plotted only for observation angles $\theta \leq 1.0$ mrad. Outside this intervall the error of the distance offset in combination with statistical errors of both, measurement and simulation calculation, increase rapidly. The error

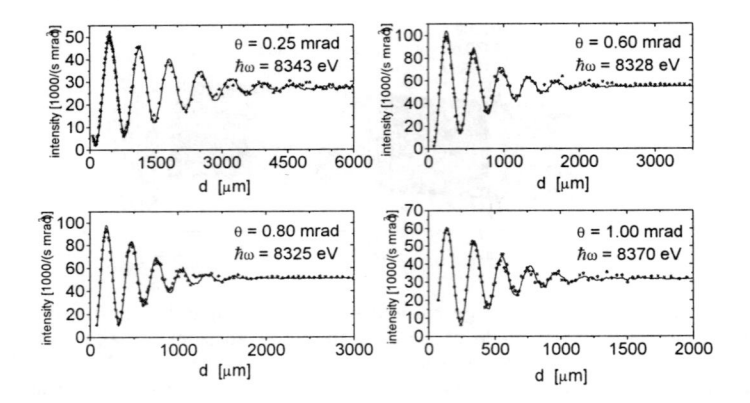

Figure 7. Intensity oscillations as function of foil distance d at fixed photon energy $\hbar\omega$ and fixed observation angle θ as indicated. Note the different scales at the abscissa. The measurements (dots) are well met by the simulation calculation (full line).

bars include, in addition, the errors of the angular divergence of the electron beam and of the foil thicknesses.

The measured dispersion $\delta_1(\omega)$ agrees at the K–absorption edge within the errors of $\Delta\delta_1/\delta_1 \leq 1.5\%$ with the Kramers–Kronig transformation of the absorption $\beta_1(\omega)$ of refs.[13,14,15], see Fig. 8(b). In comparison, the measurements of $\delta_1(\omega)$ by Bonse and co–workers[21,22] with Bonse–Hart interferometers[3] are systematically too low by about 1.5%. This small deviation may originate from a systematical error in the foil thickness measurements in these experiments. In conclusion, the good agreement within experimental errors of our measurements with that of Bonse et al.[21,22] proves that a fully operational novel x–ray interferometer has been developed in this work.

As a first application of our interferometer the calculations of Henke et al.[13] were tested with our measurents at photon energies around $\hbar\omega$ =9929 eV, well above the K–absorption edge where EDAFS in the dispersion spectra are absent. The results are shown in Fig. 8(d). It can be concluded that our interferometric measurement agrees within the experimental errors $\Delta\delta_1/\delta_1 \leq$1.4 % with the calculation of Henke et al.[13].

Acknowledgements

This work has been supported by the Deutsche Forschungsgemeinschaft DFG under contract BA 1336/1–3 and the Bundesministerium für Bildung, Wis-

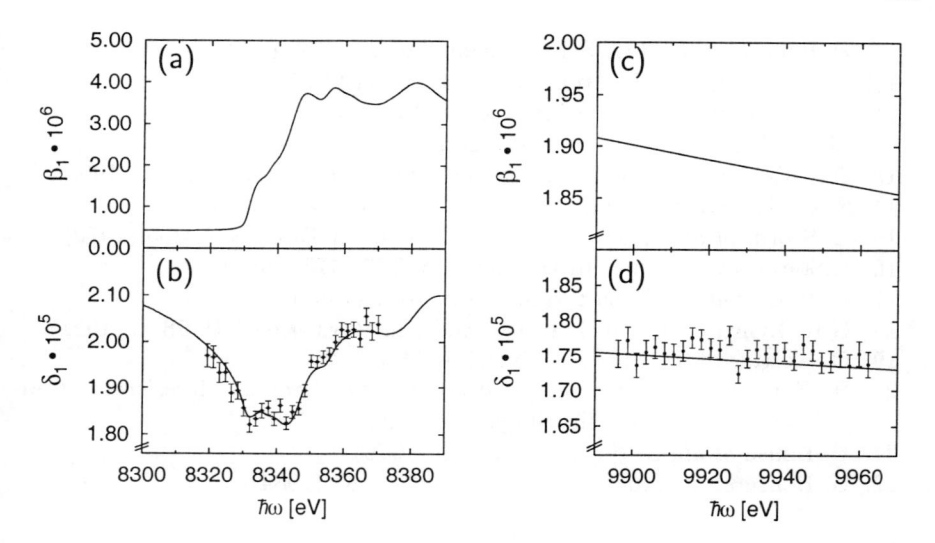

Figure 8. Complex index of refraction of nickel. Left panels: (a) absorption $\beta_1(\hbar\omega)$ and (b) dispersion $\delta_1(\hbar\omega)$ at the K–absorption edge of nickel. The results of this work are shown with error bars. The full line in panel (a) is taken from ref.[14], the full line in panel (b) is a Kramers–Kronig transform of absorption data[13,14,15]. Right panels: (c) Absorption $\beta_1(\hbar\omega)$ and (d) dispersion $\delta_1(\hbar\omega)$ at energies around 9929 eV, well above the K–absorption edge. The solid lines are calculations of Henke et al.[13].

senschaft, Forschung und Technologie under contract 06 MZ 863 I/TP 2.

References

1. *Resonant Anomalous X-ray Scattering*, edited by G. Materlik, C.J. Sparks, and K. Fischer (North Holland, Amsterdam, London, New York, Tokyo 1994).
2. B. Lengeler, in [1], p. 35.
3. U. Bonse and M. Hart, Appl. Phys. Lett. **7**, 238 (1965).
4. W.K. Warburton and K.F. Ludwig, Phys. Rev. **B 33**, 8424 (1986).
5. M. Deutsch and M. Hart, Phys. Rev. **B 30**, 643 (1984).
6. A. Freund, in *Anomalous Scattering*, eds. R. Ramaseshan, S.C. Abrahams, Munksgaard Copenhagen p. 69 (1975).
7. N. Kato and S. Tanemura, Phys. Rev. Lett. **19**, 22 (1967).
8. R.L. Blake *et al.*, in [1], p. 79 (1994).

9. F. Polack *et al.*, Rev. Sci. Instrum. **66** (2), 2180 (1995).
10. S. Marchesini *et al.*, Appl. Opt. **39**, 1633 (2000).
11. S. Dambach *et al.*, Phys. Rev. Lett. **80**, 5473 (1998).
12. M.L. Cherry *et al.*, Phys. Rev. **D 10**, 3594 (1974).
13. B.L. Henke *et al.*, Atom. Data and Nucl. Data Tabl. **54**, 181 (1993).
14. S. Kraft, private communication, PTB Braunschweig, 1995.
15. E. Storm *et al.*, Atom. Data and Nucl. Data Tabl. **A 7**, 565 (1970).
16. H.Soltau *et al.*, Nucl. Instr. Meth. **A 377**, 340 (1996).
17. E. Keil *et al.*, Z. Naturforsch. **15 a**, 1031 (1960).
18. G.R. Lynch *et al.*, Nucl. Instr. Meth. in Phys. Reas. **B 58**, 6 (1991).
19. O. Kettig, Ph.D. theses, Universität Mainz, 2001.
20. H. Euteneuer, in *Proceedings of the Fourth European Accellerator Conference* (World Scientific, Singapore, 1994), p. 506.
21. U. Bonse *et al.*, Z. Phys. **B 24**, 189 (1976).
22. U. Bonse *et al.*, Nucl. Instr. Meth. **222**, 185 (1984).

SCIENTIFIC OPPORTUNITIES AT THIRD- AND FOURTH-GENERATION X-RAY SOURCES

ANDREAS K. FREUND

European Synchrotron Radiation Facility, B.P. 220, 38043 Grenoble Cedex, France
e-mail: freund@esrf.fr

The advent of very bright X-ray sources based on synchrotron radiation has given access to new scientific research worldwide in the fields of condensed matter physics, chemistry, biology, medicine and others. Third-generation sources are electron storage rings equipped with several meter long insertion devices such as undulators and wigglers generating very well collimated, pulsed X-ray beams with a spectral brightness or brilliance of up to 10^{22} photons/s/mrad^2/mm^2/0.1% bandwidth. Examples for high-energy sources are the European Synchrotron Radiation Facility (6 GeV) in Grenoble, France, the Advanced Photon Source (7 GeV) in Argonne, USA and the SPring-8 facility (8 GeV) in Himeji, Japan. In the experimental hall around the ESRF's storage ring of 1 km circumference and even outside, about 40 beamlines are presently installed providing X-rays between 10 eV and a few hundred keV to thousands of users per year. Among the techniques are X-ray elastic and inelastic scattering, absorption spectroscopy, fluorescence, microscopy, coherent and incoherent imaging with resolutions of about 100 nanometers in space, 10 nanoradians in angle, sub-milli-eV in energy and a few tens of picoseconds in time. A selection of the presently existing instrumentation and some scientific results will be given. While third-generation sources have come into operation only a few years ago, a different type of machine, the free electron laser (FEL) in the X-ray range (8 keV) is presently being considered as the fourth-generation source. Two major projects are proposed at DESY (Hamburg) and SLAC (Stanford). These sources would produce laterally fully coherent radiation collimated to within 1 micro-radian with several orders of magnitude higher brightness than that of third generation machines. Moreover, overcoming the 50 picosecond limit of electron bunches in storage rings, these LINAC based machines should permit to deliver a few femtoseconds long X-ray pulses that will open up new experimental possibilities in ultrafast science. It is obvious that optics and sample survival will become major issues in these very intense beams. As to optics, diamond is the material with the highest performance for both storage ring and LINAC based beamlines and its importance will be highlighted, last but not least because it is very closely related to the origin of this conference.

1 Introduction

More than a century ago, in November 1895, Röntgen discovered the X-rays and he won the first Nobel prize in 1901. Since then an enormous amount of scientific work has been carried out using this type of penetrating radiation and many Nobel prizes have been awarded to scientists that utilized X-ray scattering techniques. Very important milestones in these activities were the discovery of X-ray diffraction by crystals by von Laue and his collaborators Friedrich and Knipping in 1912 and the following work by the Braggs on crystal structures derived from X-ray diffraction. In 1923 Compton observed the specular reflection of X-rays by surfaces. Medical imaging techniques were developed very early starting with simple shadowgraphs and culminating in computed axial tomography in 1979 (Cormack and Newbold). During the more recent past X-rays have been increasingly applied to the solution of large molecules starting with the famous work on DNA by Watson, Wilkins and Crick in 1962 and by Deisenhofer, Huber and Michel (1968) on protein structures. All this work was done using conventional X-ray sources, namely X-ray tubes, either sealed-off or rotating. These types of source could have been named *the 0-th generation*.

In 1947 a new type of photon sources was discovered in the research laboratories of the General Electrics Company by Blewitt and collaborators: synchrotron radiation. It is simply generated when an electric charge is accelerated and the emission process can be described in the framework of classical electrodynamics. If the electrons or positrons are relativistic, the radiation is strongly collimated in the forward direction into a narrow cone of opening angle 1/□ where □ is the electron energy. The losses of energy due to this radiation must be given back to the particles circulating in the synchrotron, which is done by radio-frequency cavities and costs electricity. Therefore, for quite some time synchrotron radiation was considered to be a nuisance by the high-energy community. Several years later, people started thinking about possible uses of this nuisance and first experiments in the VUV and the soft X-ray range were conducted using synchrotron beams. They confirmed the very exciting experimental possibilities that are due to the many interesting properties of the radiation such as high brightness, narrow collimation, linear polarization and a very wide spectral range. These first studies used radiation from dipoles or bending magnets. In the fifties a growing number of scientists, mainly physicists, made parasitic use of synchrotron beams extracted out of particle accelerators built by the high-energy physics community. These were the *first-generation sources*.

With the increasing number of people and experimental success rate, many thoughts were given to ways of increasing the flux of the synchrotron beams. At synchrotron facilities in Stanford, Hamburg, Daresbury and elsewhere new, periodic magnetic structures were inserted into straight sections of the electron or positron storage rings. These insertion devices are called *wigglers* if the magnetic field is so strong that the induced excursion of the particles exceeds the natural opening angle of synchrotron radiation, and *undulators* for smaller fields and deviations. They allowed one to obtain gains of several orders of magnitude with respect to simple dipoles. Many storage rings were equipped with such devices and then dedicated to the use as synchrotron radiation sources for an increasingly wide field of research. These were the *second-generation sources*. Many of them have been upgraded in the recent years and we could call them 2^+ *sources*.

Buoyed with the success obtained and the new research tools opened up at these facilities, in the early seventies the still rapidly growing community decided to envisage the construction of even larger, fully dedicated facilities, based on the optimised installation of insertion devices. These *third generation facilities* can be roughly subdivided into two categories:

Medium size/low energy electron storage rings with a circumference of 100 to 300 m and electron energies from 1 to 3 GeV giving X-ray energies between 10 eV and 30 keV. Existing facilities are: Elettra (Trieste, Italy), MaxII (Lund, Sweden), BessyII (Berlin, Germany), ALS (Berkeley, USA), PLS (Pohang, Korea), SRRC (Taipeh, China)... Projects under construction or planned in Europe are: SLS (Villigen, Switzerland), Anka (Karlsruhe, Germany), Soleil (Saclay, France), Diamond (Abingdon, England). Other projects worldwide are envisaged or under way in Canada (Saskatoon), China (Shanghai), Australia (?) and Japan (Himeji).

Large size/high energy storage rings with a circumference of 800 to 1300 m and electron energies from 6 to 8 GeV giving X-ray energies between 10 eV and 300 keV and even higher. Three high-energy facilities are operational at present: ESRF (Grenoble, France), APS (Chicago, USA) and SPring-8 (Himeji, Japan).

All in all, today we count more than 70 synchrotron radiation sources in operation (53), under construction, being commissioned or envisaged (19) in the world. An interactive introduction to the world of synchrotron radiation can be found in [1] and a recent review on the present state is given in [2]. The only continent that has no modern X-ray source of this kind is Africa.

2 Properties of third generation sources

The first third generation source was the European Synchrotron Radiation Facility (ESRF) in Grenoble, France. This source and the research conducted at this facility will be taken as an example for the activities taking place worldwide. First light in this machine was seen in 1991. Its electron energy and current are 6 GeV and 200 mA (for 2/3 filling), respectively, and the ring circumference is about 1 km. The principal figure of merit defining the performance of a source is the *brilliance* that is given by the number of photons per second, unit source size, unit solid angle and 0.1% relative bandwidth. It is usually defined in the following units: photons/s/mm^2/mrad^2/0.1% RBW. Other important features are the *emittance*, the *lifetime*, typically 40 hours for 2/3 filling and the *filling pattern* or the *time structure*. The emittance is the source size times the emission angle that is fundamentally limited by diffraction according to $\lambda/4\pi$ where is λ the X-ray wavelength. At most storage rings this limit is now reached in the vertical plane (a few pm.rad), but not in the horizontal plane where it is substantially higher (a few nm.rad). The source is thus quite *anisotropic*.

In the vertical plane the minimum source size is 10 μm (this also depends on the location in the storage ring, the minimum beam divergence is 10 μrad (= 2 arcsec) and the brilliance is 10^{22} photons/s/mm^2/mrad^2/0.1% RBW. Figure 1 shows the spectral distribution of the brilliance of undulator radiation as it is presently produced at the ESRF and as it can be expected for a future ring with fully optimised properties and new in-vacuum undulator and damping wiggler

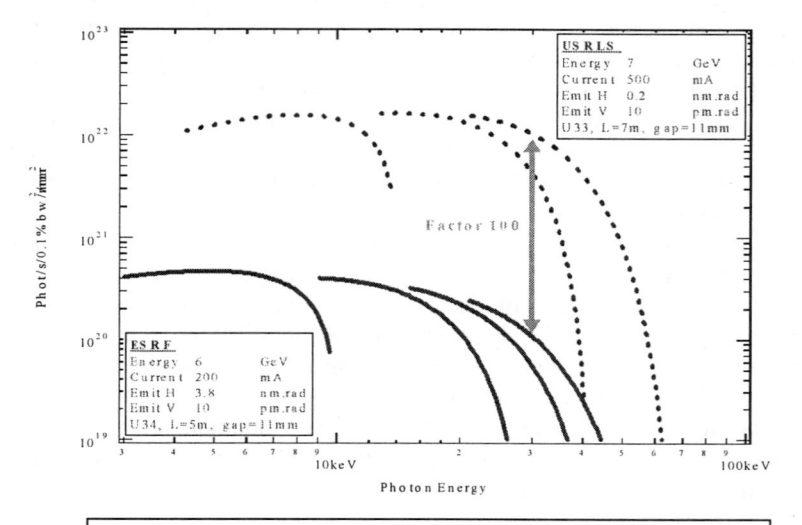

Figure 1: Present X-ray brilliance produced at the ESRF (full lines) and ultimate storage ring performance (broken lines). The different curves correspond to the undulator harmonics (1st, 3rd, 5th, ...).

technology [3]. The electrons circulate in short bunches, typically 50 ps long, corresponding to 1.5 cm spatial length. Several filling modes are possible, from a single bunch per revolution spaced a few µs (18 mA maximum current) to about 1000 bunches spaced by a few ns (200 mA maximum total current). This machine has a reliability of >96% and produces about 5000 hours of beamtime per year.

The extremely high brilliance generated by the source is accompanied by an extremely high heat load on the first elements in an X-ray beamline. Typical power densities exceed the one at the surface of the sun and thus cooling is absolutely necessary. The total X-ray power radiated by a wiggler exceeds several kW. We will come back to this problem later. Of course, all kind of optical elements must be of very high quality to be able to preserve the beam quality. For X-ray mirrors serving as high-pass filters and focusing devices this translates into a surface slope and shape error of less than 1 µrad and 10 nm, respectively, while the surface roughness should stay below 0.2 nm, *i.e.* on the atomic scale. While normally the *polarization* is 100% linear in the horizontal plane, special insertion devices can be built to generate *circular polarization* that is very important for X-ray magnetic scattering.

The low emittance and the high brilliance allow us to take advantage of the partial *coherence* of the synchrotron beams. Particularly in the recent two or three years this possibility has led to new and very exciting imaging techniques with investigations such tomographic studies of tissues and materials based on phase contrast instead of absorption contrast, also for many industrial applications. Such an outstanding source enables of course an immense variety of experimental investigations to be conducted with very high resolution in *space* (microscopy, microfocusing down to 50 nm: 100 times smaller than the diameter of a human hair!), in *angle* (grazing incidence techniques, surface diffraction, reflectometry with μrad capabilities), in *energy* (inelastic X-ray scattering with meV resolution) and in *time* (studies of dynamics at the pico-second level). The X-ray research is carried out on beamlines installed tangentially to the storage ring after a thick shield wall for biological protection. Many of them, but not all, are separated from the machine ultra-high vacuum by a beryllium and/or a diamond window. The experimental hall around the storage ring can accept up to 70 long experiments and two longer beamlines (topography and medical imaging) have their experimental stations outside the ring building.

3 Examples of applications at third generation sources

Figure 2 shows how 30 public beamlines are grouped around the ESRF's storage ring. Each beamline utilizes mainly one, sometimes two or even more combined techniques such as diffraction, absorption spectroscopy, diffuse scattering, nuclear resonance, medical imaging, microscopy, topo- and tomography, inelastic scattering. They are based on the various physical interaction processes and applied to many fields such as physics, chemistry, life and geosciences, medicine, surface and interface studies, materials science and biology. Only a few of the very many experiments installed around the ring can be described here. The selection of the examples has been guided by what the author thought would be appropriate to this conference, either through Prof. J.P.F. Sellschop's personal areas of interest, or by the conference venue. They have been chosen from research carried out on four beamlines. Much more information can be obtained directly from the ESRF either through its website http://www.esrf.fr or by asking for the annual highlights or the quarterly ESRF Newsletter. Similar information is also available from other synchrotron radiation facilities, for the addresses, see also the ESRF website under *"world links"*.

3.1 High-pressure inelastic X-ray scattering using diamond anvil cells

Inelastic X-ray scattering fully needs the very high brilliance of the third generation synchrotron source, because the spectral window is limited to a few meV energy resolution obtained by high-order reflections of a silicon perfect crystal in backscattering. Even sub-meV resolution can be reached. The experimental set-up

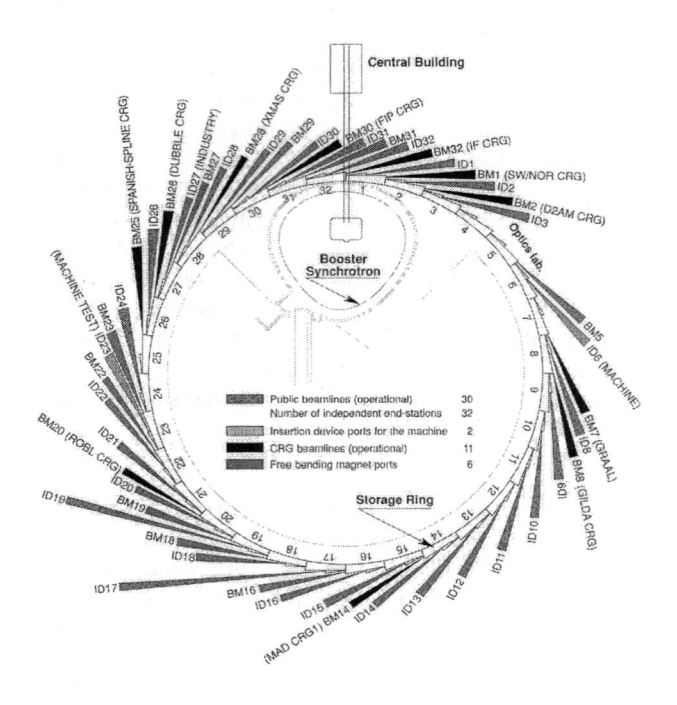

Figure 2: Beamlines around the storage ring of the European Synchrotron Radiation Facility in Grenoble.

of the beamline ID28 at the ESRF is shown in Figure 3. The beam emitted by a powerful undulator is pre-monochromatized by a double-crystal low-resolution monochromator (A) using the 111-reflection so that the high resolution monochromator (B) is not exposed to the high heat load. The back-reflected, highly monochromatic beam is doubly focused by a gold coated toroidal mirror (C) to a spot of 80 µm x 210 µm (FWHM) on the sample (D). The energy of the radiation scattered by the sample contained in the diamond anvil pressure cell is analysed by a spherically bent crystal and recorded by a solid-state detector (F). The whole beam path covers a length of almost 100 m and alignment and stability are vital for the success of such experiments. ID28 is one of the most complex and heavily oversubscribed beamlines at the ESRF.

The study of iron under megabar pressures [4] was motivated by the importance iron plays in the planetary cores. It makes up 70 to 90 weight percent of the cores and the knowledge of the elastic constants is essential for comparison with the global dynamics processes inside meV at an energy of 15.618 keV. A 99.999% pure iron

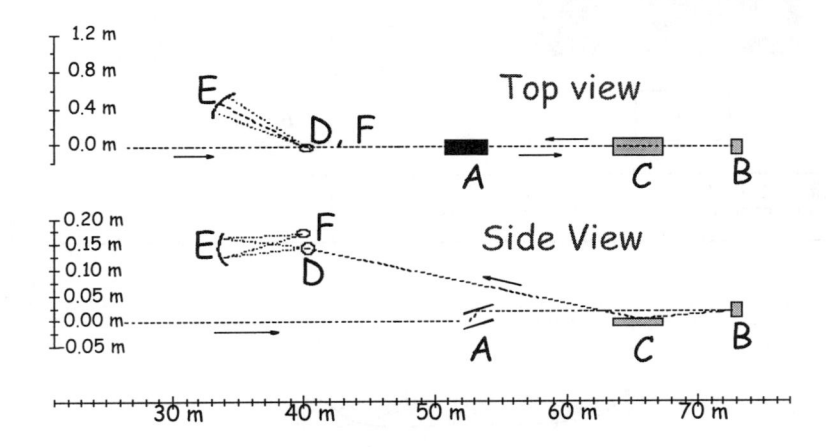

Figure 3: Experimental scheme of the inelastic scattering beamline ID28 at the ESRF. For explanation see text. Courtesy: M. Krisch.

powder sample was loaded into the pressure cell (Figures 4 *a* and *b*) and phonon dispersion curves were taken at pressures from 19 until 110 gigapascals. A typical spectrum is displayed in Figure 5. The low count rate of about 0.2 Hz is the limiting factor for this very difficult experiment. It was shown that hcp iron followed a Birch law (linear relation between density and sound wave velocity). From an extrapolation and comparison with seismic data it was concluded that the inner core must be 4 to 5% lighter than pure hcp iron. This means that low Z impurities (such as carbon) are present in the iron of the core.

3.2 X-ray holo-tomography as a new tool to reveal microstructural properties of materials

It goes almost without saying that a gain of many orders of magnitude in brilliance of a source brings surprises: the *unprecedented* creates the *unexpected*. The human mind is unable to anticipate everything. No wonder that the brilliance gain of eight orders of magnitude between a conventional X-ray tube and a third generation synchrotron X-ray source allowed us to observe completely new effects and to open unforeseen (though *a posteriori* predictable) possibilities. One of them is based on

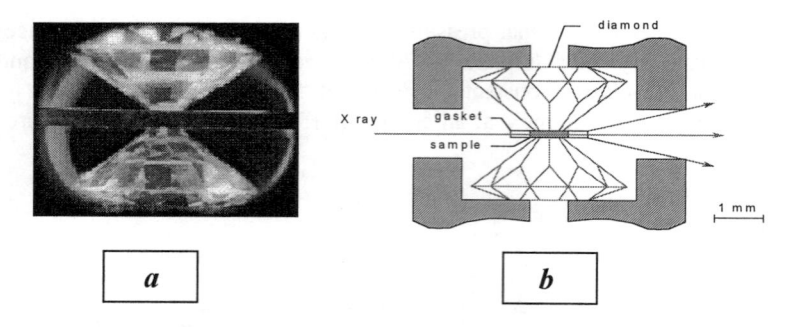

Figure 4: Diamond anvil pressure cell for the MBar pressure range. Left: photograph, right: schematic view. Courtesy: M. Krisch.

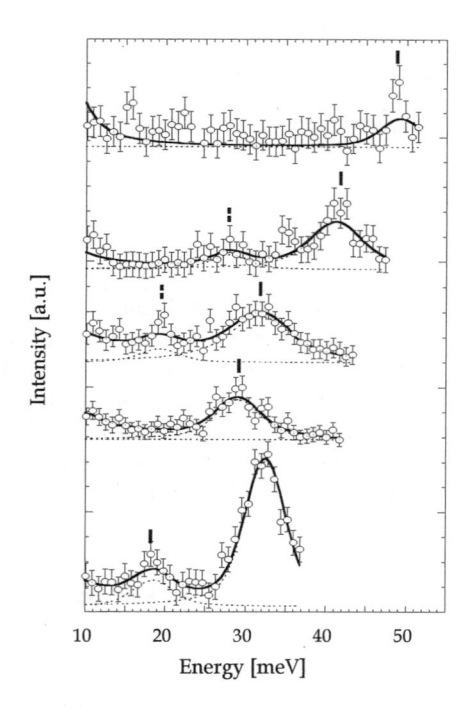

Figure 5: Dispersion of the iron LA phonon with increasing Q values: 4, 6.16, 8.31, 10.46 and 12.62 nm^{-1}. LA phonons of iron are indicated by ticks. A TA mode detected at 8.31 nm^{-1} and 10.46 nm^{-1} are indicated by broken ticks. The accuracy of the energy position of the phonons was typically 3%. From ref. [4].

coherence. First, there was a negative aspect, like when synchrotron radiation was discovered: the coherence produced by the small source sized and the excellent natural beam collimation generated speckle patterns when transmitted through beryllium windows that had a surface roughness exceeding about 1 µm. Therefore, all windows and filters in a beamline had to be polished to avoid parasitic structures in images such as topographies taken to study crystal defects. The quality of X-ray optical elements such as mirrors and multilayers (artificial nanometric Bragg reflectors) had to be drastically improved necessitating substantial investments. Then the positive side was discovered: phase contrast could be used to create pictures of materials that were very difficult to obtain using the traditional absorption contrast only, in particular for samples composed of materials with close atomic numbers. Phase contrast could also be applied to ultra-high resolution tests of optical surfaces, *i.e.*, coherent X-ray metrology.

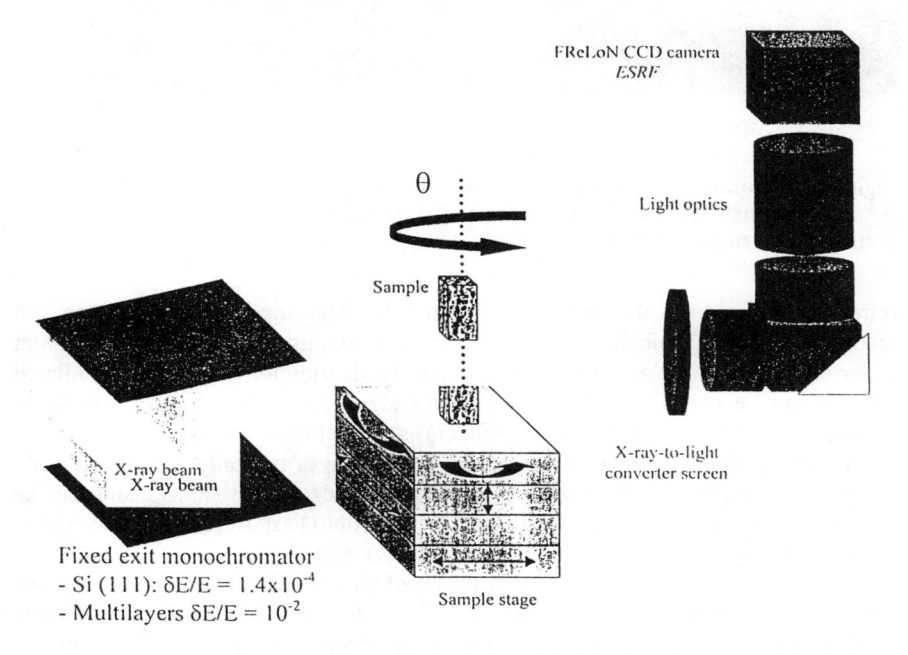

Figure 6: X-ray tomography setup on beamline ID19 of the ESRF. Courtesy: M. Salomé.

Figure 6 shows the experimental set-up for micro-tomography on beamline ID19 at the ESRF. The sample, mounted and aligned on a goniometer, is illuminated by a monochromatic X-ray beam of typically 19 keV energy. The picture recorded in transmission depends on the structure of the sample and on the contrast. The latter is generated either by absorption if the materials in the sample have strongly differing absorption properties, or by a phase lag introduced by different real parts of the

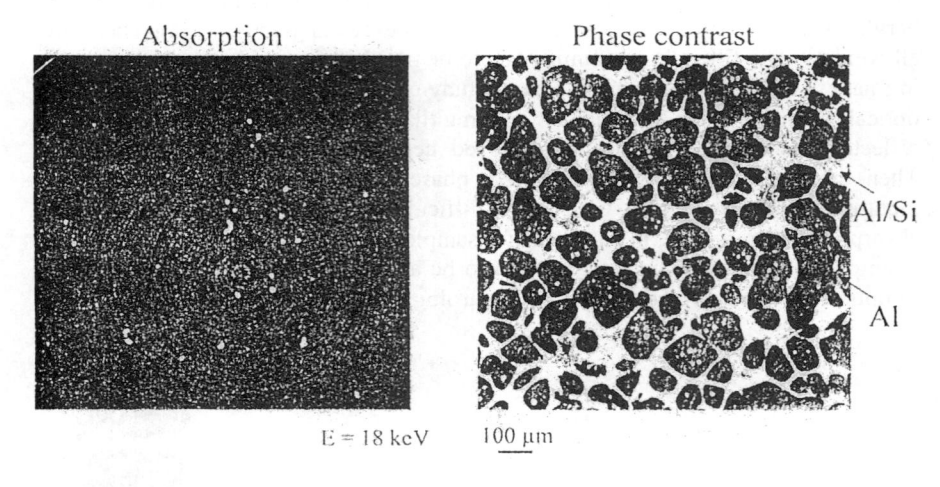

Figure 7: Absorption contrast image (left) and phase contrast image (right) of an Al-Al/Si system. The bright spots are spurious images. 800 frames were taken to reconstruct the section of the sample. Courtesy: P. Cloetens.

refractive indices of the constituent materials. The transmission patterns are recorded for many, typically several 100 angular positions, rotating the sample about a vertical axis. The three-dimensional intensity distribution inside the irradiated volume can be reconstructed by a computer algorithm. Absorption or amplitude contrast is readily obtained when the camera is very close to the object (amplitude-tomography) whereas phase contrast needs a distance of typically 1 meter to build up (holo-tomography). The spatial resolution depends on the camera and can be adjusted between 1 µm and 10 µm for the FReLoN (**Fast Readout** – **low Noise**) camera recently developed at the ESRF that has 2096x2096 pixels.

As an example, the two-dimensional structure of an Al-Al/Si system of high interest for metallurgy is invisible in absorption mode (left side of Figure 7), but a whole wealth of details are clearly revealed in phase contrast (right side of Figure 7). Three-dimensional reconstructions of ice and of polystyrene foam are presented in Figures 8 and 9. Many important structural parameters such as the connectivity can be derived from these pictures [5]. There are many other examples that could be shown here, in particular of medical interest (bone structures...) and this new technique based on phase contrast has attracted a great number of new users also from industry.

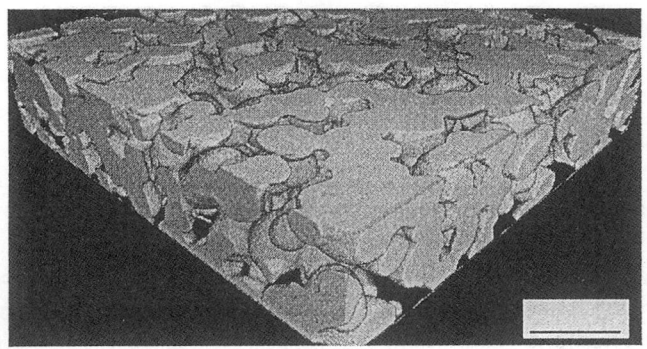

Figure 8: 3D tomographic reconstruction of a snow particle recorded with 1000 projections and 12 keV X-rays. The pixel size was 10 µm. Courtesy: W. Ludwig.

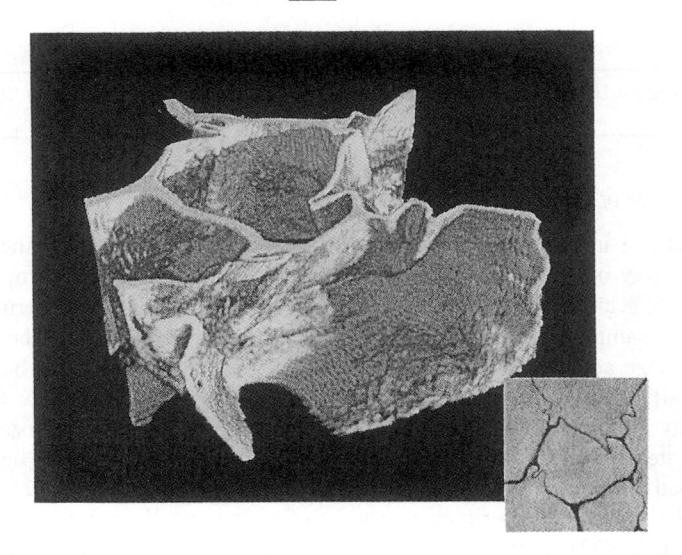

20 µm

Figure 9: Holo-tomographic reconstruction of polystyrene foam recorded with 400 projections at 4 distances. Resolution: 2 µm. 24 hours measuring time. The insert shows a tomographic slice revealing details of the connectivity. Courtesy: P. Cloetens.

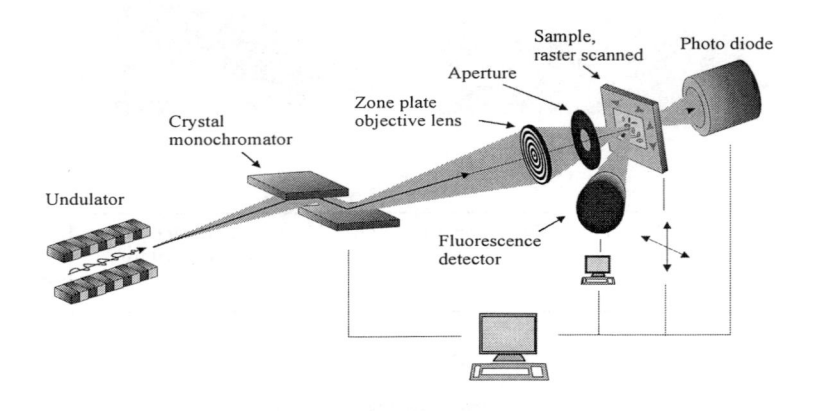

Figure 10: Scanning X-ray microscope at beamline ID21 of the ESRF. Courtesy: J. Susini.

3.3 The role of sulfur in biomineralization

In the above imaging technique the spatial resolution is given by the camera. A different way of obtaining structural details of samples on a microscopic and even nanoscopic scale is by focusing the beam to a small spot and then rastering the beam across the sample area of interest. Here the resolution is limited by the smallest X-ray spot size achievable. Many techniques of focusing X-rays have been invented and installed at modern synchrotron radiation facilities. Optical devices with sub-µm capability are tapered glass capillaries, curved mirrors and multilayers, and Fresnel zone plates. Even X-ray waveguides and refractive lenses have been recently developed. A review of modern synchrotron X-ray optics can be found in [6].

Figure 10 shows the experimental scheme of a scanning X-ray microscope installed on beamline ID21 of the ESRF. The X-ray beam emitted by an undulator is first made monochromatic with a crystal monochromator and then focused with a Fresnel zone plate to a spot of 0.25×0.25 µm^2. The fluorescence radiation is recorded with a germanium solid-state detector. By tuning the energy to a given chemical element (for example sulfur) and recording the XANES spectra (**X**-ray **A**bsorption **N**ear

Edge Structure) in fluorescence mode it is possible to map out the distribution of the chemical element with high spatial resolution. This technique was applied to the study of a seashell called *Pinna Nobilis* (Pelecypod) [7]. Several samples were prepared by cutting slices in direction perpendicular or parallel to the prisms of the outer calcitic layer and then polishing. To determine by comparison which one of the three principally possible binding modes of sulfur was present in the samples, XANES data from three pure reference samples containing sulfur in those binding states were recorded first.

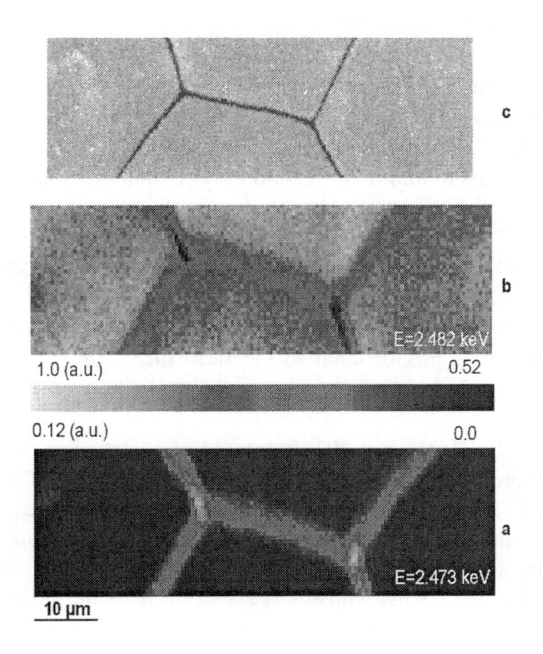

Figure 11: Chemical mapping of nodes of the polygonal organic matrix of *pinna nobilis* at the characteristic energies of of amino-acid type sulfur (a) and at the peak energy of sulfate type sulfur (b). Resolution: 0.5 µm, field: 19 x 70 µm. For comparison, an electron microscope image is shown too (c). The contrast inversion between the intra- and interprismatic areas is spectacular. From ref. 7.

The intensity mapping of Figure 11 at two energies close to the sulfur K edge showed clearly the presence of amino-linked sulfur in the inter-prismatic organic phase. The upper image is a scanning electron microscope picture from which it is impossible to obtain the information about the chemical state of sulfur. This example of XANES scanning micro-spectroscopy nicely demonstrates the strength of combining several techniques to access new and very specific information on a microscopic scale. Correlations with mappings of other elements such as calcium, magnesium or strontium are planned. They are of interest to better understand the interaction between mineral and organic phases in a great variety of biomineral materials made by nature.

3.4 The importance of diamond for synchrotron X-ray optics

Diamond single crystals have many very attractive and even unique properties as optical elements for synchrotron radiation beamlines [8]:

1. They have the highest heat conductivity κ of all materials known to man.
2. They have a very low thermal expansion coefficient α (twice less than silicon).
3. The absorption coefficient μ for X-rays is very small.
4. Their structure permits to eliminate second order contamination.
5. They are very stiff and extremely hard: an advantage for mounting and handling.
6. They resist very well to very high doses of X-rays in a wide energy range.

The first three properties are very important for cooling without creating too much thermal deformation that is proportional to the ratio $\mu\alpha/\kappa$. In Table 1 the data of diamond are compared with data for other materials silicon being the most important one.

There are unfortunately also some disadvantages, in particular with respect to silicon. The first is simply that the lattice parameter of diamond is by a factor 1.5 smaller than that of silicon, which limits the range of energies. The second shortcoming is the maximum integrated reflectivity that is twice smaller for diamond. Finally, the size of presently available synthetic stones (about 1 cm^2) is just sufficient for undulator beams whose cross section is of the order of one mm^2, but for bigger (wiggler) beams they are too small. As to crystalline perfection, diamond usually exhibits a more or less pronounced defect structure consisting of dislocations, growth bands and stacking faults that increase the width of the angular resolution curve by about one arcsec or 5 µrad, which is acceptable for a limited number of experiments. However, because this defect structure is distributed in non-uniform way across the crystal, the use of such monochromators is prohibited for all applications that require full coherence preservation, for example imaging. Two major efforts have been undertaken worldwide to prepare high quality synthetic crystals for synchrotron X-ray monochromators by:

1. The company De Beers Industrial Diamond in South Africa through a collaboration with the University of the Witswatersrand in Johannesburg (Prof. J.P.F. Sellschop) and the ESRF (A.K. Freund).
2. The company Sumitomo in Japan through a collaboration with the Japanese third generation source SPring-8 (T. Ishikawa).

The state-of-the-art of these developments has been recently reviewed [9,10]. At present a matter of debate is the influence of nitrogen impurities on the crystalline

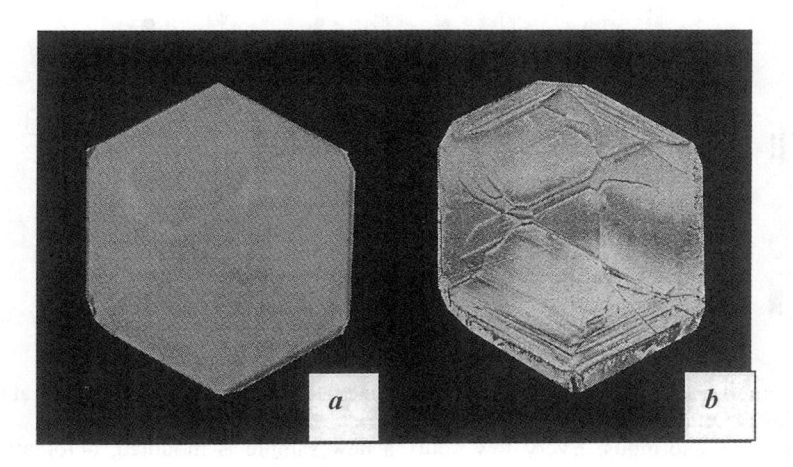

Figure 12: Close-up photograph of a diamond type Ib crystal 10 mm wide, 12 mm long and 0.4 mm thick showing the nitrogen impurity distribution (a). Lattice tilt distributions inside the same crystal determined by a μm-resolved rocking curve technique (b). The correlation of the structure in both pictures is clearly visible. From ref. 11.

perfection. A more recent comparative study showed that a type IIa specimen was completely defect-free in a quite substantial part of its volume whereas type Ib crystals containing nitrogen always gave some structures in their X-ray topographs [11]. A strong correlation between the nitrogen distribution evidenced by the yellow colour distribution in Figure 12a and the defect structure could be established [12] by taking spatially resolved rocking curves whose width distribution is shown in Figure 12b. Further efforts to improve the quality of diamond are being made.

There are other applications of diamond, for example as highly resistant windows for high power beamlines and as anvils of high pressure cells. Another interesting utilization is as a device to create or analyse polarization. In fact, diamond single crystal plates can change the ratio of $\pi-$ to σ-polarization in a photon beam. Circular polarization is very important to study magnetic scattering of X-rays or X-ray circular magnetic dichroism.

The low absorption of diamond has led to the concept of serial beam multiplexing. In this scheme the radiation of an undulator is used by several successive stations along a beamline. Each diamond crystal monochromator reflects a given wavelength (or energy) to a station while the beam transmitted (about 80%) is used further downstream for other station. Two beamlines have been built at the ESRF according to this principle: *Troika* [13] and *Quadriga*. The scheme of the latter beamline for protein crystallography is shown in Figure 13. The data production rate of this

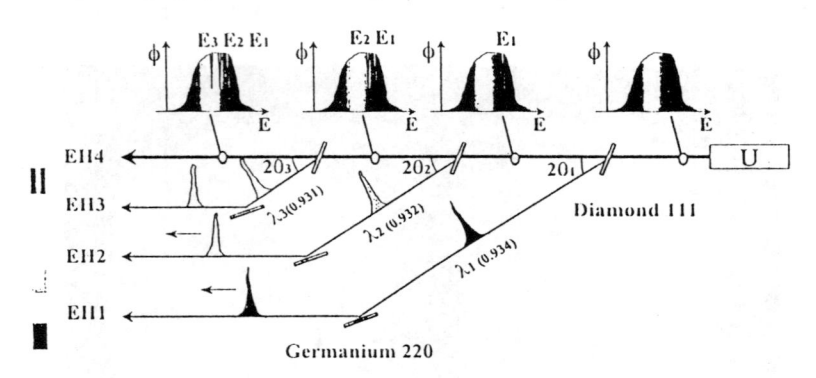

Figure 13: The *Quadriga* four-station beamline ID14 for protein crystallography at the ESRF. Courtesy: H. Belrhali

beamline is enormous. Every few hours a new sample is mounted, automatically aligned and studied. Recently a major step forward has been achieved by fully solving the structure of the 30S ribosomal subunit [14] leading to a better understanding of its interactions with antibiotics [15]. Data collected at the Quadriga beamline have been used to reach this success. Protein crystallography is a very important tool in structural genomics and about 20% of the experiment proposals at the ESRF are submitted in this field. A big center for structural genomics is planned at the Advanced Photon Source at Argonne National Laboratory.

Figure 14 shows the progress of the number of users, proposals submitted and experiments carried out per year since the beginning of ESRF operation. It is clear that third-generation sources are very productive and oversubscribed, and that this situation is not going to change in the short and the medium term. But a new type of source, much stronger than, but complementary to the present one, is already under discussion. This future development will be described briefly in the next section.

4   Properties of fourth-generation sources and possible scientific applications

Fourth-generation X-ray light sources are defined by the following properties:

1. Ultrahigh brilliance: peak brilliance 10 orders of magnitude above third-generation sources, time-averaged brilliance 2-4 orders of magnitude above third-generation sources.
2. Full lateral (transverse) coherence.
3. Sub-picosecond pulse length.
4. Wavelength around 1 Å.

They are based on single pass free lasers (FEL) where electrons produced by a low

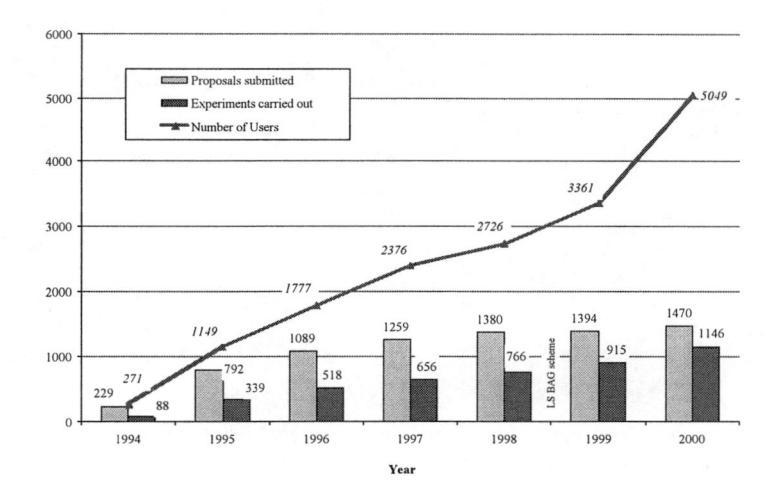

Figure 14: Summary of user operation from 1994 until 2000. Shown are the number of proposals submitted and carried out and the number of users visiting the ESRF per year.

Courtesy: R. Mason.

emittance photocathode are accelerated to energies of several GeV and bunched to about 200 fs (femtoseconds) pulse length corresponding to a spatial length of 60 μm. These bunches are sent through a long undulator (of the order of 100 m) where they are amplified by a process called SASE (Self Amplified Spontaneous Emission). It must be said that this type of machine is not a true laser (the emission is spontaneous instead of being stimulated and there is no cavity), and the longitudinal coherence is very weak compared to that of "true" lasers. The spectral bandpass is determined by the electron energy resolution convolved with the inverse number of effective periods in the undulator, a value for $\Delta E/E$ of about 0.1% is planned. But the source size is smaller than 100 μm and the divergence is as small as

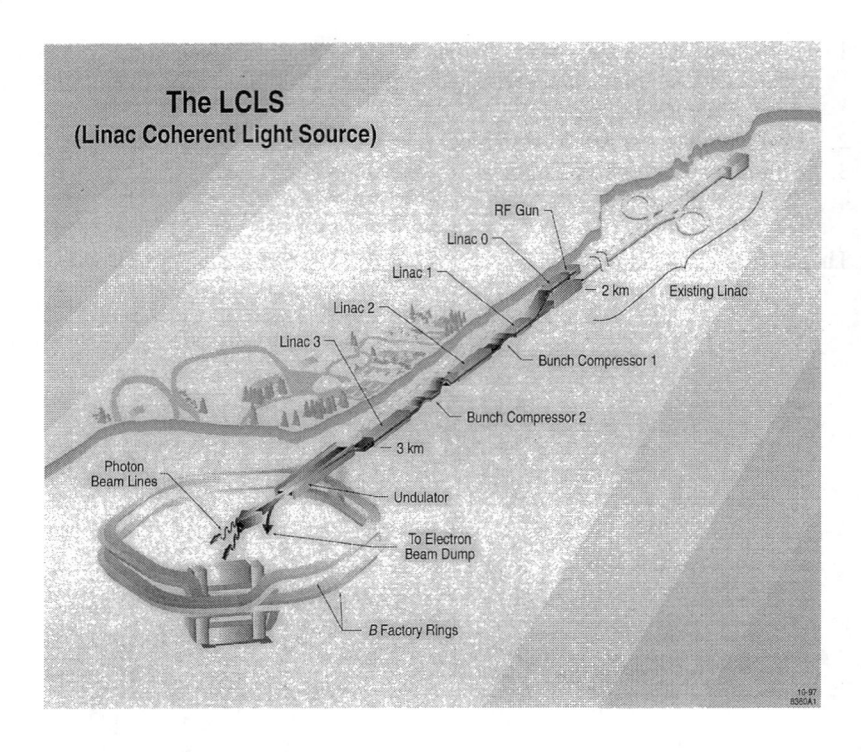

Figure 15: Conceptual view of the Linac Coherent Light Source at the Stanford Linear Accelerator Center. From ref. 16.

1 μrad, which ensures full transverse coherence. It is not possible here to describe all the details of the process and the interested reader is referred to a report describing the SLAC-LCLS project in Stanford [16] and to a website where information on a similar project at DESY in Hamburg can be obtained [17]. A main difficulty with such a single pass FEL is to reach the saturation at X-ray energies. In a first step it has been shown recently that the SASE principle works at a wavelength of about 500 nm where theoretical performances could be reached [18]. Next steps are to reach 100 nm at the APS in 2001, 6 nm at DESY in 2002 and 0.1 nm at SLAC in 2005. Figure 15 shows a an artist's view of the Stanford LCLS project. The last 1 km of the existing 3 km long LINAC will be used for creating, accelerating and compressing the electron bunches that will give rise to ultrabright coherent radiation after having passed the undulator. Several short and one about 1 km long beamlines are envisaged for installing various experiments. The Hamburg project is even more ambitious. It foresees a switchyard where the electron pulses can be directed to different undulators with different characteristics and thus several experiments can

be conducted simultaneously. The expected performance characteristics of the LCLS in terms of time-averaged and peak brilliance (or spectral brightness), is compared with other existing or projected facilities in Figure 16. It is clear that nothing but the

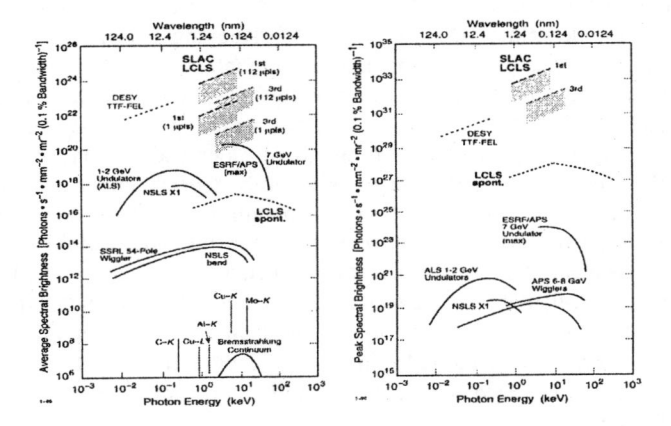

Figure 16: Performance characteristics of the planned LCLS X-ray FEL in Stanford: peak and time averaged brilliance compared with that of other facilities. The data for rotating anode X-ray tubes are also shown. From ref. 16.

spontaneous LCLS radiation is much brighter than that of any other source.
In contrast to third-generation sources whose performances can be safely predicted, an X-ray FEL has to be considered as a first experiment in itself. The second experiment will of course be concentrating on the interaction of these extremely intense (electric fields of 1 V/Å) and ultra-short (a few fs) pulses with matter. Whereas high power lasers in the visible wavelength can generate similar electric field strengths and pulse lengths, the interaction processes are completely different, because long-wavelength photons strip off the valence band electrons first. In contrast, the absorption process of X-rays empties first the K-shell, then the L and so on, because the intensity is so high that the recombination processes are not fast enough. One can even think of *true* X-ray lasers by population inversion. Totally new objects such as hollow atoms could be created. X-ray spectroscopy could be performed on these objects by sending first a pump pulse and a subsequent, time delayed, probe pulse. Such kinds of experiments have been identified [16] as being part of *atomic physics* and *plasma and warm dense matter physics*. Here the important advantage of X-rays is that they can penetrate the plasma and see its inner structure. Other studies concern *nanoscale dynamics in condensed matter* using coherent interference spectroscopy and taking full advantage of the coherence properties, and *femtosecond chemistry* where the short pulse length is the important

parameter. Pump-probe experiments can be envisaged where either an X-ray or a laser pulse can be used for changing the state of the sample and for studying the excitation and relaxation processes by following pulses with variable time delays.

Pump-probe experiments have already been carried out at several third-generation sources such as on beamline ID09 at the ESRF studying molecular motions in myoglobin [19] or chemical dissociation and recombination processes [20], but here only on the picosecond time scale and with severe flux limitations. One of the most exciting experiments at fourth-generation sources would be *structural studies of single particles and biomolecules*. The progress of biocrystallography is presently hampered by the fact that the protein crystals severely degrade under intense X-ray flux. A focused X-FEL beam would provide enough photons to collect the data needed to reconstruct the molecule by computer algorithms in several single shots from single molecules. The reconstruction procedure has been validated and molecular dynamics calculations performed to predict the following experiment. 3×10^{11} photons in a 200 fs long pulse are focused on a spot of $100 \times 100 \text{nm}^2$ and scattered off a lysozyme molecule [21]. In Figure 17 the molecule is shown before (at rest) and after having been struck by the full X-ray pulse where a Coulomb explosion will happen after about 70 fs. Therefore, for this experiment it is necessary to produce even less than 70 fs long pulses that could be obtained by special pulse slicing or compression techniques.

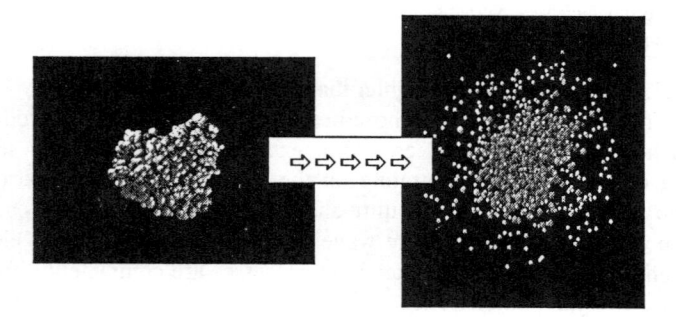

Figure 17: Coulomb explosion of a lysozyme molecule after having been hit by a 100 fs long X-ray FEL beam. From ref. 16.

Again, similar to the enormous jump in brilliance from X-ray tubes to third-generation facilities, the unprecedented experimental conditions that will be created at X-ray FELs will bring with it unexpected phenomena and new science. Until then, major technological efforts are required, but there are no *fundamental* limits that would prevent the realization of such sources and its instrumentation. It has to be said in passing that there are intermediate, less powerful, but more economic and thus less expensive projects such as the ERL (**E**lectron **R**ecirculating **L**inac) machine presently proposed at Cornell [22]. The ultimate (3^+ - *generation*) storage ring will have a brilliance increased by "only" a factor 200, but never be able to produce sub-picosecond pulses that are only possible at linac-based facilities.

It is evident that only a few materials will be capable of withstanding these enormous power densities and electric fields. Such materials are needed for beam conditioning devices. Low Z materials are naturally better suited and of course diamond is the absolute champion. Recent model calculations for the TESLA X-ray

Material	Be	C*	Si	Ge
Atomic number, Z	4	6	14	32
Atomic weight, A	9	12	28	73
Crystal structure	hcp	diamond	diamond	diamond
Lattice constant(s), a (Å)	2.286	3.567	5.431	5.658
c (Å)	3.583			
Debye temperature, T_D, at 297 K, (K)	1188	1860	543	290
Absorption coefficient, μ, at 8 keV (cm^{-1})	1.8	7.5	141	402
Conductivity, κ, at 297 K (Wcm^{-1}K^{-1})	1.93	23	1.5	0.64
Expansion coefficient, α, at 297 K (10^{-6} K^{-1})	7.7	1.1	2.4	5.6
Figure-of-merit, $\kappa/\mu\alpha$, at 297 K, (MW):	0.14	2.78	$4.4 \cdot 10^{-3}$	$2.8 \cdot 10^{-4}$
..........at 77 K....	11	120	0.20	$6.6 \cdot 10^{-3}$

Table 1: Comparison of some important X-ray monochromator materials: beryllium, diamond (C*), silicon and germanium, from [8].

FEL facility have shown that cryogenic cooling of diamond would be needed to meet the survivability requirements. At low temperatures the thermal conductivity of diamond and other crystals such as silicon and germanium can be increased by at least one order of magnitude and this effect is presently used very successfully with silicon at third-generation facilities [23]. Whereas the size of actually available diamond crystal is sufficient, it is crucial to still improve the quality and the preparation techniques so that there will be no strains left at all after growth, cutting, polishing and mounting. There are many more problems related to X-ray FEL instrumentation, already known and to be discovered, before fourth-generation facilities will reach a level of reliability and user-friendliness comparable to that of third-generation sources, but no longer than ten years ago the community had to face and to solve many unprecedented problems raised by the earlier machines and it succeeded. Then the final outcome of new developments will mainly depend on the efforts devoted to these new developments. An amount of 170 M$ US is requested for the LCLS project. We can anticipate that the discoveries visible at the horizon and beyond will be more than simply rewarding and worth the efforts for a very wide spectrum of research in fundamental and applied science. This very wide range of applications has been and will be typical for the research conducted at modern X-ray sources.

5 Summary and conclusions

The properties of present and future hard X-ray sources and their utilization at big user facilities have been described. Only a few examples out of the very wide range of experimental methods and recently achieved results could be mentioned in this paper, but the author hopes that he has given a few views that illustrate well the research conducted with bright and ultra-bright X-ray light, presently available and to come.

The strong recent acceleration of progress in the X-ray research domain due to synchrotron third-generation sources has led to a better understanding of many scientific problems, both fundamental and applied, in a very large number of communities. As a by-product, the fact that these experiments have been carried out at the same places, has stimulated a great deal of cross-talk and cross-fertilization between these communities. The general level of knowledge has raised significantly over the past few years and many outstanding research highlights can be identified. However, this has not (yet?) led to a singular discovery that would have been distinguished by a Nobel prize, maybe because it is very difficult to recognise an ensemble of outstanding research over a wide area conducted by many users at facilities whose main goal is to provide outstanding service to these users. But this is a typical ingredient of life in synchrotron research around the world and the community can live with it.

As to the new, fourth-generation sources, the scientific cases and the conceptual technical designs being readily available, it is expected that the means will be provided for their construction so that they can see – and *emit* – ultrabright light with exceptional quality in not too far future. The vision is there, the seeing is still to come. Finally, the author very much hopes that the African continent will see, also in not too far future, a synchrotron radiation facility where many communities will be able to meet through common research activities. Without any doubt there are many positive and attractive features and effects that such a facility would have, not only of scientific, but also of political and social nature.

Acknowledgements

The author is indebted to many colleagues at the ESRF who helped by providing information for writing this paper and by giving their agreement for showing their results of sometimes very recent and not yet published work. In particular, thanks go to H. Belrhali, P. Cloetens, J.M. Filhol, M. Krisch, R. Mason, M. Salomé, J. Susini and M. Wulff. I also would like to thank K. Hodgson (SSRL) for illustrations of the LCLS project. Finally, I would like to take this opportunity to express my sincere gratitude to J.P. Friedel Sellschop for both his encouragement and concrete help regarding high quality diamond crystals and for the immense pleasure of our collaboration over many years.

References

1. For an interactive and detailed introduction to synchrotron radiation, the reader is referred to the publication on CD-Rom *"Synchrotron Light to explore matter"* (bilingual: in English and French language), Springer, Berlin-Heidelberg, 2000. To purchase: contact orders@springer.de.
2. M. Altarelli, F. Schlachter and J. Cross, *"Making ultrabright X-rays"*, Sci. American **279**, 66-73 (1998).
3. J.M. Filhol, *"Third-generation synchrotron radiation sources: present performances and ultimate expectations"*, Presentation at the Symposium *"Biological Physics & Synchrotron Radiation"*, ESRF, Grenoble, 11-14 October 2000. Further information can be obtained from filhol@esrf.fr.
4. G. Fiquet, J. Badro, F. Guyot, H. Requardt and M. Krisch, *"Sound velocities in iron to 110 gigapascals"*, Science **291**, 468-471 (2001).
5. P. Cloetens, W. Ludwig, J.P. Guigay, M. Schlenker, C. Morawe, E. Ziegler and O. Hignette, *"Coherent hard X-ray imaging of and with multilayers"*, Presentation at the International Conference on Multilayers, Chamonix, January 2000. Further information can be obtained from cloetens@esrf.fr.
6. A.K. Freund, *"Synchrotron X-ray beam optics"*, in: Proceedings of the 6th Summer School on Neutron Scattering: "Complementarity Between Neutron

and Synchrotron X-Ray Scattering", A. Furrer, ed., 329-349, World Scientific (1998).
7. M. Salomé, Y. Dauphin, J. Susini, J. Doucet, B. Fayard and J.P. Cuif, *"The role of sulfur in biomineralization explored by X-ray absorption spectroscopy with submicron resolution: the relationship between sulfur chemical states and shell micro-architecture in Pinna Nobilis"*, Proc. Natl. Acad. Sci. USA – Biological Sciences/Biophysics, to be published. Further information can be obtained from susini@esrf.fr.
8. A.K. Freund, *"Diamond single crystals: the ultimate monochromator material for high power X-ray beams"*, Opt. Eng. **34**, 432-440 (1995).
9. H. Sumiya, N. Toda and S. Satoh, *"High quality large diamond crystals"*, New Diamond and Frontier Carbon Technology **10**, 233-251 (2000).
10. J.P.F. Sellschop, S.H. Connell, R.W.N. Nilen, C. Detlefs, A.K. Freund, J. Hoszowska, R. Hustache, R.C. Burns, M. Rebak, J.O. Hansen, D.L. Welch and C.E. Hall, *"Synchrotron applications of synthetic diamonds"*, New Diamond and Frontier Carbon Technology **10**, 253-281 (2000).
11. J. Hoszowska, A.K. Freund, E. Boller, J.P.F. Sellschop, G. Level, J. Härtwig, R.C. Burns, M. Rebak and J. Baruchel, *"Characterisation of synthetic diamond crystals by spatially resolved rocking curve measurements"*, Proc. Conf. XTOP2000, Ustron-Jaszowiec, Poland, September 13-15 (2000), J. Phys. D, *in press*.
12. J. Hoszowska, A.K. Freund, J.P.F. Sellschop, C. Detlefs, R.C. Burns, M. Rebak, J.O. Hansen, D.L. Welch, C.E. Hall and T. Ishikawa, *"Characterization of high-quality synthetic diamond crystals by µm-resolved X-ray diffractometry and topography"*, to be presented at the SPIE Annual Meeting 2001, conference AM134, San Diego, July 31 (2001).
13. J. Als-Nielsen, A.K. Freund, G. Grübel, J. Linderholm, M. Nielsen, M. Sanchez del Rio and J.P.F. Sellschop, *"Multiple station beamline at an undulator X-ray source"*, Nucl. Instr. Meth. Phys. Res. **B94**, 306-378 (1994).
14. B.T. Wimberly, D.E. Brodersen, W.M. Clemons Jr., R.J. Morgan-Warren, A.P. Carter, C. Vonrhein, T. Hartsch and V. Ramakrishnan, *"Structure of the 30S ribosomal subunit"*, Nature **407**, 327-339 (2000).
15. A.P. Carter, W.M. Clemons, D.E. Brodersen, R.J. Morgan-Warren, B.T. Wimberley and V. Ramakrishnan, *"Functional insights from the structure of the 30S ribosomal subunits and its interactions with antibiotics"*, Nature **407**, 340-348 (2000).
16. *"Linac Coherent Light Source (LCLS) Design Study Report"*, Report SLAC-R-521, UC-414 (1998). Stanford Linear Accelerator Center, CA 94025, USA. For scientific applications, see the Report *"LCLS – First Experiments"*, SLAC Report, October 2000. See also the SSRL website http://www-ssrl.slac.edu.
17. TESLA Project, see the website http://www-hasylab.desy.de. A Technical Design Report is being published in March 2001.

18. E. Gluskin, presentation at the Advanced Photon Source, Argonne National Laboratory, Argonne, Illinois, January 2001.
19. D. Bourgeois, T. Ursby, M. Wulff, C. Pradevand, V. Srajer, A. LeGrand, W. Schildkamp, S. Laboure, C. Rubin, T.-Y. Teng, M. Roth and K. Moffat, J. Synchrotron Radiation **3**, 65 (1996); V. Srajer, T.-Y. Teng, T. Ursby, C. Pradevand, Z. Ren, S. Adachi, W. Schildkamp, D. Bourgeois, M. Wulff and K. Moffat, Science **274**, 1726-1729 (1996).
20. S. Techert, F. Schotte and M. Wulff, *"Picosecond X-ray diffraction probed transient structural changes in organic solids"*, Phys. Rev. Letters, *in press* (2001).
21. R. Neutze, R. Wouts, D. van der Spoel, E. Weckert and J. Hajdu, Nature **406**, 752-757 (2000).
22. See the Cornell University website: http://www-chess.cornell.edu.
23. D.B. Bilderback, A.K. Freund, G.S. Knapp and D.M. Mills, *"The historical development of cryogenically cooled monochromators for third–generation synchrotron radiation sources"*, J. Synchrotron Radiation **7**, 53-60 (2000).

WAVE PACKET MOLECULAR DYNAMICS SIMULATIONS OF THE EQUATION OF STATE OF HYDROGEN AND DEUTERIUM UNDER EXTREME CONDITIONS

M. KNAUP, P.-G. REINHARD

Institut für Theoretische Physik II, Universität Erlangen-Nürnberg,
Staudtstr. 7, D-91058 Erlangen, Germany
e-mail: knaup@theorie2.physik.uni-erlangen.de

C. TOEPFFER

LPGP, Université Paris-Sud, Bât 210,
F-91405 Orsay, Ledex, France

Permanent address:
Institut für Theoretische Physik II, Universität Erlangen-Nürnberg
Staudtstr. 7, D-91058 Erlangen, Germany
e-mail: toepffer@theorie2.physik.uni-erlangen.de

Recent laser shock-wave experiments by Da Silva et al.[1] with deuterium in a regime where a plasma phase-trasition has been predicted[5] are a topic of many current discussions (e.g.[2,3,4,12,13]). In this paper we apply "Wave Packet Molecular Dynamics" (WPMD) simulations to the equation of state of hydrogen at constant temperature $T = 300$ K and of deuterium at constant hugoniot $E - E_0 + \frac{1}{2}(V - V_0)(p + p_0) = 0$ and compare them with experiments and several theoretical approaches. The WPMD method was originally used by Heller for a description of the scattering of composite particles like simple atoms and molecules[7]; later it was applied to Coulomb systems by Klakow et al.[8,9]. In the present version of our model the protons are treated as classical point-particles, whereas the electrons are represented by a completely anti-symmetrized Slater-sum of periodic Gaussian wavepackets.

1 Introduction

Classical molecular dynamics (MD) simulations are a well established tool to study equilibrium properties as well as the dynamics of manybody systems like strongly coupled plasmas. However, under extreme conditions of temperature and density quantum effects like the wave nature of the particles and their indistinguishability become important. For a Coulomb system with oppositely charged particles these ensure in fact the stability of the system as the classical equilibrium distribution has essential singularities. These can be mollified by employing effective potentials containing parameters which are chosen with more or less physical reason. On the other hand, a complete quantum treatment is an extremely demanding task as the $6N$ conjugate coordinates of the N-particle system have to be replaced by a full N-body quantum wavefunction. Here we employ an approximate quantum treatment which comes from the classical MD side, thus maintaining its simplicity as well as its correlation content.

This is achieved by blowing up the classical point particles to a localized wave packet with a simple analytical form. Such a wave packet molecular dynamics (WPMD) based on the time-dependent variational principle is a good approximation if the width of the wave packet is smaller than the typical length scale of variation of the potential. The wave packet approach is able to reproduce many dynamical properties of many-body systems as has been shown by Heller for the scattering of composite particles like atoms and molecules[7], by Feldmeier for heavy ion scattering[14] and by Klakow et al. for Coulomb systems[8,9]. Usually Gaussian wave packets are employed reducing the time evolution of a complex wavefunction to the evolution of a few relevant parameters, as position, momentum and width of the wave packet. This reduces the amount of numerical work from the solution of a partial differential equation to the much simpler case of a set of ordinary differential equations. Nevertheless there remain formidable tasks.

In the present paper we will present in section 2 the theory of WPMD, paying particular attention to the antisymmetrisation of the many-body wavefunction. In section 3 we compare results for the equation of state of hydrogen with experiments[6] and present our calculations of the Hugoniot of deuterium together with experiments[1] and other theories[1,2,3,13].

2 Theory of Wave Packet Molecular Dynamics

In the current application of WPMD to a Coulomb system of N electrons and protons, the protons are described classically, i.e. by their positions $\vec{R}_I$ and momenta $\vec{P}_I$, whereas the electrons are represented by an anti-symmetrized product of one-particle wavefunctions $\varphi_s(\vec{x}_s)$:

$$\Psi(\vec{x}_1, \ldots, \vec{x}_N) = \frac{1}{\sqrt{N! \det(\mathbf{O})}} \sum_{\sigma \in \mathcal{P}} \text{sgn}(\sigma) \prod_{s=1}^{N} \varphi_{\sigma_s}(\vec{x}_s) . \qquad (1)$$

Here, $\mathcal{P}$ is the set of permutations of order N, and $\mathbf{O}$ is the overlap-matrix

$$O_{kl} := \langle \varphi_k | \varphi_l \rangle . \qquad (2)$$

For the one-particle wavefunctions $\varphi_s(\vec{x}_s)$ we make an ansatz of periodic

Gaussian wave packets[12]:

$$\varphi_s(\vec{x}_s) \propto \sum_{\vec{n} \in \mathbb{Z}^3} \exp\left[-\left(\frac{3}{4\gamma_s^2} + \frac{ip_{\gamma_s}}{2\hbar\gamma_s}\right)(\vec{x}_s - \vec{r}_s - \vec{n}L)^2 + \frac{i}{\hbar}\vec{p}_s(\vec{x}_s - \vec{r}_s - \vec{n}L)\right]. \quad (3)$$

with the 8 variational parameters $\{\vec{r}_s, \vec{p}_s, \gamma_s, p_{\gamma_s}\}$.

Since the wavefunctions φ_s are periodic with a period of one boxlength L, one can rewrite them in Fourier representation

$$\varphi_s(\vec{x}_s) = \sum_{\vec{\nu} \in \mathbb{Z}^3} w_{\vec{\nu}}^s \exp\left(\frac{2\pi i}{L}\vec{\nu}\vec{x}_s\right). \quad (4)$$

In this formula, the complex Fourier coefficients $w_{\vec{\nu}}^s$ depend only on the variational parameters and satisfy the normalization-condition

$$1 = \int_{L^3} \varphi_s^\star(\vec{x})\,\varphi_s(\vec{x})\,\mathrm{d}^3 x = L^3 \sum_{\vec{\nu} \in \mathbb{Z}^3} |w_{\vec{\nu}}^s|^2. \quad (5)$$

As Hamilton operator we use $\hat{H} = \hat{H}_{\text{cou}} + \hat{H}_{\text{kin}} + \hat{H}_{\text{ext}}$ with

$$\hat{H}_{\text{cou}} = \frac{e^2}{4\pi\varepsilon_0}\left(\sum_{I<J}\frac{1}{|\vec{R}_I - \vec{R}_J|} + \sum_{i<j}\frac{1}{|\hat{\vec{x}}_i - \hat{\vec{x}}_j|} - \sum_{I,j}\frac{1}{|\vec{R}_I - \hat{\vec{x}}_j|}\right) \quad (6)$$

$$\hat{H}_{\text{kin}} = \sum_I \frac{P_I^2}{2M} + \sum_i \frac{\hat{p}_i^2}{2m}\;;\quad \hat{H}_{\text{ext}} = \frac{9\hbar^2}{8m\gamma_0^4}\sum_i\left(\hat{\vec{x}}_i - \langle\hat{\vec{x}}_i\rangle\right)^2. \quad (7)$$

Here, the capital variables I and J run over all protons, whereas the small variables i and j run over all electrons in the system. M and m are the masses of protons and electrons, respectively. To avoid an infinite growth of the widths γ_s of unbound electrons, we confine every wave packet in an external harmonic-oscillator potential $\hat{H}_{\text{ext}}$ which moves together with the center of mass of the electron and thus does not influence its motion. It can be interpreted physically as constraint acting on the variance $\langle(\hat{\vec{x}}_s - \langle\hat{\vec{x}}_s\rangle)^2\rangle$ of the wave packet. The constant Lagrangian parameter γ_0 adjusts the mean width of *unbound* electrons. To minimise the influence on *bound* wave packets, γ_0 is chosen much larger than the typical width of a bound electron, e.g. in an atom or molecule.[10]

The expectation value of the kinetic energy of the Slater-sum (1) can be written in terms of the inverse overlap matrix (2) and the matrix elements $E_{kl} := \langle\varphi_k|\hat{H}_{\text{kin}}|\varphi_l\rangle$:

$$\langle\Psi|\hat{H}_{\text{kin}}|\Psi\rangle = \text{Tr}\left(\mathbf{O}^{-1}\mathbf{E}\right). \quad (8)$$

If we calculate the matrix elements O_{kl} and E_{kl} from the Fourier representation (4) of the one-particle wavefunctions, we obtain

$$O_{kl} = \langle \varphi_k | \varphi_l \rangle = L^3 \sum_{\vec{\nu} \in \mathbb{Z}^3} (w_{\vec{\nu}}^k)^* w_{\vec{\nu}}^l \qquad (9)$$

$$E_{kl} = \langle \varphi_k | \hat{H}_{\text{kin}} | \varphi_l \rangle = \frac{L}{2m} (2\pi\hbar)^2 \sum_{\vec{\nu} \in \mathbb{Z}^3} \vec{\nu}^2 (w_{\vec{\nu}}^k)^* w_{\vec{\nu}}^l . \qquad (10)$$

Formula (8) was already presented in Ref.[14]. It reduces the computational effort from calculating $N!$ summands to a simple matrix inversion which scales like N^3. Furthermore, the matrix inversion is numerically very stable and no problem with the alternating signs of the permutations occur. For $N \lesssim 250$ electrons, most of the CPU time is spent in calculating the matrix elements O_{kl} and E_{kl} (which scales like N^2). Thus the antisymmetrisation does not essentially increase the computation time. For the results presented in this paper, each data point required about 1 day CPU-time on a state-of-the-art single-processor personal computer.

However, already the antisymmetrisation of the electron-proton interaction, which is a one-body operator requires N-times the effort for the kinetic energy because the sum over the protons. Therefore, in the present version of our model we use the pure Hartree product-wavefunction Ψ_H for the calculation of the expectation value of the potential energy:

$$\langle \Psi_\text{H} | \hat{H}_{\text{cou}} | \Psi_\text{H} \rangle = \frac{1}{2} \frac{e^2}{4\pi\varepsilon_0} \sum_{kl} V_{kl} \qquad (11)$$

$$V_{kl} = \iint \frac{\tilde{\rho}_k(\vec{x}_1) \tilde{\rho}_l(\vec{x}_2)}{|\vec{x}_2 - \vec{x}_1|} \, \mathrm{d}^3 x_1 \, \mathrm{d}^3 x_2 = \frac{L^5}{\pi} \sum_{\vec{\nu} \in \mathbb{Z}^3 \setminus \{0\}} \frac{1}{\vec{\nu}^2} (d_{\vec{\nu}}^k)^* d_{\vec{\nu}}^l . \qquad (12)$$

Here, $\tilde{\rho}_s(\vec{x}_s)$ is the difference of the one-particle density and a neutralizing background and $d_{\vec{\nu}}^s$ are the corresponding Fourier coefficients:

$$\tilde{\rho}_s(\vec{x}_s) := \varphi_s^*(\vec{x}_s) \varphi_s(\vec{x}_s) - \frac{1}{L^3} = \sum_{\vec{\nu} \in \mathbb{Z}^3 \setminus \{0\}} d_{\vec{\nu}}^s \exp\left(\frac{2\pi i}{L} \vec{\nu} \vec{x}_s\right) . \qquad (13)$$

For small widths $\gamma_{kl} \ll L$ of the wave packets the Fourier sum (12) converges very slowly. However, for this case one can derive the approximatic formula

$$V_{kl} \approx V_{kl}^0 - \frac{1}{r_{kl}} \operatorname{erfc}\left(\sqrt{\frac{3}{2}} \frac{r_{kl}}{\gamma_{kl}}\right) + \frac{2\pi \gamma_{kl}^2}{3L^3} \quad \text{for} \quad \gamma_{kl} \lesssim \frac{L}{10} , \qquad (14)$$

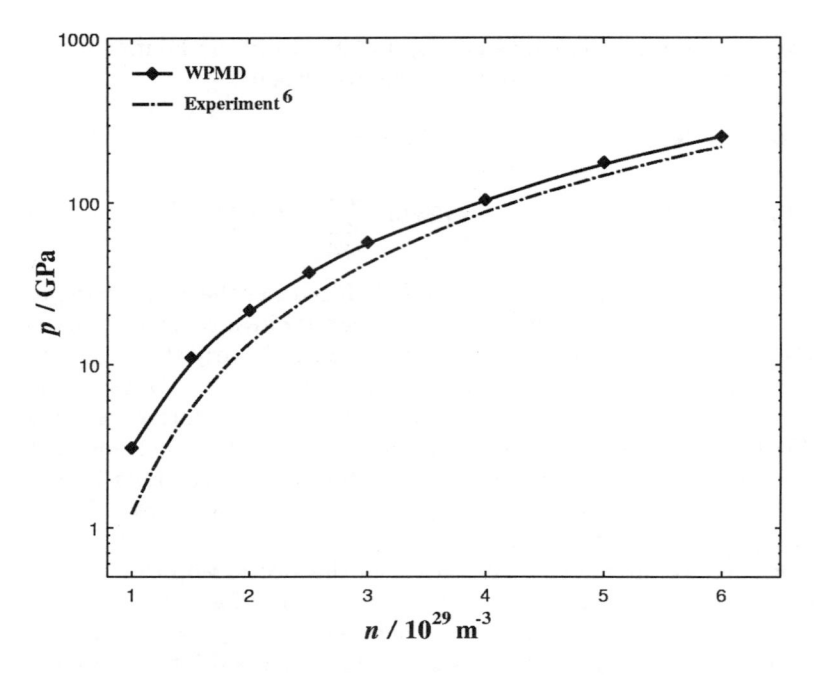

Figure 1. Equation of state of hydrogen at constant temperature $T = 300$ K. The WPMD result is compared with experiments.

with $r_{kl} = |\vec{r}_l - \vec{r}_k|$, $\gamma_{kl} = \sqrt{\gamma_k^2 + \gamma_l^2}$. V_{kl}^0 is the well-known Ewald-sum for classical point-particles[15,16].

3 Results

We applied the WPMD method to the equation of state of hydrogen and deuterium. As we are interested here in equilibrium properties only, we did not propagate the system in time explicitly, but used a force-bias Monte Carlo algorithm[17] to simulate a canonical ensemble characterized by the Gibb's distribution. The number of particles was $2N = 512$ (256 protons and 256 electrons). The Lagrangian parameter γ_0 (see sect. 2) was chosen as $\gamma_0 = \lambda_{th}$ with the thermal wavelength $\lambda_{th} = \hbar/\sqrt{mk_BT}$.

In figure 1 we compare WPMD results with experiments by Loubeyre et al.[6] who measured the pressure $p(n)$ at constant temperature $T = 300$ K. Since we apply full antisymmetrisation to the kinetic energy, our method works es-

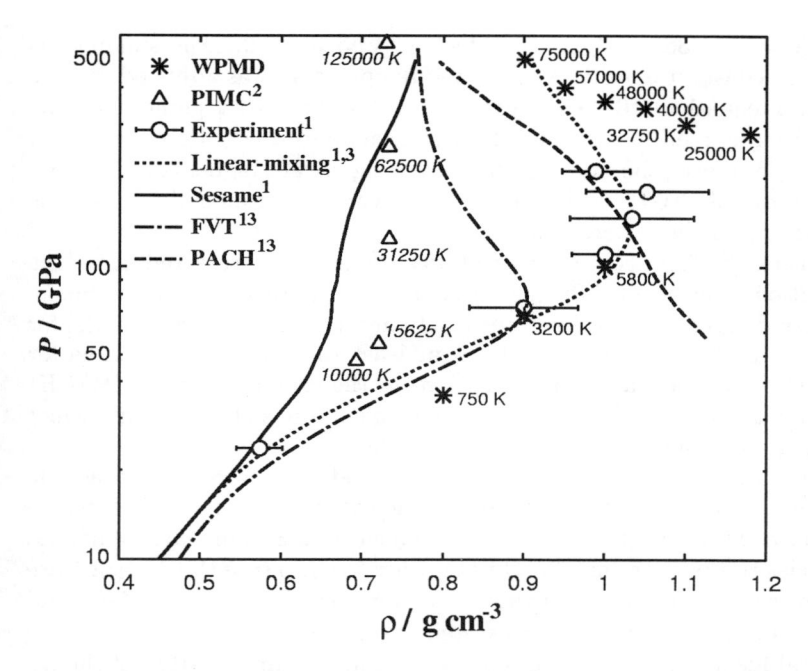

Figure 2. Equation of state of deuterium at constant hugoniot. The WPMD result is compared with experiments and other calculations.

pecially well in the regimes of weak interaction. Therefore the agreement with the experiment is quite good with increasing density. Then degeneration becomes more important than the Coulomb interaction and the system behaves more and more like an ideal Fermi gas. On the other hand, we slightly overestimate the potential energy when we neglect the antisymmetrisation. For that reason the pressure p, which is calculated via the virial theorem[10,11]

$$p = \frac{n}{3}\left(2E_{\text{kin}} - 2E_{\text{ext}} + E_{\text{cou}}\right) \qquad (15)$$

is systematically too high in the regime of low density. The next step to improve our model will be an approximate treatment of the antisymmetrisation in the calculation of the potential energy.

In figure 2 we present simulations of the Hugoniot

$$E - E_0 + \frac{1}{2}(V - V_0)(p + p_0) = 0, \qquad (16)$$

with $E_0/N = -15.886\,\text{eV}$, $\rho_0 = 0.171\,\text{g/cm}^3$ and $p_0 = 0$. The simulations were done at constant density, whereas the temperature was adjusted to fulfil the Hugoniot equation (16). We present the WPMD results together with the shock-wave experiments[1] and several theoretical approaches. Compared with the Sesame data base, the experiments show a significant higher compressibility in the range of 73 GPa to 210 GPa. The linear mixing model[1,3] shows the best agreement to the experiments.

Recent results from two other theoretical models are presented in[13]. The fluid variational theory (FVT) was developed for the regime of dense fluids. It shows good agreement to the experiments and the linear mixing model for pressures up to 70 GPa and reproduces at least the shape of the curve for higher pressures. The Padé approximations in the chemical picture (PACH) are valid for the fully ionized plasma. This calculations show an agreement to the experiments for pressures above 150 GPa.

The path-integral monte carlo (PIMC) model[2], which is just as the WPMD model "ab initio" in the sense that no external parameter is fitted to some experiment, shows a quite good agreement with Sesame but significant discrepancies to the experiments. On the other hand, our WPMD results are in excellent agreement with the experiments for pressures below 100 GPa and above 400 GPa. In these regimes the system is either close to the degenerate or the classical ideal regime, respectively, and the antisymmetrisation of the kinetic energy is sufficient. In between the compression is slightly overestimated in comparison to the experiment.

4   Conclusions

We developed a semi-classical ab initio MD-model which takes into account essential quantum effects like the wave nature of the electrons and their indistinguishability. In its present realisation it works especially well for systems where the kinetic energy is dominant, i.e. at high temperatures as well as at ultra high densities. For systems with dominant interaction the potential energy and the resulting pressure are systematically too high. An improvement of the model to remedy this drawback is in progress. We made use of the WPMD method to calculate the equation of state of hydrogen and deuterium. Comparing with the experiments we see quite good agreement and reproduce the effect of the increased compressibility of the shock-wave experiments compared to the Sesame data base.

References

1. I. B. Da Silva et al., Phys. Rev. Lett. **78** (1997) 483
2. B. Militzer and D. M. Ceperley, Phys. Rev. Lett. **85** (2000) 1890
3. M. Ross, Phys. Rev. B **58** (1998) 669
4. T. J. Lenosky, S. R. Bickham, J. D. Kress, and L. A. Collins, Phys. Rev. B **61** (2000) 1
5. D. Saumon and G. Chabrier, Phys. Rev. A **46** (1992) 2054
6. P. Loubeyre, R. LeToullec, D. Hausermann, M. Hanfland, R. J. Hemley, H. K. Mao, and L. W. Finger, Nature **383** (1996) 702-704
7. E. J. Heller, J. Chem. Phys. **62** (1975) 1544
8. D. Klakow, C. Toepffer, and P.-G. Reinhard, Phys. Lett. A **192** (1994) 55
9. D. Klakow, C. Toepffer, and P.-G. Reinhard, J. Chem. Phys. **101** (1994) 10766
10. M. Knaup, P.-G. Reinhard, and C. Toepffer, Contrib. Plasma Phys. **39** 1-2, (1999) 57
11. M. Knaup, P.-G. Reinhard, and C. Toepffer, Proceedings HIF 2000, to be published in Nuclear Instruments and Methods for Physics Research A
12. M. Knaup, P.-G. Reinhard, and C. Toepffer, Proceedings PNP-10, subm. to Contrib. Plasma Phys.
13. H. Juranek, R. Redmer, and W. Stolzmann, Proceedings PNP-10, subm. to Contrib. Plasma Phys.
14. H. Feldmeier, Nucl. Phys. A **515** (1990) 147-172
15. P. P. Ewald, Ann. Phys. **64** (1921) 253
16. J. P. Hansen, Phys. Rev. A **8** (1973) 3096
17. M. P. Allen and D. J. Tildesley, *"Computer Simulation of Liquids"*, Clarendon Press, Oxford 1987

HEAVY IONS STOPPING IN PLASMAS

GÜNTER ZWICKNAGEL

Institut für Theoretische Physik II, Universität Erlangen Staudtstr. 7, D-91058 Erlangen, Germany
E-mail: zwicknagel@theorie2.physik.uni-erlangen.de

For the stopping of highly charged heavy ions strong coupling effects in the energy transfer from the projectile–ion to a target plasma become important. A theoretical description of this nonlinear ion stopping has to go beyond the standard approaches like the dielectric linear response or the binary collision model which are strictly valid only at weak ion–target coupling. Here we outline an improved treatment which is based on a suitable combination of binary collision and linear response contributions. This approach well reproduces, up to moderate coupling strengths, the essential features of nonlinear stopping as observed and investigated in numerical simulations, in particular a dependence on the ion charge state like Z^x, with $x \lesssim 1.5$, in contrast to the usual Z^2 behavior.

1 Introduction

The interaction of charged particles with matter is a long-standing issue of experimental and theoretical investigations. Starting with the fundamental work of Bohr [1] and Bethe [2] the now available theoretical descriptions provide a good basic understanding and accurate quantitative predictions in many situations of interest, in particular for highly energetic projectiles with low charge state. However, in cases of highly charged, slow ions in a plasma and a strong projectile-target coupling perturbative approaches are not applicable. Here the well established standard treatments like the linear response description or the binary collision model fail. These nonlinear effects are crucial in at least two examples of applications. For the target design in a heavy ion driven inertial confinement fusion (HIF) scenario [3] and the electron cooling of highly charged heavy ion beams in the cooling section of a storage rings [4], which can be basically viewed as ion stopping in an electron plasma. In both cases a precise knowledge of the energy transfer rate due to the interaction of the ion beam with the target is of central importance. To investigate this nonlinear regime we concentrate on the many–body aspects of ion stopping at strong coupling and focus on the energy loss of heavy ions in a classical free electron plasma. This requires the development and use of advanced theoretical concepts and numerical tools as it has been carried out in the framework of kinetic theory [5,6] and computer simulations [7,8] during the last years. As a particular feature these investigations revealed a dependence of the energy loss on the ion charge Z like Z^x with $x \approx 1.7 \ldots 1.2$ in the nonlinear stopping regime in contrast to the Z^2-scaling at linear coupling. This behavior has been found as well in experiments on electron cooling [9,10]. We discuss the theoretical description of this nonlinear energy loss, in particular the combined scheme worked out in [5,6,8], which suitably joins binary collisions and linear response to account for both the strong short range collisions and the long range dynamic screening. The domain of validity of such extended stopping descriptions is detected by comparison with the results of computational treatments of nonlinear stopping based on a numerical solution of the nonlinear Vlasov–Poisson equation using a test–particle method [7,8].

2 Standard approaches

One standard approach for calculating the stopping power on an ion of charge Ze and (non-relativistic) velocity $\mathbf{v}$ is based on the linear response theory or dielectric description. In this treatment the stopping power $dE/ds = -Ze\hat{\mathbf{v}} \cdot \nabla\phi(\mathbf{r} = \mathbf{v}t)$ results from the electric field $\mathbf{E} = -\nabla\phi$ which is created by the induced target polarization and is here assumed to depend linearly on the perturbing potential $\phi_p(\mathbf{r}, t) = Ze/4\pi\epsilon_0|\mathbf{r} - \mathbf{v}t|$ of the moving ion. The corresponding relation in terms of the dielectric function ε reads in Fourier space $\phi(\mathbf{k}, \omega) = 2\pi Ze\delta(\omega - \mathbf{k} \cdot \mathbf{v})/\epsilon_0 k^2 \varepsilon(\mathbf{k}, \omega)$ and yields the linear response stopping power

$$\left(\frac{dE}{ds}\right)_{lr} = \frac{Z^2 e^2}{\epsilon_0 (2\pi)^3} \int d^3k \, \frac{\mathbf{k} \cdot \hat{\mathbf{v}}}{k^2} \, \text{Im}\left[\frac{1}{\varepsilon(\mathbf{k}, \mathbf{k} \cdot \mathbf{v})}\right]. \quad (1)$$

In the linear response stopping (1) many–body effects in the plasma, resulting e.g. in dynamic screening and the excitation of plasma waves, are properly taken into account through the dielectric function $\varepsilon(\mathbf{k}, \omega)$. The genuine validity of a linear response treatment is, however, restricted to the regime where a first order Born approximation can be applied. This requires for Coulomb interaction a small averaged Coulomb parameter $\langle \eta \rangle = |Z|e^2/4\pi\epsilon_0\hbar\langle v_r \rangle \ll 1$, where $\langle v_r \rangle$ represents an average over the ion-electron relative velocities $\mathbf{v}_r$. It can be easily checked that this condition is usually violated for highly charged ions in an electron cooler and for heavy ions in the last part of its range in a HIF-target where their remaining energy is relatively low.

A widely used approximation to extend the linear response expression (1) towards higher coupling is based on the introduction of appropriate cutoff parameters. In many cases the perturbation caused by the ion remains truly weak in the largest part of the ion–target interaction region and violates linear coupling only in a small vicinity of the ion $r < \langle b_0 \rangle \ll \lambda$, corresponding to high transfered momenta. Here $\langle b_0 \rangle = |Z|e^2/4\pi\epsilon_0 m \langle v_r \rangle^2$ is the classical collision diameter averaged over the ion-electron relative velocities. This suggest to restrict the k–integration in (1) by an upper limit $k_m \approx 1/\langle b_0 \rangle$, i.e.

$$\left(\frac{dE}{ds}\right)_{\mathrm{lr}}^{k_m} = \frac{Z^2 e^2}{\epsilon_0 (2\pi)^3} \int_{k<k_m} d^3k \, \frac{\mathbf{k}\cdot\hat{\mathbf{v}}}{k^2} \, \mathrm{Im}\left[\frac{1}{\varepsilon_R(\mathbf{k},\mathbf{k}\cdot\mathbf{v})}\right]. \qquad (2)$$

The cutoff parameter k_m has to be defined suitably in order to exclude the region $r < \langle b_0 \rangle$ of strong perturbation *and* to add remaining effects of the close collisions in the average. This can be achieved by comparison of the stopping power (2) with a binary collision approach (see below) using an appropriate screened potential. The details of this procedure, the resulting k_m and the whole concept of using cutoffs is outlined in [8] and the references therein.

The second standard approach describes stopping as the result of binary collisions. Starting from the momentum transfer $\Delta\mathbf{p}$ in an elementary ion-electron collision, the stopping power $dE/ds = \hat{\mathbf{v}} \cdot \langle \Delta\mathbf{p}/\Delta t \rangle$ is obtained by calculating the change of momentum per unit time as an average over all possible scattering angles and relative velocities, This leads to

$$\left(\frac{dE}{ds}\right)_{bc} = mn \int d^3v_r \, f(\mathbf{v}_r + \mathbf{v}) \, v_r \, \hat{\mathbf{v}} \cdot \mathbf{v}_r \, \sigma_{tr}(v_r), \qquad (3)$$

where m and n are the mass and density of the electrons, f is their velocity distribution and $\sigma_{tr}(v_r) = 2\pi \int_0^\pi d\vartheta \, \sin(\vartheta)[1 - \cos(\vartheta)] \, d\sigma(v_r,\vartheta)/d\Omega$ denotes the transport cross section for binary collisions. Since $(dE/ds)_{bc}$ (3) is based on a pure two-body approach the medium effects like polarization and screening must be approximated by an effective ion-electron interaction, e.g. $V_{ei}(r) = -(Ze^2/4\pi\epsilon_0)\exp(-r/\lambda)/r$. Here no restriction to $\langle\eta\rangle < 1$ arises as in the linear response theory, provided that σ_{tr} is evaluated properly e.g. from the scattering phase shifts δ_l for the potential V_{ei} obtained by solving the corresponding Schrödinger equation. The binary collision approach is, however, insofar incomplete as the effective interaction V_{ei} has to be supplied as an external parameter whose appropriate choice is the crucial issue. In a widely used simplification the required transport cross section for the screened effective interaction is approximated by a pure Coulomb scattering with a truncation of impact parameters larger than the screening length λ. This results in the appearance of a Coulomb logarithm $\ln\Lambda = \ln(\lambda/\langle b_0\rangle)$ in the final expressions. This type of approximation as well as the use of a cutoff in the linear response result, Eq.(2), can be successfully used as long as the argument of the Coulomb logarithm or $k_m\lambda$ are sufficiently large, that is, for

$$\frac{\langle b_0 \rangle}{\lambda} = \frac{|Z|e^2\omega_p}{4\pi\epsilon_0 m \langle v_r \rangle^3} \approx \frac{\sqrt{3}|Z|\Gamma^{3/2}}{(1+(v/v_{th})^2)^{3/2}} \ll 1. \qquad (4)$$

Here the screening length was estimated by $\lambda \approx \langle v_r \rangle/\omega_p$, the relative velocity by $\langle v_r \rangle \approx (v_{th}^2 + v^2)^{1/2}$ in terms of the ion velocity v and the thermal velocity of the electrons $v_{th} = (k_B T/m)^{1/2}$, while $\Gamma = (e^2/4\pi\epsilon_0 k_B T)(4\pi n/3)^{1/3}$ is the classical plasma parameter and ω_p is the plasma frequency. The parameter $\langle b_0 \rangle/\lambda$ or equivalently $|Z|\Gamma^{3/2}$ represents the ion–electron or projectile–target coupling parameter. Equation (4) defines weak coupling. Strong, nonlinear coupling corresponds to $|Z|\Gamma^{3/2} \gtrsim 1$. With increasingly nonlinear coupling, however, an improved theoretical treatment for the energy loss is required.

3 Combined model

Considerable progress in this direction has been achieved during the last years. Footing on ideas already introduced over thirty years ago by Gould and De-Witt [11] in the framework of kinetic theory, these improvements are mainly based on a specific combination of the linear response and the binary collision approach. This combined expression $(dE/ds)_{\text{com}}$ is composed by three terms through [6,5,8]

$$\left(\frac{dE}{ds}\right)_{\text{com}} = \left(\frac{dE}{ds}\right)_{\text{lr}} + \left(\frac{dE}{ds}\right)_{\text{bc}} - \left(\frac{dE}{ds}\right)_{\text{lr}}^0. \tag{5}$$

Here $(dE/ds)_{\text{lr}}$ represents the stopping power (1) which involves a fully dynamically screened interaction in linear response, $(dE/ds)_{\text{bc}}$ equals expression (3) where the transport cross section σ_{tr} is determined from a static, spherically symmetric effective ion-electron interaction $V_{\text{ei}} = -e\phi_{\text{ei}}(k)$ which is also used in the linear response of a *non-interacting* target

$$\left(\frac{dE}{ds}\right)_{\text{lr}}^0 = \frac{1}{(2\pi)^3} \int d^3k \; \hat{\mathbf{v}} \cdot \mathbf{k} \; |V_{\text{ei}}(k)|^2 \; \text{Im}[\chi_0(k, \omega(\mathbf{k}))]. \tag{6}$$

Here χ_0 is the free density-density response function. By construction the combined expression (5) guarantees the correct asymptotic behavior of stopping at high and low velocities [8]. Two different choices for V_{ei} have actually been considered. The first one [5,6] uses the static limit of the dynamically screened linear response interaction $\phi_{\text{ei}}(k,\omega) = \phi_p(k)/\varepsilon(k,\omega)$, i.e. $V_{\text{ei}}(k) = -Ze^2/\epsilon_0 k^2 \varepsilon(k, \omega = 0)$. In the limit of classical plasma targets this static interaction just represents the Debye-screened Coulomb potential

$$V_{\text{ei}}(k) = -\frac{Ze^2}{\epsilon_0} \frac{1}{k^2 + 1/\lambda_D^2} \quad \leftrightarrow \quad V_{\text{ei}}(r) = -\frac{Ze^2}{4\pi\epsilon_0 r} \exp(-\frac{r}{\lambda_D}), \tag{7}$$

where λ_D is the Deybe length.

For the second choice [8,12] of an effective interaction V_{ei}, a velocity dependent screening length $\lambda(v)$ is introduced and determined from linear response in the framework of the combined expression (5). Here the effective interaction

$$V_{ei}(r) = -\frac{Ze^2}{4\pi\epsilon_0 r}\exp(-\frac{r}{\lambda(v)}) \quad\leftrightarrow\quad V_{ei}(k) = -\frac{Ze^2}{\epsilon_0}\frac{1}{k^2 + 1/\lambda^2(v)}, \quad (8)$$

is employed to evaluate the contributions $(dE/ds)_{lr}^0$ and $(dE/ds)_{bc}$. To determine $\lambda(v)$ the linear response part $(dE/ds)_{lr} - (dE/ds)_{lr}^0$ is viewed as a correction to the binary collision term $(dE/ds)_{bc}$ which vanishes by demanding $(dE/ds)_{lr} - (dE/ds)_{lr}^0 = 0$, that is,

$$\left(\frac{dE}{ds}\right)_{lr} \stackrel{!}{=} \frac{1}{(2\pi)^3}\int d^3k\,\hat{\mathbf{v}}\cdot\mathbf{k}\left|\frac{Ze^2}{\epsilon_0(k^2 + 1/\lambda^2(v))}\right|^2 \text{Im}[\chi_0(k,\omega(\mathbf{k}))]. \quad (9)$$

Eq.(9) has to be resolved numerically for $\lambda(v)$. This concept aims to include dynamic effects by mapping the dynamically screened interaction contained in the linear response description into a spherically symmetric velocity dependent effective interaction. In both schemes for the combined expression $(dE/ds)_{com}$ no external parameter appears in contrast to all cutoff approaches and a pure binary collision treatment.

In Fig. 1 the different theoretical expressions are compared for the stopping power of a positively charged ion in a classical electron plasma as a function of the ion velocity and for different coupling parameters $|Z|\Gamma^{3/2}$. The dashed curves represent the combined expression (5) for the statically screened interaction (7), while the dotted curves show the linear response stopping power (2) with cutoff [8] $k_m = 2/\gamma\langle b_0\rangle$, where $\ln\gamma = 0.577\ldots$ is Euler's constant. The solid curves result from $(dE/ds)_{com}$ (5) with velocity dependent screening length $\lambda(v)$ provided from Eq. (9), that is, $(dE/ds)_{lr} - (dE/ds)_{lr}^0 = 0$. For the lowest displayed coupling strength $Z\Gamma^{3/2} = 0.11$ all expressions are almost identical except of small deviations around the maximum of the stopping power and at very low v. These deviations vanish quickly for still smaller coupling but strongly increase for growing coupling strengths, where we reach the fringe to the nonlinear regime at $Z\Gamma^{3/2} = 0.36$ and very low ion velocities, while for the highest shown coupling strength $Z\Gamma^{3/2} = 11.2$ nonlinear coupling with $\langle b_0\rangle/\lambda = 3^{1/2}|Z|\Gamma^{3/2}/(1 + v^2/v_{th}^2)^{3/2} \geq 1$ is expected for $v \lesssim 2.5 v_{th}$. Here the combined expression with $\lambda = \lambda_D$ (dashed) stays always above the approach with velocity dependent screening $\lambda = \lambda(v)$ (solid) with an increasing deviation at medium velocities for growing $Z\Gamma^{3/2}$. Both descriptions well agree in all cases at low and sufficiently high velocities. The linear response description with cutoff (dotted curves) considerably deviates at low velocities and for increasing coupling where it exhibits a drastically smaller stopping power. The presented combined approaches for the stopping power are, however, essentially based on the assumption that the collective, static as well as dynamic, screening effects, are well described by linear response. This can be justified only a posteriori and was checked by a comparison with simulation results.

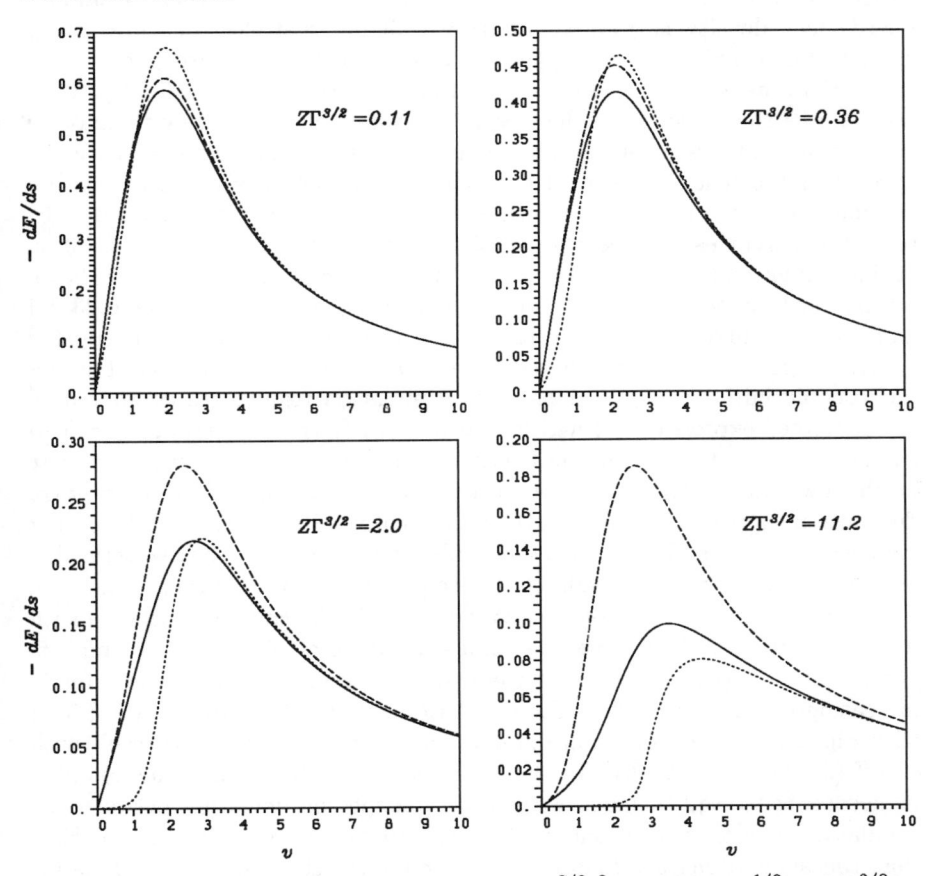

Figure 1. Normalized stopping power $dE/ds/(Z\Gamma^{3/2})^2$ in units of $3^{1/2}k_BT/\Gamma^{3/2}\lambda_D$ as function of the ion velocity v in units of the thermal velocity $v_{th} = (k_BT/m)^{1/2}$ for a positively charged ion at various $Z\Gamma^{3/2}$. The dashed curves represent the combined expression (5) for the screened interaction (7) with $\lambda = \lambda_D$. The dotted curves are the linear response stopping power with cutoff (2) and the solid curves exhibit the second version of the combined expression where the velocity dependent screening length $\lambda(v)$ from Eq.(9) is used for the screened interaction (8).

4 Comparison with numerical treatments

The stopping power on a heavy ion moving through an ideal target plasma of classical electrons can be described by the nonlinear Vlasov–Poisson equations. These are solved numerically by test–particle simulations, where a smooth phase-space density is approximated by splitting real electrons into many charged test–particles. Analogous to the well-known standard particle-in-cell (PIC) schemes the test–particles act as sources for a mean–field and the evolution of the phase–space density of the system is governed by the motion of the test–particles in this selfconsistent mean–field and, in our case, in the bare Coulomb field of the projectile-ion. An elementary cubic simulation volume, which is periodically continued in all directions, contains some 10^5 test-particles corresponding to about 500 real electrons. A detailed description of the simulation technique is given in [8]. Typical simulation results are shown by the filled diamonds in Fig. 2 for coupling strengths $Z\Gamma^{3/2} = 0.11, 0.23, 1.1$ and 4.6. To guide the eye the simulation results are fitted by the dotted curves. The errorbars are related to an ensemble averaging over different microscopic initial configurations. The simulation results are compared to the combined expression (5) with either $\lambda = \lambda_D$ (dashed curves) or $\lambda = \lambda(v)$ from Eq. (9) (solid curves) in the screened interaction V_{ei}. We first concentrate on the low and medium velocities up to $v/v_{th} = 3\ldots 4$. Here both theoretical descriptions agree very well with the numerical results for $Z\Gamma^{3/2} = 0.11$. For increasing coupling $Z\Gamma^{3/2} = 0.23$ the combined expression with fixed screening length $\lambda = \lambda_D$ provides to high values around the stopping power maximum. This trend proceeds for increasing coupling, while the combined scheme with $\lambda = \lambda(v)$ still excellently fits the simulation data at $Z\Gamma^{3/2} = 1.1$ and 4.6. For $Z\Gamma^{3/2} > 5$ (not shown) also the combined expression with $\lambda(v)$ finally increasingly fails to reproduce the simulation results. A comparison of Fig. 1 with Fig. 2 also reveals the pure prediction of the linear response result with cutoff (2) for $Z\Gamma^{3/2} > 0.1$. In the high velocity regime where the coupling is, according to Eq.(4), much weaker, an unexpected deviation between the simulations and the theoretical predictions appears. Here it is to be remarked that the simulations are inapplicable at high velocities when the dynamic screening length $\propto v$ exceeds the finite length L of the simulation box. To allow nevertheless some comparison at high v, one can incorporate the finite simulation box in the theoretical predictions [8] which results in the long-dashed curves in Fig. 2. They are in a similar good agreement with the combined scheme at high velocities as found for low and medium ones. The more or less pronounced humps which can be seen in the simulation results are also related to the finite size and periodicity of the simulation cube.

We now focus on the stopping power at low velocities and investigate the dependence on the ion charge Z. Here dE/ds becomes linear in v and can be parameterized in terms of a dimensionless friction coefficient $R(Z\Gamma^{3/2})$ as

$$\frac{dE}{ds}\frac{\Gamma^{3/2}\lambda_D}{\sqrt{3}\,k_B T} = -\left(\frac{v}{v_{\text{th}}}\right) R(Z\Gamma^{3/2})\,, \qquad v \ll v_{\text{th}}\,. \tag{10}$$

The stars in Fig. 3 show the $Z\Gamma^{3/2}$ dependence of R computed with test-particle simulations for positively charged ions. The solid curve is again the

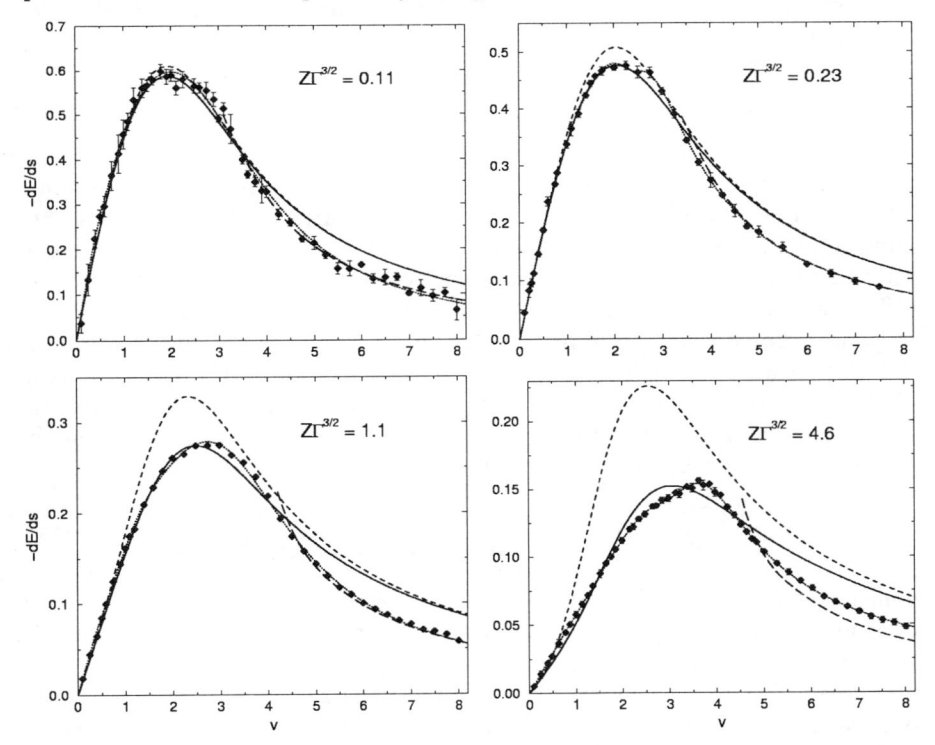

Figure 2. Normalized stopping power $dE/ds/(Z\Gamma^{3/2})^2$ in units of $3^{1/2}k_B T/\Gamma^{3/2}\lambda_D$ as function of the ion velocity v in units of $v_{\text{th}} = (k_B T/m)^{1/2}$ for $Z > 0$ and different $Z\Gamma^{3/2} = 0.11, 0.23, 1.1$ and 4.6. The filled diamonds with errorbars are the results of Vlasov-Poisson (test-particle) simulations, which have been fitted by the dotted curves. The dashed and the solid curves are respectively, as in Fig. 1, the predictions of the combined expression (5) for $\lambda = \lambda_D$ or $\lambda = \lambda(v)$. The long-dashed curves represent the high velocity limit of the solid curves when taking into account the finite size of the simulation box.

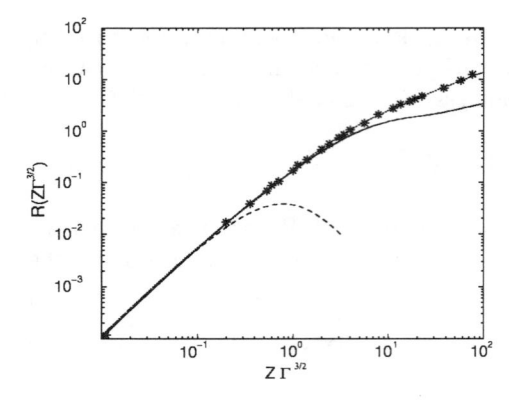

Figure 3. Dimensionless friction coefficient $R(Z,\Gamma)$ (10) as function of $Z\Gamma^{3/2}$. The dashed curve represents the linear response description with cutoff (2) and the solid curve the combined expression (5) for the screened interaction (8). The stars are the results of Vlasov simulations which have been fitted by the dotted curve, Eq. (11).

combined expression (5). It fits the simulation results quite well up to around $Z\Gamma^{3/2} \approx 5$ in agreement with the previous observations. The remaining differences at higher coupling parameters can be mainly explained by nonlinear screening effects [8] where the selfconsistent potential in the Vlasov calculations differs from the screened effective interaction (8) as determined by the linear response. The dashed curve represents the prediction of the linear response description with cutoff (2). For a given plasma parameter Γ the dependence of R on the coupling parameter $Z\Gamma^{3/2}$ just reflects the dependence of the low velocity stopping power on the charge state Z. It can be read off from the fit (dotted curve) through the simulation data

$$R \approx \frac{(Z\Gamma^{3/2})^2}{8} \left[\ln\left(1 + \frac{0.14}{(Z\Gamma^{3/2})^2}\right) + \frac{1.8}{1 + 0.4(Z\Gamma^{3/2})^{1.3}} \right], \quad (11)$$

which reveals a gradual decrease of the power $x = x(Z)$ of a local charge dependence like Z^x. Starting from a $R \sim Z^2 \ln(\text{const}/Z)$ behavior close to the linear regime, $Z\Gamma^{3/2} \ll 1$, the charge dependence changes towards the very weak $R \sim Z^{0.7}$ reached at the highest studied $Z\Gamma^{3/2} \gg 1$. Such a weaker dependence on Z has also been observed in experiments on electron cooling [9,10]. A comparison with the experimental data can be found in [8].

5 Conclusions

In agreement with experiments, the charge dependence of nonlinear stopping is considerably weaker than in the linear regime. The friction coefficient R scales locally like Z^x with $x = x(Z)$ ranging from $x \approx 2$ in the weak coupling regime $Z\Gamma^{3/2} \ll 1$ to $x \approx 0.7$ at strongly nonlinear coupling $Z\Gamma^{3/2} \gg 1$.

Standard approximations like the linear response stopping with cutoff are applicable only for $Z\Gamma^{3/2} \lesssim 0.1$, while the presented combined expression (5) with a velocity dependent effective interaction, i.e. $\lambda = \lambda(v)$ as provided from linear response by Eq. (9), gives very good predictions up to $Z\Gamma^{3/2} \leq 5$. At even higher coupling additional efforts are necessary to obtain a theoretical description of the stopping which also takes into account the nonlinear screening.

References

1. N. Bohr, Phil.Mag. **25** (1913) 10; Phil.Mag. **30** (1915) 581.
2. H. Bethe, Ann.Physik **5** (1930) 325.
3. Proceedings of the 12th Int. Symp. on Heavy Ion Inertial Fusion and the 8th Int. Workshop on Atomic Physics for Ion-Driven Fusion, eds. I. Hofmann and H.J. Bluhm, Heidelberg, Germany, Sept. 1997, Nucl.Instr. and Meth. A **415** (1998) Nos. 1-3.
4. Proceedings of the workshop on electron cooling and related topics, Uppsala, May 1999, ed: T. Bergmark, Nucl.Instr. and Meth. A **441** (2000) 1-304.
5. K. Morawetz and G. Röpke, Phys.Rev E **54** (1996) 4134.
6. D. O. Gericke, M. Schlanges and W.D. Kraeft, Phys.Lett. **222A** (1996) 241; Laser and Particle Beams **15** (1997) 523; D.O. Gericke and M. Schlanges, Phys.Rev. E **60** (1999) 904.
7. G. Zwicknagel, Nucl. Inst. and Meth. A **415** (1998) 680.
8. G. Zwicknagel, C. Toepffer and P.-G. Reinhard, Phys.Rep. **309** (1999) 117.
9. A. Wolf et al.,*Beam Cooling and Related Topics*, J.Bosser ed., CERN 94–03, Genf, 1994, p. 416.
10. Th. Winkler et al., Hyp.Int. **99** (1996) 277; Nucl.Instr.Meth in Phys.Res. A **391** (1997) 12.
11. H.A. Gould and H.E. DeWitt, Phys.Rev. **155** (1967) 68.
12. G. Zwicknagel, Nucl.Instr. and Meth. A **441** (2000) 44.

HEAVY-ION STOPPING: BOHR THEORY REVISITED

PETER SIGMUND

Physics Department, Odense University (SDU), DK 5230 Odense M, Denmark

ANDREAS SCHINNER

Inst. f. Experimentalphysik, J. Kepler-Universität, A-4040 Linz-Auhof, Austria

Niels Bohr's classical theory of charged-particle stopping from 1913 is known to be a more appropriate starting point for a theory of heavy-ion stopping than Bethe's quantum theory, but various obstacles have prevented its application in quantitative estimates. This paper provides an overview of recent progress achieved in estimating stopping forces for a wide range of ion-target combinations over a broad velocity range with standard input characterizing excitation spectra of the target and charge states of the projectile without use of adjustable parameters. Comments are made on the familiar effective-charge postulate, and a resolution is offered to a longstanding problem in stopping of channeled heavy ions.

1 Introduction

The range of scientific and technological problems which can be attacked successfully with beams of swift heavy ions is remarkable and still seems expanding. To go through any number of examples for the present audience would be like carrying sand into the Namib desert, but let me mention ion beam analysis of materials as a prime area of expertise of the group at Witwatersrand and its leader. That field is rapidly moving toward depth resolution in the atomic-monolayer range, and this has generated renewed interest in the classical problem of the energy loss of swift ions in matter, not the least heavier ions such as carbon, nitrogen and oxygen. It is the energy loss which determines the depth scale in analysis and hence needs to be known accurately for any improved experimental resolution to make an impact in practice.

Those in need of stopping data have become accustomed to look into Jim Ziegler's comprehensive tabulations which are updated continuously as new experimental data come along. However, considering $\sim 10^4$ elemental ion-target systems, a wide range of ion energies and a multitude of ionic charge states of potential interest, coverage with experimental data ranges from incomplete over scarce to nonexistent and will do so for many years to come. In other words, theory is indispensable.

While the theory of stopping for light particles such as electrons and protons is one of the most developed areas in atomic physics, pioneered by Bethe's master piece in 1930 and followed up by hundreds, perhaps thousands of more detailed studies, stopping of heavier ions from lithium to uranium is an area dominated by fitting and interpolating experimental data. This, at least, is the impression you get when scanning the current literature, and there is a good reason for it since there is a manifold of electronic processes contributing to stopping which are difficult to grasp theoretically, at least by any first-principles quantum theory.

On the other hand, those of us who have had an opportunity to listen to Jens Lindhard lecturing on particle penetration have all heard about Niels Bohr's kappa criterion which divides up particle stopping into a Bethe regime for $\kappa = 2Z_1 v_0/v \ll 1$ where the Born approximation is valid, and conversely a classical regime where the de Broglie wavelength is small enough compared to the Coulomb collision diameter to allow construction of wave packets following classical orbits. Here, Z_1 and v are the atomic number and speed of the projectile and v_0 the Bohr velocity. The limiting velocity is equivalent to an energy of $Z_1^2 \cdot 100$ keV/u. This implies that ions heavier than helium at not too high energies typically fall into the classical regime.

2 Bohr Theory

Bohr's classical stopping theory[1] divides up interactions between projectile and target electrons into close collisions obeying Rutherford's law and distant collisions that are treated as in classical dispersion theory. While superceded by Bethe's theory as a description of α-particle stopping the theory has made its way into textbooks because of the good physical insight it provides. Moreover, Bohr made use of his model[2] when estimating the range of fission fragments, but subsequent attempts at establishing scaling laws for heavy-ion stopping were based on Bethe's rather than Bohr's theory, ignoring the kappa criterion.

Figure 1 illustrates a good reason for this on the case of oxygen ions in aluminium which is here going to serve as a standard of reference because of an exceptionally good coverage with experimental data[3]. Writing Bohr's formula in the form

$$-\frac{dE}{dx} = \frac{4\pi Z_1^2 Z_2 e^4}{mv^2} N \sum_\nu f_\nu \log \frac{Cmv^3}{Z_1 e^2 \omega_\nu}, \qquad (1)$$

where ω_ν and f_ν are resonance frequencies and dipole oscillator strengths of the target, N the number density and $C = 1.1229$, we may identify $\sum_\nu f_\nu \log \omega_\nu = \log I/\hbar$ with the experimental value $I = 164$ eV for the mean excitation energy of aluminium. The top graph shows that Bohr's prediction is indeed superior to Bethe's for $E \gtrsim 0.5$ MeV/u, but at lower velocities it turns negative which is unphysical. The Bethe prediction, on the other hand, overestimates the stopping force for $E \gtrsim 0.1$ MeV/u. That feature is commonly ascribed to screening by projectile electrons and adjusted empirically. We shall come back to this point.

Inspection of Bohr's theoretical procedure reveals application of an asymptotic expansion of Bessel functions which breaks down for heavy ions at low velocities. Avoiding that expansion[4] replaces the logarithm in eq. (1) by a positive-definite function and leads to the second graph in figure 1 which still underestimates the stopping force significantly at low speed. The target has here been characterized by just one single resonance frequency $I/\hbar$. Replacing this by an excitation spectrum of three resonances for the principal shells of aluminium yields the third graph. The theory now overestimates stopping, but this is to be expected for a completely stripped projectile. However, the necessary screening correction now appears significantly reduced in comparison to what would be expected from the Bethe curve in the top graph.

Further refinement requires modifications and generalizations of the Bohr model which we have performed recently.

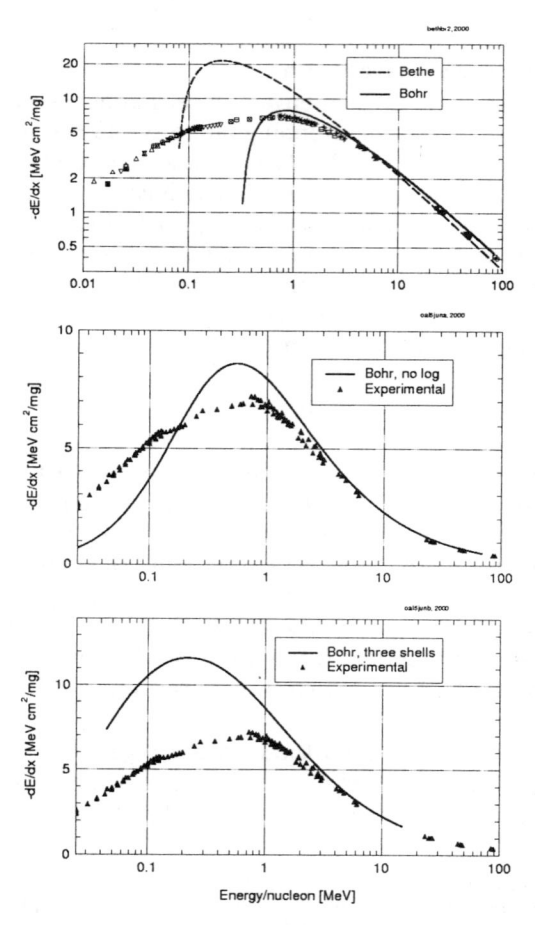

Figure 1. Stopping of oxygen in aluminium. Top: Straight Bohr and Bethe; Middle: Asymptotic expansion avoided; Bottom: Three target resonances. Experimental data (15 sets) compiled by H. Paul.

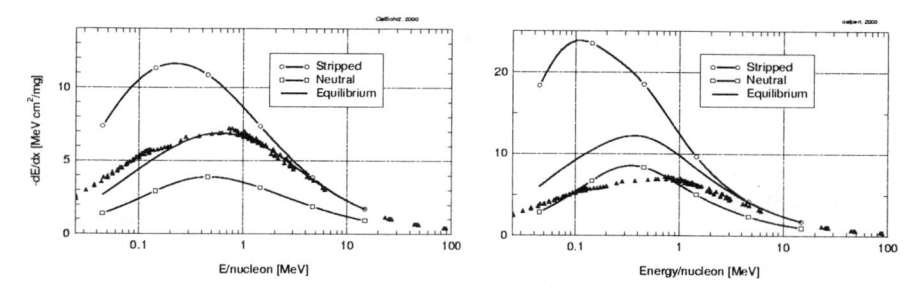

Figure 2. Stopping of oxygen in aluminium. Left: Bohr model including screening; Right: Same including Z_1^3 perturbation.

3 Modified Bohr Theory

Inclusion of screening due to electrons bound to the projectile turns out to fit well into the analytical framework of the Bohr model[5]. The Coulomb interaction has been replaced by a partially-screened Yukawa-type potential governed by a suitable screening radius a. On the basis of the Thomas-Fermi-Amaldi model an approximate expression $a = a_{\mathrm{TF}}(1 - q_1/Z_1)$ was derived, where a_{TF} is the Thomas-Fermi radius of a neutral projectile atom and q_1 the ionic charge. Inclusion of these features into the Bohr model delivers charge-dependent stopping forces. The stopping force in charge equilibrium can then be determined approximately by insertion of a mean equilibrium charge in standard Thomas-Fermi form,

$$\langle q_1 \rangle = Z_1 \left(1 - e^{-v/Z_1^{2/3} v_0}\right). \qquad (2)$$

Figure 2 (left graph) shows surprisingly good agreement with experimental data. Predictions for stripped and neutral oxygen projectiles have been included for comparison.

We could perhaps have stopped at this point[6], but in addition to screening several other low-speed corrections are known which cannot just be ignored. The first is the Z_1^3 or Barkas effect which depends on the sign of the charge and may become quite sizable even for lighter projectiles. This correction is conventionally found from the next term in a perturbation series in terms of the interaction potential. We have evaluated[7] the Z_1^3 correction in accordance with well-established schemes but allowing for projectile screening. The result in figure 2 (right graph) shows a ~ 100 % correction to the leading contribution! This suggests that the good agreement found in the lowest order was fortuitous. The obvious conclusion is that for heavy ions, perturbation theory has to be abandoned as a quantitative tool.

4 Binary Theory

To explain how we arrived at a non-perturbative theory[8] we need to discuss a few specifics of the Bohr theory, going back for a moment to unscreened Coulomb interaction. In distant interactions, target electrons are assumed to be bound harmonically, and this is seen to limit the range of the Coulomb interaction to the so-called adiabatic radius $a_{\rm ad} = v/\omega_\nu$. Hence the effect of binding is equivalent with a kind of screening of the interaction. It now turns out that for the Bohr model the effect of harmonic binding on the Coulomb interaction can be modelled *rigorously* by a Yukawa potential with screening radius $a_{\rm ad}$. Specifically, Bohr's result for the kinetic-energy transfer in distant collisions may be found from free binary scattering by adoption of such a screened potential.

However, also potential energy is transferred to a harmonically-bound electron. In order to model this effect in binary scattering we need the distance r_0 from the origin (figure 3 top). That distance was found by comparing the

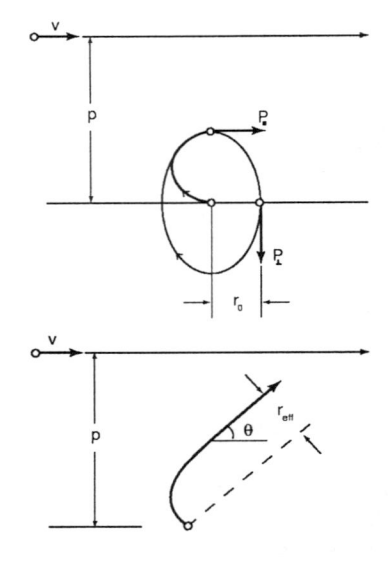

Figure 3. Orbit of Coulomb-excited electron in Bohr model (top) and in equivalent binary model (bottom); the same angular momentum is transferred in distant collisions; the momentum arm turns out to equal the distance determining the transfer of potential energy[8].

angular-momentum transfer in the two geometries (figure 3). Skipping some details here we may summarize that the essentials of the Bohr theory are contained in a binary scattering problem which may be solved rigorously. We have explored the properties of this model in detail and found that higher-order Z_1 corrections are described accurately in the limit where they are small, and results obtained for large perturbations are more credible than those found by perturbation expansion.

At this point we incorporated Thomas-Fermi screening again by means of a combined screening radius $(1/a^2 + 1/a_{\rm ad}^2)^{-1/2}$ and evaluated stopping

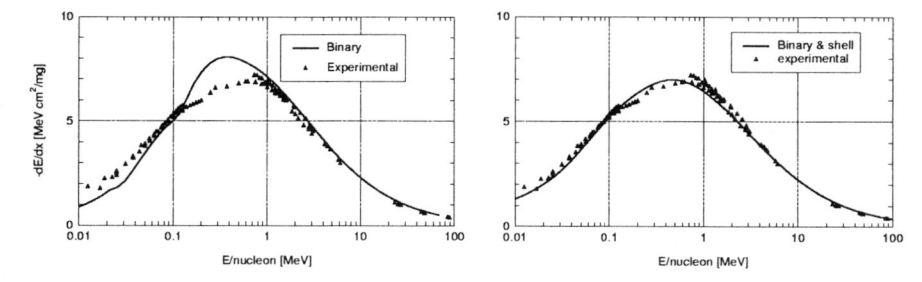

Figure 4. Stopping force for oxygen in aluminium predicted by binary model without (left[8]) and with (right[11]) shell correction.

forces. Results are shown in figure 4 (left graph). The agreement with the data is almost as good as in figure 2 (left), but you may ask where the Z_1^3 effect has gone. Well, it is still there around the stopping maximum, although much smaller than in figure 2, but at low velocities the prediction of the binary model falls slightly *below* that of the Bohr theory. The solution of this seeming paradox lies in the division into close and distant collisions that is necessary in the Bohr model. At low speed there is a broad interval where neither the dipole approximation nor Rutherford's law are valid. This gives rise to a considerable error margin in the unmodified Bohr theory[8]. That problem has been eliminated in the binary theory where a division into regimes is avoided altogether.

Another wellknown low-speed effect is the shell correction which accounts for internal electron motion in the target. A rigorous evaluation of this correction is a nontrivial matter in both classical and quantal stopping theory[9], but in binary scattering it reduces to a transformation between moving reference frames (kinetic theory[10]). Insertion of atomic velocity distributions into this transformation yields the result shown in figure 4 (right). Here the agreement with the experimental data has come within experimental error[11].

The above results all refer to the classical regime. In order to extend the range of validity of the theory into the Bethe regime we applied what we call an 'inverse Bloch correction'[4]. The standard Bloch correction, wenn added to the Bethe logarithm, leads to the Bohr logarithm in the low-velocity limit. The inverse Bloch correction serves the same purpose in the reverse direction. This is important for lighter ions. Figure 5 shows two examples. The agreement with experiment is quite good but can most likely be improved by replacing eq. (2) by a more accurate expression for the mean charge.

There are a few open ends in the theoretical description as it stands

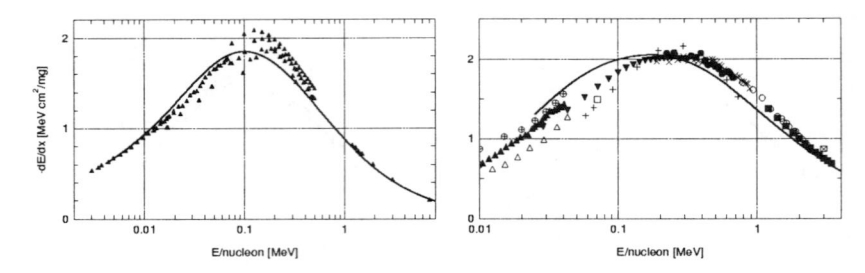

Figure 5. Stopping force on He in C (left[14]) and Li in Al (right) including inverse Bloch correction. Experimental data compiled by H. Paul.

which we are currently trying to straighten up. We have not yet allowed for projectile processes such as energy loss by charge exchange and projectile excitation. More important is improved input for the equilibrium charge eq. (2) which is known to show a gas-solid effect that needs to be accounted for. For the lightest projectiles it may be necessary to separate the stopping force into weighted contributions from individual charge states. On the other hand we have not so far found it necessary to characterize target excitation by anything going beyond the rather corse spectra distinguishing merely between the principal target shells. We emphasize that one and the same procedure has been utilized for all graphs shown over the entire velocity range, and that the use of adjustable parameters has been avoided.

5 Screening and Effective Charge

The effective charge is a concept that you will inevitably meet in the literature on heavy-ion stopping, originally designed[12] to generate unknown stopping forces on heavy ions by scaling known stopping forces on hydrogen ions at the same velocity, i.e.,

$$\left(\frac{dE}{dx}\right)_{Z_1} = (\gamma Z_1)^2 \left(\frac{dE}{dx}\right)_{Z_1=1}. \qquad (3)$$

Here the velocity-dependent effective-charge fraction γ, found by interpolation between experimental data, is supposed to account for projectile screening. Figure 6 (left) shows γ calculated by the binary theory for argon on aluminium, albeit with helium as a reference instead of hydrogen which is one of two alternatives found in the literature[13]. We show argon here instead of oxygen in order to enhance screening.

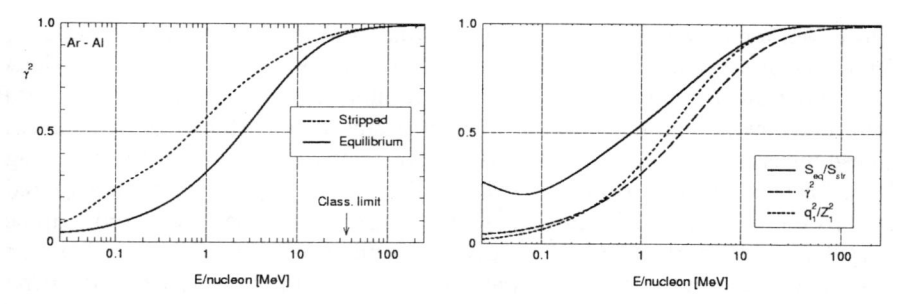

Figure 6. Left: Calculated effective-charge fraction on argon in aluminium, defined with reference to helium instead of eq. (3)[14]. Solid line: Both ions in charge equilibrium; dotted line: Both ions stripped. Right: Calculated equilibrium stopping force on argon in aluminium normalized to stopping force on stripped ion S_{eq}/S_{str} (solid curve). Also included are the square of the equilibrium charge fraction, $\langle q_1 \rangle^2/Z_1^2$ and the effective-charge fraction from the figure to the left.

The solid curve in figure 6 (left) shows the characteristic S-shape known from tabulations of experimental stopping data, but it is striking that the same qualitative shape is found by plotting the ratio of the two stopping forces for completely stripped ions, i.e., in the absence of projectile screening[14]. As a matter of fact, the two curves merge into each other before merging into unity at high speed. Evidently the qualitative shape of the two curves is unrelated to screening but must reflect the transition from the Bethe to the Bohr regime. The upper end of the 'S' is defined by $\kappa = 1$ for argon (marked in the graph).

The actual effect of projectile screening is illustrated in figure 6 (right) which primarily shows the ratio between stopping forces on dressed and stripped ions. It is seen that this curve does not approach zero at low speed, reflecting the fact that also neutral projectiles undergo electronic stopping via close collisions. This graph illuminates the general experience that effective charges determined from experimental data do not exhibit the scaling properties of the ion charge as expressed by eq. (2). In the literature this has led to a distinction between ion charge and stopping charge[15]. A more appropriate implication would seem to abandon the effective charge as a tool in ion stopping.

6 Frozen-Charge Paradox in Channeling

Finally we want to comment on a problem in channeling of heavy ions that has persisted for over 20 years. In a series of classical experiments Datz and coworkers[16], and later Golovchenko and coworkers[17] measured stopping forces on heavy ions in crystals under planar or axial channeling conditions as a function of charge state. This is possible for channeled ions because charge states are largely frozen over the range of foil thicknesses involved. These measurements revealed a fairly accurate q_1^2 dependence of the stopping force for several ions in both gold and silicon. The experiments were geared toward elucidating the Barkas (Z_1^3) effect which was expected to show up as a q_1^3-dependent correction. The puzzling result was that deviations from q_1^2 scaling, when found at all, either were negative instead of positive or increasing with energy instead of decreasing.

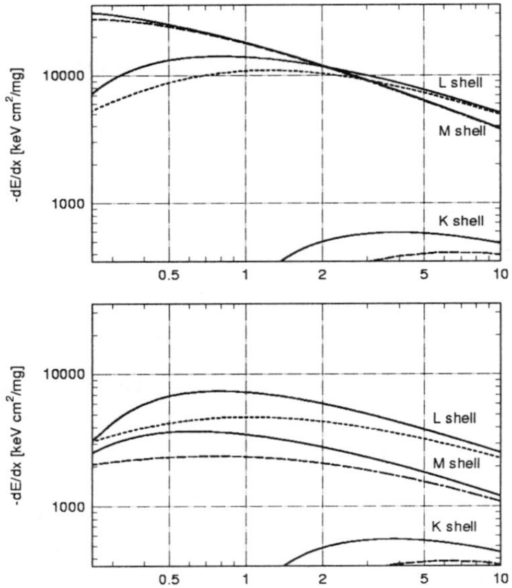

The top graph[18] in figure 7 shows that random stopping of Cl^{17+} in silicon is predominantly due to target L and M electrons while the Barkas correction increases dramatically from the M to the K shell. The bottom graph shows that for a neutral ion the contribution from the M shell has dropped while that of the K shell has remained constant. Thus, the significance of the Barkas correction increases with increasing screening. Since the two effects have opposite signs the net result depends on their relative magnitude. Figure 8 shows that for channeling the contribution to stopping from the K shell is negligible, causing an overall reduction of the Barkas correction. This suggests screening to dominate, resulting in a seemingly negative Barkas effect. Figure 8 also shows a rapid increase in stopping due to L electrons for stripped ions. This is the likely causes of an *increasing* Barkas correction with increasing velocity.

Figure 7. Stopping of chlorine in silicon[18]. Stripped ion in random target; dotted lines: Bohr theory; solid lines: Corrected for Barkas effect; Bottom: Same for neutral ion.

This work has been supported by the Danish Natural Science Research Council. Thanks are due to H. Paul for his continuous interest in this project.

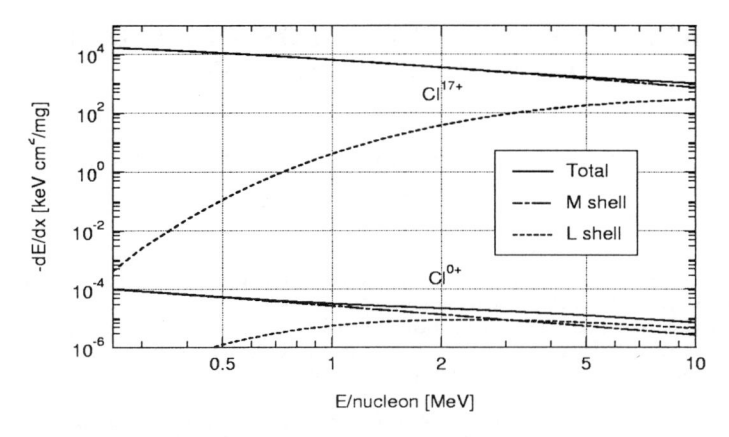

Figure 8. Stopping of chlorine in silicon[18]: Stripped and neutral ion channelled axially ($\langle 110 \rangle$); uncorrected for Barkas effect.

References

1. N. Bohr, Philos. Mag. **25**, 10 (1913).
2. N. Bohr, Phys. Rev. **59**, 270 (1941).
3. H. Paul, *Stopping power for light ions*, www2.uni-linz.ac.at/fak/TNF/atomphys/STOPPING/welcome.htm.
4. P. Sigmund, Phys. Rev. A **54**, 3113 (1996).
5. P. Sigmund, Phys. Rev. A **56**, 3781 (1997).
6. H. Paul, A. Schinner, and P. Sigmund, Nucl. Instrum. Methods B **164-165**, 212 (2000).
7. A. Schinner and P. Sigmund, Nucl. Instrum. Methods B **164-165**, 220 (2000).
8. P. Sigmund and A. Schinner, Europ. Phys. J. D **12**, 425 (2000).
9. P. Sigmund, Europ. Phys. J. D **12**, 111 (2000).
10. P. Sigmund, Phys. Rev. A **26**, 2497 (1982).
11. P. Sigmund and A. Schinner, Phys. Scr. in press.
12. L. C. Northcliffe, Ann. Rev. Nucl. Sci. **13**, 67 (1963).
13. R. Bimbot et al., Nucl. Instrum. Methods **153**, 161 (1978).
14. P. Sigmund and A. Schinner, Nucl. Instrum. Methods B, in press.
15. W. Brandt and M. Kitagawa, Phys. Rev. B **25**, 5631 (1982).
16. S. Datz et al., Phys. Rev. Lett. **38**, 1145 (1977).
17. J. A. Golovchenko et al., Phys. Rev. B **23**, 957 (1981).
18. P. Sigmund and A. Schinner, Phys. Rev. Lett. 86, 1486 (2001).

RADIATION EFFECTS MICROSCOPY AND CHARGE TRANSPORT SIMULATIONS

K.M.HORN, D.S.WALSH, P.E.DODD and B.L.DOYLE
Sandia National Labs, Albuquerque, NM, USA

The passage of energetic ions through semiconductor material results in the creation of electron-hole pairs that can introduce electrical charge into sensitive nodes of electronic devices. This charge, if of sufficient magnitude and duration, can cause the transient or permanent failure of a device. The measurement and modeling of charge collection processes is vital for validating predictive design simulation codes. Radiation effects microscopy is used to experimentally replicate the conditions that are simulated by three dimensional charge transport calculations of a circuit's response to a single ion strike at a specific circuit location. This paper describes two radiation microscopy techniques used in studying integrated circuits, the effect of ion-induced damage on these techniques, and the use of these techniques in validating three dimensional charge transport simulations. Examples of simulation verification are shown for the TA788 16K SRAM; this device is a 16K test version of the radiation hardened SA3953 256K SRAM fabricated using 0.5 micron design rules i1n Sandia's CMOS 6R process.

1. Nuclear Microprobes and Radiation Effects Microscopy

Over the last decade, focused, ion microbeams have been used for radiation testing at labs in Japan, Europe, Africa, Australia and the United States. A conceptual rendering of a microprobe-based test system is illustrated in Figure 1. An energetic beam of ions is magnetically focused to a sub-micron-sized spot at the surface of a sample. (The world record for the smallest beam spot is currently held by the University of Leipzig with a spot size of roughly 40 nm.) The focused ion beam can be rastered across a region of the sample or directed to a specific spot. The use of ion fluences of several hundred ions/s or less permit the effect of each ion to be measured and analyzed individually.

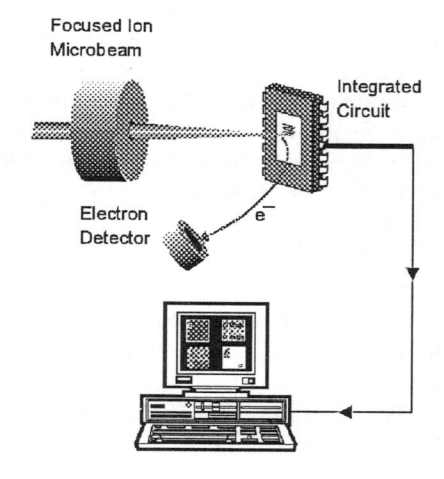

Figure 1. Schematic drawing of microbeam radiation testing.

As ions penetrate into the circuit, they lose energy, creating electron-hole pairs; this is the underlying cause of single event effects (SEE). A single ion can potentially deposit enough charge at a sensitive node within a circuit to cause a change of logic state (single event upset – SEU), an inability to change state (single event latchup – SEL), or even a destructive power short (single event burn-out – SEB). In the case of SEU, the total area of the circuit over which an ion strike can cause upset, measured in units of cm^2, is called the upset cross-section.. Linear Energy Transfer, (LET), measured in units of $MeV/mg/cm^2$, is the amount of energy deposited by the incident ion per unit track length. A circuit is 'hardened' to radiation by reducing its upset cross section, increasing the threshold LET at which an incident ion can cause SEE, or both. Three dimensional charge transport simulations of circuit response to ion strikes are used to evaluate and predict the radiation hardness of circuit designs. Experimental measurements of upset cross sections and threshold LETs for single event effects are used to verify radiation hardness specifications.

1.1 Single Event Upset Imaging

In October of 1990, the nuclear microprobe was first used to directly image single event upsets (SEU) in SRAMs fabricated with 1.25 micron technology []. It has since been used to measure and image upsets in DRAMS, EEPROMS, buffers, and other semiconductor devices. Prior to this application to the study of radiation effects, nuclear microprobes had predominantly been used for analytical measurements of the elemental composition of micron-sized regions of various sorts of targets (biological, mineral, man-made, etc.). In the field of radiation effects, researchers had previously used apertured systems to localize the exposure of integrated circuits (IC) to ionizing radiation [,,]. By focusing rather than aperturing the incident ions, higher ion fluences are obtained than with apertured systems and it is possible to perform two-dimensional scans quickly and with flexible control of the scan area. Upset cross sections can also be measured directly from the upset-image, (such as Figure 2), rather than inferred from the statistics of whole-die exposures. By processing the data live-time, the experimenter can directly image and identify circuit structures

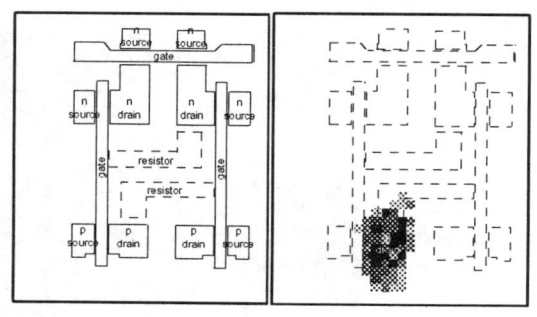

Figure 2. The circuit layout for a single memory cell of the TA670 16K SRAM is shown next to the corresponding single event upset image recorded using 30 MeV Cu.

susceptible to upset – thus the whole system acts as a sort of *radiation microscope* to "view" upsets and even the underlying charge collection occurring at different circuit structures.

We have published descriptions of the operation of the upset imaging system elsewhere [6] and shall only briefly summarize here. Once the ion beam has been focused, the size of the scan is calibrated by imaging TEM (transmission electron microscope) grids of known dimensions and pitch. The full charge generation of the incident ions is then measured in a fully depleted silicon *p-i-n* diode whose depletion depth exceeds the range of the incident ion. This measurement is used to calibrate the signal electronics (charge sensitive pre-amp, amplifier and digitizer) for subsequent, quantitative, charge collection measurements. In order to measure the circuit's functional response to the incident ions, two computers act in tandem to record the upset image. One computer controls the positioning of the focused ion beam and the dwell time of the beam at each pixel of the xy-scan while a second computer exercises the target circuit and 'notifies' the first computer whenever a change of logic state is detected. The first computer then records the X and Y position of the beam, the occurrence of the upset, and the elapsed exposure time for the measurement. The resulting data file constitutes an "event file" which records the position and time dependence of each upset occurrence. This event file can be rendered as an upset image, such as shown in Figures 2.

1.2 Ion Beam Induced Charge Collection Imaging

The underlying physical cause of single event upset is the collection of electrical charge at sensitive nodes of the integrated circuit. This charge is produced by creation of electron-hole pairs as the incident ion loses energy while penetrating the integrated circuit. Measurements of charge collection using apertured ion exposures were previously done in the 1980's and 90's [,.]. In these measurements, a charge sensitive pre-amplifier is connected to the V_{DD} or V_{SS} pins of an integrated circuit. The network of metalizations that bias the circuit structures of the device are in this way also used to collect charge from the structures. Breese *et al.* first applied scanned, focused ion microbeams to the imaging of charge collection within integrated circuits [], using this same methodology. This application constitutes another form of *radiation microscopy* for directly viewing charge collection magnitudes within a device in almost real-time. When

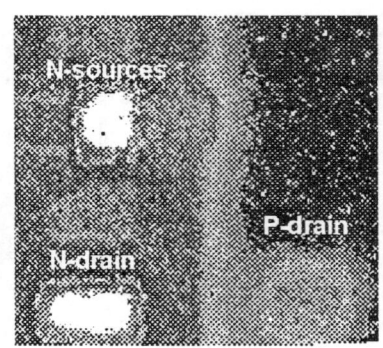

Figure 3. IBICC image of two memory cells in a TA670 16K SRAM.

combined with SEU-Imaging, these two techniques yield complementary information: (1) How much charge is collected at a specific site **and** (2) does that charge cause a circuit malfunction. In the charge collection image shown in Figure 3, (where lighter pixels correspond to higher charge collection), the N-drain and N-sources exhibit the highest charge collection within the memory cell, while from Figure 2 it is seen that it is the p-drain which actually causes circuit upset when exposed to 30 MeV Cu ions.

2.0 Beam Induced Damage Effects

Unlike an optical microscope, the use of the ion microprobe as a radiation microscope is not totally benign. The ions incident on the circuit, which produce the upset or IBICC signals, also produce displacement damage resulting in defects within the device itself.

2.1 Damage Effects During Charge Collection Imaging

The defects created during the irradiation of a device can act as trapping sites for mobile carriers and greatly influence the amount of charge collected into a sensitive node. This effect is explicitly shown by a sequence of IBICC images collected from an isolated test FET structure on the TA670 16K SRAM shown in Figure 4. The scan size of these IBICC images is 40x40 microns; the p-drain (denoted with the dashed white line in the first and last panels) is 12x5 microns.

0 ions/pixel   50 ions/pixel   100 ions/pixel   150 ions/pixel

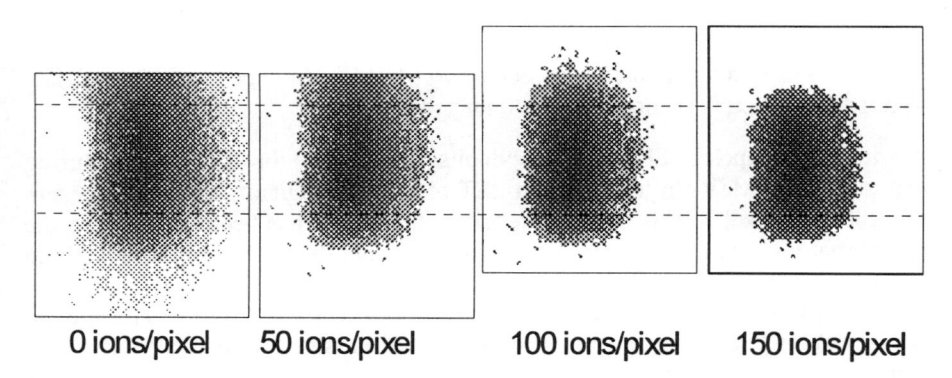

Figure 4. Displacement damage from 14 MeV C exposure reduces charge collection into an isolated FET test structure after only 50 ions/pixel. The p-drain of the FET is outlined with a white dashed line in the first and last panels. The p-drain was centered within the scan area after the second panel.

A maximum of 200 14 MeV C ions were delivered to each pixel for the measurement shown in Figure 4. The effect of the accumulated ion damage on charge collection is apparent in the gradual decrease of the region from which charge is collected into the centrally located p-drain. This effect can be successfully modeled as a decrease in the effective diffusion length for charge collection into the drain as displacement damage, and hence trap sites, accumulate over the entire ion exposure area [].

2.2 Damage effects during upset imaging

The effect of ion induced damage on SEU measurement is most clearly illustrated in the very initial stages of exposure, as shown in Figure 5. In this instance, the

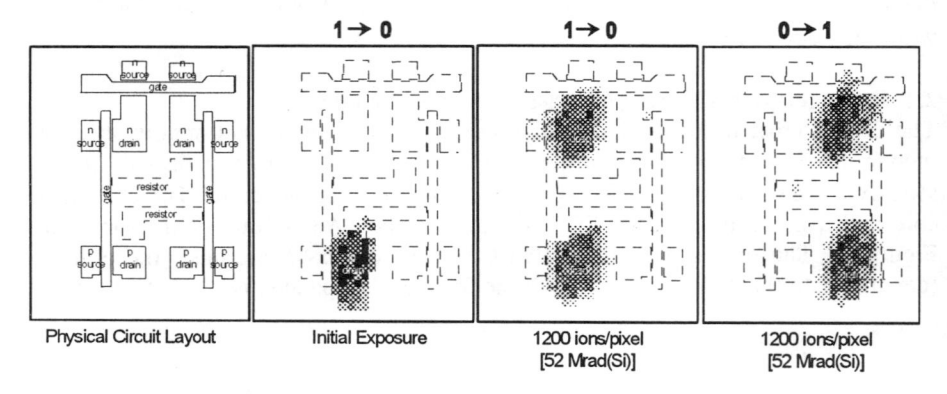

Figure 5. Displacement damage caused by 30 MeV Cu ion exposure increases the measured upset cross section after 1200 ions/pixel.

previously unexposed SRAM cell exhibits upsets only in the p-drain during exposure to 30 MeV Cu ions with an LET of 27 MeV/mg/cm^2. After an exposure of 1200 ions/pixel in this 64x64 pixel scan, the device has been damaged to the extent that ion strikes to the corresponding n-drain of the SRAM cell also begin to produce upsets. This condition also exists in the opposite logic state, as shown in the right-most panel. The dose of 1200 ions/pixel over the 40x40 square micron scan range of this measurement corresponds to a massive 52 Mrad(Si) total dose! This underscores the fact that even using a beam current of only 0.5 fA (~3000 ions/sec), the very small exposure area of the incident focused microbeam results in a very large local total dose damage effect. Consequently, we routinely use ion fluences of only 100 ions/sec or less and limit exposures to the minimum required to make a measurement.

3.0 Simulations of Charge Collection

As shown in the preceding sections, energetic particles incident on an electronic device create both electron-hole pairs and displacement damage as they transit the device. Electronic systems exposed to the radiation environments of space are bombarded by highly energetic particles from solar, galactic and intergalactic origins. Laboratory testing, using broad area ion exposures, can determine whether a device will continue to function under the irradiation of their intended space environment and over their intended mission duration.

Computer modeling of the three dimensional charge transport that occurs within the device during passage of an energetic ion through the circuit helps designers to understand the complex interplay among device design, operation, and the introduction of charge into the device at arbitrary locations due to irradiation. The detailed calculation of charge transport within a semiconductor structure is performed using finite-element charge transport calculations that solve Poisson's equation and the continuity equation for each volume element comprising the

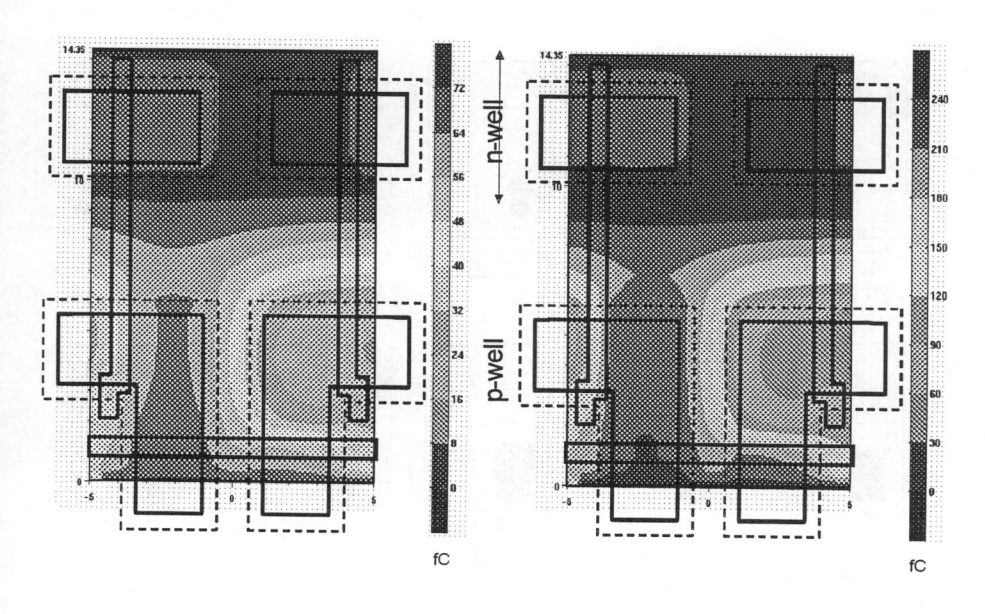

3 MeV Helium   20 MeV Carbon

Figure 6. Charge collection simulations for 3 MeV and 20 MeV Carbon ion exposures of the TA788. No upsets occur for single ion strikes of either ion. (N.B. A different numeric scale is used for each charge collection color scale.)

modeled circuit structure. Using these results, full-scale circuit simulations are performed in order to determine the effect of a specific ion strike at a specific circuit location on overall circuit operation.

3.1 Davinci 3D Charge Transport Calculations of the TA788

Shown in Figure 6 are the results of charge transport simulations for 3 MeV He and 20 MeV C ion exposures of one memory cell of the TA788 SRAM. The integrated values for the time-dependent current transients calculated throughout the memory cell are rendered with a color scale to produce a charge collection image comparable to the microbeam-based IBICC image. The color scale for both images is calibrated in units of femtoCoulombs of collected charge. Note that the charge collection scale for the 20 MeV C exposure is approximately three times greater than that for the 3 MeV He exposure in order to accommodate the greater charge generated by the larger, more energetic, C ion. In the case of both ions, the p-on and n-off drains collect the greatest amount of charge when struck by the ion. However, full circuit simulations of the memory cell predict that neither 3 MeV He nor 20 MeV C generate sufficient electrical charge from single ion strikes to

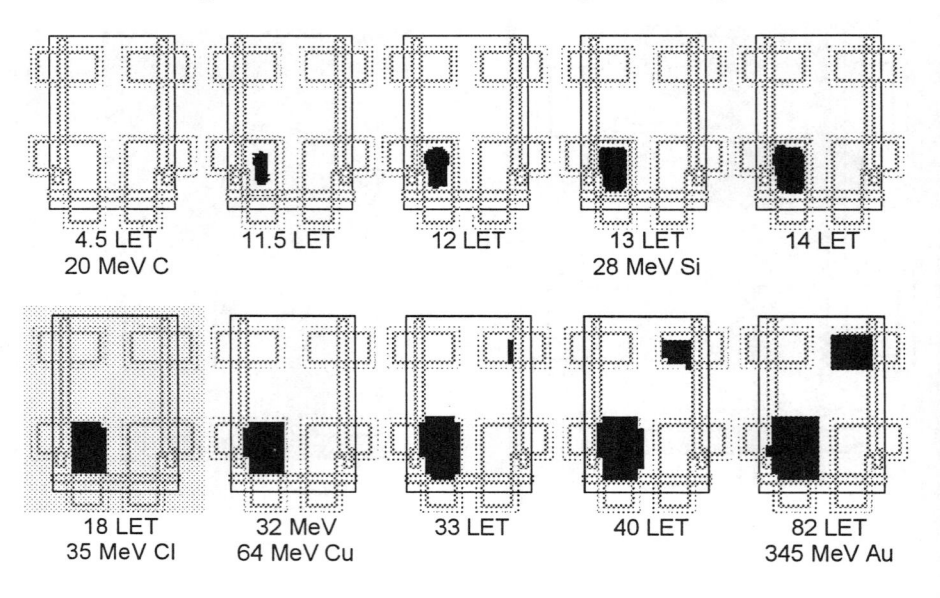

Figure 7. Progression of expanding upset regions calculated with Davinci simulations of the TA788 memory cell for increasing values of linear energy transfer (LET). Upsets occur first in the n-off drain at LETs above 11.5 MeV/mg/cm^2. Above 33 LET the p-off drain begins to also upset.

induce upset in the SRAM; this is confirmed by both ion microbeam and broad area ion exposures.

3.2 Simulations of Single Event Upset

The occurrence of simulated upset in the SRAM cell is determined by introducing a time-dependent current transient at the site of the ion strike into a circuit-level simulation of the SRAM cell. (Recall that the time-resolved current transient is calculated by a finite element calculation within a confined region of the circuit, e.g. drain, source). Both the magnitude and duration of the collected charge transient are important in determining whether circuit malfunction occurs. For any given memory cell, the magnitude of the collected charge must be greater than a critical amount, (Q_c), and persist for a time longer than the characteristic feedback time of the memory cell. The magnitude of the ion-generated charge

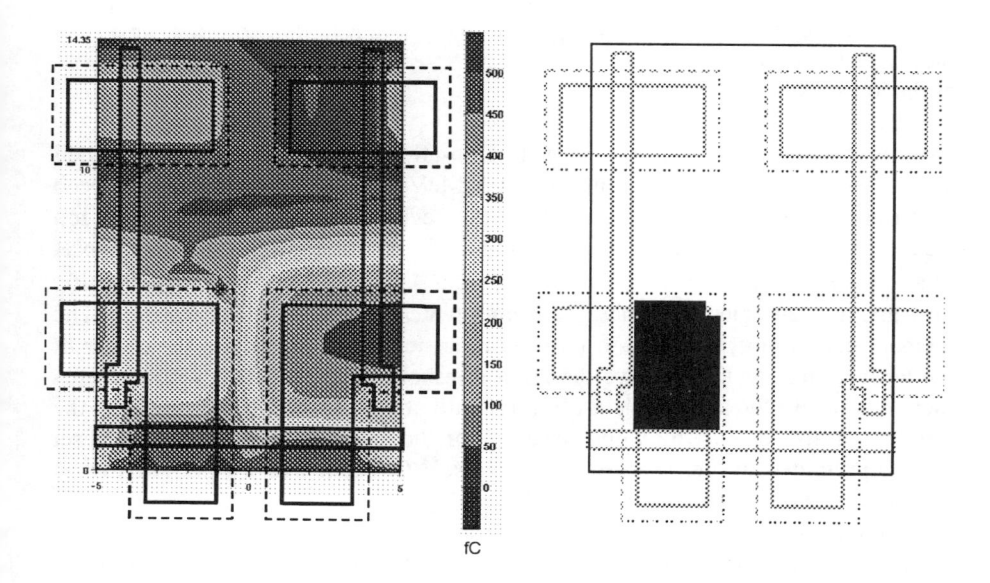

35 MeV Chlorine   **35 MeV Chlorine**

Figure 8. Davinci simulation of charge collection measured from V_{DD} (left) and ion impact sites that induce upset (right). The exposure sites producing upset are shaded in black; all fall within the boundary of the n-off drain.

varies greatly, depending on the ion and energy; this effect is visible by comparing the collected charge magnitudes in figures 6 and 8. It can also be seen in figures 6 and 8 that the amount of charge collected by the circuit varies significantly depending upon the exact location of the ion strike. Simulations indicate that single event upset occurs in the n-off drain of the TA788 memory cell at a threshold LET of approximately 11.5 MeV/mg/cm^2. This result has been confirmed by broad beam upset testing of the device. The effect of increasing ion LET on upset cross section is shown in Figure 8. In this figure, a simulated upset image is generated for ion LETs ranging from 4.5 to 82 MeV/mg/cm^2. Specific ions and ion energies are listed for beams typically used in upset testing. The memory cell is most sensitive to upset from an ion strike directly to the center of the n-off drain. At LETs above the upset threshold, the upset cross sectional area increases within the susceptible n-off drain until at an LET of 33 MeV/mg/cm^2, the p-off drain also begins to exhibit upset. For the experimental microbeam upset images presented in the next section, a 35 MeV Cl beam with an LET of 18 was used and only n-off drain upsets were observed.

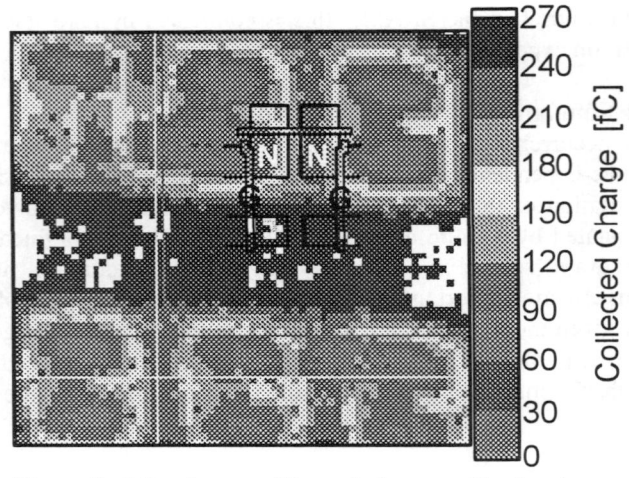

Figure 9. Microbeam calibrated charge collection image measured with 20 MeV carbon from the V_{DD} lines of the TA788.

The simulated IBICC and upset images for a 35 MeV Cl ion exposures are shown in Figure 8. The region shaded black in the upset image denotes those locations that produce circuit upset when struck by 35 MeV Cl ions.

4.0 Radiation Microscopy Measurements of the TA788

Precise simulation of the radiation response of an integrated circuit requires accurate knowledge of the as-fabricated circuit design parameters, (e.g. dopant concentrations and profiles, feature boundaries and sizes). In reality these parameters may differ slightly from the as-designed specifications but greatly change the radiation response. Simulation results are therefore validated by experiment. Focused ion beams are used to replicate the simulation conditions of a single ion strike at a specific circuit location in order to test the validity and predictive ability of the simulated circuit response.

4.1 Charge Collection Imaging

An experimentally measured IBICC image of the TA788, recorded using 20 MeV C ions, is shown in Figure 9. An overlay shows the circuit layout for one memory cell. N-drains, p-drains and gates are labeled N, P and G respectively. The substrate of the TA788 is lightly doped p-type Si. The n-drains sit in more heavily doped p-wells (visible as the bright red horizontal bands across the top and bottom of the image). The p-drains are formed in the n-well (visible as the brown horizontal band across the center of the image). The logic states of the memory cells vary within the image, resulting in the presence of different charge collection states for some drains in adjacent memory cells.

Quantitative charge collection spectra can be extracted from specific circuit structures in the image and compared to simulation results. Figure 10 presents a comparison of Davinci simulation results and experimental measurement of the charge collection from the p-off drain of one memory cell. The three dimensional charge transport calculations predict a total collected charge of 255 fC from the p-off drain struck by 20 MeV C. Experimental measurement of this site using the ion microbeam yields a collected charge of 271 fC - approximately 6% above the simulated value and within the uncertainties of the simulation's prediction and the experiment's accuracy. The full energy peak of the incident 20 MeV ions occurs at 885 fC; this charge calibration peak is measured in a fully depleted PIN diode whose depletion depth exceeds the range of the incident 20 MeV C ion.

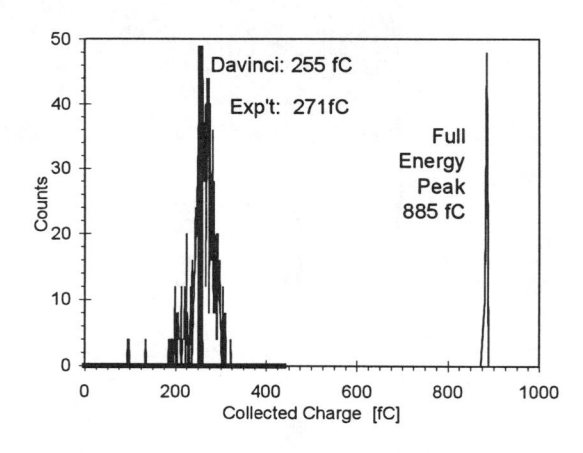

Figure 10 Calculated and measured charge collection spectra for 20 MeV C ion strikes to the p-drain of the TA788.

We have used the charge collection results from the 20 MeV C exposures for comparison to simulation in order to display charge collection spectra whose statistics are not limited by damage effects. 35 MeV Cl charge collection measurements are an accurate simulation of the single event process only until accumulated damage begins to affect collection efficiency. Therefore, only the first few hundreds of ions accurately replicate the simulation conditions. Other circuit structures can be compared to simulation results by digitally extracting spectra from other regions of the graphically rendered charge collection data.

4.2 Single Event Upset Imaging
A measured upset image from the TA788 is shown in the lower right of Figure 11.

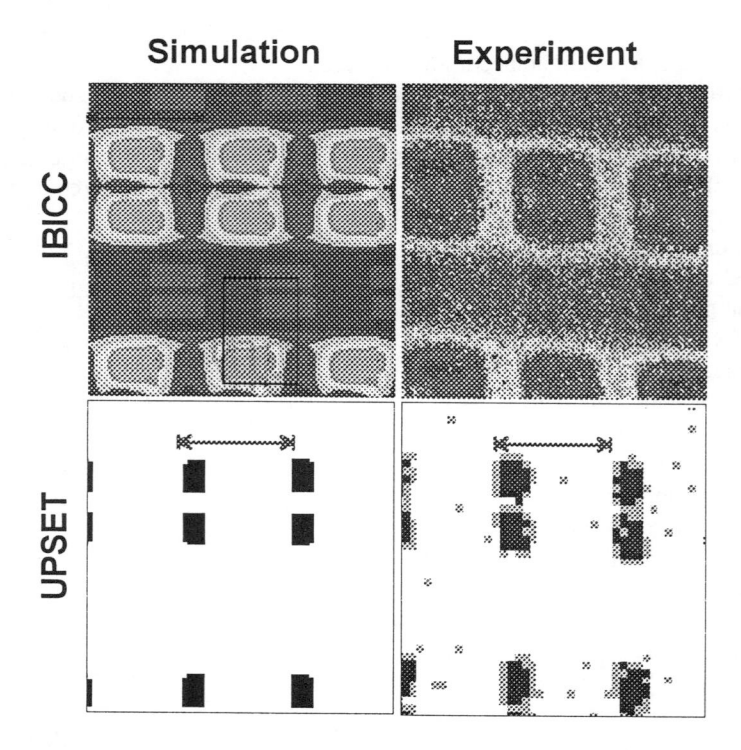

Figure 11. 35 MeV Cl experiment measurements and Davinci simulations of the charge collection and upset of TA788 SRAM.

The dark pixels in the upset images reflect the occurrence of 0-to-1 logic-state transitions in the memory cells during 35 MeV Cl irradiation. The corresponding simulated upset image is shown in the lower left of Figure 11. A simulated charge collection image is shown in the upper left panel and the experimentally measured IBICC image is shown in the upper right. Scale bars of equal length are shown in each panel, representing the 10 micron pitch of the memory cells.

The highly localized and periodic clusters of upset sites in the measured upset image agrees well with the prediction. The isolated, scattered upset points in the experimental upset image are attributed to a very low intensity halo that can sometimes be present around the focused ion beam.

5.0 Conclusion

The use of scanned, focused ion microbeams for radiation testing has proved to be a useful tool in simulation verification and design evaluation of integrated circuits. However, ion microbeam measurements of radiation sensitivity must avoid beam damage effects to be accurate. This paper has briefly described the radiation microscopy techniques used to measure the radiation sensitivity of integrated circuits, illustrated the technical obstacles posed by beam-induced damage effects with examples from past microbeam experiments, and shown comparisons between three dimensional charge transport simulations of charge collection and ion microbeam experimental measurements of the TA788 16K SRAM.

6.0 References

1. K.M.Horn, B.L.Doyle, D.S.Walsh and F.W.Sexton, "Application of the Nuclear Microprobe to the Imaging of Single Event Upsets in Integrated Circuits", Scanning Microscopy, Vol. 5, No. 4, 1991, pp. 969-976.
2. A.B.Campbell and A.R.Knudson, "Use of an ion microbeam to study single event upsets in microcircuits," IEEE Trans. Nucl. Sci., vol. NS-28, pp. 4017-4021, 1981.
3. F.J.Henley and W.G.Oldham, "Soft error studies using a scanning source", Proc. 20th IEEE Reliability Physics Symp., 1982, pp.88-91.
4. D.F.Heidel, U.H.Bapst, K.A.Jenkins, L.M.Geppert, and T.H.Zabel, "Ion microbeam radiation system", IEEE Trans. Nucl. Sci., vol. 40, pp.127-134, 1993.
5. A.R.Knudson and A.B.Campbell, "Charge collection Measurements for Energetic Ions in Silicon", IEEE Trans. Nucl. Sci., vol. NS-29, pp.2067-2071, 1982.

6. P.J.McNulty, W.J.Beauvais, D.R.Roth, J.E.Lynch, A.R.Knudson, and W.J.Stapor, "Microbeam analysis of MOS circuits", RADECS 91: First European Conf. On Radiation Effets on Devices and Systems 1991, pp.435-439.
7. T.J.Aton, J.A.Seitchik, S.D.Hantz and H.Shichijo, "Accurate measurements of small charges collected on junctions from alpha particle strikes using an accelerator-produced microbeam", Proc. Intl. Reliability Physics Symp., 1995, pp.303-310.
8. M.B.H.Breese, P.J.C.King, G.W.Grime, and F.Watt, "Microcircuit imaging using an ion-beam induced charge", J.Appl. Phys., vol. 72, no. 6, pp. 2097-2104, 1992.
9. M.B.H.Breese and K.M.Horn, "The Influence Of Ion Induced Damage On Lateral Charge Collection And IBIC Image Contrast", ", Nucl. Instrum. And Methods B, vol. 138, pp. 1349-1354, 1998.

BOMBARDMENT-INDUCED TOPOGRAPHY ON SEMICONDUCTOR SURFACES

JOHAN B MALHERBE & QUINTIN ODENDAAL

Department of Physics, University of Pretoria, Pretoria, South Africa
E-mail: malherbe@scientia.up.ac.za

This paper presents a review of low energy (E < 20 keV) bombardment-induced topography on semiconductor surfaces. The importance of such topography studies is outlined. The advantages and disadvantages of the different techniques used for the characterisation of the topography are discussed. This shows that the modern tendency is towards using scanning probe microscopy (SPM) techniques such as atomic force microscopy (AFM). The advent of these quantitative techniques coincided with a movement towards the study of regular features, such as ripples, in the bombardment-induced topography. A short discussion is given on the main features of the bombardment-induced topography on Si, InP and GaAs surfaces. In contrast to Si and GaAs where the topography is relatively small and can only really be investigated by SPM, InP exhibits spectacular topographical features that are easily discernible in an ordinary scanning electron microscope.

1 Introduction

The systematic investigation of bombardment-induced topography on semiconductor surfaces such as Si [1], Ge and several compound semiconductors substrates started in the late 1960's [2, 3]. Scientific curiosity driven research was the main reason for these studies. In the middle 1970's more practical reasons appeared for studying bombardment-induced topography. It was shown [4, 5] that the sputter depth resolution in ion gun sputter depth profiles in surface analytical techniques such as Auger electron spectroscopy (AES) and X-ray photoelectron spectroscopy (XPS), depends critically on the topography of the sample before [4] and after [5] ion sputtering. A major advance in the study of this topography and in sputter depth profiling was the discovery by Sykes *et al.* [6] of the tremendous improvement in depth resolution and the reduction of bombardment-induced topography when using dual ion guns. This led to the development of sample rotation [7] during sputter depth profiling. Bombardment-induced topography can also affect other surface sensitive analytical techniques. The accuracy of surface techniques such as secondary ion mass spectroscopy (SIMS) employing ion beams as their excitation and analysis sources, can be detrimentally influenced by the development of topography during the ion bombardment. The onset of surface topography is sometimes accompanied by a change in the sputter yield [8]. An example illustrating both effects (i.e. deterioration of the depth resolution and quantitative compositional analysis) is shown in fig. 1 depicting the effect of sample rotation and

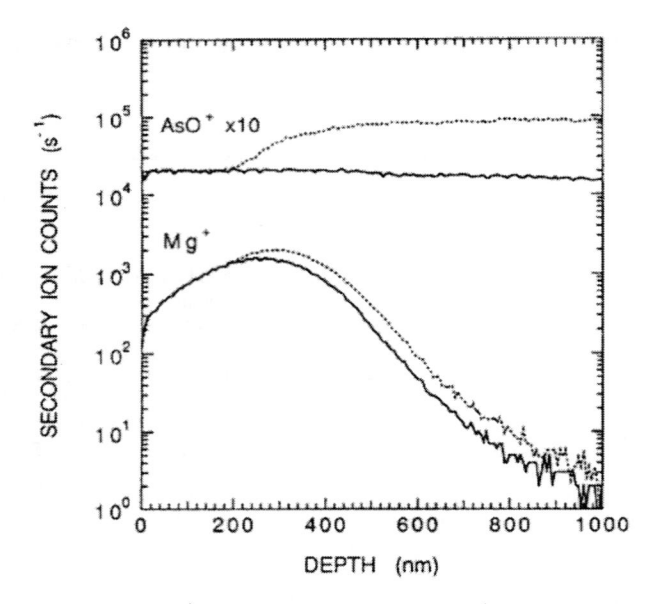

Figure 1. A SIMS profile of Mg^+-implanted GaAs with 3keV O_2^+ at 40° showing AsO^+ and Mg^+ secondary ion yields with (solid line) and without (dashed line) rotation (after Cirlin [8]).

non-rotation on depth discussed in section 4.2, GaAs develops a ripple structure on its surface during oxygen ion bombardment. With sample rotation these ripples disappear.

The movement towards smaller dimensions in IC devices and the increasing use of ion beams makes the study of surface topography of increasing importance for the semiconductor science and technology. A recent development is the possible use of bombardment-induced topographical features for the manufacture of semiconductor quantum dots [9]. The thin needle-like cones appearing on ion bombarded InP has been suggested as a possible characterising tool for the shape of SPM tips [10]. Ion bombardment of semiconductor surfaces sometimes lead to a smoothing effect of originally rough surfaces [11]. It has also been suggested to be used as a micro-machining tool in the manufacturing of small devices such as micro lenses [12].

Apart from the above short outline of the importance of studying bombardment-induced topography on semiconductor surfaces, we also present in this paper a short overview of the different techniques used for studying topography. A historical development of the topography theories is given. The main features of bombardment-induced topography on Si, InP and GaAs surfaces are very briefly outlined.

2 Topography analysis techniques

Traditionally surface topography has been studied using scanning electron microscopy (SEM) and transmission electron microscopy (TEM). Although a variety of other techniques have also been used, the three main techniques have been SEM, TEM and SPM. The use of the latter with its quantitative capabilities will probably lead to big advances in the study of this kind of topography. A summary of the main advantages and disadvantages of these three techniques are given in table 1.

Table 1. A summary of the main advantages and disadvantages of the popular topography techniques.

Technique	Advantages	Disadvantages
SEM	Ease of operation. Little experimental convolution.	Poor resolution. Poor contrast. Few quantitative measurements (cone angles, heights). Subjectivity in choosing area of investigation.
TEM	Good resolution. Little experimental convolution. Crystallographic information.	Complicated and time consuming sample preparation. Few quantitative measurements. Subjectivity in choosing area of investigation.
SPM	Quantification of topography: rms roughness; dimensions of sputter craters, cones, etc.; ratio of surface area to projected area; height correlation function; Atomic resolution. Individual ion-crater / solid interactions. Crystallographic information. Sometimes compositional information.	Tip artifacts. Destruction of features on very soft samples.

3   Theoretical models

Because of the complexity of the ion/solid interaction process, there are more qualitative models than quantitative ones [13]. Two phases can be distinguished in the development of quantitative theories. Initially the emphasis was on the explanation of irregular topographical features and sputter depth resolution dependence. Lately the trend is to explain the development of regular features such as microscopic ripple-like structures on the surfaces reminiscent of the ripples developing on sandy surfaces due to the influence of wind.

The main quantitative theory belonging to the first phase was developed by Carter and his co-workers since 1969 [14]. This ion flux theory uses the equation

$$\frac{\partial h}{\partial t} = -\frac{J\,Y(\theta)}{N}$$

to describe the time evolution of local height h in terms of the ion flux density J, sputter yield Y as a function of the incidence angle θ of the ions and the atomic density N. The next major development was due to Sigmund [15] who considered the spatial dependence of energy deposition by the bombarding ion inside the substrate together with a curvature dependent sputter rate.

The SLS (sequential layer sputter) model by Hofmann [5] was very successful to link sputter depth resolution with the development of surface roughness on an atomic scale due to the stochastic sputtering process. Erlewein and Hofmann [16] was also first to introduce atomic transport into a quantitative topography theory.

The Bradley-Harper theory [17] was a major advance in the explanation of regular features, such as ripples, and described it in terms of a competition between the surface curvature-dependent sputter rate (leading to surface roughening) and surface diffusion which has a smoothing effect. Chason et al. [18] and several others extended this theory by the inclusion of viscous flow relaxation resulting in a noisy Kuramoto-Sivashinsky partial-differential equation. Carter [19], collected all these processes in the a stochastic partial-differential equation, which includes the random or noisy arrival of ions and the statistical variation of the sputtering rate, to define the spatiotemporal evolution of local-surface height h, viz.

$$\frac{\partial h}{\partial t} = -\left(\left|\frac{\partial h}{\partial x}\right|+\left|\frac{\partial h}{\partial y}\right|\right) + \sum_m A_m\left(\frac{\partial h}{\partial x}\right)^m + B_m\left(\frac{\partial h}{\partial y}\right)^m + \sum_2^4 C_n\frac{\partial^n h}{\partial x^n} + D_n\frac{\partial^n h}{\partial y^n} + \eta(x,y,t)$$

In this equation the first term on the right-hand side represents viscous relaxation, the first summation term illustrates the fact that sputtering causes erosion along the local-surface normal with a rate dependent on a trigonometric function of the surface gradients, the second summation term includes the effects of the curvature-dependent sputtering rate (n = 2), surface (n = 4), and bulk (n = 3) diffusion, and the final term represents the noise.

In an interesting, different approached Rudy and Smirnov [20] explained the ripples using a hydrodynamic model (Navier-Stokes). The amorphised layer due the ion bombardment is considered as a Newtonian fluid on a hard surface.

Apart from the above analytical models it is also possible to obtain ripple formation by computer simulation in a Monte Carlo code that employs binary collisions between the bombarding ion species and the substrate atoms coupled with surface diffusion [21]. The results thus obtained agree with some general experimental observations such as the ripple orientation dependence on incident ion angle.

4 Topography of different semiconductors

In the examples and discussion given below the main emphasis will be on bombardment-induced ripple structures. The other topographical features that follow ion bombardment have been discussed in numerous other review papers [13, 14, 22].

4.1 Silicon

It is no surprise that silicon is the semiconductor surface most extensively studied for bombardment induced topography and thus also for ripple formation. The ripples that develop on silicon is generally so small that it is difficult to detect them with normal SEM. With the introduction of SPM techniques and their superior resolution, a big expansion happened in the ripple studies.

A few general observations can be made on bombardment-induced ripple formation on Si. As can be seen from fig. 2, the onset of the ripples is sputter depth dependent, i.e. ion dose dependent [23]. This example further indicated that the development of surface topography can have a detrimental effect on quantitative surface analytical techniques, such as AES. The formation of the periodic ripples is ion angle and ion energy dependent. For normal or near normal incident ions, no ripples appear, even at high doses. The ripples are easiest formed when the angle of incidence θ of the ions is in the range $30° < \theta < 70°$ and also appear for a large variety of ions. The orientation of the ripples with respect to incident ions also depends on the angle of incidence. For $30° < \theta < 70°$ the wave vector of the ripples

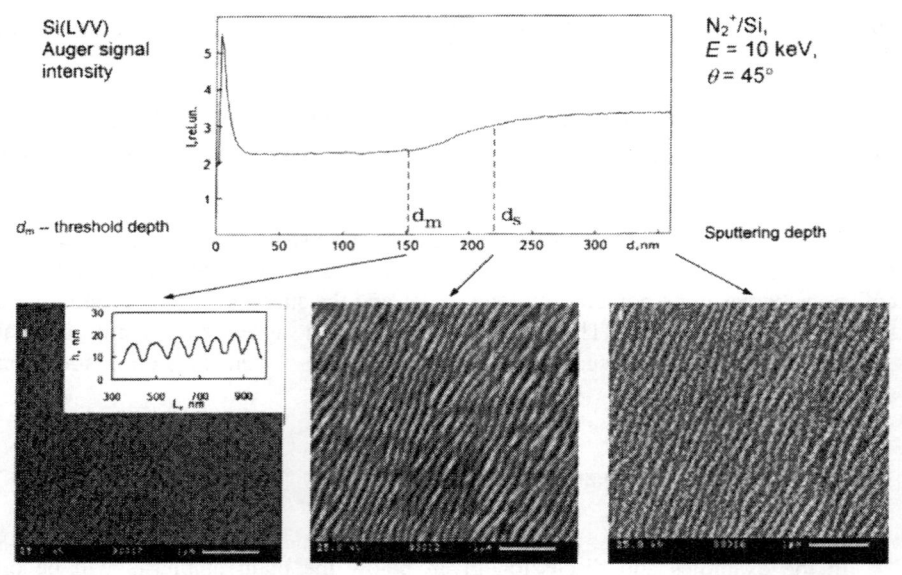

Figure 2. Auger Si (LVV) signal intensity as a function of sputtering depth showing the evolution of ripples on the silicon surface (after Smirnov *et al.*, [23]).

is parallel to the projection of the ion flux. At glancing angles of incidence it becomes perpendicular.

Ripples are readily produced when using reactive ion species such as oxygen and nitrogen ions. It also appears when using noble gas ion bombardment. In the latter case bombardment with lower mass ion species (like Si^+, Ne^+) produce patchy ripple structures in contrast to higher mass species (like Ar^+ and Xe^+). The low mass species produced these ripples only at low temperatures (even as low as 120 K) while with Xe^+ bombardment the ripples appear at room temperature. This low temperature has significant implications for bombardment-induced ripple theories. As was seen in the previous section, most ripple theories use a surface diffusion mechanism. At very low temperatures (such as 120 K) one would expect little surface diffusion to take place. Vajo et al. [24] compared their ripple results on Si obtained by O_2^+ bombardment with several theories. This comparison indicated the need for a better quantitative theory.

4.2 GaAs

The bombardment-induced topography on GaAs is very similar to that on silicon. Ripples appear with the same general features and dependencies as with silicon. For more details the reader is referred to reference [22].

4.3 InP

Spectacular topography in the form of cones, needle-like cones and ripples develop on InP after ion bombardment. A large number of publications dealing with the topography has appeared. Much attention has centered on the composition of cones that develop. At low doses the topography appears in the form of small protrusions or cones far apart from each other (see fig. 3(a)). With higher doses, these develop into denser and larger cones (fig. 3(b)) going over to a spectacular dense needle-like structure (fig. 3(c)). At high doses ripples are formed (fig. 3(d)). In a series of papers one group used AFM to quantitatively measure the topography after bombardment of ions from the noble gas family (excluding the radioactive radon) [25 - 28]. Their results in general disagreed with the predictions of the Bradley-Harper theory [17].

5 Summary

It has been shown that the study of bombardment-induced topography on semiconductor surfaces has progressed from a scientific curiosity driven research phase to one which is of practical importance. In recent times great progress has been made in the development of quantitative theories. They are currently being tested experimentally and, thus, being refined. Bombardment-induced topography on semiconductor surfaces depends primarily on the semiconductor species. In the case of Si and GaAs the topography is small ripples while for InP a variety of larger structures are formed. The bombarding ion parameters play a secondary, but important, role in the resulting topography.

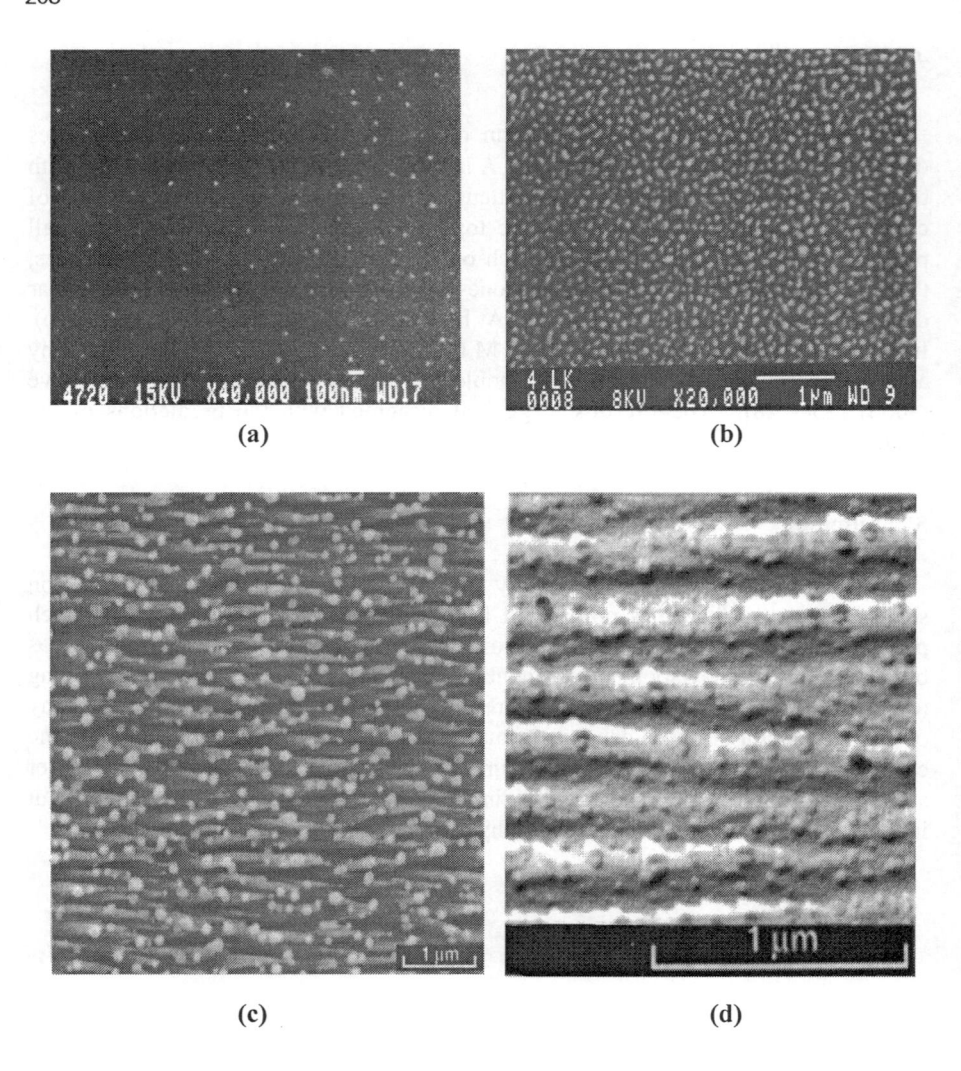

Figure 3. SEM images (a)-(c) and a TEM micrograph of a Cr-C replica (d) depicting the development of Ar$^+$ bombardment topography on In P as a function ion dose: (a) 3×10^{15} Ar$^+$/cm^2; (b) 1.5×10^{16} Ar$^+$/cm^2; (c) 5×10^{16} Ar$^+$/cm^2; (d) 2×10^{18} Ar$^+$/cm^2.

References

1. Stewart A. D. G. and Thompson M. W., Microtopography of surfaces eroded by ion-bombardment, *J. Mater. Sci.* **4** (1969) pp. 56-60.
2. Wilson I. H., The topography of sputtered semiconductors, *Radiat. Eff.* **18** (1973) pp. 95–103.
3. Barber D. J., Frank F. C., Moss M., Steeds J. W. and Tsong I. S. T., Prediction of ion-bombarded surface topographies using Frank's kinematic theory of crystal dissolution, *J. Mater. Sci.* **8** (1973) pp. 1030-1040.
4. Hofmann S., Erlewein J. and Zalar A., Depth resolution and surface roughness effects sputter depth profiling of NiCr multilayer sandwich samples using Auger electron spectroscopy, *Thin Solid Films* **43** (1977) pp. 275-283.
5. Hofmann S., Evaluation of concentration-depth profiles by sputtering in SIMS and AES, *Appl. Phys.* **9** (1976) pp. 59-66.
6. Sykes D. E., Hall D. D., Thurstan R. E. and Walls J. M., Improved sputter-depth resolution using two ion guns, *Appl. Surf. Sci.* **5** (1980) pp. 103-106.
7. Zalar A., Improved depth resolution by sample rotation during Auger electron spectroscopy depth profiling, *Thin Solid Films* **124** (1985) pp. 223–230.
8. Cirlin E.-H., Auger electron spectroscopy and secondary ion mass spectrometry depth profiling with sample rotation, *Thin Solid Films* **220** (1992) pp. 197–203.
9. Facsko S., Dekorsy T, Koerdt C., Tappe C., Kurz H., Vogt A. and Hartnagel H. L., Formation of ordered nanoscale semiconductor dots by ions sputtering, *Science* **285** (1999) pp. 1551-1553.
10. Seah M. P., Spencer S. J., Cumpson P. J. and Johnstone J. E., Cones formed during sputtering of InP and their use in defining AFM tip shapes, *Appl. Surf. Sci.* **144-145** (1999) pp. 151-155.
11. Menzel R., Bachman T, Machalett F, Wesch W., Lanf U., Wendt M., Musil C. and Mühle R., Surface smoothing and patterning of SiC by focussed ion beams, *Appl. Surf. Sci.* **136** (1998) pp. 1-7.
12. Wada O., Ion-beam etching of InP and its application to the fabrication of high radiance InGaAsP/InP light emitting diodes, *J. Electrochem. Soc.* **131** (1984) pp. 2373–2380.
13. Malherbe, J.B., Sputtering of compound semiconductor surfaces. II Compositional changes and radiation-induced topography and damage, *Critical Rev. Solid State Mater. Sci.* **19** (1994) pp. 129-195 and references therein.
14. Carter G., Navinsek B. and Whitton J. L., Heavy ion sputtering induced surface topography development. In *Sputtering by particle bombardment II*, ed. by Behrisch R. (Springer-Verlag, Berlin, 1983) p. 640.
15. Sigmund P., A mechanism of surface micro-roughening by ion bombardment, *J Mater. Sci.* **8** (1973) pp. 1545-1553.

16. Erlewein J. and Hofmann S., A model calculation of the influence of surface transport on the depth resolution in sputter profiling, *Thin Solid Films* **69** (1980) pp. L39-L42.
17. Bradley R. M., and Harper J. M. E., Theory of ripple topography induced by ion bombardment, *J. Vac. Sci. Technol. A* **6** (1988) pp. 2390-2395.
18. Chason E., Mayer T. M., Kellerman B. K., McIlroy D. T. and Howard A. J., Roughening instability and evolution of the Ge(001) surface during ion sputtering, *Phys. Rev. Lett.* **72** (1994) pp. 3040-3043.
19. Carter G., Effect of surface-height derivative processes on ion-bombardment-induced ripple formation, *Phys. Rev. B* **59** (1999) pp. 1669-1672.
20. Rudy A. S. and Smirnov V. K., Hydrodynamic model of wave-ordered structures formed by ion bombardment of solids, *Nucl. Instrum. Meth. B* **159** (1999) pp. 52-59.
21. Koponen I., Hautala M. and Sievänen O. –P., Simulations of ripple formation on ion-bombarded solid surfaces, *Phys. Rev. Lett.* **78** (1997) pp. 2612-2615.
22. Malherbe J. B. and Odendaal R. Q., Ion sputtering, surface topography, SPM and surface analysis of electronic materials, *Appl. Surf. Sci.* **144-145** (1999) pp. 192-200, and references therein.
23. Smirnov V. K., Kibalov D. S., Krivelevich S. A., Lepshin P. A., Potapov E. V., Yankov R. A., Skorupa W., Makarov V. V. and Danilin A. B., Wave-ordered structures formed on SOI wafers by reactive ion beams in physics research, *Nucl. Instrum. Meth. B* **147** (1999) pp. 310-315.
24. Vajo J. J., Doty R. E. and Cirlin E.-H., Influence of O energy, flux, and fluence on the formation of sputtering-induced ripple topography on silicon, *J. Vac. Sci. Technol. A* **14** (1996) 2709.
25. Demanet C. M., Malherbe J. B., van der Berg N. G. and Sankar K. V., Atomic force microscopy investigation of argon-bombarded InP: Effect of ion dose density, *Surf. Interface Anal.* **23** (1995) pp. 433-439.
26. Demanet C. M., Sankar K. V. and Malherbe J. B., Atomic force microscopy investigation of ion-bombarded InP: Effect of angle of ion bombardment, *Surf. Interface Anal.* **24** (1996) pp. 503-510.
27. Demanet C. M., Sankar K. V., Malherbe J. B., van der Berg N. G. and Odendaal R. Q., Atomic force microscopy investigation of ion-bombarded InP: Effect of ion energy, *Surf. Interface Anal.* **24** (1996) pp. 497-502.
28. Sankar K. V., Demanet C. M., Malherbe J. B. and van der Berg N. G., Atomic force microscopy investigation of ion-bombarded InP: Effect of ion species, *Surf. Interface Anal.* To be published

THE ACTIVATION VOLUME FOR SHEAR

F.R.N. NABARRO

*Material Physics Institute,
University of the Witwatersrand, Private Bag 3, WITS 2050
and
Division of Manufacturing and Materials, CSIR,
P.O.Box 395, Pretoria 0001, South Africa*

Becker (1925) and Mott and Nabarro (1948) gave different expressions for the dependence of activation energy on stress when the stress approaches that at which plastic deformation can occur without thermal activation. We show that, when allowance is made for the inevitable elastic non-linearity in the activated state, Becker's model leads to the same dependence as that of Mott and Nabarro. This limiting expression is likely to be valid in many, but not all processes of thermally activated plastic deformation. We then consider the influence of the sharpness of the dependence of the strain rate on the applied stress. Following Clough and Simmons (1975), we observe that the activation volume under a general stress is a tensor, representing the volume integral of the strain, which occurs during the process of activation. Activation of plastic deformation at one site sheds loads on to other sites which may then activate athermally, producing a multiplication factor which is itself stress dependent and may diverge before the applied stress reaches that at which activation occurs without thermal activation.

1. Introduction

This paper summarizes the results of a study in collaboration with J.W. Cahn of the U.S. National Institute of Standards and Technology. A full account will be submitted to Philosophical Magazine A. In 1925, R. Becker analyzed the plastic flow of solids under shear stress. He distinguished two limiting cases, the flow of amorphous materials, where the shear stress biases thermally – induced atomic migrations, and the plastic deformation of crystalline materials, which will deform only when the local stress reaches a critical value σ_o. If the applied stress σ_a is less than σ_o, deformation can be initiated if thermal activation raises the local stress to σ_o in an activated volume V. He showed that the activation free energy for this process was proportional to $(1-\sigma_a/\sigma_o)^2$.

In 1948, after the establishment of the dislocation mechanism of the plastic deformation of crystals, Mott and Nabarro analyzed the specific model of a dislocation segment overcoming a barrier, and found that, as σ_a approached σ_o, the activation free energy vanished as $(1-\sigma_a/\sigma_o)^{3/2}$.

The first aim of the present work was to resolve this discrepancy. The next was to bridge the theoretical treatments which are appropriate in the limits $\sigma_a \to \sigma_o$ and $\sigma_a \to 0$. This leads to a discussion of the way in which the sharpness of the dependence of strain rate on stress affects the geometry of deformation, and then to an analysis of the nature of the activation volume.

A distinction is drawn between driving stresses for plastic deformation and modulating stresses. At low stresses, the activation volume tends to a constant value which, however, cannot be determined by the usual analysis of experimental observations.

Experimental observations are usually analyzed by assuming that the strain rate is a direct measure of the rate of release of dislocations from obstacles. Section 5 discusses the validity of this assumption.

2. The Becker and Mott-Nabarro formulae.

The Becker model calculates the free energy of activation as the elastic energy of the volume V in the activated state, less the elastic energy under the applied stress σ_a, and less the work done by σ_a as the material in V deforms when the stress in it increases from σ_a to σ_o. The Mott-Nabarro model assumes that σ_a is close to σ_o. Since σ_o is the maximum stress with which the obstacle repels the dislocation segment, this repulsive stress will depend parabolically on the coordinate of the dislocation segment when the segment is close to the critical position. The variation of activation free energy as $(1-\sigma_a/\sigma_o)^{3/2}$ follows.

An argument similar to that of Mott and Nabarro must hold for almost all models in which the configuration is described by a single reaction coordinate. The system becomes elastically soft in the reaction coordinate as the critical configuration is approached, and Becker's use of linear elasticity is necessarily invalid.

3. High and low stresses: the Schmid Law

At low stresses the strain rate $\dot{\varepsilon}$ is a linear function of the stress σ. For an anisotropic material this may be written.

$$\dot{\varepsilon}_{ij} = r_{ijk\ell}\, \sigma_{k\ell} \tag{1}$$

More generally, changes in strain rate are related to changes in stress by

$$\delta\dot{\varepsilon}_{ij} = r_{ijk\ell}(\sigma)\delta\sigma_{k\ell} \tag{2}$$

For the relatively rapid deformation of a single crystal the situation is entirely different. For almost all orientations of the applied stress tensor σ_{ij} with respect to the crystal axes, the deformation occurs by glide on a low-index plane (usually close-packed) with normal $\hat{n}_j$ along a single glide direction (also usually close-packed) in a direction $\hat{b}_i$, with a unit displacement b_i. The Schmid Law states that glide will begin when the shear stress resolved on to the plane i in the direction j, which may be written $\hat{b}_i \sigma_{ij} \hat{n}_j$, reaches a critical value. This is a scalar quantity, the pressure tending to expand a loop of dislocation of Burgers vector b_i on the plane $\hat{n}_j$.

The symmetry of the crystal produces a family of equivalent glide systems $\{ij\}$. It follows from eq. (1) that glide must occur on all of these systems. If, as is observed experimentally, the glide rate increases very rapidly with the resolved shear stress on a particular glide system, one of the equivalent systems will begin to glide rapidly as the stress is increased before the glide rate on the other systems becomes observable (unless the stress system has a symmetrical orientation with respect to the crystal axes).

For a polycrystal with randomly oriented grains, Clough and Simmons (1975) showed that, when the glide rate depends linearly on the stress as in eq. (1), the plastic deformation will be determined by the von Mises condition that $\sigma'_{ij}\sigma'_{ij}$ reaches a critical value, where

$$\sigma'_{ij} = \sigma_{ij} - \tfrac{1}{3}\delta_{ij}\,\sigma_{kk} \tag{3}$$

is the deviatoric stress tensor. When the strain rate on each system is a rapidly increasing function of the corresponding resolved shear stress, each grain will, if unconstrained, glide on a single system, and the behaviour of the ensemble will be dominated by those grains for which one glide system is maximally stressed, leading to the Tresca criterion that the onset of plasticity is determined by the maximum difference of principal stresses.

4. The activation volume

The thermodynamic activation volume $V^\dagger$ is defined in terms of the activation free energy $F^\dagger$ by

$$V^\dagger(\sigma_a) = -\left(\frac{\partial F^\dagger}{\partial \sigma_a}\right)_T . \tag{4}$$

It is not Becker's volume V of the region in which the activation process occurs, but the change in this volume during the activation process. If the attack frequency is v_o, the rate of the activation process is

$$r(\sigma_a) = v_o \exp[-F^\dagger(\sigma_a)/kT] . \tag{5}$$

With the further assumption that the observed rate of strain $\dot{\varepsilon}(\sigma)$ is a direct measure of $r(\sigma_a)$, we have

$$\dot{\varepsilon}(\sigma_a) = \dot{\varepsilon}_o \exp[-F^\dagger(\sigma_a)/kT] , \tag{6}$$

and $V^\dagger(\sigma)$ may be determined from experimental observations by the relation

$$V^\dagger(s) = kT\left(\frac{\partial \ln\dot{\varepsilon}(s)}{\partial s}\right)_T , \tag{7}$$

where $\quad s = \hat{b}_i \, \sigma'_{ij} \, \hat{n}_j . \tag{8}$

This formalism neglects the fact that in plastic deformation σ is not a scalar pressure but a tensor shear stress σ_{ij}. If we are concerned with single glide, so that $\dot{\varepsilon}$ is a scalar, we may define a tensor activation volume

$$V^\dagger_{ij}(\sigma) = kT\left(\frac{\partial \ln\dot{\varepsilon}(\sigma)}{\partial \sigma_{ij}}\right)_T. \qquad (9)$$

If the change in strain at an arbitrary point in the body during the process of activation is $\delta e^\dagger_{ij}$, then $V^\dagger_{ij}$ is given by the volume integral

$$V^\dagger_{ij} = \int \delta e^\dagger_{ij} d\tau. \qquad (10)$$

More generally, we may define an experimentally observable non-tensorial quantity $V^\dagger_{ijk\ell}$ by

$$V^\dagger_{ijk\ell} = kT\frac{\partial \ln\dot{\varepsilon}_{ij}}{\partial \sigma_{k\ell}}. \qquad (11)$$

This approach fails at low stresses, where, as in eq. (1), $\dot{\varepsilon}$ is linear in σ and so, from (8),

$$V^\dagger = kT/\sigma, \qquad (12)$$

and tends to infinity as $\sigma \to 0$.

It is now necessary to take into account backward as well as forward jumps. Assuming that the crystal symmetry does not influence the relative probability of backward and forward jumps, we replace eq. (6) by

$$\dot{\varepsilon} = \dot{\varepsilon}_o \exp[-F^\dagger_+/kT] - \dot{\varepsilon}_o \exp[-F^\dagger_-/kT] \qquad (13)$$

where
$$F^\dagger_+ = \int_{\sigma_a}^{\sigma_o} V(\sigma) d\sigma \qquad (14a)$$

and
$$F^\dagger_- = \int_{-\sigma_o}^{\sigma_a} V(\sigma) d\sigma. \qquad (14b)$$

This leads to

$$\left(\frac{\partial \ln \dot{\varepsilon}(\sigma_a)}{\partial \sigma_a}\right)_T = \frac{V(\sigma_a)}{kT} \coth\left[\int_o^{\sigma_a} \frac{V(\sigma)d\sigma}{kT}\right] \quad (15)$$

As $\sigma_a \to \sigma_o$, the expression in square brackets becomes large, the hyperbolic cotangent tends to unity, and we recover eq. (7).

As $\sigma_a \to O$, $\dot{\varepsilon}$ becomes proportional to σ_a according to eq. (1), the left-hand side of eq. (14) tends to $1/\sigma_a$, and (15) is satisfied by any constant value of $V^\dagger(O)$. This indeterminacy of $V^\dagger(O)$ corresponds to the physical situation that in the linear regime the observed strain rate involves the product of $V^\dagger(O)$ and an unknown mobility.

5. Driving stresses and modulating stresses

For a single crystal in single glide on the system b_i n_j, only the stress component $\hat{b}_i \sigma'_{ij} \hat{n}_i$ does work in the glide process. This component is the driving stress. The remaining stresses

$$\sigma^m_{pq} = \sigma_{pq} - \hat{b}_p \hat{n}_q s \quad (16)$$

are the modulating stresses. They determine the deviations from the Schmid Law. Thus σ^m_{pp}, the hydrostatic stress, causes small deviations from the Schmid Law. Other components cannot induce flow on the ij system, but they can strongly modulate the rate of flow by altering the core structure of the dislocation. A simple example, though one which produces only weak effects, is that of a dislocation lying between two close-packed planes. The dislocation is dissociated into two partial dislocations. The driving stress acts equally on the two partials. The component of stress resolved onto the glide plane at right angles to the glide direction alters the separation of the two partial dislocations, and so alters the dislocation mobility.

6. Site activation rate and strain rate

The passage from eq. (5) to eq. (6) involves the assumption that a single thermally activated event leads to an increment of strain which is independent of the applied stress σ_a. This assumption is not likely to be true. Firstly, the area A of glide plane swept by a dislocation when it is thermally released from an obstacle is likely to be an increasing function $A = A(\sigma_a)$. Secondly, as pointed out by Mott (1953), Kuhlmann-Wilsdorf (1987) and more recent by Cottrell (1996a,b, 1997), the thermally-activated release of a dislocation from an obstacle sheds load on to dislocations held up at neighbouring obstacles. Each thermally activated release immediately triggers athermally $\mu(\sigma_a)$ secondary events, where μ is an increasing function of σ_a. These in turn trigger athermally μ^2 tertiary events, so that the contribution of a single thermally triggered event is multiplied by a factor

$$M(\sigma_a) = 1 + \mu + \mu^2 + \ldots = 1/[1-\mu(\sigma_a)]. \tag{17}$$

Equation (6) is then replaced by

$$\dot{\varepsilon}(\sigma_a) = \dot{\varepsilon}_o \, A(\sigma_a) M(\sigma_a) \exp[-F^\dagger(\sigma_a)/kT]. \tag{18}$$

Experiments other than the measurement of $\dot{\varepsilon}(\sigma_a)$ are then required to relate the macroscopically observed $\dot{\varepsilon}(\sigma_a)$ to the microscopic $F^\dagger(\sigma_a)$. Such experiments could study internal friction at very low amplitudes either in the absence of a biasing stress or as a function of a small biasing stress, and amplitude-dependent internal friction, possibly as a function of a biasing stress. Further information could be obtained by measuring these quantities over a range of temperatures, and by measurements of stress relaxation at constant strain.

REFERENCES

[1] Becker, R., 1925, *Phys. Z.* **26**, 919.

[2] Clough, R.B. and Simmons, J.A., in Rate processes in Plastic Deformation of Metals, ed. J.C.M. Li and A.K. Mukherjee, *American Society for Metals*, 1975, p. 266.

[3] Cottrell, A.H., 1966a, *Phil. Mag. Lett.*, **73**, 35; 1996 b, *Phil. Mag.* A **74**, 1041; 1997, Phil. Mag. Lett. **75**, 227.

[4] Mott, N.F., 1953, *Phil. Mag.* **44**, 741.

[5] Mott, N.F. and Nabarro, F.R.N., 1948, Report on Strength of Solids *(London: The Physical Society)*, p.1.

3. Elementary Particle Physics

CHALLENGES AND OPPORTUNITIES IN PARTICLE PHYSICS

JOHN ELLIS

Theoretical Physics Division, CERN, CH-1211 Geneva, Switzerland
E-mail: John.Ellis@cern.ch

The Standard Model of particle physics describes very accurately all confirmed experimental data from accelerators. It leaves unanswered the origin of particle masses, the unification of the fundamental interactions, and the nature of a quantum theory of gravity. Particle masses may originate from a Higgs boson, and data from CERN's LEP accelerator have provided a first hint for its existence. Data on solar and atmospheric neutrinos indicate that neutrinos have masses, as suggested by grand unified theories. Astrophysics and cosmology may provide other hints about physics at very high energies. Central to the quest for the Higgs and other new particles will be CERN's next accelerator, the LHC.

1 The Standard Model of Particle Physics

Particle physics not only addresses the fundamental structure of matter, its constituents and the forces between them, but also gives us insights into the very early history of the Universe. When the age of the Universe was less than about one second old, its temperature was so high that the typical kinetic energies of elementary particles were above a few MeV. At earlier times, nuclear physics is inadequate to describe their behaviour. Perhaps it is not surprising that particle physicists need large accelerators to recreate the conditions of the Big Bang: the largest CERN accelerator (LEP) has a circumference of about 27 km and collides electrons and positrons with energies around 100 GeV each. It thereby reproduces, in some sense, conditions when the Universe was about 10^{-10} seconds old. CERN's next accelerator, the LHC, will recreate conditions when it was only 10^{-12} seconds old.

What have experiments at LEP and elsewhere been able to reveal so far about the elementary particles and their interactions? It has been established that nuclear matter is fabricated out of quarks, of which there are six types (or flavours) in total: the *up, down, strange, charm, bottom* and *top*, in order of increasing mass, and that the electron is followed by two heavier charged siblings (leptons), the *muon* and *tau*, each of which is accompanied by its own neutrino, for a total of three neutrino flavours. Four fundamental forces between these elementary particles have been identified: the familiar electromagnetic and gravitational forces, and the strong and weak nuclear forces.

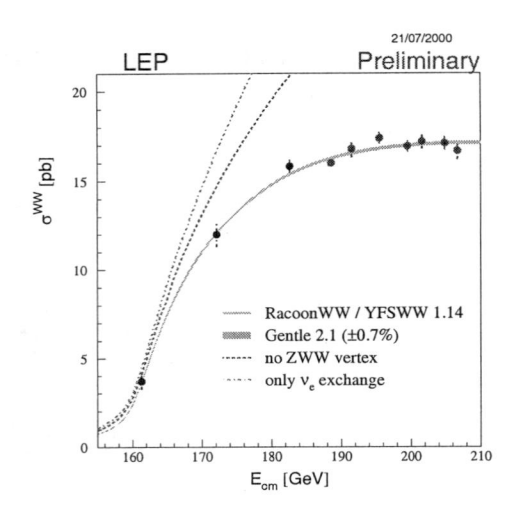

Figure 1. The cross section for $e^+e^- \to W^+W^-$ measured at LEP, as a function of the centre-of-mass energy, compared with theoretical calculations. The dotted lines show that the data require not only neutrino exchange, but also photon and Z^0 exchanges, with the couplings predicted in the Standard Model [4].

All these fundamental forces are mediated by the exchanges of some carrier particles: the photon for electromagnetism, eight gluons for the strong interactions, the $W^\pm$ and Z^0 for the weak interactions, and (it is firmly believed) the graviton for gravity. The massless photon has been known since 1900, from the work of Planck and Einstein. The massless gluons were discovered at the DESY laboratory in Hamburg in 1979 [1], and the massive $W^\pm$ and Z^0 were discovered at CERN in 1983 [2]. The massless graviton remains to be discovered.

All the quarks and charged leptons are massive, and it now seems that the neutrinos may also have very small masses. The $W^\pm$ and Z^0 weigh as much as medium-sized nuclei, and the heaviest quark, the top [3], weighs as much as a large nucleus. The most pressing problem in particle physics is to understand where these masses come from, and why some particles, such as the photon and gluons, get away without acquiring masses.

This Standard Model has been tested with high precision by experiments at CERN elsewhere. For example, using LEP it was possible to measure the mass of the Z^0 with an accuracy of about two parts in 10^5, and many other properties of the Z^0 and its decays agree with the Standard Model with an accuracy around 0.1% [4]. Recently, with increased energies at LEP, it has become possible to measure the mass of the $W^\pm$ with an accuracy around five parts in 10^4, and to verify that the photon and Z^0 boson couple to the $W^\pm$ just as predicted by the Standard Model, as seen in Fig. 1.

All this sounds very impressive, but many important questions are raised by the Standard Model, and await answers from the next generations of accelerator experiments. Where do the *particle masses* come from? Are the fundamental forces *unified*? Why are there so many different *flavours* (types) of matter particles? Above all, how to explain all the parameters of the Standard Model: six quark masses, three charged lepton masses, two boson masses, four weak mixing angles and phases, three gauge couplings and a CP-violating phase in the strong interactions, for a total of 19 parameters, not to mention the masses and mixing angles of neutrinos?

In the following sections, the prospects for addressing some of these fundamental questions are reviewed.

2 The Problem of Mass

2.1 Broken Symmetry

In the basic formulation of the Standard Model as a gauge field theory, all particles are born massless, reflecting the symmetries of the underlying theory. However, in the real world, only a few of these particles are still massless, namely the photon and the gluons, reflecting the persistence of unbroken local 'gauge' symmetries, $U(1)$ in the case of electromagnetism and $SU(3)$ in the case of the strong interactions. The other elementary particles, namely the quarks, leptons, $W^\pm$ and Z^0, not to mention you and I, are now massive. Presumably this is because other underlying symmetries of the Standard Model are broken, and the game is to discover how.

The mechanism favoured by particle physicists envisages different particles as corresponding to different directions in some internal space. Initially, the equations of the theory do not distinguish between these different degrees of freedom, that can be imagined as different directions in some analogue of a flat plane. In the unbroken case, there is no obstacle to moving in any direction in the plane, corresponding to all the particles being massless. However, the symmetry may be broken by folding the plane, rather like a rift valley created by plate tectonics, or more prosaically by folding a piece of paper, as seen in Fig. 2. After folding once a piece of paper, there is one flat direction and a perpendicular direction that rises and falls. The flat direction might correspond to a particle that remains massless, whereas the 'hilly' direction might correspond to a massive particle.

These 'hills' and 'valleys' are provided in the Standard Model by a new sort of field that permeates all space, much like the more familiar electromagnetic and gravitational fields. Just as the electromagnetic and gravitational

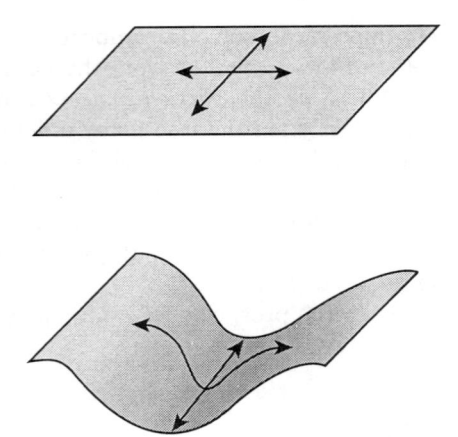

Figure 2. The symmetric phase of theory is analogous to a flat plane, in which there is no obstacle to motion in any direction - all particles are massless - whereas a broken phase has the analogue of a 'rift valley', with obstacles to motion in some directions but not others - corresponding to the appearance of massive particles.

fields have quantum fluctuations called the photon and graviton, so also the quantum fluctuations of this new mass-giving field should yield at least one observable particle, called the Higgs boson [5].

It is a classical vacuum expectation value of the Higgs field that enables other particles to gain mass, by breaking the gauge symmetry of the Standard Model. For example, it is this new Higgs field that provides an extra polarization state for each of the $W^{\pm}$ and Z^0, enabling them to become massive. Remember that a massless spin-one particle such as the photon has just two polarization states, whereas a massive spin-one particle such as the $W^{\pm}$ or Z^0 requires three polarization states: one longitudinal state as well as two transverse states. In addition to these 'hidden' Higgs states, even in the minimal version of the Standard Model there is one physical Higgs boson in the spectrum, and more complicated models, such as the supersymmetric extensions discussed below, have more Higgs bosons awaiting discovery.

Although the Higgs mechanism might look at first like a dubious trick, one can in fact show that it is the only way of providing the particles of the Standard Model with masses, if one wants a theory with results that can be calculated reliably [6]: what a field theorist calls renormalizable.

Many such calculations have been made in the Standard Model, and the results agree very well with experiment, as already mentioned. This good

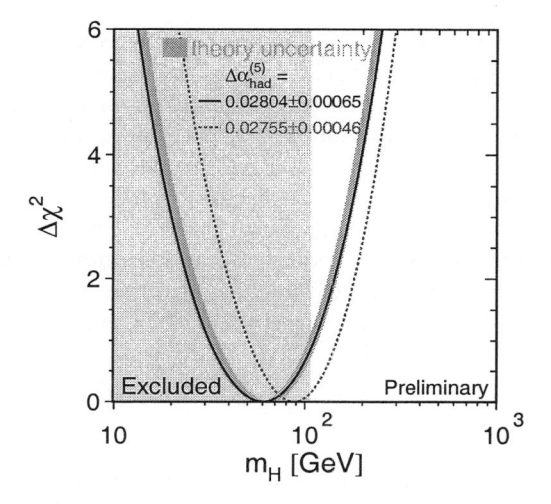

Figure 3. The likelihood function for the mass of the Higgs boson, as estimated from a global analysis of data on the electroweak interactions. The solid and dotted lines correspond to different estimates of the effective value of the fine-structure constant at the Z^0 energy [4].

agreement was used some years ago to predict successfully the mass of the top quark, which had not at that time been detected directly by any experiment [7]. More recently, similar calculations have been used to predict successfully the mass of the $W^\pm$. These calculations are quite sensitive to the mass of the infamous Higgs boson [8], and the agreement between the data and the predictions can be used to estimate its mass [4]:

$$m_H = 62^{+53}_{-30} \text{ GeV} \tag{1}$$

with a one-sided 95% confidence level upper limit of 170 GeV, as seen in Fig. 3.

2.2 A Supersymmetric Conspiracy?

Although the Higgs boson provides a possible source for particles masses, there is another problem associated with their values [9]. If one calculates quantum corrections to the masses of bosons, such as those of the $W^\pm$, the Z^0 and the Higgs itself, one finds that they are so large as to risk destabilizing them. There is apparently nothing to prevent the electroweak scale from becoming very large, perhaps as great as the Planck scale $G_N^{-1/2} \simeq 10^{19}$ GeV. In principle, one could evade this problem by fine-tuning the classical values of the boson masses to very large and almost opposite numbers, so that their net values are around 100 GeV. However, this strikes many theorists as unnatural.

This instability may be avoided by postulating equal numbers of bosons and fermions with equal couplings, as in supersymmetric theories [10]. In this case, the positive corrections due to the quantum effects of bosons are cancelled naturally by the corresponding negative quantum corrections due to fermions. Supersymmetry is a very beautiful theory, and is required at some energy scale in string theory, which, as discussed later, is the only candidate we have for a quantum Theory of Everything (TOE), including gravity. However, this aesthetic argument does not tell us what masses supersymmetric particles might have. The only hint comes from the naturalness (or fine-tuning) argument summarized above, which suggests that some supersymmetric particles should weigh less than about 1 TeV.

2.3 The Higgs Quest

In order to verify these ideas, it is clear that the highest priority is to discover experimentally the Higgs boson. The precision electroweak data mentioned above suggest that it might be tantalizingly close. Compared with the estimate (1), the direct searches at LEP have established a lower limit of 113.5 GeV, and have attained a sensitivity to a Higgs mass of 115 GeV. Much excitement has been caused recently by the appearance of an apparent excess of Higgs candidates with a mass around 114 to 115 GeV. These include events reported in particular by the ALEPH collaboration, principally in the $H \to \bar{b}b + Z \to \bar{q}q$ final state, as seen in Fig. 4 [11].

Interesting events have also been reported by the L3 collaboration, including one in the $H \to \bar{b}b + Z \to \bar{\nu}\nu$ final state, seen in Fig. 5 [12]. Overall [13], there is an excess of about 2.9 standard deviations corresponding to $m_H = 115^{+1.3}_{-0.7}$ GeV [14]. During the course of this meeting, it was decided not to continue running LEP during 2001 in an effort to confirm or refute this possible signal. Instead, it was decided to push ahead as rapidly as possible with the preparations for the LHC accelerator, which should be able to discover definitively the Higgs boson, whatever its mass. In the mean time, the Tevatron $\bar{p}p$ collider may have a chance to confirm the discovery of the Higgs boson, if its mass really is around 115 GeV.

In addition to the Higgs boson, many experiments are looking for supersymmetric particles, which should have the same internal quantum numbers as the known particles, but spins differing by half a unit. Unsuccessful indirect searches at LEP indicate that many sparticle species must have masses greater than around 100 GeV. If they are to do their job of stabilizing the masses of bosons in the Standard Model, these sparticles should weigh less than about 1 TeV. The LHC will be able to discover strongly-interacting sparticles, the

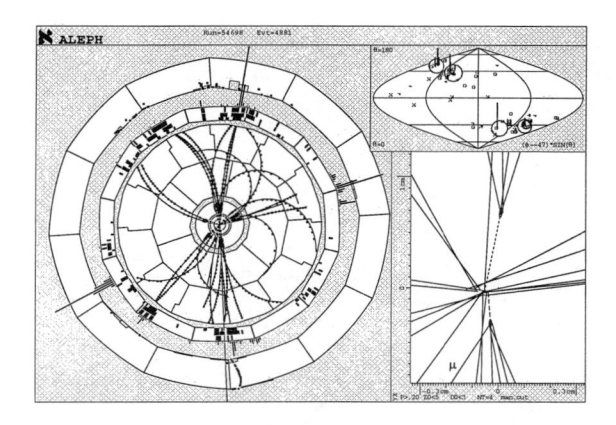

Figure 4. Display of a candidate for $e^+e^- \to (Z^0 \to \bar{q}q)$ + Higgs seen by the ALEPH collaboration [11].

squarks and gluinos, if they weigh less than about 2 TeV, and will also have interesting sensitivities to other sparticles [15]. In the mean time, in addition to direct searches at the Tevatron, there are many direct and indirect searches for the dark matter that astrophysicists believe to abound in the Universe, much of which may be composed of supersymmetric particles.

3 The Problem of Unification

The Standard Model treats the fundamental forces in a rather messy way, unifying partially the weak and electromagnetic interactions, but leaving the strong nuclear interactions off on one side. Following Einstein, it has been the dream of many ambitious physicists to find a single unified field theory that encompasses them all. If such a Grand Unified Theory (GUT) exists, the unifying symmetry must be broken at very high energy scale, if the manifest differences in strength between the different interactions are to be understood, and if new interactions are not to cause protons to decay more rapidly than is allowed by experiment: the present lower limit on the proton lifetime is about 10^{33} years [16].

The quantum numbers of the quarks and leptons fit very neatly with the assignments suggested by the simplest GUTs, and there are experimental hints that may favour some GUT scenarios. For example, the strengths of

a) Run # 933204   Event # 4704   Total Energy : 112 GeV

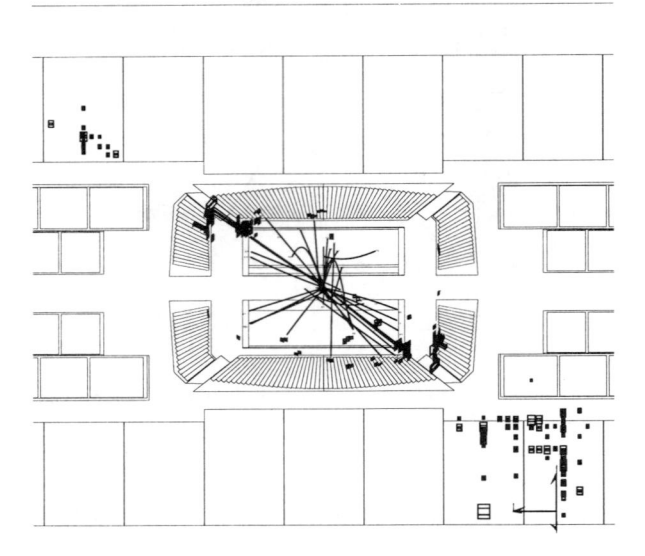

Figure 5. Display of a candidate for $e^+e^- \to (Z^0 \to \bar{\nu}\nu) +$ Higgs seen by the L3 collaboration [12].

the different fundamental interactions vary as functions of the effective energy scale, in a manner calculable once one knows the spectrum of matter particles. If one assumes just the particles in the Standard Model, the different interaction strengths would become almost equal at an energy around 10^{14} to 10^{15} GeV as seen in Fig. 6, and the proton lifetime would be about 10^{30} years. Close, but no banana! Both these difficulties are removed by postulating the appearance of supersymmetric particles at energies below about 1 TeV. In this case, the different measured interaction strengths are in perfect agreement with unification at an energy scale around 10^{16} GeV [17], as seen in Fig. 6, and the proton lifetime exceeds the present experimental lower limit. Such a supersymmetric GUT not only gets the different interaction strengths right, but also offers tantalizing relations between the masses of quarks and leptons.

The key to testing GUTs is surely to find characteristic new forces or particles. One generic prediction of GUTs is that *neutrinos should have small but non-zero masses,* and the first experimental evidence pointing in the GUT

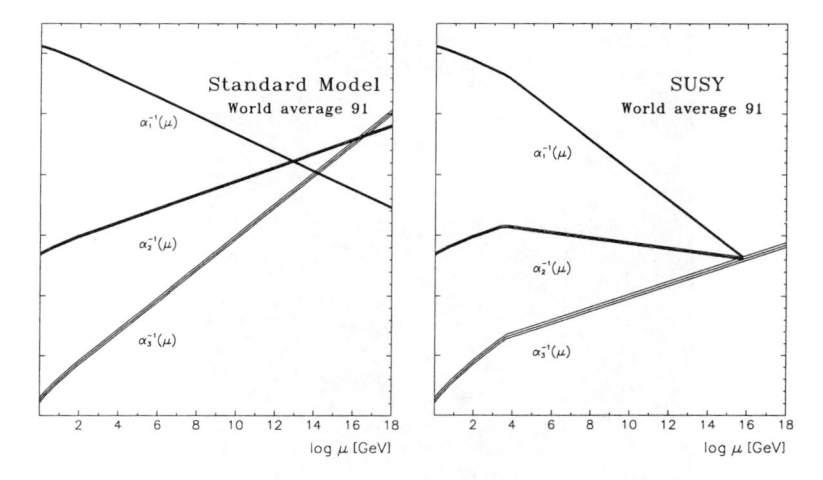

Figure 6. The energy variations of the gauge coupling strengths in theories without and with supersymmetry. The latter are consistent with grand unification at an energy around 10^{16} GeV [17].

direction has been provided by indications that neutrinos may indeed be massive. These hints have been furnished by measurements of neutrinos from the Sun [18] and cosmic-ray interactions in the atmosphere [19]. In both cases, fewer neutrinos are seen than would be expected in the Standard Model, and these experiments are interpreted as evidence that the neutrinos produced by the astrophysical sources - electron neutrinos from the Sun and muon neutrinos from the atmosphere - change (oscillate) into other neutrino flavours before reaching the detectors. Such neutrino oscillations are predicted if the different neutrino species mix among themselves like quarks do, and if the mass eigenvalues are different. Experiments are now underway in Japan and being prepared in the United States and Europe to test these ideas with neutrino beams produced by accelerators, which are well-understood and controllable.

If neutrinos do have masses, they might constitute a significant fraction of the missing dark matter postulated by astrophysicists and cosmologists. However, the minimum masses required to interpret the neutrino oscillation experiments are too small to be of great cosmological interest [20], and astrophysicists generally prefer much heavier particle candidates for dark matter, such as supersymmetric particles.

The search for *proton decay* has already been mentioned. If proton decay were to be seen, it would constitute very direct and convincing evidence for

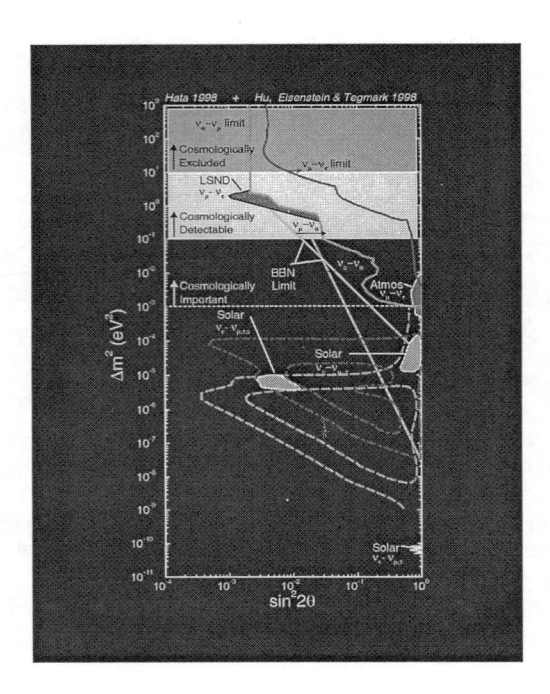

Figure 7. A compilation of information about neutrino masses. The regions of mass differences and mixing angles indicated by different solar, atmospheric and accelerator neutrino oscillation experiments are compared with cosmological limits. Apart from the LSND accelerator experiment, the data favour small mass differences, corresponding to masses that are unlikely to dominate the cosmological dark matter, though they may have some astrophysical significance [20].

GUTs : after all, other sources of neutrino masses could be envisaged. For this reason, discussions are underway about the possibility of building a very large next-generation experiment that could continue the search for proton decay.

As for new particles, *magnetic monopoles* are predicted in most GUTs. Unfortunately, according to modern ideas of inflationary cosmology, no magnetic monopole is likely to have remained within our observable Universe.

4 Towards a Theory of Everything

The job description of a Theory of Everything (TOE) is to unify gravity with the other particle interactions. The first step in such a programme is to formulate a consistent quantum theory of gravity. This is the greatest piece of business left unfinished by twentieth-century physics, and hence the ultimate challenge for theoretical physics in the twenty-first century. The greatest successes of twentieth-century physics were quantum mechanics and general relativity, and its greatest embarrassment has been the failure to combine them.

Hope for making a quantum theory of gravity is provided by string theory [21], which also seems sufficiently powerful to incorporate matter particles and the other fundamental interactions in a TOE. However, although string theory has made tremendous advances and resolved some aspects of the problem of quantum gravity, much of its promise remains to be fulfilled.

The first challenge offered by quantum gravity is that *higher-order corrections* to the single-graviton exchange responsible for the classical gravitational field, are horribly infinite. Thus two-graviton exchange would be much greater than one-graviton exchange at, for example, a conjectural e^+e^- collider with a centre-of-mass energy near the Planck scale of 10^{19} GeV. The challenge of making sense of these quantum corrections has been met successfully by string theory.

Further challenges are offered by *black holes*, which have been found in semi-classical treatments to obey the laws of thermodynamics, with non-zero temperature and entropy. The question is raised whether quantum mechanics can survive unscathed in strong gravitational fields with event horizons across which information is apparently lost. These may be present microscopically, because of localized quantum fluctuations in the gravitational field, the so-called *space-time foam* [22]. String theory provides a successful accounting for the quantum states of macroscopic black holes, but a complete stringy treatment of space-time foam remains an open challenge.

Finally, how is one to *combine gravity with the other interactions*, given that the graviton has spin two, whereas the carrier particles for the other fundamental interactions have spin one ? This sounds like a good assignment for supersymmetry, which automatically relates particles of different spin. Indeed, string theory seems to require supersymmetry for its consistent formulation. However, no specific TOE has yet emerged convincingly from among the myriad possibilities offered by superstring theory.

5 Hints from Astrophysics and Cosmology ?

In view of the high energy scales involved in many theories of physics beyond the Standard Model, it is natural to wonder whether astrophysics and cosmology may be able to answer some of our questions.

One possible hint may come from *ultra-high-energy cosmic rays* [23]. According to conventional physics, they would have been absorbed before reaching us if they came from sources more than about 5×10^7 light-years away. On the other hand, there are no obvious local sources. Perhaps they are due to new physics, for example, the decays of massive GUT particles spread through the galactic halo ?

A major puzzle in astrophysics and cosmology is the *dark matter* that is required to explain the formation of structures in the Universe. As already mentioned, it may be composed of massive weakly-interacting particles, such as neutrinos or supersymmetric particles. Extensive experimental programmes to probe these possibilities are underway around the world.

A more speculative suggestion is to use sources of energetic photons and neutrinos to *probe fundamental physics*, such as special relativity or quantum mechanics. Distant sources that emit pulses of energetic radiation, such as gamma-ray bursters, might be able to test some models of space-time foam [24].

One cosmological observation that seems very likely to take us back to the GUT epoch is the *microwave background radiation* [25], which exhibits perturbations believed to have been generated during an early period of cosmological inflation. These may have been laid down by quantum fluctuations of some Higgs-like scalar field at an energy scale close to that estimated from the GUT unification of particle interactions.

Recent observations of the microwave background, together with observations of distant supernovae and measures of the dark matter density [26], suggest that the Universe may be dominated by *vacuum energy*. If so, this provides a second number to be confronted with any theory of gravity, in addition to Newtons constant $G_N \simeq 1/(10^{19} \text{GeV}^2)$. Calculating the value of this vacuum energy may be the ultimate challenge for string theory.

6 Prospects for the Twenty-First Century

As we have seen, the Standard Model of particle physics works very well, but leaves many questions unanswered. Many theoretical ideas have been proposed that address at least some of these questions, such as the Higgs boson, supersymmetry, grand unification and string theory. As also seen, there may be ways of testing some of these ideas using astrophysical and cosmological data. But many of the best prospects for verifying or refuting these ideas will be provided by forthcoming accelerator experiments, including those using bottom-meson factories and long-baseline neutrino beams. However, the centrepiece of the next generation of particle experiments will be the LHC.

This will collide beams of 7 TeV protons and/or lead nuclei with about

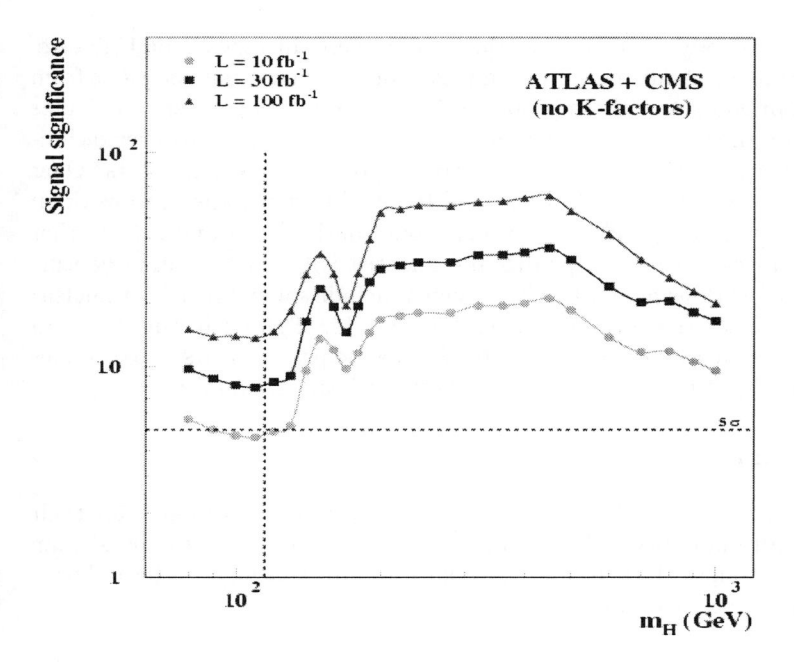

Figure 8. An illustration of the significance to be expected at the LHC for the discovery of a Higgs boson, for different amounts of accumulated collisions [15].

3 TeV per nucleon. The LHC will be installed in the tunnel previously occupied by LEP, and will have four major experiments. Two of them, ATLAS and CMS, are designed for discovering new particles such as the Higgs boson - see Fig. 8 - and/or supersymmetric particles. Another experiment, LHCb, will mainly be looking at matter-antimatter differences in the decays of particles containing bottom quarks. Finally, the ALICE experiment will mainly be studying ultra-relativistic heavy-ion collisions, to look for signatures of the quark-gluon plasma.

Several ideas for possible accelerators beyond the LHC are being proposed at CERN and elsewhere. One possibility could be a linear e^+e^- collider with a centre-of-mass energy in the TeV range. Another possibility could be a complex for producing, accelerating and storing muons, that could be used as a neutrino factory or to make high-energy muon-antimuon collisions.

We believe that the scientific problems studied at CERN are potentially of interest to everybody curious about the structure and history of the Universe. We have a number of outreach programmes aimed at answering questions from the general public about these problems [27]. These questions exert particular fascination on the young, even those who subsequently move into other technical fields, like the majority of students working on CERN experiments. Over 6000 scientists and engineers from around the world are registered to work on experiments at CERN, including groups from South Africa and a few other African countries. CERN is open to any scientist with a promising scientific programme of work using our facilities, which include some directed at nuclear physics, as well as the high-energy experiments highlighted in this talk. In particular, CERN would welcome extended participation in its programmes by more interested members of the Southern African community.

Acknowledgments

It is a pleasure to thank the organizers, particularly Simon Connell, for their kind invitation and special hospitality. I should also like to thank Marian Tredoux and Rodger Hart for their company and for sharing Southern Africa with me on the way to Lüderitz.

References

1. R. Brandelik *et al.*, TASSO Collaboration, Phys. Lett. **B86** (1979) 243; based on a proposal by J. Ellis, M. K. Gaillard and G. G. Ross, Nucl. Phys. **B111** (1976) 253.
2. G. Arnison *et al.*, UA1 Collaboration, Phys. Lett. **B122** (1983) 103; M. Banner *et al.*, UA2 Collaboration, Phys. Lett. **B122** (1983) 476; G. Arnison *et al.*, UA1 Collaboration, Phys. Lett. **B126** (1983) 398; P. Bagnaia *et al.*, UA2 Collaboration, Phys. Lett. **B129** (1983) 130.
3. F. Abe *et al.*, CDF Collaboration, Phys. Rev. Lett. **74** (1995) 2626; S. Abachi *et al.*, D0 Collaboration, Phys. Rev. Lett. **74** (1995) 2632.
4. LEP Electroweak Working Group, http://lepewwg.web.cern.ch/LEPEWWG/Welcome.html.
5. P. W. Higgs, Phys. Rev. Lett. **13** (1964) 508.
6. J. S. Bell, Nucl. Phys. **B60** (1973) 427; C. H. Llewellyn Smith, Phys. Lett. **B46** (1973) 233; J. M. Cornwall, D. N. Levin and G. Tiktopoulos, Phys. Rev. Lett. **30** (1973) 1268.
7. J. Ellis and G. L. Fogli, Phys. Lett. **B232** (1989) 139.
8. J. Ellis, G. L. Fogli and E. Lisi, Phys. Lett. **B274** (1992) 456.
9. See, for example: L. Maiani, in *Electromagnetic Interactions*, Proceedings of the Summer School On Particle Physics, Gif-Sur-Yvette, 1979 (IN2P3, Paris, 1979).
10. H. P. Nilles, Phys. Rept. **110** (1984) 1; H. E. Haber and G. L. Kane, Phys. Rept. **117** (1985) 75.
11. R. Barate *et al.*, ALEPH Collaboration, Phys. Lett. **B495** (2000) 1.
12. M. Acciarri *et al.*, L3 Collaboration, Phys. Lett. **B495** (2000) 18.
13. OPAL collaboration, G. Abbiendi *et al.*, hep-ex/0101014; DELPHI collaboration, P. Abreu *et al.*, preprint 0285, http://delphiwww.cern.ch/~pubxx/www/delsec/papers/public/papers.html.
14. For a compilation of the LEP data presented on Sept. 5th, 2000, see: T. Junk, hep-ex/0101015.
    For a compilation of the LEP data presented on Nov. 3rd, 2000, see: P. Igo-Kemenes, for the LEP Higgs working group, http://lephiggs.web.cern.ch/LEPHIGGS/talks/index.html.
15. ATLAS Collaboration, Chapters 19 and 20 of Physics Technical Design Report, http://atlasinfo.cern.ch/Atlas/GROUPS/PHYSICS/TDR/access.html; CMS Collaboration, Chapter 12 of Technical Proposal, http://cmsinfo.cern.ch/TP/TP.html.

16. D.E. Groom et al., Particle Data Group, European Physical Journal **15** (2000) 1, http://pdg.lbl.gov/.
17. J. Ellis, S. Kelley and D. V. Nanopoulos, Phys. Lett. **B249** (1990) 441; U. Amaldi, W. de Boer and H. Furstenau, Phys. Lett. **B260** (1991) 447; P. Langacker and M. Luo, Phys. Rev. D **44** (1991) 817.
18. For an overview, see: J. N. Bahcall, Phys. Rept. **333-334** (2000) 47.
19. Y. Fukuda et al., Super-Kamiokande Collaboration, Phys. Rev. Lett. **81** (1998) 1562.
20. W. Hu, D. J. Eisenstein and M. Tegmark, Phys. Rev. Lett. **80** (1998) 5255.
21. M. B. Green, J. H. Schwarz and E. Witten, *Superstring Theory* (Cambridge University Press, UK, 1987).
22. J. Ellis, N. E. Mavromatos and D. V. Nanopoulos, gr-qc/9909085 and references therein.
23. For a recent review, see: A. V. Olinto, astro-ph/0011106.
24. G. Amelino-Camelia, J. Ellis, N. E. Mavromatos, D. V. Nanopoulos and S. Sarkar, Nature **393** (1998) 763.
25. See, for example: J. R. Bond et al., MaxiBoom Collaboration, astro-ph/0011379.
26. N. Bahcall, J. P. Ostriker, S. Perlmutter and P. J. Steinhardt, Science **284** (1999) 1481.
27. See http://public.web.cern.ch/Public/ and J. Ellis, physics/0005021.

SUPERSTRINGS: WHY EINSTEIN WOULD LOVE SPAGHETTI IN FUNDAMENTAL PHYSICS

S. JAMES GATES, JR.

Department of Physics
University of Maryland
College Park, MD 20742-4111 USA
E-mail: gatess@wam.umd.edu

There are some questions in physics that until recently could not be answered due to the lack of a complete theory of gravitation. Some of these were, "How does the force of gravity work on objects a billion billions times smaller than the hydrogen atom?" or "What was the universe like, the very instant after the BIG BANG?" or "What is the complete physics of Black Holes?" In these arenas, the effects of gravity and all the other forces must be very different from those seen in everyday experience. Einstein suspected this and it led him to the belief that there must exist a "unified field theory" to describe our world at the tiniest scales. He spent the last forty years of his life unsuccessfully searching for this construction. More recently there appeared new mathematical models called "superstring theory" that have apparently succeeded in reaching his goal. This talk is an introduction to the idea of superstrings and heterotic strings as well as a progress report on the newest frontiers of this subject, "M-theory."

1 The Unification Dream

1.1 A Hidden Classical Clash

Ever since Maxwell synthesized the equations that describe all electromagnetic phenomena, there has been a detectable progress toward the unification of the fundamental laws of physics. The equations of electromagnetism provided a unified description of the electric *and* magnetic phenomena. A major proponent of extending this approach to all of the laws of physics was Albert Einstein who sought to create a theory that he called "the unified field theory." Such a construction would at once describe all the fundamental equations of physics within a single consistent mathematical framework.

Modern physics stands on *two* principle foundations; quantum theory and relativity. These two pillars have enormously extended the range of phenomena that can be given a comprehensive description. Yet for most of the twentieth century there was a hidden clash between these two basic themes. This was epitomized by lack of a consistent theory of *quantum gravity*. There was no completely consistent way to describe the physics of regimes where both quantum and relativistic behaviors must simultaneously be taken into account.

1.2 A New Paradigm

This situation began to dramatically change in the 1980's with the appearance of superstring/heterotic string theories [1,2]. These theories posit that the essential reason that traditional general relativity and traditional quantum theories are inadequate for the task of describing a quantum gravity is because both rely on a concept that has lain at the basis of ideas about physical reality for two millenia. This idea is that the geometical point particle is an appropriate construct with which to begin descriptions of matter and energy as well as space and time themselves. Newton's second law desribes the motion of a *point* particle. Events within a relativistic description are *points* in a four-dimensional space-time manifold. The wavefunction of Schroedinger's equation takes as its argument the coordinates of *points* in space and time. Point are ubiquitous in pre-string physics.

The idea of superstring/heterotic string theory is that the basic visualization of fundamental dynamical entities must be one that occurs by excising the geometrical point concept and replacing it with the idea of the one-dimensional curve, or "little pieces of spaghetti."

2 A Universe of Pasta?

2.1 Basic Ideas

In fact, the essence of string theory is to conceive of a universe in which the most fundamental entities are one dimensional curves or "strings" from which all matter and energy and even space and time themselves are just attributes. The simplest such construction is the bosonic string as originally conceived by Nambu and Goto[3] and improved upon by Polyakov[4]. This can be done by writing an action for the bosonic string

$$S = -\tfrac{1}{2} T \int d\tau \, d\sigma \, e^{-1} [\, \eta^{ab} \eta_{\underline{mn}} (e_a X^{\underline{m}})(e_b X^{\underline{n}}) \,] \quad ,$$

$$X^{\underline{m}}(\tau,\sigma) \; ; \quad \underline{m} = 0, 1, \ldots, d-1 \quad , \tag{2.1}$$
$$e_a(\tau,\sigma) \equiv e_a{}^m(\tau,\sigma) \, \partial_m \quad ,$$
$$\partial_m \equiv (\partial_+, \partial_=) \quad , \quad \partial_+ \equiv \partial_\sigma + \partial_\tau \quad , \quad \partial_= \equiv \partial_\sigma - \partial_\tau \quad .$$

The actual string is described by $X^{\underline{m}}$ which may be identified as a point in a $d-1$ dimensional Minkowski space-time. Unlike a relativistic point particle, however, it is a function of both a proper time coordinate τ and a proper space coordinate σ. The quantity T is a proposed new constant of Nature called the "string tension" and it sets the scale at which the effects of string theory depart from those predicted by a purely point particle based model. When expressed as an equivalent length its magnitudes turns out to be about 10^{-35} meters.

As the string moves, it sweeps out an area known as the "world sheet." If the string is in the form of a closed non-intersecting curve this area is more generally a tube-like structure. In either case this swept out area may be considered as being infinitely flexible, it can be bent deformed, etc. We can even punch holes into the world sheet. This is where another quantity, the quantity $e_a{}^m$, of some importance appears to play an important role in the Polyakov bosonic string action. This mathematical entity, called the "zweibein," encodes information about the punctures or possibly handles described by the world sheet.

2.2 Exactly How Strange Is Our Universe?

As early string theory evolved, it was realized that the action in Eq. (2.1) was not necessarily simultaneously consistent with the laws of special relativity and quantum theory. In fact in order to achieve this, it was shown that the number of "string coordinates" must far exceed the usual number of coordinates of space-time as envisions by Einstein. In fact for the action in (2.1), simultaneous consistency initially seemed possible only for $d = 26$. In other words, our universe would have to have twenty-five *distinct* directions (plus time) in which an object can move!

This was problematical for the use of (2.1) in an attempt to build a fundamental new theory of physics. An improvement in this situation occurs if an additional modification of the string action is undertaken. This modification was suggest by A. Neveu and J. Schwarz [5] and as well P. Ramond [6]. These researchers suggested that some additional terms could be added to Eq. (2.1) in such a way that the number of independent directions is reduced relative to the bosonic string. The key to their approach was to allow for additional dynamical variables to appear. They introduced what we now call "NSR fermions" and usually denoted by $\psi^{\underline{m}}(\tau, \sigma)$. There is a sense in which their suggestion described little pieced of spaghetti that are spinning all along their lengths. For this reason, their constructions are called "spinning strings." In our short review we will not explicitly describe the action for these theories. Suffice it to say that these models achieve simultaneous consistency not with twenty-five distinct directions but only with nine. This still cannot be considered as a viable description of our universe. The NSR construction also has another most unusual property called "supersymmetry."

This observation led Green and Schwarz to find a way in order to make this property more transparent with the construction of "superstrings [1]."

In our world, there are two distinct types of objects whose physics we have observed. One of these classes of objects contain fermions, with the electron being the most obvious example. Fermions obey the Pauli Exclusion Principle (PEP). On the other hand there are objects in Nature that do *not* obey the PEP with the photon being perhaps the premiere example. These latter such objects are called bosons.

Electrons are just one example of fermions that have been observed in particle physics experiments during the last eighty years. By now it is known that the electron is but one member of a family of six objects called "leptons." The other members of this family are; the electron neutrino, muon, muon neutrino, taon and taon neutrino. We group these particles into three pairs (electron, electron neutrino), (muon, muon neutrino) and (taon, taon neutrino) referred to as the first, second and third generation of leptons. In the heart of hadronic matter, there occur other types of fermions called "quarks." There are presently known to be eighteen such object called by the names; up, down, charm, strange, top and beauty quarks. These are also grouped into generations in the most obvious way. Collectively leptons and quarks are called "matter fields."

The photon is also a member of a larger collection of bosonic objects. This collection is responsible for conveying the forces or interactions that act on leptons and quarks. There are presently known four distinct forces that act on matter fields; gravitational, weak, electromagnetic and strong (or chromodynamic) interactions. Each one of these interactions has one or more objects that communicate these forces to matter. For the gravitation interaction there is the "graviton." For the weak interaction there are the "neutral and charged intermediate vector bosons (IVB)" and for the chromodynamic interaction there are eight "gluons."

The most intriguing property about supersymmetry is that were it to be an exact symmetry of nature, then the role of bosons and fermions would not be as sharply distinct as we have observed in the laboratory. In a supersymmetrical universe, some fermions would also be carriers of forces and other bosons would be correspond to matter upon which the forces act. One of the most important question that twenty-first century physics must address is "Does supersymmetry in a broken realization apply to our universe?"

The usual conservation laws of angular momentum, energy and linear momentum arise as consequences of the action in Eq. (2.1). However half-integer spins can only arise after the NSR modification of the bosonic string action. In 1984, a construction was found by D. Gross, J. Harvey, E. Martinec and R.

Rohm [2] (also known as the "Princeton string quartet" since this work was done at Princeton) that contains all of the types of elementary particles described in the last few paragraphs. This mathematical construction is called "the heterotic string." In its original presentation, there were given two distinct descriptions.

10D Fermionic Formulation

$$S_{HET-1} = \int d^2\sigma d\zeta^- \mathbf{E}^{-1} \left\{ -i\tfrac{1}{2} T \eta_{\underline{mn}} (\nabla_+ \mathbf{X}^{\underline{m}})(\nabla_= \mathbf{X}^{\underline{n}}) \right. \\ \left. - (\eta_-{}^I \nabla_+ \eta_-{}^I) \right\} , \quad (2.2)$$

where $I = 1, \ldots 32$.

10D Bosonic Formulation

$$S_{HET-2} = \int d^2\sigma d\zeta^- \mathbf{E}^{-1} \left\{ -\tfrac{1}{2} T [i \eta_{\underline{mn}} (\nabla_+ \mathbf{X}^{\underline{m}})(\nabla_= \mathbf{X}^{\underline{n}})] \right. \\ + i\tfrac{1}{2} [(\nabla_+ \mathbf{\Phi}_R^{\hat{a}})(\nabla_= \mathbf{\Phi}_R^{\hat{a}}) \\ \left. + \Lambda_={}^{\mp}(\nabla_+ \mathbf{\Phi}_R^{\hat{a}})(\nabla_{\mp} \mathbf{\Phi}_R^{\hat{a}})] \right\} , \quad (2.3)$$

where $\hat{a} = 1, \ldots 16$. This second formulation of the heterotic string utilized a discovery [7] that there exists action for 2D world sheets that describe bosons (called "chiral bosons") which move solely in one direction. Right moving chiral bosons may be called "rightons" and left-moving ones "leftons."

Above, we have written these actions using a special mathematical technique known as "the (1,0) superfield formulation[8]" so as to be as concise as possible. In the original analysis it was found that this theory is also a ten-dimensional one that could describe 496 distinct Yang-Mills fields associated with the Lie algebra $E_8 \otimes E_8$. However, neither one of the actions above actually possesses this symmetry. In order to accommodate all 496 gauge fields, solitonic solutions of the equations of motion are promoted to the same role as the fundamental variables in the action. Working in collaboration[10] with W. Siegel in 1988, we were able to find a third formulation for which all the basic degrees of freedom to realize the $E_8 \otimes E_8$ symmetry are present in an action *without* the use of solitons

10D Non − linear Bosonic Formulation

$$S_{HET-3} = \int d^2\sigma d\zeta^- \mathbf{E}^{-1}\Big\{ -\tfrac{1}{2}T[i\eta_{\underline{mn}}(\nabla_+\mathbf{X}^{\underline{m}})(\nabla_=\mathbf{X}^{\underline{n}})] \\
- i\tfrac{1}{4\pi}Tr\{R_+R_= + i\Lambda_={}^\ddagger R_+\nabla_+R_+ \\
+ \tfrac{2}{3}\Lambda_={}^\ddagger\{R_+, R_+\}R_+ \\
+ \int_0^1 dy\,[(\tfrac{d\widetilde{U}}{dy}\widetilde{U}^{-1})[\nabla_=((\nabla_+\widetilde{U})\widetilde{U}^{-1}) \\
+ \nabla_+((\nabla_=\widetilde{U})\widetilde{U}^{-1})]\}\Big\}\,, \qquad (2.4)$$

$$R_a \equiv U^{-1}\nabla_a U\,, \quad U \equiv exp\Big[i\Phi_R{}^{\hat a}t_{\hat a}\Big]\,.$$

From our perspective the most interesting difference between our construction and the ones in (2.2) and (2.4) is that the variables that we use to describe the "internal sector" (i.e. $\Phi_R{}^{\hat a}$) may be thought of as the coordinates of an "isotopic charge space." These are "non-abelian rightons." In the original formulations the internal sectors are described by objects that reside either on the root space or weight spaces (in the sense of Lie algebras) of the isotopic charge space.

3   Compactification Or Not

The process of truncating a higher dimensional string theory down to a four dimensional one is called "compactication." After the initial consideration of the problem of heterotic string compactification [11], there appeared to be a myriad of different techniques;

(a) *Calabi − Yau*
(b) *Calabi − Yau with torsion*
(c) *Covariant lattices*
(d) *Orbifolds*
(e) *Free fermions*
(f) *Asymmetric orbifolds*
(g) *Chiral WZNW coset models*

with which to accomplish this goal. All of these take the viewpoint that the physics we observe in our universe is due to some of the directions of a higher dimensional universe having "curled up" so tightly as to not be observable. The Calabi-Yau approach posits that the extra six directions lie on a special type of complex manifold, hence the name. This method has been particularly popular as a way to obtain non-perterbative information from string theory[12].

A recurrent theme in string theory is that it is the first mathematically based construct that seems to demand that our universe possess more than the traditional four dimension of time (1) and space (3). However, all present observation indicates that our universe is actually four dimensional. Applying Occam's razor, we investigated the question of whether heterotic string theory in particular can be formulated in such a way that there is *no* logical and mathematical need whatsoever to consider higher dimensions? Some of the methods ((b.)-(g)) above follow this philosophy. In our own research we have done this with the added caveat that we wish to restrict ourselves to theories where *all* symmetries must be present in an action. We suceeded in constructing such models [10] that are genuine four dimensional heterotic string actions

$$\text{4D Non-Abelian Bosonic Formulation}$$

$$\begin{aligned}
S_{HET-4} = \int d^2\sigma d\zeta^- \mathbf{E}^{-1} \Big\{ & -\tfrac{1}{2}T\left[i\eta_{\underline{mn}}(\nabla_+\mathbf{X}^{\underline{m}})(\nabla_=\mathbf{X}^{\underline{n}})\right] \\
& - i\tfrac{1}{4\pi}Tr\{R_+R_= + i\Lambda_=^{\neq}R_+\nabla_+R_+ \\
& + \tfrac{2}{3}\Lambda_=^{\neq}\{R_+,R_+\}R_+ \\
& + \int_0^1 dy\,[(\tfrac{d\widetilde{U}}{dy}\widetilde{U}^{-1})[\nabla_=((\nabla_+\widetilde{U})\widetilde{U}^{-1}) \\
& + \nabla_+((\nabla_=\widetilde{U})\widetilde{U}^{-1})]\} \\
& - i\tfrac{1}{4\pi}Tr\{L_+L_= + \Lambda_+^{=}L_=L_= \\
& + \int_0^1 dy\,[(\tfrac{d\widetilde{L}}{dy}\widetilde{L}^{-1})[\nabla_=((\nabla_+\widetilde{L})\widetilde{L}^{-1}) \\
& - \nabla_+((\nabla_=\widetilde{L})\widetilde{L}^{-1})]\} \Big\} \quad , \\
R_a \equiv U^{-1}\nabla_a U \quad , & \quad U \equiv exp\left[i\Phi_R{}^{\hat{a}}t_{\hat{a}}\right] \\
L_a \equiv L^{-1}\nabla_a L \quad , & \quad L \equiv exp\left[i\Phi_L{}^{\hat{a}}\widetilde{t}_{\hat{a}}\right] \quad .
\end{aligned} \qquad (3.5)$$

We empahsize that this action makes absolutely no reference to higher dimensions. The string coordinates here are only *four* in number. The device by which these models work is that they consistently use *both* non-abelian leftons and non-abelian rightons The leftons and rightons may belong to different groups or cosets. For the sake of completeness, we should say that whenever chiral bosons are used, there is also an important but non-dynamic set of fields "notons" that play an important kinematic role.

4 The Quantum/Gravity Resolution and Einstein's Pasta Surprise

One of the marvelous results of accepting the superstring paradigm is that it almost automatically "cures" the clash between perturbative quantum theory and general relativity. This is most obvious if the technique of studying the conformal invariance to the string in the presence of massless condensates is adopted. Such an approach leads to the consideration of 2D non-linear σ-models. An example of this technique appears in the case of the string moving through a background of gravitons described by $g_{mn}(\mathbf{X})$ and in this case the appropriate action is

$$S_\sigma = \int d^2\sigma d\zeta^- \mathbf{E}^{-1} \left\{ -\tfrac{1}{2} T \left[ig_{mn}(\mathbf{X}) (\nabla_+ \mathbf{X}^m)(\nabla_= \mathbf{X}^n) \right] \right\} + \ldots . \quad (4.6)$$

Using standard methods of the background field techniques it was first shown by Friedan [8] that this action leads to equations of the form

$$R_{mn}(\mathbf{X}) = \hbar \mathcal{F}_{mn}(R) + \ldots \quad (4.7)$$

where $\mathcal{F}_{mn}(R)$ is a calculable function of the curvature tension and dependent on the string tension parameter. This function is calculable to any order in $\hbar$.

The σ-model techniques were extended to the other massless condensates[9] of the string, particularly in the case of the four dimensional theory, and the Yang-Mills, Maxwell and Klein-Gordan equations all emerge as did the case of the Einstein equation in Eq. (4.7). These calculation also reveal that these matter fields properly make contributions to the energy-momentum tensor as the sources in Einstein's equation of motion. We thus have explicit proofs that string theory provides a realization of Einstein's long sought after dream of the construction of a "unified field theory." Superstrings/heterotic strings provide a single mathematical structure from which *all* equations of motion for fundamental particles can be derived in a uniform manner.

In the non-perturbative regime, superstring/heterotic string theory has had a remarkable perhaps even spectacular success in unexpected areas. In the area of black hole physics, a highly non-perturbative regime, it has been found that the Hawking-Bekenstein "gray-body factors" appear[13] naturally within the context of superstring/heterotic string theory. Another striking non-perturbative advance[14] occured in the strong coupling regime of Yang-Mills gauge theories. Here it has been shown that the strong coupling regime of a Yang-Mills type theory can, under certain circumstances, be described by a gravitational theory! Correlation functions in the former can be found by looking at equivalent quantities in the latter.

One of the powers of supersting/M-theory is that it allows us new ways to think about the universe. The idea that there may be more than the traditional four directions is one example of this. With such mathematical machinery in hand, whole new possibilities open before us. There has been the recent example of the "Randall-Sundrum" scenario [15]. In this approach, we imagine that the four dimensional universe sits like a pane of glass in a bigger universe of more directions. Such seemingly fanciful an initially unlikely ideas have implications for particle physics!

5 Myth or Reality: M-theory

By the end of the eighties, it was recognized that the low-energy limits of superstrings/heterotic strings did not seem to include the 11D supergravity theory. At present eleven dimensions is the largest dimension in which it is known how to describe a supergravity theory. The M-theory conjecture[16] is generally interpreted as implying the existence of a meta-theory that includes all known superstring/heterotic string theories in different limits and as well as possessing one low energy approximation in which its effective action takes the form (κ^2 is proportional to the 11D Newton constant below)

$$-\kappa^2 S \; = \; \int d^{11}x \; e^{-1} \, [\; R \; + \; \tfrac{1}{2 \cdot 4!} F^{abcd} F_{abcd} \;] \; + \; \tfrac{1}{6} \int A \wedge F \wedge F \; + \; \ldots \; . \tag{5.8}$$

This is the well form of 11D supergravity[17]. M-theory is really a conjecture. Although remarkable progress has been made in confirming many non-trivial tests, no truly fundamental description of this putative meta-theory exists. It is known to be connected via an intricate series of duality maps and compactifications to superstring/heterotic string theories.

At the level of a microscopic viewpoint, M-theory seems to arise from consideration of cutting out a piece of a string and exploring the dynamics of this small bit (sometimes called an "M-parton"). This approach has been most successfully used in the "$D - 0$ brane/M(atrix) model" description[18] of M-theory. There are two major drawback in proceeding this way, however. Firstly to extract the actually eleven dimensional dynamics, a certain parameter must be first sent to infinity. This leaves open the question of the possibility of emergent behavior. Secondly, this approach is not 11D Lorentz covariant.

6 An Unfinished Dream

Superstring/M-theory has shown us a tantalizing new way in which to envision our universe. It has given us a glimpse at mathematical constructs that seem capable of describing all matter and energy and even space and time as arising from a single origin. However, this new paradigm is an unfinished work.

At the APS Centennial meeting in March of 1999, I gave a lecture entitled "A Supersymmetrist Looks at the New Fundamental Physics A'Borning" in which I pointed out that despite the diligent activities by a large fraction of the world's community of theoretical physicists, superstring/M-theory has in a deep sense yet to be 'born.' By this I mean that all of the work of the last decade and a half has failed to produce a true theory. Presently, a more accurate description of "superstring/M-theory" is that it is a growing collection of consistent facts looking for a reason for its existence. Let me try to make this point by the analog I used at the APS meeting.

Imagine that we could build a time machine and go back to the year 1919 and meet with the world's most accomplished collection of physicists to ask a single question, "What is quantum theory?" The answers we would receive would be most interesting from our modern perspective. The introduction of the basic idea of quantum theory began with the Bohr hypothesis (around 1912) as exemplified by the Bohr-Sommerfeld rule. But this is not a *true* theory. Instead this is an ad hoc assumption that is grafted onto classical Newtonian mechanics. It is not until the work of Schröedinger and Heisenberg (around 1925) that it can be said that a true quantum theory was created. Thus, in 1919 most of the world's leading physicists would agree that there was some new paradigm emerging but no one would have a clear vision of its final comprehensive form. This is an analogy to where fundamental physics of superstring/M-theory is today.

Instead of experimental input, the pioneering works by a large hosts of physicists that was initiated by the work in [19] provided *mathematical* input. These are also the analog to the observations of the atomic spectra preceeding Bohr's revelation. Superstring/M-theory is very unusual as an effort to describe the universe around us because it has received *no* experiment input up to this point. This, of course, is very troubling given that physics *must* always look to Nature as the absolute arbitor. What should we expect from a true theory of superstrings/M-theory? Since there is no realiable way to answer this, I can here only report on my intuitions.

There several ingredients that I would expect in our final theory;

(a.) String-like fundamental degrees of freedom that are the argument of a string field functional (analog to the Schröedinger wavefunction.)

(b.) Superspace pre-potential gauge theory-like formulation similar to SUSY YM and supergravity, i.e. manifest supersymmetry and NSR-like (or GS-like) infinite dimensional Lie algebra symmetry.

(c.) Manifest gauge symmetry with the appearance of extended field strengths, gauge transformations and Bianchi-like identities.

(d.) A Maxwell/Schröedinger/Dirac/ Einstein like equation describing dynamics.

All of these are essentially a description of "covariant superstring field theory." Although this is an area whose investigation has not been thoroughly and consistently undertaken during the last fifteen years, I remain convinced that it is only this area that will ultimately provide the tools that will permit us to reach the same level of comprehensive understanding as was provided by Einstein via the equivalence principle for the theory of general relativity. Recently there has been increased activity in this area due to the use of D-branes as probes of the string field potential [20] function.

Finally I wish to close by saying my intuition also leads me to expect that progress in covariant superstring/M-theory field theory is likely to hinge on a very old and almost forgotten problem in the class of supersymmetrical field theory. This is the notorious "off-shell or auxiliary field" problem. If there is to be a successful covariant superstring/M-theory field theory it must contain the solution to this problem. Some of my research of recent [21] has been directed to developing new tools addressed at this problem. We have presented at least one system in which the off-shell problem has been completely solved. It is my hope these will prove of value in the wider realm.

Acknowledgments

I wish to acknowledge the support of the U.S. National Science Foundation (PHY-98-02551) as well as the endowment of the John S. Toll Professorship of Physics at the University of Maryland. Finally I wish to extend my thanks to the organizers (Simon Connell and the entire organization committee) of this meeting held in Lüderitz, Namibia for their kind invitation to attend and be part of the celebration of the career of a remarkable scientist, Friedel Sellschop. Finally I wish to thank Prof. Rudolph Tegen for consideration of during the period of the preparation of this manuscript.

References

1. M. B. Green and J. H. Schwarz, *Nucl. Phys.* **B181**, 502 (1981); idem. *Nucl. Phys.* **B198**, 252 (1982); idem. *Nucl. Phys.* **B198**, 441 (1982); idem. *Phys. Lett.*, **B109** 444 (1982).

2. D. J. Gross, J. A. Harvey, E. Martinec, and R. Rohm, *Phys. Rev. Lett.* **54**, 502 (1985); idem. *Nucl. Phys.* **B256**, 253 (1985).
3. T. Goto, Prog. Theor. Phys. **46**, 1560 (1971); Y. Nambu, Symmetries and Quark Models, eed. R. Chaud, (Gordan and Breach) 269 (1970).
4. A. Polyakov, *Phys. Rev. Lett.* **103**, 207 (1981); ibid. *Phys. Rev. Lett.* **103**, 211 (1981).
5. A. Neveu and J. Schwarz, *Nucl. Phys.* **B31**, 86 (1971)
6. P. Ramond, Nuovo Cim. **A4**, 544 (1971); *Phys. Rev.* **D3**, 2415 (1971).
7. W. Siegel, *Nucl. Phys.* **B238**, 307 (1984).
8. D. Friedan, *Phys. Rev. Lett.*, **45** 1057 (1980); idem. Ann. Phys. **163**, (1985) 318.
9. S. J. Gates, Jr., C. M. Hull, and M. Roček, *Nucl. Phys.* **B248**, 157 (1984); T. Curtright and C. Zachos, *Phys. Rev. Lett.* **53**, 1799 (1984); P. Howe and H. Sierra, *Phys. Lett.* **B148**, 175 (1984); E. S. Fradkin and A. A. Tseytlin, *Phys. Lett.* **B158**, 316 (1985); idem. *Nucl. Phys.* **B261**, 1 (1985); A. Sen, *Phys. Lett.* **B166**, 300 (1986); D. A. Depireux, S. J. Gates, Jr. and Q-Han Park, *Phys. Lett.* **B244**, 364 (1989); S. Bellucci, S. J. Gates, Jr., and D. Depireux, *Phys. Lett.* **B232**, 67 (1989).
10. S. J. Gates, Jr. and W. Siegel, *Phys. Lett.* **B206**, 631 (1988).
11. P. Candelas, G. Horowitz, A. Strominger, and E. Witten; *Nucl. Phys.* **B258**, 46 (1985), L. Dixon, J. Harvey, C. Vafa and E. Witten, *Nucl. Phys.* **B261**, 678 (1985); K. S. Narain, *Phys. Lett.* **B169** 41 (1986); I. Antoniadis, C. Bachas, C. Kounnas and P. Windey, *Phys. Lett.* **B171**, 51 (1986); H. Kawai, D. C. Lewellyn, and S.-H.-H. Tye, *Phys. Rev. Lett.* **57**, 1832 (1986).
12. S. T. Yau, *Proc. of Symposium on Anomalies, Geometry and Topology*, eds. W. A. Bardeen and A. R. White, (World Scientific, Singapore, 1985); A. Strominger and E. Witten, Commun. Math. Phys. **101**, (1986) 341; B. R. Greene, K. H. Kirklin, P. J. Miron and G. G. Ross, *Nucl. Phys.* **B278**, 341 (1986); T. Hübsch, Commun. Math. Phys. **108**, (1987) 291; P. S. Green and T. Hübsch, Commun. Math. Phys. **109**, (1987) 99; ibid. **115**, (1988) 231; R. Schimmrigk, *Phys. Lett.* **B193**, 175 (1987); P. S. Aspinwall, B. R. Greene, K. H. Kirklin and P. J. Miron, *Nucl. Phys.* **B294**, 193 (1987); P. Candelas, A. M. Dale, C. A. Lütken and R. Schimmrigk, *Nucl. Phys.* **B298**, 493 (1988); P. Candelas, C. A. Lütken and R. Schimmrigk, *Nucl. Phys.* **B306**, 113 (1988).
13. A. Strominger and C. Vafa, *Phys. Lett.* **B379**, 99 (1996).
14. J. Maldacena, Adv. Theor. Math, Phys., **2**, 231 (1998).
15. L. Randall and R. Sundrum, *Phys. Rev. Lett.* **83**, 3370 (1999).
16. E. Witten, *Nucl. Phys.* **B443**, 85 (1995); E. Witten and C. Vafa, *Nucl.*

Phys. **B447**, 264 (1995); M. J. Duff, J. T. Liu and R. Minasian, *Nucl. Phys.* **B452**, 261 (1995)
17. E. Cremmer, B. Julia and J. Scherk, *Phys. Lett.* **B176**, 409 (1978); E. Cremmer and B. Julia, *Phys. Lett.* **B78**, 48 (1978); idem. *Nucl. Phys.* **B159**, 141 (1979).
18. T. Banks, W. Fischler, S. H. Shenker and L. Susskind, *Phys. Rev.* **D55**, 5112 (1997).
19. G. Veneziano, Nuovo Cim. **57A**, 190 (1968); idem. Phys. Rep. **C9**, 199 (1974); J. A. Shapiro, *Phys. Rev.* **D179**, 1345 (1969); idem. *Phys. Rev.* **D179** 1945 (1972); Z. Koba and H. B. Nielsen, *Nucl. Phys.* **B10**, 633 (1969); C. Lovelace, *Phys. Lett.* **B34**, 500 (1971).
20. A. Sen, JHEP **9810**, 021 (1998), hep-th/9809111, idem. JHEP **9812**, 021 (1998) hep-th/9812031, idem. JHEP **9812**, 021 (1999) hep-th/9911116; N. Berkovits, hep-th/0008145; JHEP **0009**, 046 (2000) hep-th/0006003; N. Berkovits, A. Sen and B. Zweibach, *Nucl. Phys* B **587**, 147 2000.
21. S. J. Gates, Jr., L. Rana, *Phys. Lett.* **B352**, 50 (1995); idem. *Phys. Lett.* **B369**, 262 (1996); idem. *Phys. Lett.* **B438**, 80 (1998); C. Curto, S. J. Gates, Jr. and V. G. J. Rodgers, *Phys. Lett.* **B480**, 337 (2000).

COMMON FEATURES OF PARTICLE MULTIPLICITIES IN HEAVY ION COLLISIONS.

F. BECATTINI

Università di Firenze and INFN Sezione di Firenze, Largo E. Fermi 2, I-50125, Florence, Italy

J. CLEYMANS

Department of Physics, University of Cape Town, Rondebosch 7701, Cape Town, South Africa

A. KERÄNEN, E. SUHONEN

Department of Physical Sciences, University of Oulu, FIN-90571 Oulu, Finland

K. REDLICH

Institute of Theoretical Physics, University of Wroclaw, PL-50204 Wroclaw, Poland.

Results of a systematic study of fully integrated particle multiplicities in central Au–Au and Pb–Pb collisions at beam momenta of 1.7A GeV, 11.6A GeV (Au–Au) and 158A GeV (Pb–Pb) using a statistical-thermal model are presented. The close similarity of the colliding systems makes it possible to study heavy ion collisions under definite initial conditions over a range of centre-of-mass energies covering more than one order of magnitude. We conclude that a thermal model description of particle multiplicities, with additional strangeness suppression, is possible for each energy. The degree of chemical equilibrium of strange particles and the relative production of strange quarks with respect to u and d quarks are higher than in e^+e^-, pp and p$\bar{p}$ collisions at comparable and even at lower energies. The average energy per hadron in the comoving frame is always close to 1 GeV per hadron despite the fact that the energy varies more than 10-fold.

1 Introduction

It is becoming more and more clear that results from relativistic heavy ion collisions at many different energies[1] show striking common traits. Statistical-thermal models are able to reproduce particle multiplicities in a satisfactory manner by using a very small number of parameters: temperature, volume, baryon chemical potential and a possible strange-quark suppression parameter, γ_s [2]. We report here the results[3] of an analysis of data from collisions at several different energies, with emphasis on the similarity of the colliding system. We have focussed our attention on central Au–Au collisions at beam momenta of 1.7A GeV (SIS) [4], 11.6A GeV (AGS) [5] and on central Pb–Pb collisions at 158A GeV (SPS) beam momentum [6]. As far as the choice of data (and, consequently, colliding system) is concerned, our leading rule is the availability of full phase space integrated multiplicity measurements because a pure statistical-thermal model analysis of particle yields, without any consideration of dynamical effects, *may* apply only in this case [7]. Such data, however, exist only in a few cases and whenever legitimate we have extrapolated spectra measured in a limited rapidity window to full phase space. The use of extrapolations is more correct than using data over limited intervals of rapidity, especially in the framework of a purely statistical-thermal analysis without a dynamical model. Moreover, the usually employed requirement of zero strangeness ($S = 0$) demands fully integrated multiplicities because strangeness does not need to vanish in a limited region of phase space.

In order to assess the consistency of the results obtained, we have performed the statistical-thermal model analysis by using two completely independent numerical algorithms whose outcomes turned out to be in close agreement throughout. Similar analyses have been recently made by other authors (see e.g. [9,10]); however, both the model and the used data set differ in several important details, such as the assumption of full or partial equilibrium for some quark flavours, the number of included resonances, the treatment of resonance widths, inclusion or not of excluded volume corrections, treatment of flow, corrections due to limited rapidity windows etc. Because of these differences it is difficult to trace the origin of discrepancies between different results. We hope that the present analysis, covering a wide range of beam energies using a consistent treatment, will make it easier to appreciate the energy dependence of the various parameters such as temperature and chemical potential.

2 Data set and model description

As emphasized in the introduction, in the present analysis we use the most recent available data, concentrating on fully integrated particle yields and discarding data that have been obtained in limited kinematic windows. We have derived integrated multiplicities of π^+, Λ and proton in Au–Au collisions at AGS by extrapolating published rapidity distributions [12,13,14] with constrained mid-rapidity value (y_{NN}=1.6). For proton and Λ we have fitted the data to Gaussian distributions, whilst for π^+ we have used a symmetric flat distribution at midrapidity with Gaussian-shaped wings on either side; the point at which the Gaussian wing and the plateau connect is a free parameter of the fit. The fits yielded very good χ^2's/dof: 0.27, 1.24 and 1.00

for π^+, proton and Λ respectively. The integrated multiplicities have been taken as the area under the fitted distribution between the minimal y_{min} and maximal y_{max} values of rapidities for the reactions NN → πNN, NN → ΛK for pions and Λ's respectively; the difference between these areas and the total area has been taken as an additional systematic error. The area between y_{min} and y_{max} amounts to practically 100% of the total area for pions and about 95% for Λ's.

We have not included data on deuteron production because of the possible inclusion of fragments in the measured yields. This is particularly dangerous at low (SIS) energies where inclusion or not of deuterons modifies thermodynamic quantities like ϵ/n [15].

The data analysis has been performed within an ideal hadron gas grand-canonical framework supplemented with strange quark fugacity γ_s. In this approach, the overall average multiplicities of hadrons and hadronic resonances are determined by an integral over a statistical distribution:

$$\langle n_i \rangle = (2J_i + 1)\frac{V}{(2\pi)^3} \int d^3p \, \frac{1}{\gamma_s^{-s_i} \exp\left[(E_i - \boldsymbol{\mu} \cdot \mathbf{q}_i)/T\right] \pm 1} \qquad (1)$$

where $\mathbf{q}_i$ is a three-dimensional vector with electric charge, baryon number and strangeness of hadron i as components; $\boldsymbol{\mu}$ the vector of relevant chemical potentials; J_i the spin of hadron i and s_i the number of valence strange quarks in it; the + sign in the denominator is relevant for fermions, the − for bosons. This formula holds in case of many different statistical-thermal systems (i.e. clusters or fireballs) having common temperature and γ_s but different arbitrary momenta, provided that the probability of realizing a given distribution of quantum numbers among them follows a statistical rule [8,16]. In this case V must be understood as the sum of all cluster volumes measured in their own rest frame. Furthermore, since both volume and participant nucleons may fluctuate on an event by event basis, V and $\boldsymbol{\mu}$ (and maybe T) in Eq. (1) should be considered as average quantities [8].

The overall abundance of a hadron of type i to be compared with experimental data is determined by the sum of Eq. (1) and the contribution from decays of heavier hadrons and resonances:

$$n_i = n_i^{\text{primary}} + \sum_j \text{Br}(j \to i) n_j \qquad (2)$$

where the branching ratios $\text{Br}(j \to i)$ have been taken from the 1998 issue of the Particle Data Table [17].

It must be stressed that the unstable hadrons contributing to the sum in Eq. (2) may differ according to the particular experimental definition. This is a major point in the analysis procedure because quoted experimental multiplicities may or may not include contributions from weak decays of hyperons and K_S^0. We have included all weak decay products in our computed multiplicities except in Pb–Pb collisions on the basis of relevant statements in ref. [18] and about antiproton production in refs. [11,19]. It must be noted that switching this assumption in Au–Au at SIS and AGS does not affect significantly the resulting fit parameters.

The overall multiplicities of hadrons depend on several unknown parameters (see Eq. (1)) which are determined by a fit to the data. The free parameters in the fit are T, V, γ_s and μ_B (the baryon chemical potential) whereas μ_S and μ_Q, i.e. the strangeness and electric chemical potentials, are determined by using the constraint of overall vanishing strangeness and forcing the ratio between net electric charge and net baryon number Q/B to be equal to the ratio between participant protons and nucleons. The latter is assumed to be Z/A of the colliding nucleus in Au–Au and Pb–Pb.

For SIS Au-Au data we have required the exact conservation of strangeness instead of using a strangeness chemical potential. This gives rise to slightly more complex calculations which are necessary owing to the very small strange particle production (Au–Au). The difference between these strangeness-canonical calculation and pure grand-canonical calculation of multiplicities of K and Λ for the final set of thermal parameters (see Table 1) turns out to be as large as a factor 15 in Au–Au at $1.7A$ GeV.

Owing to few available data points in SIS Au–Au collisions, we have not fitted the volume V nor the γ_s therein. The volume has been assumed to be $4\pi r^3/3$ where $r = 7$ fm (approximately the radius of a Au nucleus) while γ_s has been set to 1, the expected value for a completely equilibrated hadron gas. Since we have performed strangeness-canonical calculation here, the yield ratios involving strange particle are not independent of the chosen volume value as in the grand-canonical framework. Thus, in this particular case, V is meant to be the volume within which strangeness is conserved (i.e. vanishing) and not the global volume defining overall particle multiplicities as in Eq. (1). Also, in order to test the dependence of this assumption on our results, we have repeated the fit by varying V by a factor 2 and 0.5 in turn.

A major problem in Eq. (2) is where to stop the summation over hadronic states. Indeed, as mass increases, our knowledge of the hadronic spectrum becomes less accurate; starting from ≈ 1.7 GeV many states are possibly missing, masses and widths are not well determined and so are the branching ratios. For this reason, it is unavoidable that a cut-off on hadronic states be introduced in Eq. (2). If the calculations are sensitive to the value of this cut-off, then the reliability of results is questionable. We have performed all our calculations with two cut-offs, one at around 1.8 GeV (in the analysis algorithm A) and the other one at 2.4 GeV (in the analysis algorithm B). The contribution of missing heavy resonances is expected to be very important for temperatures $\geq$ 200 MeV making thermal models inherently unreliable above this temperature.

Table 1. Summary of fit results. Free fit parameters are quoted along with resulting minimum χ^2's and λ_s parameters.

	Average
Au–Au 1.7A GeV	
T (MeV)	49.6±2.5
μ_B (MeV)	813±23
γ_s	1 (fixed)
V(fm^3)	1437 (fixed)
λ_s	0.0054±0.0035
Au–Au 11.6A GeV	
T (MeV)	119.8±8.3
μ_B (MeV)	553.5±16
γ_s	0.720±0.097
$VT^3 \exp(-0.7\text{GeV}/T)$	2.03±0.34
λ_s	0.43±0.10
Pb–Pb 158A GeV	
T (MeV)	158.1±3.2
μ_B (MeV)	238±13
γ_s	0.789±0.052
$VT^3 \exp(-0.7\text{GeV}/T)$	21.7±2.6
λ_s	0.447±0.025

3 Results

As mentioned in the introduction, we have performed two analyses (A and B) by using completely independent algorithms.
In the analysis A all light-flavoured resonances up to 1.8 GeV have been included. The production of neutral hadrons with a fraction f of $s\bar{s}$ content has been suppressed by a factor $(1 - f) + f\gamma_s^2$. In the analysis B the mass

cut-off has been pushed to 2.4 GeV and neutral hadrons with a fraction f of $s\bar{s}$ content have been suppressed by a factor γ_s^{2f}. Both algorithms use masses, widths and branching ratios of hadrons taken from the 1998 issue of Particle Data Table [17]. However, it must be noted that differences between the two analyses exist in dealing with poorly known heavy resonance parameters, such as assumed central values of mass and width, where the Particle Data Table itself gives only a rough estimate. Moreover, the two analyses differ by the treatment of mass windows within which the relativistic Breit-Wigner distribution is integrated.

A summary of the final results is shown in Fig. 1. For each analysis an estimate of systematic errors on fit parameters have been obtained by repeating the fit

- assuming vanishing widths for all resonances
- varying the mass cut-off to 1.7 in analysis A and to 1.8 in analysis B
- for Au–Au at 1.7A GeV, the volume V has been varied to $V/2$ and to $2V$ (see discussion in Sect. 2)

The differences between new fitted parameters and main parameters have been conservatively taken as uncorrelated systematic errors to be added in quadrature for each variation (see Table 1). The effect of errors on masses, widths and branching ratios of inserted hadrons has been studied in analysis A according to the procedure described in ref. [8] and found to be negligible. Finally, the results of the two analyses have been averaged according to a method suggested in ref. [22], well suited for strongly correlated measurements.

4 Discussion and conclusions

From the results obtained, an indication emerges that a statistical-thermal description of multiplicities in a wide range of heavy ion collisions is indeed possible to a satisfactory degree of accuracy, for beam momenta ranging from 1.7A GeV to 158A GeV per nucleon. Furthermore, the fitted parameters show a remarkably smooth and consistent dependence as a function of centre-of-mass energy.

The temperature varies considerably between the lowest and the highest beam energy, namely, between 50 MeV at SIS and 160 MeV at SPS. Similarly, the baryon chemical potential changes appreciably, decreasing from about 820 MeV at SIS to about 240 MeV at SPS. However, since the changes in temperature and chemical potential are opposite, the resulting energy per particle shows little variation and remains practically constant at about 1 GeV per

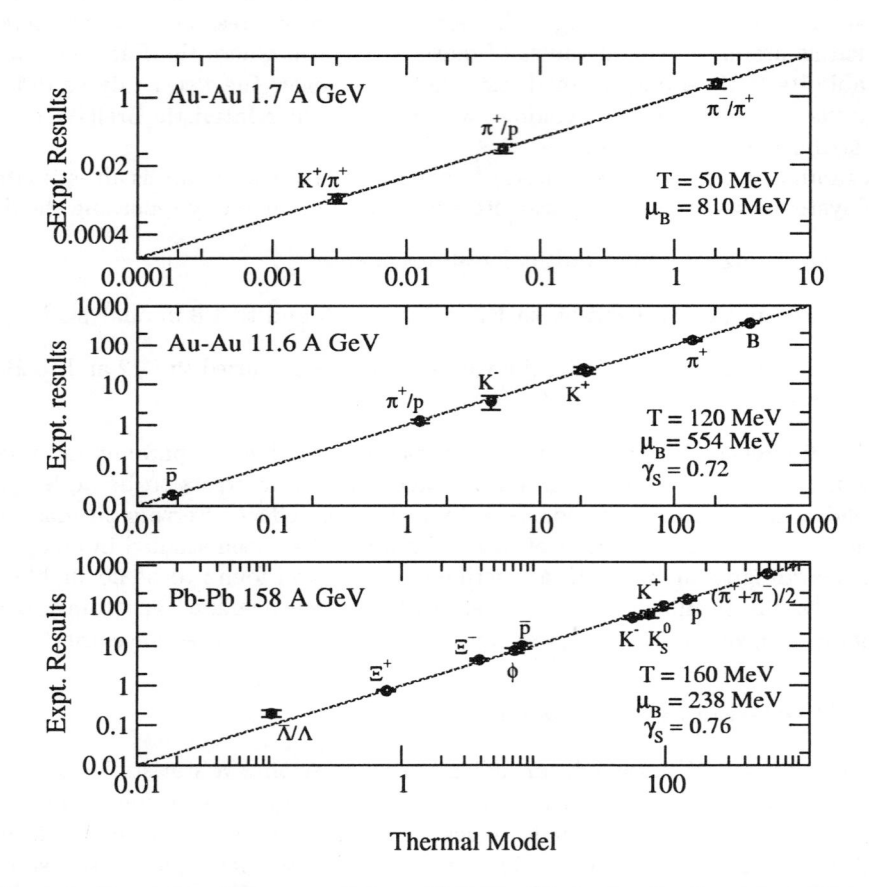

Figure 1. Comparison between particle multiplicities fitted using the thermal model and experimental results.

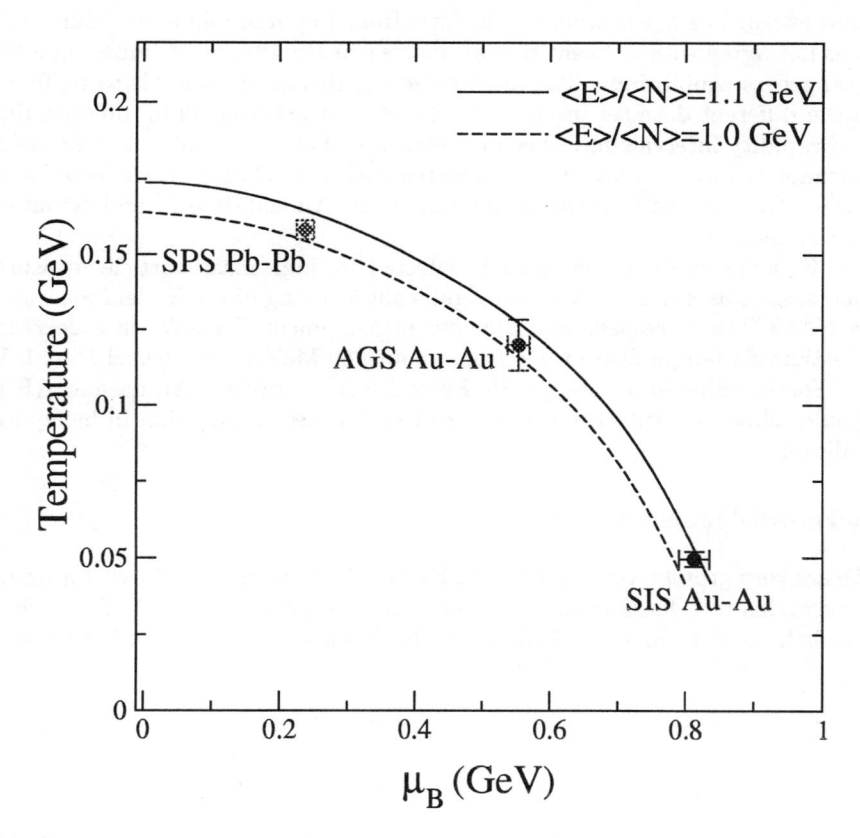

Figure 2. Fitted temperatures and baryon-chemical potentials plotted along with curves of constant energy per hadron.

particle; this is shown in Fig. 2.

The supplementary γ_s factor, measuring the deviation from a completely equilibrated hadron gas, is around 0.7 – 0.8 at all energies where it has been considered a free fit parameter. At the presently found level of accuracy, a fully equilibrated hadron gas (i.e. $\gamma_s = 1$) cannot be ruled out in all examined collisions except in Pb–Pb, where γ_s deviates from 1 by more than 4σ. This result does not agree with a recent similar analysis of Pb–Pb data [9] imposing a full strangeness equilibrium. The main reason of this discrepancy is to be found in the different data set used; whilst in ref. [9] measurements in different limited rapidity intervals have been collected, we have used only particle yields extrapolated to full phase space. The temperature values that we have found essentially agree with previous analyses in Au–Au collisions [20] and estimates 11.7 A GeV [23].

The T value in Pb–Pb is strongly affected by high mass particle measurements, such as ϕ and Ξ. A recent significant lowering of the Ξ yield measured by NA49 [21] with respect to a previous measurement [24] results in a decrease of estimated temperature value from about 180 MeV to the actual 160 MeV.

Forthcoming lower energy Pb–Pb and high energy Au–Au data at RHIC should allow to clarify the behaviour of strangeness production in heavy ion collision.

Acknowledgements

We are very grateful to N. Carrer, U. Heinz, M. Morando, C. Ogilvie for useful suggestions and discussions about the data. We especially thank H. Oeschler for his help with the GSI SIS data and R. Stock for his help with NA49 data.

References

1. U. Heinz, Nucl. Phys. A661 (1999) 140c.
2. J. Letessier, J. Rafelski, A. Tounsi, Phys. Rev. C64 (1994) 406; C. Slotta, J. Sollfrank, U. Heinz, Proc. of Strangeness in Hadronic matter, J. Rafelski (Ed.), AIP Press, Woodbury 1995, p. 462.
3. F. Becattini, J. Cleymans, A. Keränen, E. Suhonen and K. Redlich, (in preparation).
4. see e.g. H. Oeschler, Lecture Notes in Physics 516, "Hadrons in dense matter and hadrosynthesis", Springer-Verlag (1999), Eds. J. Cleymans, H.B. Geyer, F.G. Scholtz.
5. see e.g. C. Ogilvie, Nucl. Phys. A638 (1997) 57c.
6. see e.g. R. Stock, Nucl. Phys. A661 (1999) 282c.
7. J. Cleymans and K. Redlich, Phys. Rev. C60 (1999) 054908.
8. F. Becattini, M. Gaździcki and J. Sollfrank, Eur. Phys. J. C5 (1998) 143.
9. P. Braun-Munzinger, I. Heppe and J. Stachel, Phys. Lett., B465 (1999) 15.
10. J. Letessier and J. Rafelski, Nucl. Phys. A661 (1999) 497c.
11. T. Abbott et al., E-802 collaboration, Nucl. Phys. A525 (1994) 455c.
12. L.Ahle et al., E-802 Collaboration, Phys. Rev. C59 (1999) 2173.
13. S. Ahmad et al., Phys. Lett. B382 (1996) 35.
14. L.Ahle et al., E-802 Collaboration, Phys. Rev. C60 (1999) 0649001.
15. J. Cleymans and K. Redlich, Phys. Rev. Lett. 81, 5284 (1998).
16. F. Becattini, Lecture Notes in Physics 516, "Hadrons in dense matter and hadrosynthesis", Springer-Verlag (1999), Eds. J. Cleymans, H.B. Geyer, F.G. Scholtz.
17. Particle Data Group, Eur. Phys. J. C3 (1998) 1.
18. F. Sikler (NA49 Collaboration), Nucl. Phys. A661 (1999) 45c.
19. L.Ahle et al., E-802 Collaboration, Phys. Rev. Lett. 81 (1998) 2650.
20. J. Cleymans, H. Oeschler and K. Redlich, Phys. Rev. C59 (1999) 1663 and references therein.
21. R.A. Barton, NA49 Collaboration, talk given at the "Strangeness 2000" conference, Berkeley, (July 1999), to be published in the proceedings of the conference.
22. M. Schmelling, Phys. Scripta 51 (1995) 676.
23. J. Stachel, Nucl. Phys. A610 (1996) 509c.
24. H. Appelshauser et al., NA49 Collaboration, Phys. Lett. B444 (1998) 523.

THE INFLUENCE OF STRONG CRYSTALLINE FIELDS ON QED-PROCESSES INVESTIGATED USING DIAMOND CRYSTALS IN γ,γ COLLIDERS

E. UGGERHØJ

Institute for Storage Rings Facilities, ISA
University of Aarhus, Denmark
ugh@ifa.au.dk

The very recent indications of Higgs–candidates at CERN have led to a strong interest in new types of facilities like high–energy photon colliders. This again leads to a search for strong high–energy gamma sources. In the present paper it is shown that single crystals are unique radiators due to the strong crystalline fields of 10^{12} V/cm or more, in which incident particles move over very large distances ($\sim 100\mu$m). Along axes, radiation emission and energy loss is enhanced more than two orders of magnitude. This dramatic effect leads to radiation cooling followed by capture to high–lying channeling states. The radiation is emitted in the forward angular cone of 40 μrad or less. In the planar cases certain incident directions give hard photons with an intensity ~ 10 times the normal coherent bremsstrahlung. Therefore, in general, crystals turn out to be very interesting γ-sources for photo production and coming γ,γ colliders.

1 Introduction

With the possible signs of the elusive Higgs particle on the horizon, a strong interest has come up on the possibility of new techniques to produce Higgs particles. In LEP, the Higgs hints were produced by a e^+/e^- collision, giving back–to–back z and Higgs particles but the signals were difficult to disentangle from more common processes. A more clean way of producing Higgs particles would be in a γ,γ collider. Such a facility would require an efficient technique to convert the e^+/e^- beams to high–energy γ–rays. For this purpose, crystals are very interesting, as will be shown below.

In the present paper some of the most recent experimental results[1] from NA–43 are presented concerning radiation emission, photon multiplicities, radiation cooling, energy loss, and photon emission angles. For details, the reader is referred to Ref. 1 and references therein. See Ref. 2 for an introduction to channeling and strong channeling fields.[2] Last but not least, the possibility of using crystals as γ-sources for photon colliders is discussed.

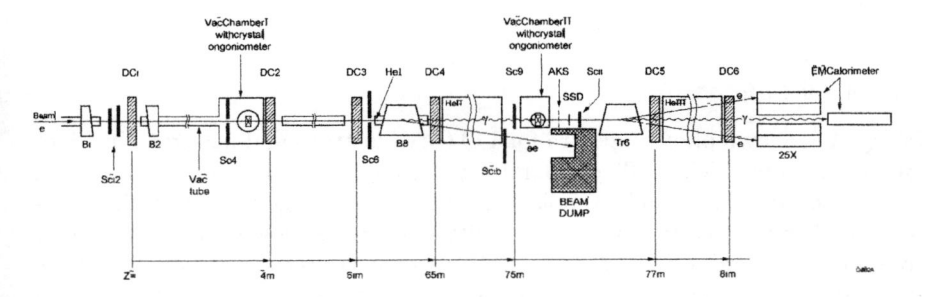

Figure 1. A schematic drawing of the setup used in NA-43. See text for details.

2 Experiment

The experiment was performed in the North Area of the CERN SPS. The beam (H_2) is a tertiary one containing electrons, positrons or pions with energies ranging from 35 GeV to 300 GeV. The experimental arrangement is shown schematically in Fig. 1. Drift chambers ($DC_{1,2,3,4}$) define incident and exit angles ($\theta_{in}, \theta_{out}$) together with exit paticle momentum. $DC_{5,6}$ and a magnet (Tr 6) act as a pair spectrometer (PS) with a minimum photon energy of 5 GeV. A fully depleted solid state detector (SSD) with a 1 mm thick Pb-foil in front measures photon multiplicities for photon energies $E_\gamma \geq 0.5$ GeV. The PS is also used for photon multiplicities.

For the pair spectrometer, the specific energy of each converting photon is measured, whereas when using the SSD, only the total energy of all photons is measured in the calorimeter. The pair spectrometer can also be used to find the direction of the emitted photons. Since the photons are emitted with $1/\gamma$, which is comparable to the angular resolution of the drift chambers, the approximate particle direction can be measured at the moment of photon emission. The energy of these photons is measured by the lead–glass array. In this way the angular distribution of the electron beam can be measured in front and behind the first crystal *but* also the direction of the electron just before emitting the photon *inside* the crystal can be detected, see below.

3 Radiation Emission and Photon Multiplicities

In the following, the experimental radiation spectra are plotted as a function of emitted energies. The emitted radiation intensity is normalized to that from an equivalent amorphous target of the same thickness – giving the plotted enhancement.

3.1 Axial Case

In Fig. 2 is shown photon spectra for 149 GeV electrons and positrons incident on the 0.7 mm thick ⟨110⟩ diamond crystal for which the channeling angle, $\psi_1 = 30\mu$rad. The incident polar angle regions are given below the spectra. Further on, two types of multiplicity curves are shown: One type (right column) measured with the solid state detector (SSD) – and another type (middle column) measured with the pair spectrometer. For very well–aligned particles ($\theta_{in} : 0-10\mu$rad) the photon spectra agree with the general channeling picture for e^+/e^-, i.e. electrons are focused around the target nuclei and emit hard photons, whereas positrons are pushed away from the strong crystalline fields and emit mostly softer photons. For increasing incident angles the $e^+/e^=$ spectra agree fairly well – apart from the very pronounced peak at $0.8 \cdot E_0$. In Fig. 3 is shown the number of particles in the incident beam giving rise to radiative energy loss in the photon peak ($E_{rad} : (0.6-1.0)E_0$). For incident–angle regions of (0-10) μrad, about 40% of all 149 GeV particles lose more than 60% of their energy E_0. For 243 GeV electrons this number goes down to 25%. In both cases these numbers are about 10 times higher than what is obtained by surface transmission. These results can be understood by assuming a very pronounced radiative cooling in diamond crystals.[3,4] In Ref. 1 is shown that all these particles exit the crystals with angles (θ_{out}) close to the Lindhard angles calculated for the final particle energies $E_f = E_0 - E_{rad}$.

3.2 Planar Case

In an earlier experiment[5] some of us found for the first time a very pronounced high–energy photon peak when 149 GeV electrons are incident along the (110) planes and at 0.3 mrad to the ⟨100⟩ axis in a diamond crystal. In Fig. 4 is shown the same type of effect incident at $\pm 10\mu$rad to the (111) planes in the 0.7 mm diamond crystal and at 0.6 mrad to the ⟨100⟩ axis.

The very pronounced peak at 110 GeV is due to this new type of coherent bremsstrahlung emitted when the electrons cross the rows of atoms forming the (111) crystal planes. In the Lindhard theory these incident directions are called 'the strings of strings region' [SOS]. The photons are expected to be nearly 100% planar polarized – like CB is in many cases. An experimental investigation of the polarization of these photons was obtained recently by NA–43[6] by comparing the pair production along two perpendicular crystal planes in a second crystal placed 40 m behind the radiator. Here, large asymmetries were found – implying a high degree of polarization. In Fig.4b the

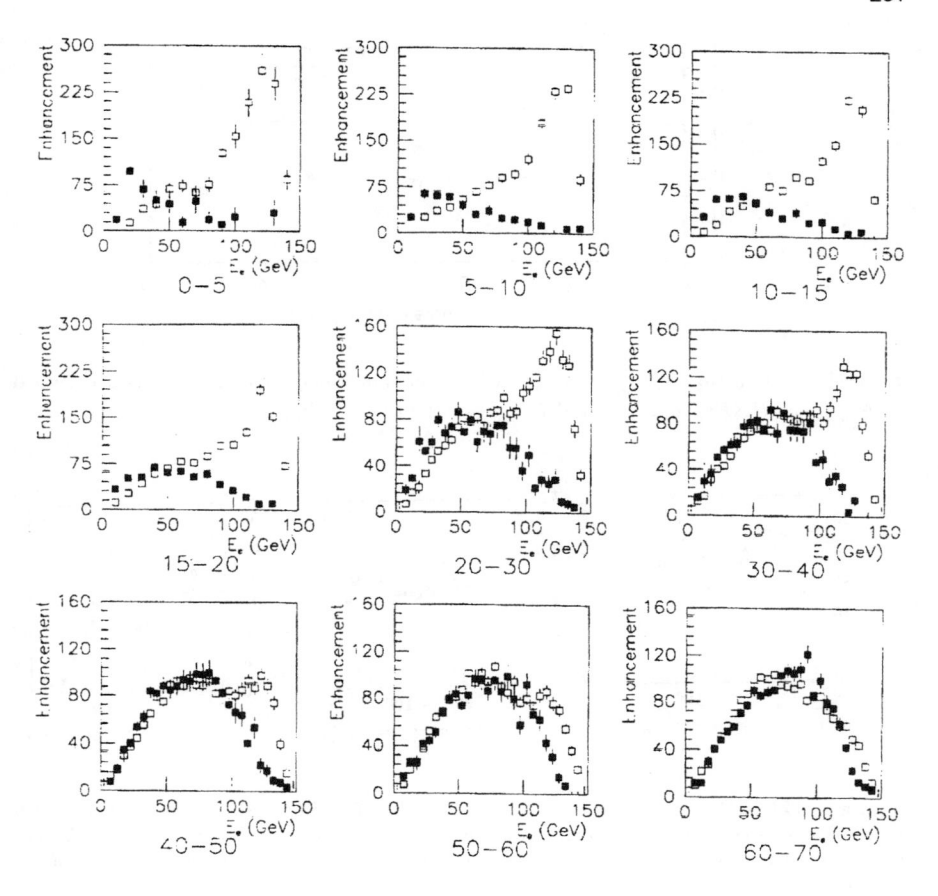

Figure 2. Enhancement (left column), pair spectrometer multiplicity)middle column) and SSD multiplicity (right column) as a function of different incident angles fiven in units of μrad. The open squares are for electrons whereas the filled ones are for positrons. Note the change of vertical scales.

corresponding multiplicity spectrum is shown from which it appears that the electron emits about 1.7 photons each with an energy above 0.5 GeV in the region of the peak. Finally, in Fig. 4c the 'single photon spectrum', i.e., the energies of the emitted photons, is shown and it is proven that the 'string of strings'–peak consists of very high energy protons followed by less energetic

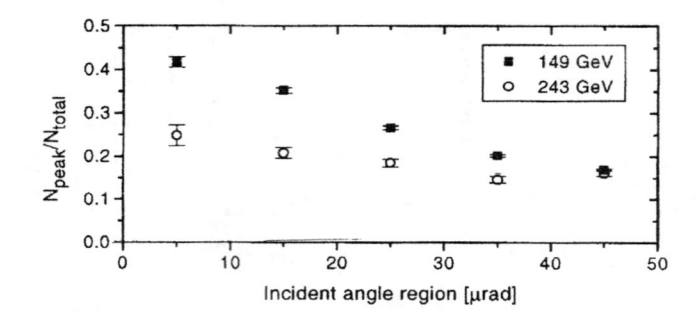

Figure 3. The fraction $N_{\text{peak}}/N\text{total}$ as a function of incident angle. The values are defined as $N_{\text{peak}} = N(\Delta E\epsilon[0.6;1] \cdot E_e$ and $E_{\text{total}} = N(\Delta E\epsilon[0;1] \cdot E_e)$

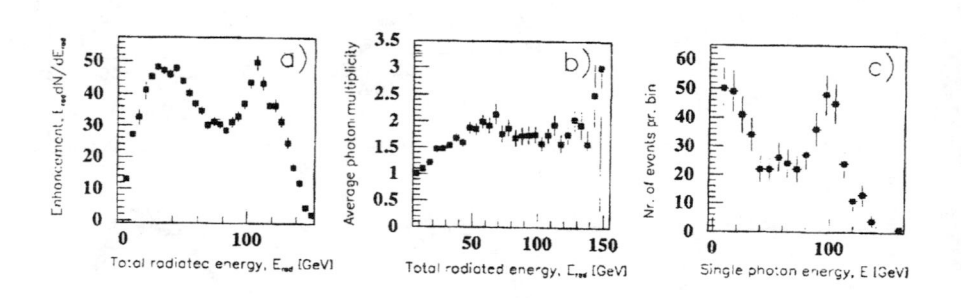

Figure 4. 149 GeV electrons on the (111) plane of 0.7 mm diamond, 0.6 mrad to the ⟨110⟩ axis. a) shows the radiation enhancement as a function of total radiation energy, b) the photon multiplicity, and c) the number of photons emitted as a function of the energy of the photon.

ones. We emphasize that Fig. 4c shows a counting spectrum, not a power spectrum; the effect is thus very strong.

This effect is now being used as a γ–source in the new CERN collaboration – NA–59.[7] Here, a second crystal is used to turn planar polarized photons into circular ones. These circularly polarized photons could open new 'windows' in high–energy physics, like measuring the contribution from gluons to the spin of the nucleon.

Further on it should be pointed out that this so–called 'string of strings' (SOS) incident angle region would be an excellent source for high–energy gamma–rays. The enhancement is a factor of 10 larger than for normal CB.

The effect could also be used in coming photon colliders.

4 Radiative Energy Loss

The energy loss for GeV electrons and positrons is practically all due to radiation emission. In Fig. 5 is shown the average radiative energy loss of 149 GeV electrons and positrons incident along the (110) axis. From the curves, the channeling effect for positively and negatively charged particles is clear. The overall radiative energy loss is dramatic, i.e., well–aligned electrons lose around half their energy in a 0.7 mm thick crystal where, for a comparison, the energy loss in an amorphous 0.7 mm foil is less than 1 GeV. The energy loss is hence enhanced almost two orders of magnitude. Secondly, it should be noticed that this strongly enhanced energy loss is not just found for the rather small channeling angles – it continues far outside the channeling angular region. Here it only decreases rather slowly for increasing incident angles, which is due to the strong crystalline fields. For increasing particle energy the potential region for distances $r_\perp \cong (4-5)a$ from the axis becomes more and more important as pointed out by Kononets.[8]

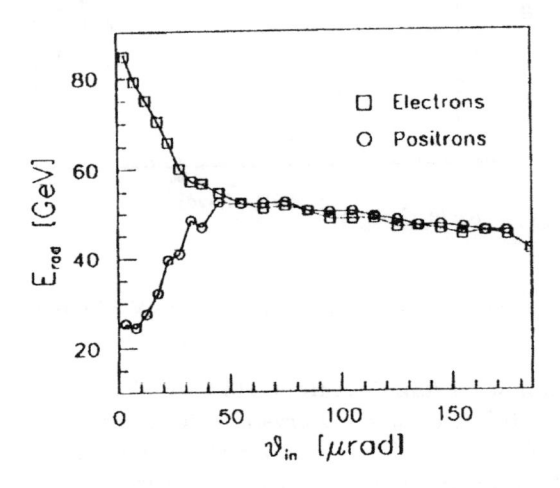

Figure 5. Average radiation energy by 149 GeV electrons and positrons.

5 Radiative Cooling and Capture

The dramatic enhancements of radiation emission from multi–GeV electrons/positrons traversing single crystals have through the years raised the question about radiative cooling – or: Is it possible to reduce the transverse energy of particles by going through a crystal and thereby obtain smaller exit angles θ_{out} than incident angles θ_{in} ? When a particle emits a high–energy photon its transverse energy $E_\perp = \gamma m v \psi^2 + U(r_\perp)$ decreases and thereby the angle θ to the crystal axis decreases as well. This so–called radiative cooling will counteract the multiple scattering and the particle might come out from the crystal with a smaller angle to the axis than the incident one. In

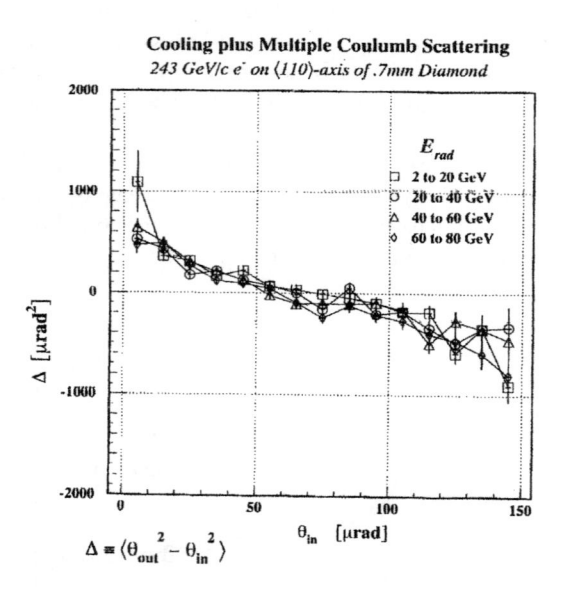

Figure 6. Radiation cooling for 243 GeV relectrons.

Fig. 6 is shown new cooling data from 243 GeV electrons incident along the $\langle 110 \rangle$ axis in the 0.7 mm diamond crystal. Here, very strong cooling effects are found as compared to 149 GeV data,[1] the cooling here is giving negative $\Delta \equiv \theta_{out}^2 - \theta_{in}^2$ already for a radiative energy loss of a few GeV. According to theory[8] the radiative cooling scales as $(\delta \langle \theta^2 \rangle / \delta L)_{rad} \propto Z^3$ for small χ and $(\delta \langle \theta^2 \rangle / \delta L)_{rad} \propto Z^2$ for $\chi \gg 1$. Compared to the multiple scattering

$(\delta\langle\theta^2\rangle/\delta L)_{\rm ms} \propto Z^2(E^2)$, this means that the net radiative cooling is expected to be very strong at 243 GeV, as observed.

For incidence within (ψ_1) the exit angles are practically all inside the channeling regime showing that nearly all incident particles experience radiative capture to channeling states – a very exceptional situation for negatively charged particles.

The strong radiation cooling gives a great advantage for producing strong γ–sources. For normal channeling radiation the incident angle region is very narrow, especially in the GeV region. The cooling effect increases this incident angle region strongly.

The very strong cooling effects in diamond are due to the fact that radiation emission is enhanced more than two order of magnitude, and multiple scattering is minimal as compared to other crystals.

6 Photon Emission Angles and Photon Colliders

Calculations on photon emission from crystals by the Frankfurt group[9] using the Dirac equation showed a rather surprising result, i.e, the hard photons are emitted with large angles to the crystalline axis – for 50 GeV electrons the photons in the interval 34–50 GeV are emitted with a typical angle of 140 μrad to the $\langle 110 \rangle$ axis of Ge. These results were obtained to determine whether or not axial radiation would be applicable for $\gamma - \gamma$ physics, e.g., the hunt for the Higgs boson.[10] Figure 7 shows the average angle of photon emission, $\theta\gamma$, as a fucntion of the photon energy, $E\gamma$, for 149 GeV electrons incident with an angle $\theta_{\rm in} \leq 20\mu$rad to the $\langle 110 \rangle$ axis. In contrast to the theoretical expectations mentioned above, the average emission angle is 36 μrad with about 55% of the photons emitted within 10 μrad to the axis. For the 243 GeV data the corresponding result is $\langle\theta_\gamma\rangle = 24\mu$rad for entry angles $\theta_{\rm in} \leq 16\mu$rad. We note that $\langle\theta_\gamma\rangle$ in both cases is close to the respective (ψ_1) values of 23 and 30 μrad, respectively. This is also what is expected from the above discussed radiative capture which will bring the electrons into high–lying channeling states.

These results are very encouraging for the application of channeling/-strong–field radiation for $\gamma - \gamma$ physics because crystals could be used to produce very narrow γ–ray beams for high–energy photon colliders. In the case of the strings–of–string radiation, the emission angles are of the same order as for axial radiation; furthermore, that hard photons are usually followed by only one photon of a few GeV and the photons could be linearly polarized.[6]

In the hunt for Higgs particles and physics beyond the Standard Model,

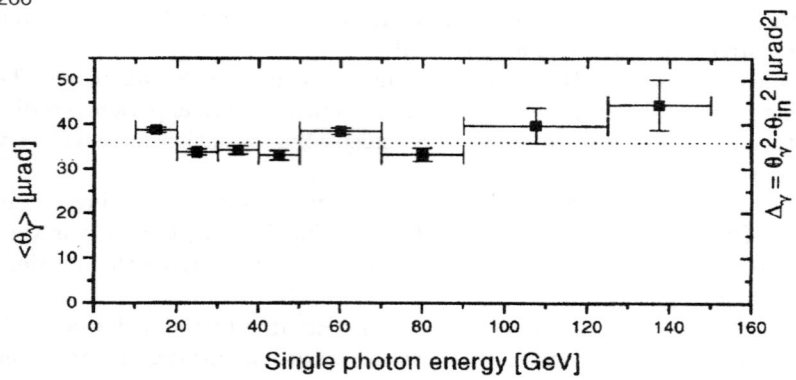

Figure 7. Average angles of emission with respect to the ⟨110⟩ axis in diamond. Single photons emitted by 149 GeV electrons incident within 20 μrad to the axis as a function of photon energy. The line denotes the average value.

a new generation colliders are being designed. They will without doubt be linear e^+/e^- colliders like NLC and Tesla, but a strong interest is coming up for having a γ,γ colliding branch in these facilities. Photon colliders give a much cleaner particle (Higgs) production. The conversion of e^+/e^- into high–energy photons could either be performed by 180° Compton scattering of laser light on the electrons/positrons. Another possibility is to use single crystals and convert e^+/e^- into axial or planar channeling radiation. The luminosity for these new colliders is 10^{33} or more. This is accomplished by focusing the e^+/e^- beams down to sub–micron diameters and with $\sim 10^{10}$ particles/bunch and repetitiion rates of more than 100. A luminosity of 10^{33} corresponds to a production of about 1000 Higgs/year.

If the above discussed SOS–radiation should be used, about 10% of the incoming particles would give off a high–energy photon in a 1 mm thick diamond crystal. But for this, two main questions arise: 1) How much will the low–energy photons smear the decay pictures? and 2) What happens to the diamond crystal when these high–intensity e^+/e^- beams hit the cystal in sub–micron spots over longer periods of time?

7 Conclusion

From the present experimental results it is clear that single crystals – especially diamonds – are unique for investigations of the influence of strong fields on QED processes. Crystal fields are 10^{12}V/cm or more, and the incident particles move in these strong fields up to 100 μm – in contrast to nucleus–nucleus collisions.

Crystals are excellent targets for producing hard gamma rays. The dramatically enhanced radiationn emission leads to strong angular cooling for electrons, which again leads to radiative capture to high–lying channeld states. Among the emitted photons from an electron or positron there is normally at least one very hard photon emitted in the first part of the crystal.

The strong cryssalline fields lead to enormous radiative energy losses. In just a 0.7 mm (110) diamond a 150 GeV and 243 GeV electron lose 60% of its total energy, which should be compared to the corresponding energy loss of less than 1% in an amorphous foil of the same thickness.

The present experiments have shown that for channeled particles the hard photons are emitted in a very narrow angular cone around the axis. Crystals would therefore be unique for future $\gamma\gamma$–colliders. Here, the production of rare particles are more favorable and clean than in e^+/e^- colliders.

It should finally be pointed out that the continued demand for higher beam energies and luminosities mean that many beam phenomena involve quantum effects like the ones described above. For colliding beams the Lorentz boost is even much more dramatic than in the present esperiments.

Acknowledgment

The author thanks all members of the NA–43 collaboration for their continuous stimulating interest through the years. Special thanks to Dr. Yu.V. Kononets for many fruitfull discussion during the evaluation of the NA–43 results. The strong support from De Beers concerning diamond crystals is highly appreciated.

References

1. K. Kirsebom, U. Mikkelsen, E. Uggerhøj, K. Elsener, S. Ballestrero, P. Sona, S.H. Connell, J.P.F. Sellschop, Z.Z. Viulakazi, Radiation emission and its influence on the motion of multi–GeV electrons and positrons incident on a single diamond crystal. Nucl. Inst. Meth. B. in print.
2. J.U. Andersen: Channeling Revisited. Present volume.

3. V.N. Baier, V.M. Katkov and V.M. Strakhovenko: Electromagnetic Processes at High Energies in Oriented Single Crystals (World Scientific, London, 1998).
4. R. Medenwaldt, S.P. Møller, A.H. Søresnen, S. Tang–Petersen, E. Uggerhøj, K. ELsener, M. Hage–Ali, P. Siffert, J.P. Stoquert and K. Maier, Phys. Rev. Lett. **63**, 2827 (1989).
5. R. Medenwaldt, S.P. Møller, S. Tang–Petersen, E. Uggerhøj, K. Elsener, M. Hage–Ali, J. Stoquart, P. Sona and K. Maier, Phys. Lett. B **242**, 517 (1990).
6. K. Kirsebom et al., Phys. Lett B 4593471999.
7. A. Apyan et al., Proposal to the CERN SPS Committee, CERN/SPSC 98–17, SPSC/P308, 1998.
8. Yu.V. Kononets in: Quantum Aspects of Beam Physics, Monterey 1998, ed: Pisin Chen (World Scientific, London, 1999).
9. J. Klenner, J. Augustin, A Schäfer and W. Greiner, Phys. Rev. A **50**, 1019 (1994).
10. A. Schäfer, S. Graf, J. Augustin, W. Greiner and E. Uggerhøj, J. Phys. G **16**, L131 (1990).

USING CRYSTALS TO SOLVE THE NUCLEON 'SPIN CRISIS' TODAY... AND LOOKING FOR PHYSICS BEYOND THE STANDARD MODEL TOMORROW

MAYDA M. VELASCO
Northwestern University, Evanston, Illinois

I will describe the ongoing R&D on aligned crystal radiation of the NA59 experiment. We are aiming at producing circularly polarized photons, starting from unpolarized electrons. For this purpose we need: (1) to prove that the birefringent capabilities of an aligned crystal allow it to behave as a quarter wave plate, as predicted by Cabibbo in the early sixties. The motivation of this is to be able to deploy the resulting photon beams to solve the 'spin crisis' by measuring the polarized gluon contribution to the nucleon spin using polarized photo-production of jets and heavy quarks. (2) to make further studies in 'strong-field (SOS)' radiation discovered by the NA43 collaboration. SOS radiation could become the source of very energetic photons where their polarization will be defined by that of the parent electron. A series of measurements of SOS radiation could lead to the realization of the first 'Higgs-Factory'.

1 Introduction and Physics Motivation

I present the results and prospects from a series of measurements and physics studies, performed by the NA59 collaboration at CERN, that could lead to the realization of the first 'Higgs-Factory'. There are strong indications that the mass of the Higgs boson is less than 200 GeV. I believe that the best way to study a light Higgs in detail and in a reasonable time scale is at a $\gamma\gamma$-collider, where the cross sections for the Higgs s-channel production are rather large and the required electron beams energies ($E_e \leq 100$ GeV) are rather modest. I propose that an efficient and cost effective way to produce, for example a 70 GeV photon beam starting from 100 GeV electrons, is an aligned crystal as a radiator, selecting angles of incidence such that the electrons experience 'strong-field' effects.

The main aim of the NA59 experiment is to test the birefringent capabilities of an aligned crystal. This characteristic allows the crystal to behave as a quarter wave plate for multi GeV photons, as predicted by Cabibbo in the early sixties. This will make it experimentally possible to have a new series of polarized photo-production experiments that can be used to solve the so called 'Nucleon Spin Crisis'.

1.1 The Nucleon Spin Crisis

The new generation of polarized DIS experiments[a] has firmly established that only 30% of the nucleon spin is carried by quarks, as first observed by the EMC Collaboration [1]. Their measurements also implied that there is a net negative polarization carried by the strange quarks. Despite their efforts it is not possible to tell what carries the rest – gluons and/or orbital angular momentum are possible candidates. There are several competing explanations based on non-perturbative QCD: (1) A large fraction of the nucleon spin is carried by polarized gluons Δg. (2) A large fraction of the nucleon spin is carried by the orbital angular momentum L_Z and not by the gluons. An attempt has been made to extract the polarized gluon contribution from all existing data on the spin structure function g_1. Unfortunately, the theoretical uncertainties are and will remain too large to make a conclusive statement. Therefore, the next step will be to understand the fraction of the nucleon spin carried by strange quarks, polarized gluons and orbital angular momentum from new dedicated experiments. It will be crucial to perform independent measurements of Δs and Δg to distinguish between the different models. As we will argue below, the first attempts to obtain Δg are not sufficient and the best way in which $\Delta g(x)$ can be measured is from high energy polarized photo-production experiments.

The first glimpse of $\Delta g = \int_0^1 \Delta g(x)\,dx$ was obtained in a model dependent way from leading-order (LO) and next-to-leading (NLO) QCD fits to the GLAP equations of all the available data on the $g_1(x, Q^2)$ spin-dependent structure function, where x could be interpreted as the fraction of the nucleon momentum carried by the struck quark, and Q^2 is the four momentum squared of the mediating virtual photon. In the simplest Quark-Parton Model, where we ignore the interaction of quarks with gluons, g_1 is given by the average of the sum of $e_i^2 \Delta q_i(x)$ over all flavors, where $\Delta q_i(x) = q_i^+(x) - q_i^-(x) + \bar{q}_i^+(x) - \bar{q}_i^-(x)$, and $q_i^+ (\bar{q}_i^+)$ and $q_i^- (\bar{q}_i^-)$ are the distribution functions of quarks (antiquarks) with spin parallel and anti-parallel to the nucleon spin, respectively, and e_i is the electric charge of the quarks of flavor i. Hence, g_1 contains information on the orientation of the quark spin with respect to the proton spin. Using the OPE description for the first moment of g_1, the 'spin content' of the nucleon $\Delta\Sigma = \Delta u + \Delta d + \Delta s$ is obtained. From a combined analysis of all published g_1 data for the proton, deuteron and neutron the SMC collaboration finds $\Delta\Sigma = 0.29 \pm 06$ at $Q^2 = 5$ GeV, see Fig. 1(a). A full description of the analysis and updated values can be found in references [2]

---

[a]The SMC Collaboration at CERN, the E142, E143, E154 and E155 Collaborations at SLAC, and the HERMES Collaboration at DESY.

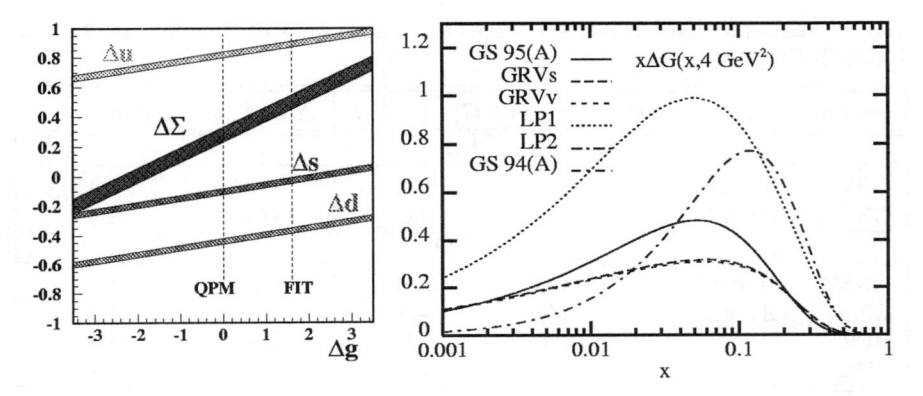

Figure 1. (a) Model dependent solution: Quark contribution to the nucleon spin from a NLO pQCD-fit of all $g_1(x,Q^2)$ data available (assuming the AB-scheme). (b) The NLO gluon distributions and their ratio as a function of the fraction of the nucleon momentum carried by the gluon.

and [3], respectively. Since the first moment is evaluated at a fixed value of Q_0^2, all data had to be evolved from the their measured Q^2 to Q_0^2. The primary reason for the NLO QCD fits was this Q^2 evolution. As a side product of this QCD analysis two things have become more obvious:

- The dependence of the singlet part, which we refer to as $\Delta\Sigma$, on the gluon cannot be neglected if we want to obtain the spin content of the nucleon. Unfortunately, the decomposition of $\Delta\Sigma$ into the 'true spin content of the nucleon', $\Delta\widetilde{\Sigma}$, and a gluon contribution is scheme-dependent [4]. For example, in the Adler–Bardeen (AB) [5] factorization scheme [8], used in several of these analyses, the decomposition is

$$\Delta\Sigma(Q^2) = \Delta\widetilde{\Sigma} - n_f \frac{\alpha_s(Q^2)}{2\pi} \Delta g(Q^2), \qquad (1)$$

where the last term was originally identified as the anomalous gluon contribution [9,10,11]. In this scheme $\Delta\widetilde{\Sigma}$ is independent of Q^2, but it cannot be obtained from the measured $\Delta\Sigma(Q^2)$ without an input value for Δg.

- The results for Δg from the LO and NLO analysis from several groups are given in Table 1. They all give positive values for Δg. The main differences are due to the data samples used, the choice made for the reference scales Q_{ref}^2 at which the parton distribution function are parametrized,

Table 1. Results for the first moment of the polarized gluon distribution from the LO and NLO pQCD analyses.

group	fit	Ref.	scheme	Q^2_{ref}(GeV^2)	$\Delta g(Q^2_{ref})$	$\Delta g(10\text{ GeV}^2)$
LO						
GRV	standard	6	$\overline{\text{MS}}$	0.34	0.44	1.9
GS	Gluon B	7	$\overline{\text{MS}}$	4.0	1.9 (fixed)	2.3
NLO						
GRSV	standard	12	$\overline{\text{MS}}$	0.34	0.5	1.7
GS	Gluon B	13	$\overline{\text{MS}}$	4.0	1.6	1.9
BFR		14	AB	1.0	1.5	2.8
SMC		3	AB	1.0	0.9	1.7

and the assumed form of the parameterization. However, the main difference comes from the chosen factorization scheme. This in conjunction with the renormalization scheme is the main source of uncertainty. For example, in the SMC analysis [3] $\Delta g = 1.6 \pm 0.3(stat.) \pm 1.0(sys.)$ at $Q^2_{ref}=5$ GeV^2. Despite the differences between the LO and NLO analyses, all these results provide a good description of the data, therefore we can conclude that the sensitivity of these fits to $\Delta g(x)$ is weak and dominated by theoretical uncertainties. The corresponding gluon distributions for the LO analyses are shown in Fig. 1(b).

1.2 The Higgs is Right Around the Corner

One of the main goals of the next set of collider experiments is the discovery of the still elusive Higgs boson, because we cannot believe that the Standard Model (SM) is correct until the Higgs has been observed. In the SM, the Higgs is responsible for the observed breaking of the $SU(2) \times U(1)$ symmetry, and it is through this mechanism that the particles acquire mass.

Within the theoretical framework of the SM, the coupling of the Higgs boson to the fermions and gauge bosons is determined by the weak coupling g, the particle masses, and the Higgs mass m_H. Consequently, all production cross sections and decay widths are specified as a function of m_H. The Higgs mass, however, cannot be predicted by the theory, not even within the confines of the SM. Supersymmetry, a popular extension of the SM, also cannot predict the exact value of the Higgs mass, but does *require* a light Higgs with mass no greater than 135 GeV.

Fits to precision electroweak observables indicate that the Higgs mass is less than about 200 GeV, see Fig. 2. Searches at LEP place a lower limit of about 109 GeV, while in the latest data there may be a hint of Higgs events corresponding to a mass of 114 GeV [16]. This mass range will be explored at the upcoming high luminosity run at the TEVATRON, to begin in 2001. For all these reasons, I believe that *the Higgs is right around the corner,* and we must prepare for the next stage: a Higgs factory.

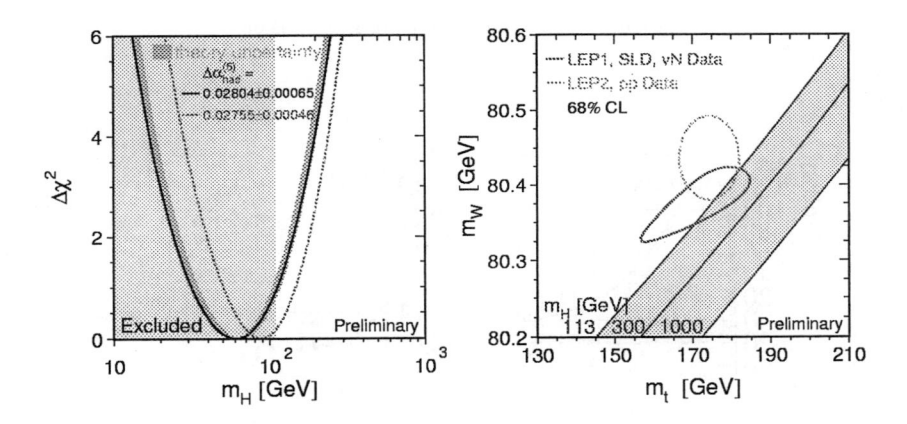

Figure 2. Constraints on the Higgs boson mass from SM fits to electroweak observable. Preliminary results from LEP EWWG [15].

2 Future Measurements Needed With High Energy Polarized γ-beams

2.1 *Photo-production of Charm and Experimental Asymmetry*

A determination of the gluon polarization can only be obtained from a process involving the gluon in leading-order. If we assume that the contribution from intrinsic charm quarks to the nucleon is negligible, then charm production from photon-gluon fusion (PGF) is the best candidate. In order to perform this measurement as cleanly as possible, we need photon-nucleon interactions at energies sufficiently above threshold and good reconstruction of charm mesons.

For real photons the photo-production cross section for charm production

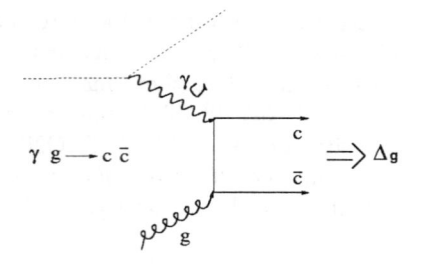

Figure 3. Diagram for photon-gluon fusion.

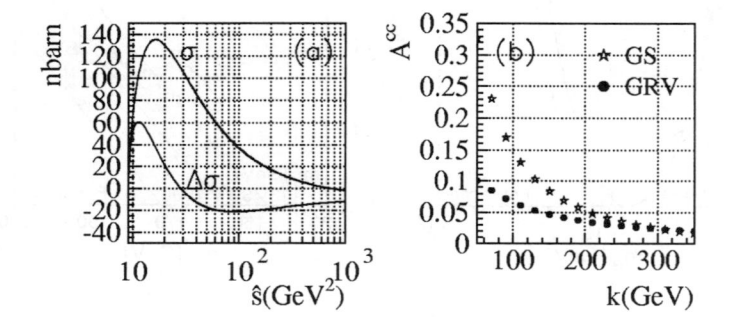

Figure 4. (a) The photon-gluon cross sections as function of the c.m. energy. (b) The photon-nucleon cross section asymmetry at LO for charm production as a function of the beam energy.

from the PGF process, $\gamma g \to c\bar{c}$, can be written as

$$\sigma^{\gamma g \to c\bar{c}} = \sigma(\hat{s}) + \lambda_\gamma \lambda_g \Delta\sigma(\hat{s}), \qquad (2)$$

where $\sigma(\hat{s})$ and $\Delta\sigma(\hat{s})$ are the spin-averaged and spin-dependent contributions, respectively, and $\hat{s} = (q + k)^2$ is the square of the c.m. energy in the photon-gluon system, while $\lambda_{\gamma,g}$ are the corresponding helicities. As shown in Fig. 3, $\hat{s}$ is given by $\hat{s} = x_g s$ where s is the square of the c.m. energy in the photon-nucleon system, and x_g represents the fraction of the nucleon momentum carried by the gluon. For large photon energies $s = 2Mk$, where M is the nucleon mass.

If we ignore non-perturbative hadronization effects for $\gamma N \to c\bar{c}$, then the photon-nucleon cross-section asymmetry $A^{c\bar{c}}_{\gamma N}$ is obtained from the PGF cross sections and the gluon distributions

$$A_{\gamma N}^{c\bar{c}}(k) = \frac{\Delta\sigma^{\gamma N \to c\bar{c}X}}{\sigma^{\gamma N \to c\bar{c}X}} = \frac{\int_{4m_c^2}^{2Mk} d\hat{s}\, \Delta\sigma(\hat{s})\, \Delta g(x_g, \hat{s})}{\int_{4m_c^2}^{2Mk} d\hat{s}\, \sigma(\hat{s})\, g(x_g, \hat{s})}, \qquad (3)$$

where the minimum x_g is defined by the charm threshold $\hat{s} = 4m_c^2$. At LO, the PGF cross sections are given by [29,30]

$$\Delta\sigma(\hat{s}) = \frac{4}{9} \frac{2\pi\alpha_e\alpha_s(\hat{s})}{\hat{s}} \left\{ 3\beta - \ln\frac{1+\beta}{1-\beta} \right\},$$

$$\sigma(\hat{s}) = \frac{4}{9} \frac{2\pi\alpha_e\alpha_s(\hat{s})}{\hat{s}} \left\{ -\beta(2-\beta^2) + \frac{1}{2}(3-\beta^4)\ln\frac{1+\beta}{1-\beta} \right\}, \qquad (4)$$

where $\beta = \sqrt{1 - 4m_c^2/\hat{s}}$ is the c.m. velocity of the charmed quarks and antiquarks. Both terms rise sharply at the threshold, and $\Delta\sigma(\hat{s})$ changes sign at about four times the threshold (Fig. 4a). In Fig. 4b we show our estimates for $A_{\gamma N}^{c\bar{c}}$ as a function of the photon beam energy. They were calculated assuming that $m_c = 1.5$ GeV, and by using the LO parameterizations for $g(x)$ and $\Delta g(x)$ already shown in Fig. 1(b). The parameterization used for these calculations are consistent with the expected behavior of $\Delta g(x)/g(x) \to 0$ as $x \to 0$ and $x \to 1$. In QCD, at small x, $g(x)$ should rise due to soft gluon singularities that are not present in the polarized case [7], while at large x they should both go as $(1-x)^4$ according to QCD counting rules [32,33].

The calculated $\sigma^{\gamma N \to c\bar{c}X}$ for 150 GeV photons is of the order of 500 nb, which is consistent with the data from photo-production experiments. At the same energy we predict the unpolarized contribution $\Delta\sigma^{\gamma N \to c\bar{c}X}$ to be around 50 nb. As the energy increases we do not see a significant change in either cross section, but the rate starts to decrease very rapidly below 75 GeV.

What we propose to measure is the spin dependent cross section asymmetry, $A_{\rm RAW}$, for charm photo-production, which is related to $A_{\gamma N}^{c\bar{c}}$ by

$$A_{\rm RAW} = \frac{N_{\uparrow\downarrow}^{c\bar{c}} - N_{\uparrow\uparrow}^{c\bar{c}}}{N_{\uparrow\downarrow}^{c\bar{c}} + N_{\uparrow\uparrow}^{c\bar{c}}} = P_B P_T f A_{\gamma N}^{c\bar{c}}, \qquad (5)$$

where $N_{\uparrow\downarrow}^{c\bar{c}}$ ($N_{\uparrow\uparrow}^{c\bar{c}}$) are the number of charm containing events when the photon and the nucleon polarization are anti-parallel (parallel) to each other, P_B (P_T) is the beam (target) polarization and f is the fraction of polarized material.

Even though NLO fits for $\Delta g(x)$ and $g(x)$ are available, we have presented the $A_{\gamma N}^{c\bar{c}}(k)$ at LO because only LO QCD calculations of the polarized PGF process exist. For the unpolarized process NLO order corrections are available [34], but with theoretical uncertainties larger than the experimental accuracy. Nevertheless, their predictions are in good agreement with the

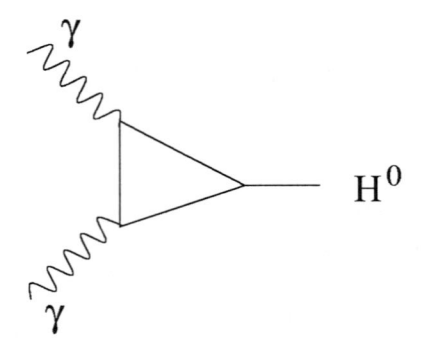

Figure 5. The coupling of the Higgs boson to two photons, $\gamma\gamma \to H^0$, proceeds through loops of heavy charged particles.

data, showing that photo-production cross sections are well described by NLO PGF and that a value of $m_c \simeq 1.5$ GeV is clearly favored. As we can see from Eq. (3) and (5), in order to obtain $\Delta g(x)/g(x)$ we need the PGF cross sections. Therefore for the final analysis of $\Delta g(x)/g(x)$ all NLO calculations will be highly desirable.

2.2 The $\gamma\gamma$-collider as a Higgs Factory

Suppose the Higgs has been observed with a mass below 200 GeV. While this observation would confirm the idea that a fundamental scalar field breaks electroweak symmetry, it would not in itself point to physics beyond the SM. New Physics will not show up in the Higgs mass value *per se*, but rather in the decay modes, cross section, and angular asymmetries. We should be planning *now* how to study the properties of the Higgs boson in as much detail as possible, for only then will we get past the minimal picture of electroweak symmetry breaking that the SM provides.

I am convinced that the best technology for a Higgs factory is the $\gamma\gamma$-collider. There are several reasons for this:

1. The energies required are relatively modest, because the Higgs would be produced as an *s*-channel resonance[18], Fig. 5.

2. The signal can be obtained in modes other than $b\bar{b}$.

3. The control of the beam kinematics and helicity provides a *unique* tool for probing production mechanisms, which happen to be particularly direct (see below).

4. There are potentially two techniques for producing high energy photon beams: laser back-scattering (LB) and aligned crystals (AC).

In this section the first three points will be reviewed briefly. The last point, concerning the production of polarized high energy photon beams with aligned crystals, is the subject of this proposal and one of the main goals of my research.

The option of pursuing frontier physics with real photon beams is often overlooked, despite many interesting and informative studies [19,20]. The high energy physics community has focussed on particle beams (mainly protons and electrons) for historical reasons, and risks missing an excellent opportunity to do exciting physics in the near future. In the context of the next generation of accelerators, most people are comfortable with the idea of colliding TeV electrons and positrons, but dismiss the idea of colliding 100 GeV photons. Yet this is a far easier and less costly enterprise, and could deliver crucial information on the Higgs sector sooner. As is well known, laser back-scattering methods have been successfully deployed at the SLC. The only barrier for using this technology for a Higgs factory is the need for a super-high power laser. An alternative method, based on the use of aligned crystals, could circumvent this sticking point and open the way to a Higgs factory in this decade. Given a Higgs mass of, say, 120 GeV, the high energy community could then consider the options of a Higgs factory at an e^+e^- machine with beams with energies around 150 GeV each, or a $\gamma\gamma$-collider with beams with energies of 60 GeV each.

While the physics possibilities at a $\gamma\gamma$-collider are still being explored, it is clear that such a machine would furnish a wealth of information:

1. The mass can be pinned down to 110 MeV, using the scanning methods described in [17,21], and an integrated luminosity of 50 fb^{-1}.

2. With the right kind of photon luminosity spectrum, the total width can be measured within $\Delta\Gamma_H/\Gamma_H$=0.06, if it is assumed that the $\Gamma(H \to \gamma\gamma)$ and $\Gamma(H \to b\bar{b})$ are predicted in the SM [21].

3. Branching ratios to $b\bar{b}$, $\gamma\gamma$, W^+W^-, and possibly $\tau^+\tau^-$ can be measured to better than 10% [22].

4. Information on the CP quantum numbers can be inferred from angular correlations in the $t\bar{t}H^0$ final state, and possibly from $\gamma\gamma \to H^0 \to W^+W^-$ [23].

We plan to study these possibilities in more detail – see the section on 'Related

Activities'. Since, however, this proposal is aimed at R&D for the generation of the photon beams, we discuss the physics only briefly.

In $\gamma\gamma$ collisions the Higgs boson will be produced as a single resonance in the s-channel. The cross section can be written simply as:

$$\sigma(\gamma\gamma \to H^0 \to X) = z\frac{dL_{\gamma\gamma}}{dz}\frac{4\pi^2}{M_{H^0}^3}\Gamma(H^0 \to \gamma\gamma)BR(H^0 \to X)(1+\lambda_1\lambda_2), \quad (6)$$

where $z = \sqrt{s_{\gamma\gamma}}/2E_e$, λ_i is the helicity of the ith photon, and $L_{\gamma\gamma}$ is the effective $\gamma\gamma$ luminosity. The crucial parameters are the energy spread of the individual γ beams needed to obtain an optimum $L_{\gamma\gamma}$ and a high degree of polarization.

In contrast to hadron colliders or e^+e^- machines, the fact that the Higgs is produced through the s-channel allows us measure the cross section as a function of $\sqrt{s}$ across the Higgs threshold, thereby gaining information on the Higgs mass.

However, since there is no tree-level coupling of real photons to the Higgs, this s-channel process proceeds through a triangle diagram as shown in Fig. 5. Note that all charged particles participate in this loop, in contrast to the gluon fusion process $gg \to H^0$ found in hadron colliders, in which only quarks participate. The cross section $\sigma(\gamma\gamma \to H^0)$ is sensitive to physics beyond the SM through the new particles flowing in this loop.

For the same reason, the partial width $\Gamma_{\gamma\gamma}$ for $H^0 \to \gamma\gamma$ would be sensitive to heavy new charged particles circulating in the triangle. Deviations from the SM value can be as large as 20% in the MSSM. The total cross section on threshold gives a measurement of $\Gamma_{\gamma\gamma}$ which could be as good as 3%. A measurement of the total width also relies on the photon energy spectrum.

The fact that the Higgs is a scalar and most SM processes involve pairs of fermions or pairs of vector bosons means that backgrounds can be strongly suppressed by a suitable choice of beam polarizations. Some detailed studies have been published, and they indicate that clear signals can be obtained in the $b\bar{b}$ and W^+W^- final states. Since the yield would be several thousands for 10 fb^{-1} (one year's running), a measurement at the couple of percent level is feasible. There is also hope for the $\tau^+\tau^-$ decay mode, and possibly also the $\gamma\gamma$ and γZ states, although the number of events would be modest.

If a $\gamma\gamma$-collider runs concurrently with an e^+e^- machine, then the combination of measurements would powerfully constrain – or elucidate – physics beyond the standard model.

One of the current 'hot' topics in Higgs theory is the possibility of non-trivial CP violation in the Higgs sector. Preliminary studies indicate that the CP quantum numbers could be inferred from the $t\bar{t}H^0$ and W^+W^- final

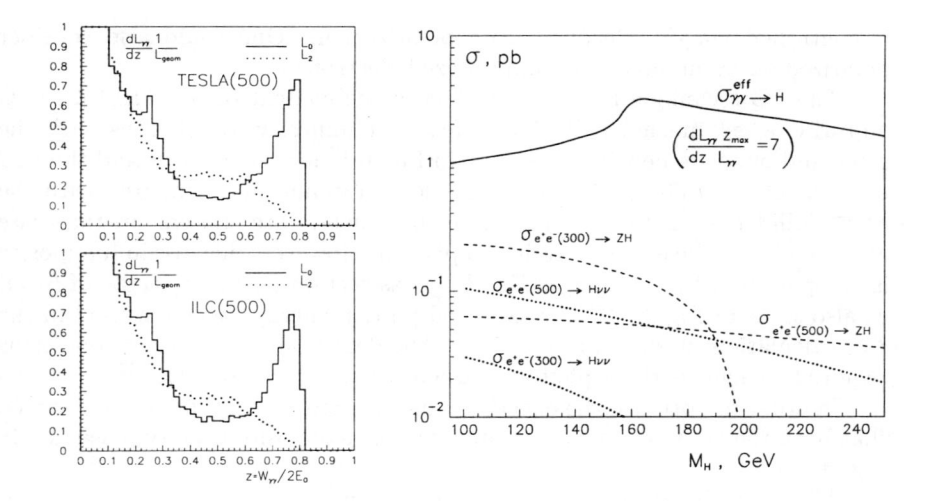

Figure 6. (a) Effective $\gamma\gamma$ luminosity for photons with equal and opposite helicity. (b) Comparison of the Higgs production cross sections for $\gamma\gamma \to H^0$ and e^+e^--collisions.

states. The control of the beam polarization is key to these studies.

3 How do we make the high energy photon beams?

The sketch of physics possibilities in the last section shows that the energy spectrum and polarization of the photon beam is crucial. The worthiness of any plan to build a $\gamma\gamma$-collider would depend on the R&D of beam techniques. (The demands on the detector technology, while not trivial, are easily within the capabilities of existing labs.) I will discuss first the known technology of laser back-scattering, and then propose an alternative technology using aligned crystals.

3.1 Laser Back-scattering (LB) (Standard design)

High energy photons can be produced by Compton scattering photons from a laser off a high energy electron beam. This technique has been used very successfully at SLAC, where single photons with a high fraction of the electron energy form a tightly collimated beam. The polarization of the photon beam is controlled via the electron beam polarization. One could also consider polarized lasers impinging on unpolarized electrons.

The LB technique has been studied in the context of the TESLA, NLC and JLC e^+e^- machines [24]. The effective luminosity for photons with the same and opposite helicities is shown in Fig. 6 [17] for the TESLA and the ILC e^+e^- machines. There is about a 40% loss in luminosity compared with the e^+e^- collider when the optimum laser parameters are chosen. In that case about 65% of the electrons produce a photon. However, this loss in luminosity is compensated by the increase in the cross section (about a factor of 6–30) as also show in Fig. 6 [17]. The expected photon energy spectrum used in the $L_{\gamma\gamma}$ calculations is shown in Fig. 7(a). The details about the laser conditions required to achieve these photon characteristics can be found in [25].

In order to optimize the effective cross sections the authors have assumed that both the electron and the photon will have opposite helicity (case (a)) in Fig. 8.

The main challenges of these designs are the complicated optical schemes they require around the interaction region, and the very high power laser to provide the input photons. These are currently under study by the TESLA group [26,27] and the NLC group [28]. In this proposal I discuss an alternative scheme based on aligned crystals.

3.2 Aligned Crystals (AC) (New design)

The LB technique has received most of the attention in designs for $\gamma\gamma$-colliders. The general conclusion is that such a design would deliver the physics, *if* an extremely high power laser was used. Unfortunately, such a laser based optical system presents a formidable challenge. It is for this reason that I propose the development of aligned crystal techniques.

My proposal is based on the work I have done at CERN concerning the production of polarized photons from electrons passing through Silicon crystals. I formed the NA59 Collaboration [35] in 1998 and we have had two data taking runs, in 1999 and 2000. I will give a brief overview of the principles of the technique before detailing some of the studies we are conducting.

Why Aligned Crystals are Special Radiators?

The Bethe-Heitler (BH) energy loss formula pertains to amorphous materials. When electrons pass through a periodic solid such as silicon, however, the *coherent* interaction of atomic nuclei with the electrons can lead to a very large increase in the energy loss [36]. One manifestation of this phenomenon is called "coherent bremsstrahlung," (CB) and is well known since decades. A

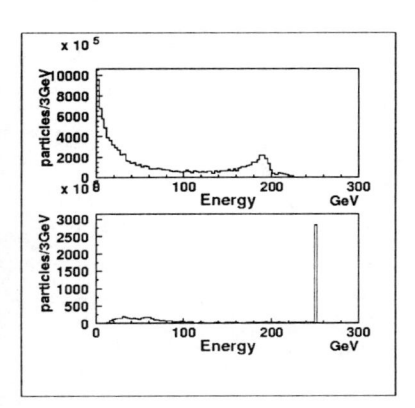

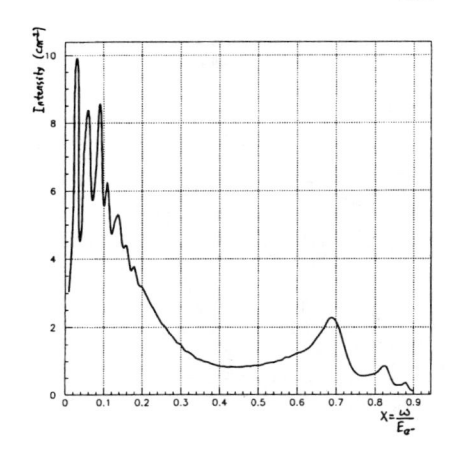

Figure 7. Left: expected photon energy spectrum from LB used in $L_{\gamma\gamma}$ calculations shown in Fig. 6 [17] and energy of the out going electron (35% of the e^- do not convert). Right: the plot show the single photon energy spectrum expected from an aligned crystal in the proposed (SOS) configuration. The plot is for a 1.5cm Si crystal, where electrons penetrate at 0.3 mrad from the $\langle 100 \rangle$ axis on the (110) plane. In this case 10% of the e^- do not convert. This were the operating conditions of the NA59 experiment, not the one proposed for a $\gamma\gamma$ collider where we will aim at having thinner diamond crystals.

newer theoretical development is called the "string of strings," (SOS) and it appears to be more promising for the purposes of beams in a $\gamma\gamma$-collider. A simple drawing showing under which experimental condition we will obtain CB or SOS radiation enhancement is shown in Fig. 9

Cabbibo realized in 1962 that the radiation produced in coherent bremsstrahlung would be polarized [37]. This arises from the difference in the probability to radiate perpendicular and parallel to the crystallographic plane. Macroscopically speaking, the indices of refraction $n_\perp$ and $n_\parallel$ are different. This property of aligned crystals can be exploited to turn a linearly polarized beam into circularly polarized one, or in the construction of efficient polarimeters. I will not discuss these possibilities in detail, however.

Strong Field Effects: SOS

Theoretical Background

The phenomena behind the SOS radiation that we want to use can be described as follows. Let us consider a charged particle penetrating a single

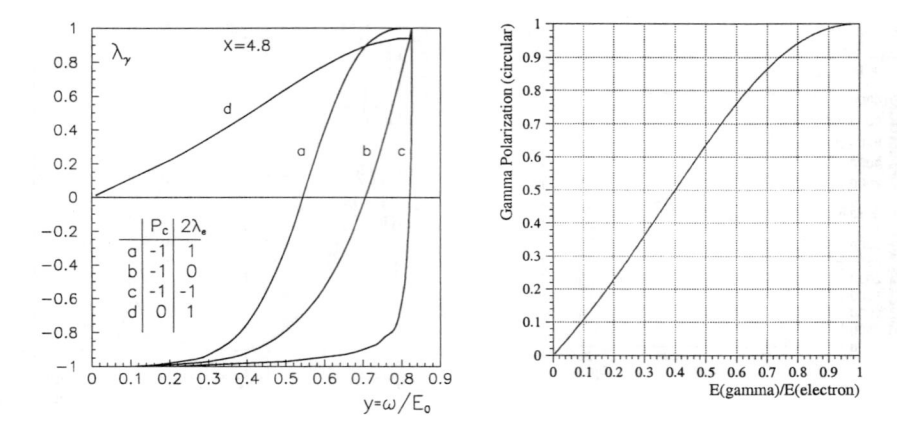

Figure 8. (a) Expected photon circular polarization produced from laser backscattering for different laser and incoming electron conditions, and (b) polarization for photons produced from a 100% longitudinally polarized electron radiating off an amorphous material, or crystal configurations in which no net linear polarization is obtained in the case of unpolarized radiating electrons.

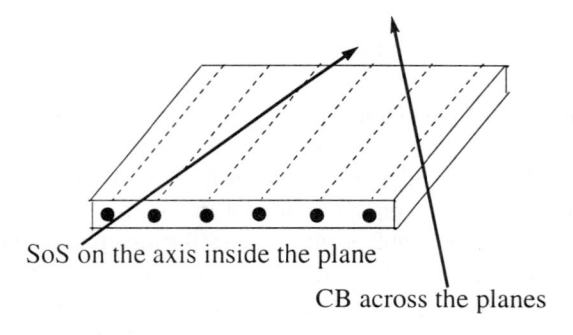

Figure 9. Drawing of incident electron direction with respect to a crystal plane and axis. For the 'Strong-Field-Effects (SOS)' to occur we need to be in the plane and experience the coherent effects of the strings of atoms that are sitting in the plane. For the 'conventional' coherent bremsstrahlung the electron experiences coherence as it goes across planes.

crystal (Fig. 10 (a)). For a sufficiently small angles of incidence to a crystallographic plane direction, the coherent scattering off the atomic constituents

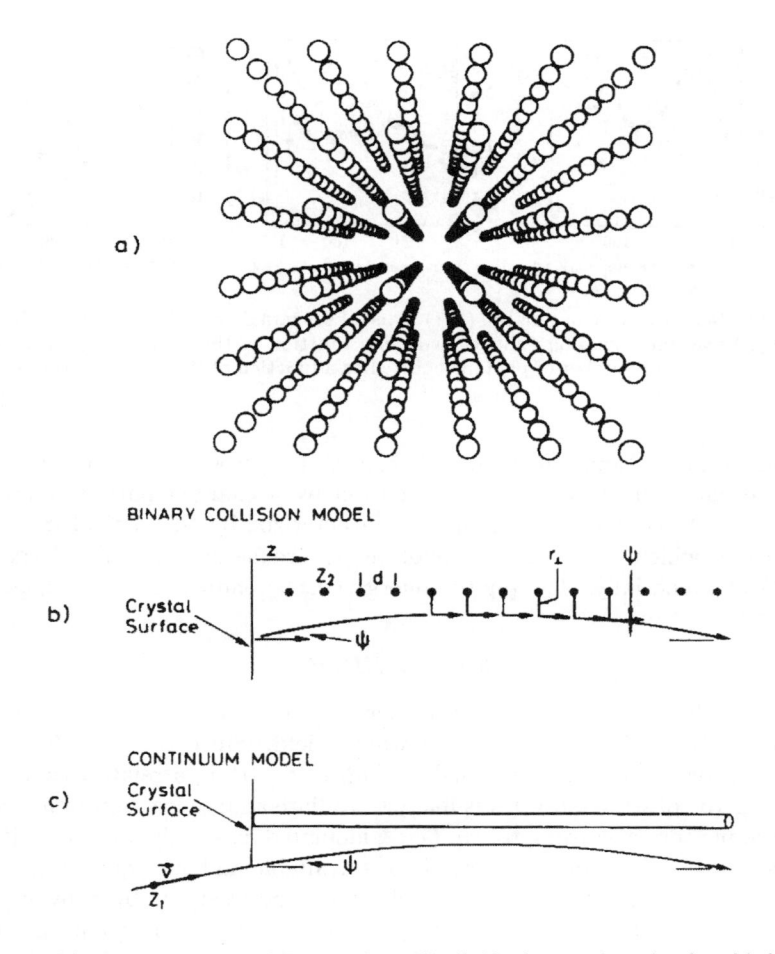

Figure 10. (a) Perspective view of the positions of individual atoms in a simple cubic lattice. Below, the deflection of an incident particle by a string of atoms in (b) the binary collision picture and (c) the continuum picture.

(Fig. 10 (b)) acts on the particle as if the charges of the screened nuclei were smeared along this direction (Fig. 10 (c)). Thus, the particle is deflected by a 'continuous string of charge'. Even though the forces are purely electric in the laboratory frame, in the electron plane the deflection will have the same character as that of the deflection by a magnetic field. Furthermore, in the

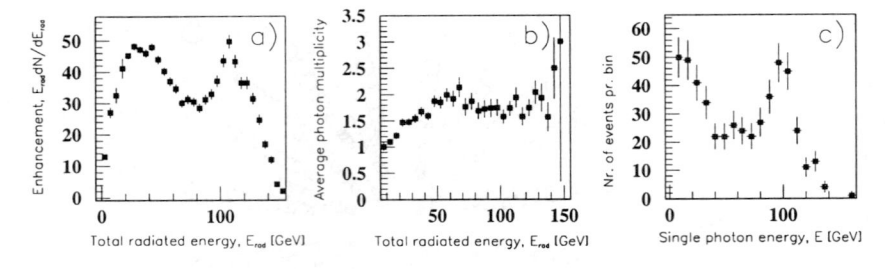

Figure 11. 149 GeV electrons on the (111) plane of 0.7 mm diamond, 0.6 mrad to the ⟨110⟩ axis. (a) Show the radiation enhancement as a function of the total energy, (b) photon multiplicity, and (c) number of photons emitted as a function of the energy of the photon.

continuum approximation the field becomes of macroscopic extension.

The emission of synchrotron radiation by a charged particle passing a magnetic field is among other things characterized by the critical frequency, ω_c, beyond which the number of photons per frequency interval, $dN/d\omega$, decreases exponentially. The photon energy corresponding to this frequency is given by

$$\hbar\omega_c = 3\gamma EB/B_0, \qquad (7)$$

where γ is the Lorentz factor, E the energy of the electron, B is the magnetic field and $B_0 = 4.4 \cdot 10^9$ T is the Schwinger field. Since the axial fields in a crystal correspond to magnetic fields of up to $\simeq 10^5$ T, a relativistic particle with $\gamma \simeq 10^5$ passing a crystal is likely to radiate a large fraction of its energy into one photon, according to Eq. (7). Calculated classically as above, it may even seem to radiate more energy than is available and thus quantum corrections become important for the calculation as pointed out by Schwinger [41]. The invariant parameter, $\chi = \gamma B/B_0$, is a measure of the quantum effects in synchrotron radiation and already when χ is around 0.1 the corrections become significant. This is the basis of the so-called strong field effects (SOS) which have been investigated in detail by the NA43 collaboration at CERN led by Uggerhøj [42], and treated theoretically by several groups, e.g., the groups led by Baier in Novosibirsk and Kononets in Moscow. This new type of coherent bremsstrahlung is emitted when the electrons cross the rows of atoms forming a plane.

Experimental Background

Fig. 11 shows NA43 collaboration results on SOS radiation using 149 GeV electrons incident at a $\pm 10\mu$rad to the (111) planes in a 0.7 mm diamond crystal and with an angle of incidence of 0.6 mrad to the $\langle 110 \rangle$ axis. The random yield for the 0.7 mm diamond crystal is 0.58% of a radiation length (X_0), but as shown in Fig. 11(a) a factor of 50 in enhancement is observed for this type of crystal alignment compared to the random case. Larger degrees of enhancement are expected as the angle with respect to the plane is reduced. Also shown is the single photon spectrum in Fig. 11(c), which is the energy of the emitted photon. As shown, most of the time a photon with $E_\gamma = 0.7 E_e$ is produced along with a low energy photon. These are conditions that are highly desirable for beam line construction.

Now, contrary to reports in [43] concerning SOS producing a large degree of linearly polarized high energy photons, we have NA59 data to show that this is not the case. Just as predicted by Strakhovenko we find SOS radiation not to be polarized for energies above 70 GeV if we start with 180 GeV electrons (see Fig. 16). As a consequence, if the incoming electron is longitudinally polarized, then the circular polarization of the emitted photon is described by the function already shown in Fig. 8(b).

Coherent Bremsstrahlung

Another type of radiation obtained from the penetration of electrons in crystals is the more familiar 'coherent bremsstrahlung' (CB). This type of radiation has a maximum when the inverse formation length[b] of the photon coincides with the projection of a vector of the reciprocal lattice on the direction of the electron. Equivalently, it is a resonance phenomenon which appears when the waves emitted from the passage of subsequent planes are in phase. Earlier studies, with the Omega spectrometer and an 80 GeV electron beam radiating off a Si-crystal, have shown that linear polarizations of 10-60% can be obtain with this method [36]. We will use this type of radiation in our NA59 experiment as further evidence in our conclusions on the SOS polarization, but believe it is not as feasible for a $\gamma\gamma$ collider because of the smaller enhancement resulting in a lower number of energetic photons.

---

[b]The formation length can be considered as the length over which the photon and the emitting electron become separated by one wavelength of the emitted radiation. In the case of pair creation, it is the length required to separate the e^+ and e^- by a Compton wavelength.

3.3 Why do I think that a crystal based $\gamma\gamma$-collider could be feasible?

The reason why I think that an aligned crystal scheme based on SOS radiation look promising is because of the following advantages and characteristics over the LB method:

- the rate of useful photons compared to the number of low energy photons is better. The energy spectrum for the produced photons in the LB is shown in Fig. 7 and Fig. 11(c) show data on the single photon energy for the proposed SOS crystal based design.

- a broader region of the energy spectrum is useful because the degree of polarization for the photon is still high for photons that are carrying only 50% of the initial electron energy (see second plot in Fig. 8)[c],

- relative suppression of the low energy photons, see Fig. 11(a) and (c),

- the requirements on the stability of the beam are already part of the main e^+e^--designs, like small e beam divergence and beam position (the photons are emitted within $1/\gamma$, therefore the incoming electron will define the photon beam spot),

- the setup requirements around the IR are simpler,

- the required technology (crystals, goniometer, radiation hard controllers, etc.) are already available.

4 What have we learnt from the NA59 Experiment?

The CERN-SPS NA59 experiment is composed of several of the members of the NA43 collaboration, who pioneered the SOS radiation and many other radiation phenomena at high energies, and members of the SMC collaboration, who are interested in performing the required R&D in order to be able to produced circularly polarized photons, starting from unpolarized electrons. The motivation for that is to be able to solve the 'spin crisis' by measuring the polarized gluon contribution to the nucleon spin using polarized photo-production of jets and heavy quarks (produced via photon-gluon fusion) [29,30,31].

The stated purpose[35] of CERN-NA59 was the study of the birefringence of crystals to show that they could be used as a 'quarter-wave plate' ($\lambda/4$-plate)

---

[c]This is to be compared with LB, where the polarization is already zero.

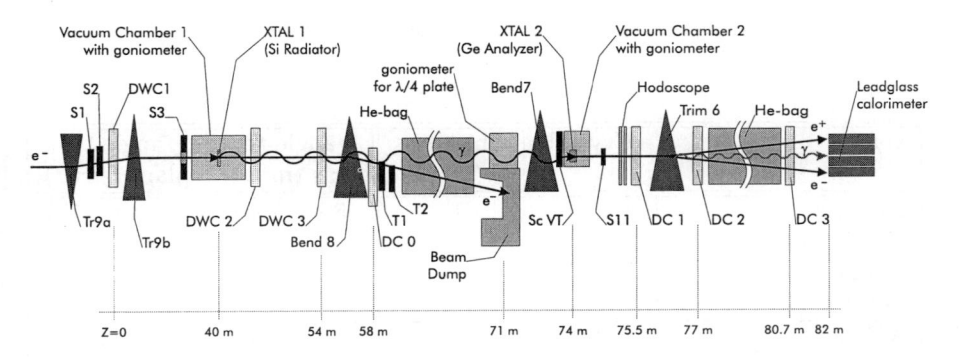

Figure 12. Experimental setup used for the pair production asymmetry polarization method.

to convert linear polarization into circular. In the course of our investigations we have learned many things that are important in the context of $\gamma\gamma$-colliders:

- measurement of the enhancement in radiation by unpolarized electrons in thick aligned crystals and the resulting photon linear polarization, for both SOS and CB radiation,
- multiplicity and single photon energy measurements for both type of radiation in thick crystals,
- development of new polarimetry techniques using an aligned crystal as an analyzer,
- production of large arrays of essentially perfect ultra large diamond crystals that will have multiple purposes for any further R&D that we might make with crystals. (This were produced by deBeers, our collaborators from Scholand Research Center at the University of the Witwatersrand, and the pre-alignment was performed at the ESRF in Grenoble.)
- test of the quarter wave plate, which will allow us to produce low maintenance beam polarization monitors for circularly polarized photons.

The NA59 experiment finished taking data last year, and we are in the process of analyzing the data. The setup used is shown in Fig. 12. Electrons penetrating the first crystal generate the linearly polarized photon beam. The polarization of these photons is 'rotated' in a second crystal, which acts as a

Table 2. *Crystal thickness and beam angle of incidence.*

Usage	Crystal	Thickness (mm)	angle to axis (mrad)	angle to plane (μrad)
Radiator (Xtal-1, $e^- \to \gamma e^-$)	Si (CB)	15	5 to $\langle 100 \rangle$	180 to (110)
Radiator (Xtal-1, $e^- \to \gamma e^-$)	Si (SOS)	15	0.3 $\langle 100 \rangle$	0 to (110)
$\lambda/4$-plate ($\gamma \to \gamma$)	Si	100	2.29 $\langle 110 \rangle$	0 to (110)
Analyzer (Xtal-2, $\gamma \to e^+ e^-$)	Ge	10	3 $\langle 110 \rangle$	0 to ($1\bar{1}0$)
Analyzer (Xtal-2, $\gamma \to e^+ e^-$)	C	4	6.2 $\langle 001 \rangle$	560 to (110)

$\lambda/4$−plate. After this point we have an admixture of linearly and circularly polarized photons and the linear polarization is analyzed with an analyzing crystal.

An unpolarized electron beam with an energy of 180 GeV was provided. The available beam intensity was more than 2×10^5 particles/spill for 4.9×10^{12} proton on the T2 target at a 3 mrad production angle, and a beam spot with a diameter of 4 cm. We ran at 8×10^4 e^-/ppp. The angular divergence of the beam was about 50 μrad in both vertical and horizontal projections.

The setup was composed of two spectrometers. The upstream spectrometer (Bend-8) was used to determine the energy lost by the electron in the crystal radiator (XTAL 1), while the downstream spectrometer (Trim-6) was used to analyze the $e^+ e^-$ pairs produced by conversion of the photon beam in the crystal analyzer (XTAL 2).

The material, thickness and angle of incidence of the electron (photon) beam with respect to a plane and a crystal axis are summarized in Table 2 for the crystal radiator (analyzer) and labelled as XTAL 1 (XTAL 2) in Fig. 12.

4.1 Results

The year 2000 data taking can be divided into three parts: (1) the evaluation of the linear polarization for CB and SOS radiation were made with a two crystal setup, one as a Si-radiator and a second one with a Ge-crystal as an analyzer, (2) the Si-$\lambda/4$−plate crystal was added and tested, (3) the linear polarization measurements were repeated with a diamond array as the analyzer instead of Ge. The motivation for changing to diamond is that the measured asymmetry from which the polarization is inferred is three times larger.

Enhancement measurements:

The first checks of the data were made by looking at the energy loss by the electron beam with the aligned Si-radiator, and comparing them with an independent data sample taken with the Si-radiator at a 'random' position [d]. When the radiator is at a random position there is no net polarization generated for the photon beam, and the energy loss follows the usual $1/E_{loss}$ energy spectrum, as a consequence the power spectrum, $E_{loss} * (dN/dE_{loss})$, should be equal to unity through the full range of E_{loss} (see Fig. 13). We can see that the most reliable region in the calibration and resolution of the leadglass calorimeter is between 50 to 160 GeV. For this reason we have limited the analysis of the data to that region.

The results on energy loss are shown in Fig. 14, for the Si crystal aligned in a SOS and a CB configuration, along with the enhancement plots. An enhancement factor of 15 and 25 for electron radiation due to the CB and SOS effects compared to the case for radiation when the electron goes through an unaligned crystal, is found.

In Fig. 15 we show the comparison of data with the prediction for the energy loss spectrum in the CB, however this prediction is not available for the SOS case at this kind of crystal thickness. However, we do have a comparison for the single photon already shown in Fig. 7 for the SOS and the expected polarization shown in Fig. 16.

Multiplicity measurements:

The multiplicity measurements are consistent with the predictions for the CB configuration, namely 2-3 photons per incoming electron. However, for SOS we find much larger multiplicities that the previous measurement shown in Fig. 11. This is believed to be due to the effect of having a thicker radiator and in a configuration where we are on the plane. The average photon multiplicity was about 2.5 times large than in the CB configuration.

---

[d] When the crystal is at a random orientation (and away from crystalline planes) with respect to the incoming beam direction, then it behaves as an amorphous material and the energy loss by the electron follows the usual Bethe-Heitler behavior.

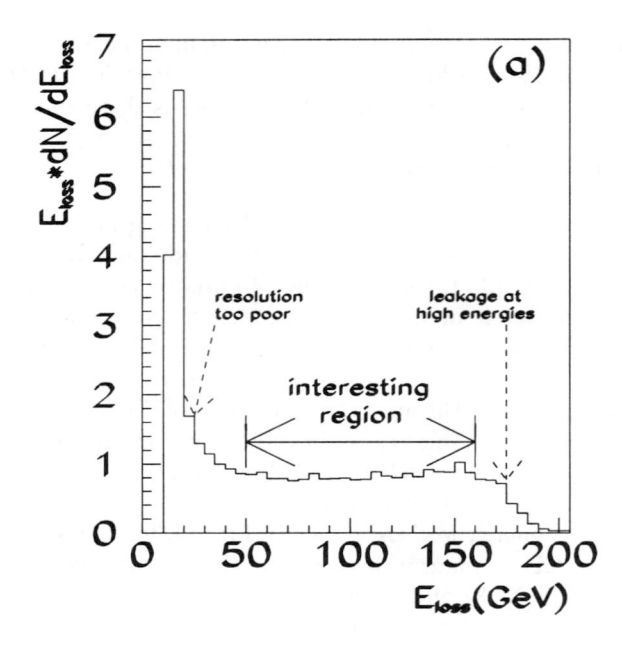

Figure 13. Power spectrum for data for unaligned or random crystal Si-crystal setting.

Large diamond array

Our South African colleagues have succeeded in persuading the de Beers Diamond Research Laboratory in Johannesburg, South Africa to engage in a program of research and development in the production of large high quality synthetic diamonds specifically for the purposes of high energy research and (their) research at the ESRF in Grenoble.

This program is paying special attention to better control of the crystal growth parameters. The phase diagram of carbon is well-known, but what is less well-known is how to maintain conditions in the diamond-stable region of this diagram (extremes of pressure and temperature, with critical gradients and control of growth orientation) and how to emerge out of this region of extremes back to normal temperature and pressure without total or even significant partial dissolution of the diamond.

An array composed of 4 diamonds was prepared using these newly produced stones (See Fig. 17) and they were mutually aligned by Freund's group in Grenoble. Each tile was 8mm×8mm×4mm resulting in an analyzer with

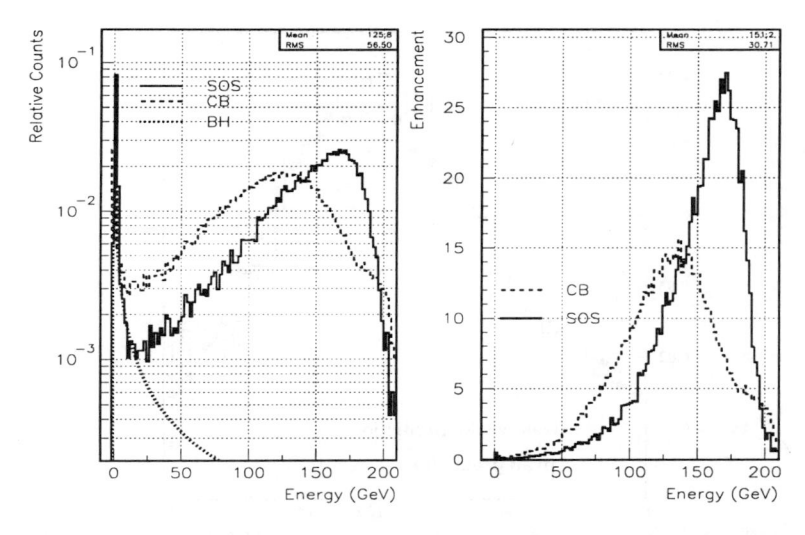

Figure 14. (a) Data showing the energy loss by the 180 GeV electron beam in the Si-radiator (15% of a X_0) with the crystal alignment in an SOS (solid) and a CB (dashed) radiation configuration, (b) measured enhancement with respect to the crystal in an unaligned configuration. As shown, about 10% of the electrons going into the crystal do not radiate more than 2-3 GeV, but the rest radiate more than 50 GeV. This is to be contrasted with the expected mean energy loss in case that the crystals is unaligned. In that case the 180 GeV electron will have lost 17 GeV on average. These results are based on a minimum bias trigger and the lead glass electromagnetic calorimeter information.

three times more analyzing power than the Ge crystal used previously. We aim at having all diamond tiles, mutually oriented to a polar angle of within 0.002 degrees.

From an alignment scan made by us, once the array was mounted in its final position to perform our polarization measurements, we found that the quality of the crystal was high (See Fig. 18(a)). From detailed scanning we could see that there was a small azimuthal misalignment between the tiles (See Fig. 18(b)) of more than the 2 mdegrees. The alignment procedure is now improved.

This is the beginning of an R&D to achieve large diamond arrays, and it will have a significant number of applications in high energy beamline construction. For example, a $\lambda/4$-plate in the future could be made out of consecutive mutually aligned crystals. The motivation being that the photon survival rate for equivalent 'rotation' efficiency compared to our 10cm-Si $\lambda/4$-plate is eight times better. This requires a total length of 2cm of diamond.

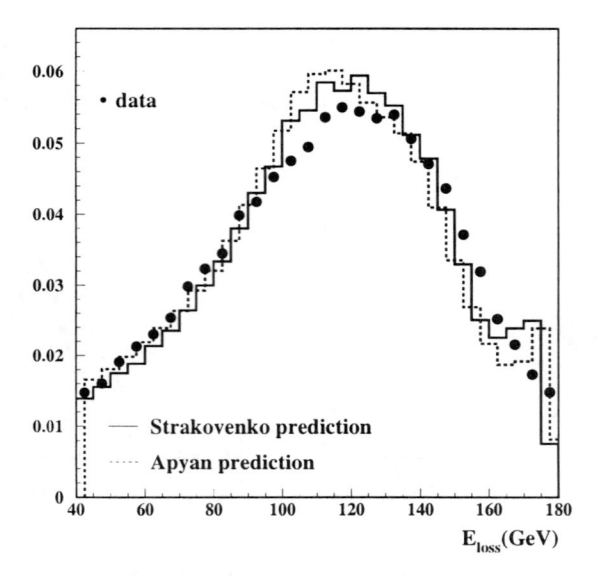

Figure 15. Data showing the energy loss by the 180 GeV electron beam in the Si-radiator while the crystal is in a CB configuration compared with two theoretical predictions [38]-[39]. These results are based on a minimum bias trigger and the lead glass electromagnetic calorimeter information.

New HE polarimetry technique and polarization measurements:

With the crystal-polarimeter ('pair' method), we take advantage of the fact that the pair production probability for linearly polarized photons penetrating into an aligned crystal is larger when the polarization is parallel to the crystallographic plane, than when the polarization is perpendicular to the crystallographic plane. Therefore, in this method we simply use another aligned crystal as an 'analyzer' and measure the asymmetry between the pair production probability of photons with 'parallel' and 'perpendicular' linear polarization. The expected asymmetry after taking into account the chosen radiator and analyzer is shown in Fig. 19 (a) for CB configuration. For SOS the asymmetry is expected to be small for $E_\gamma > 70$ GeV because the polarization is essentially zero Fig. 16 (a).

We took eight sets (0°, ± 45°, ± 90°, ± 135°, 180°) of data with the radiators at a fixed position and the analyzer crystal oriented to be at the

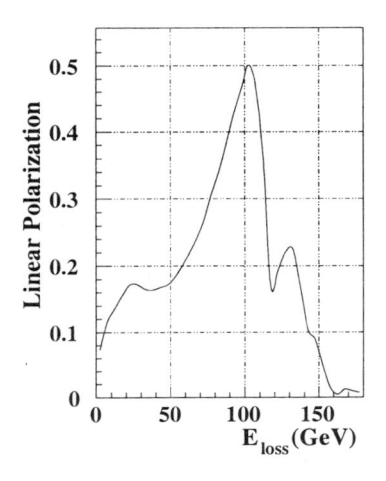

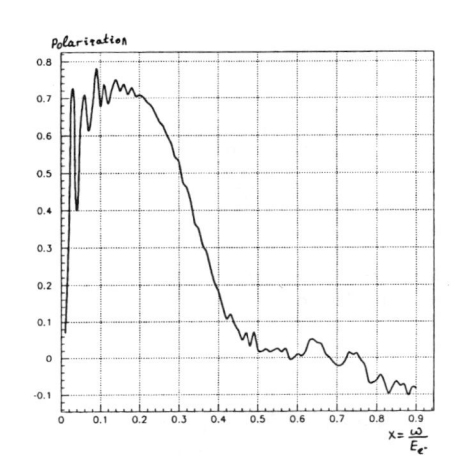

Figure 16. (a) Predictions of the photon polarization produced by 180 GeV electrons penetrating a 15 mm thick Si-radiator. In this case the crystal is aligned with respect to the electron beam axis at an angle of 5 mrad from the $\langle 011 \rangle$ axis and 180 μrad from the (110) plane, CB-configuration. (b) Expected polarization for the SOS-configuration. In this case the crystal is aligned with respect to the electron beam axis at an angle of 0.3 mrad from the $\langle 100 \rangle$ axis and 0 μrad from the (110) plane. The predictions were provided by [38], and [39], respectively.

Figure 17. Left: Example of synthetic diamonds produced for our experiment. Right: Mounted essentially perfect diamonds in the four tile array used in our experiment.

required angle between the crystallographic plane and the linear polarization plane (perpendicular to the crystallographic plane of the radiator), to analyze the polarization of the photon beam produced using the aligned Si-radiator from the 'pair' method.

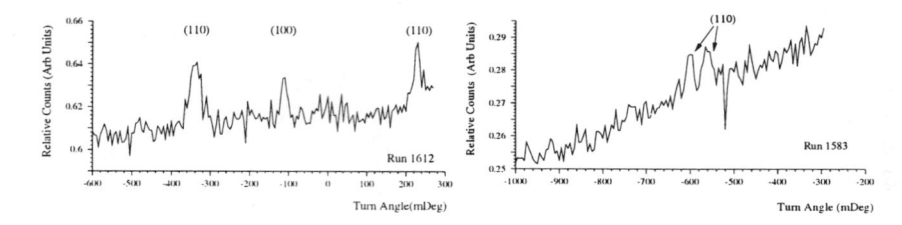

Figure 18. An alignment scan for the diamond tile array where (a) index planes are clearly identified, and (b) a small misalignment among tiles can be detected.

The radiator was kept at a fixed position, and therefore the linear polarization was not changed though the data taking, except when we were changing from CB to SOS configurations. The difference between the configurations will appear when we make a requirement on the pair conversion of the photons in the Ge-analyzer by requiring: (1) no tracks in the veto counter located in front of the Ge-crystal, (2) two minimum ionizing signals in the S11 counter, which is located right after the Ge-analyzer, and (3) by an (e^+e^-)-pair detection with the pair spectrometer.

The asymmetry that we measure is $(\sigma_\parallel - \sigma_\perp)/(\sigma_\parallel + \sigma_\perp)$, where σ is the pair production cross section $(\gamma \to e^+e^-)$ measured with the Ge-analyzer. The expected asymmetry from the combined Si-radiator and Ge-analyzer dual crystal setup is predicted to be around 5-6% asymmetry in the region of interest. In order to determine $\sigma_{\parallel(\perp)}$ we will use the minimum bias trigger to determine the flux, while the $\gamma \to e^+e^-$ events are given by a dedicated trigger. In Fig. 19 we show preliminary results for the 90° and 180° configuration and compare it to the prediction. This preliminary result is completely consistent with expectation, therefore we can conclude that the degree of linear polarization is as expected for CB.

For SOS we have no prediction available other than that the linear polarization goes to zero for energies above 75 GeV and increases very quickly for energies below 75 GeV (See Fig. 16). In Fig. 19 we show the measured asymmetries between the 0° and 90° orientations and using the diamond analyzer that is more sensitive than the Ge. We find fractions of a percent asymmetry at higher photon energies, while the asymmetry grows very quickly at low energies just like the polarization. Therefore, we conclude that the polarizations obtained for $x = E_\gamma/E_e > 0.5$ are indeed very small if any.

The asymmetry analysis for all configurations is well advanced and the final results will be available in the near future.

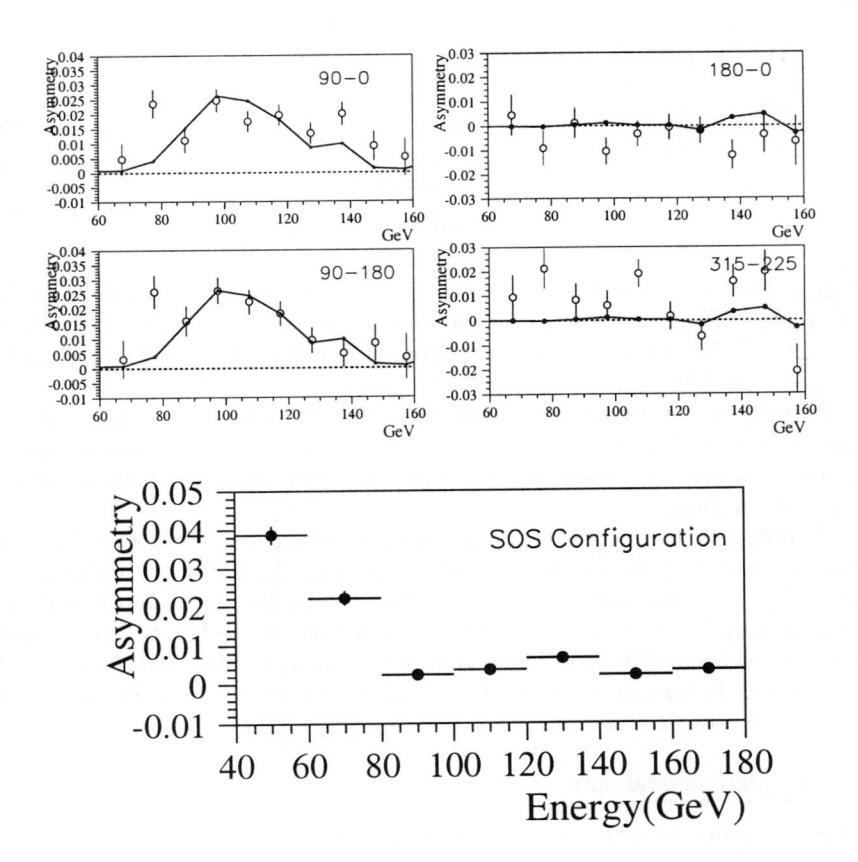

Figure 19. Measured linear polarization asymmetry for the CB and SOS radiation. The CB data is compared with the theoretical prediction which find this size of asymmetry consistent with a 50% polarization, while the SOS is consistent with low to no polarization for energies above 80 GeV. The energy range in the SOS was expanded to confirm the trend observed, which follows the predicted polarization shown in Fig. 16. Only statistical error are included at the moment.

Birefringence: $\lambda/4$-plate and linear polarization analyzing capabilities of crystals

In the early sixties Cabibbo and collaborators [37] proposed the use of a crystal as a $\lambda/4$−plate for energetic photons. This relies on the difference between the indices of refraction, $\Delta n = n_\perp - n_\parallel$, for photons polarized perpendicular and parallel to a crystallographic plane:

$$\Re(n_\perp - n_\|)x_{\lambda/4} = \lambda/4, \tag{8}$$

$$\Re(n_\perp - n_\|)\frac{\omega}{c}x_{\lambda/4} = \pi/2, \tag{9}$$

$$\Delta k x_{\lambda/4} = \pi/2, \tag{10}$$

where ω is the incoming photon beam energy and $x_{\lambda/4}$ the thickness of the $\lambda/4$-plate. What this means is that we need a non-zero wave-vector difference Δk. Under this condition the degree of circular polarization is given by $P_c(x_{\lambda/4}) = \sin(\Delta k x_{\lambda/4})$.

In order to determine the optimum crystal thickness for the $\lambda/4$-plate we take two factors into account: (1) $\Re(n_\perp - n_\|)$ for a given material after a relative alignment between the beam and a crystallographic plane, and (2) beam attenuation.

To obtain $\Re(n_\perp - n_\|)$ we calculate the pair conversion ($\gamma \to e^+e^-$) probability for photons polarized perpendicular and parallel to the crystallographic plane, $W_\perp, W_\|$. The difference between the two probabilities is proportional to the absorptive term of the index of refraction, $W_\| - W_\perp \propto \Im(n_\perp - n_\|)$. Therefore, this quantity can then be related to the dispersive part, $\Re(n_\perp - n_\|)$, through the dispersion relation. The number of surviving photons after the $\lambda/4$-plate crystal is $N(x_{\lambda/4}) = N_0 exp\{-x_{\lambda/4}(W_\perp + W_\|)/2\}$.

<u>$\lambda/4$-plate crystal optimization:</u>

A big difference in $\Re(n_\perp - n_\|)$ makes the crystal also act as a polarizer, therefore we must minimize the absorption probability simultaneously such that $W_\| - W_\perp/W_\| + W_\perp << 1 \to 2\Delta k/(W_\| + W_\perp) << 1$.

In general, we find that for a 90 mm thick Si $\langle 110 \rangle$ crystal at room temperature we can obtain a 70-75% efficient $\lambda/4$-plate, if the beam arrives at 1.5 mrad off the axis on the (100) plane. That is, the resulting photon beam will be 70-75% circularly polarized if the incident photon beam was originally 100% linearly polarized. Under these conditions about 15-20% of the beam survives, see Fig. 20.

A photo of the $\lambda/4$-plate crystal and the annular stage is shown in Fig. 21. As shown in Fig. 20, the chosen $\lambda/4$-plate crystal reduces the linear polarization by 30% and 35% of circular polarization will be produced. This will be induced from a reduction in the linear polarization asymmetry measurement. This analysis is in progress. The first look of the data using the $\lambda/4$-plate is shown in Fig. 22.

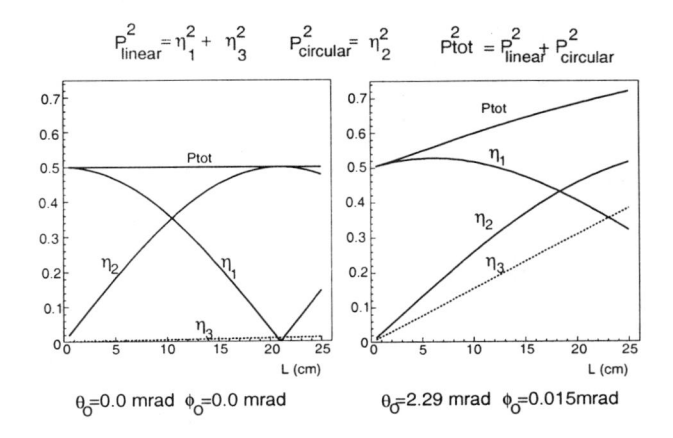

Figure 20. Polarization conversion in Si-$\lambda/4$ plate as a function of crystal thickness. Here we are expressing the polarization in terms of the Stokes parameters [40].

Figure 21. The Si $\lambda/4$-plate crystal mounted in the annular stage. The crystal has a diameter of 5 cm and a length of 10 cm.

5 Conclusion

The case for a polarized photo-production experiment and a Higgs factory at a $\gamma\gamma$-collider are clear. The crucial design element in both cases is the photon

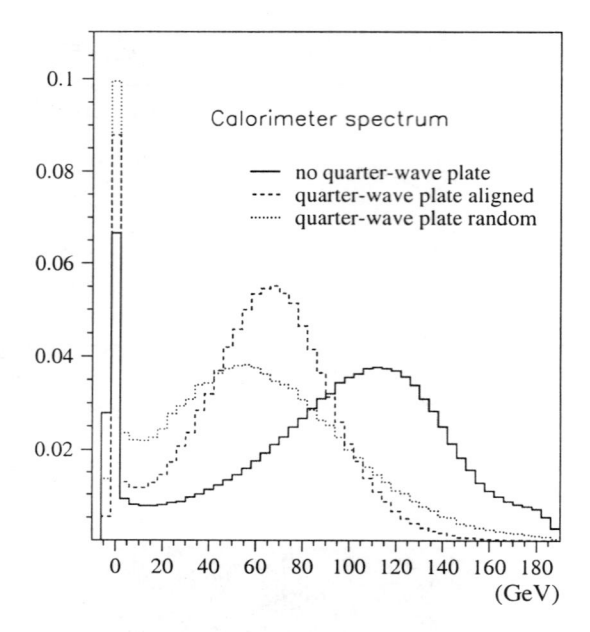

Figure 22. First look at $\lambda/4$-plate.

beam.

The work of NA59 shows that the generation of suitable photon fluxes from aligned crystals is feasible. These photons are mostly unpolarized, therefore the resulting magnitude and helicity of the circularly polarized photon beam will be defined by the longitudinal polarization of the radiating electron. These two facts will now be used to make estimates of the expected luminosity of the proposed $\gamma\gamma$-collider Higgs factory.

In the case of unpolarized electrons like those available in recent photo-production experiments, a sizable degree of linear polarization can be achieved as shown by the NA59 experiment if a proper alignment for the crystal-radiator is chosen. Once we have a linearly polarized beam, we will need the equivalent of a '$\lambda/4$−plate' for high energy photons to convert their linear polarization into circular. The data taken by the NA59 collaboration will be able to prove Cabibbo's proposal to use a crystal as a $\lambda/4$−plate for energetic photons. If the birefringence capability of crystals is proven, a new generation of polarized photo-production experiments will be possible with already existing facilities at Fermilab.

References

1. EMC, J. Ashman et al., Phys. Lett. **B206**, 364 (1988); Nucl. Phys. **B328**, 1 (1989).
2. SMC, D. Adams et al., accepted for publication by Physical Review D, **D56** (1 Nov. 97), **hep-ex/9702005**.
3. SMC, D. Adams et al., Phys. Lett. **B396** 338 (1997).
4. H.Y. Cheng, Preprint IP–ASTP–25–95, Academia Sinica (Taipei, Taiwan, Dec. 1995), **hep-ph/9512267**; Int. J. Mod. Phys. **A11**, 5109 (1996)
5. S. Adler and W. Bardeen, Phys. Rev. **182**, 1517 (1969).
6. M. Glück, E. Reya,, W. Vogelsang, Phys. Lett. **B359** 201 (1995).
7. T. Gehrmann and W.J. Stirling, Z. Phys. **C65**, 461 (1995).
8. R.D. Ball, S. Forte, and G. Ridolfi Nucl. Phys. **B444**, 287 (1995).
9. G. Altarelli and G.G. Ross, Phys. Lett. **B212**, 391 (1988).
10. A.V. Efremov and O.V. Teryaev, J.I.N.R. Preprint E2–88–287, Dubna (1988).
11. R.D. Carlitz, J.C. Collins, and A.H. Mueller, Phys. Lett. **B214**, 229 (1988).
12. M. Glück, E. Reya, M. Stratmann, W. Vogelsang, Phys Rev. **D53**, 4775 (1996).
13. T. Gehrmann and W.S. Stirling, Phys. Rev. **D53**, 6100 (1996).
14. R.D. Ball, S. Forte, and G. Ridolfi, Phys. Lett. **B378**, 255 (1996).
15. EWWG, http://lepewwg.web.cern.ch/CERN/LEPEWWG.
16. LEP-Fest, http://CERN.web.cern.ch/CERN/Announcements/2000/LEPFest.
17. V.Telnov, Int. J. Mod. Phys. A 13 (1998) 2399, e-print:hep-ex/9802003.
18. D.L. Borden,et. al., Phys. Rev. **D48**, 4018 (1993).
19. M.Battaglia, HU-P-264, Apr 1999, To be published in the proceedings of 4th International Workshop on Linear Colliders (LCWS 99), Sitges, Barcelona, Spain, 28 Apr - 5 May 1999, e-print: hep-ph/9910271.
20. G.Jikia, S.Soldner-Rembold, To be published in the proceedings of 4th International Workshop on Linear Colliders (LCWS 99), Sitges, Barcelona, Spain, 28 Apr - 5 May 1999. e-print: hep-ph/9910366.
21. T.Ohgaki, Feb. 8, 2000, e-print: hep-ph/0002083.
22. T.Ohgaki, T. Takahasgi, I. Watanabbe, Mar. 12, 1997, e-print: hep-ph/9703301.
23. E.Asakawa, May 31, 2000, e-print: hep-ph/0005313.
24. Proc.of Workshop on $\gamma\gamma$ Colliders, Berkeley CA, USA, 1994, Nucl. Instr. &Meth. A **355**(1995).
25. V.Telnov,Nucl.Instr.&Meth.A **294** (1990)72.
26. V.Telnov, SLAC-PUB-7337, Phys.Rev.Lett., **78** (1997) 4757, erratum ibid 80 (1998) 2747, e-print: hep-ex/9610008.
27. V.Telnov, Proc. Advanced ICFA Workshop on Quantum aspects of beam physics, Monterey, USA, 4-9 Jan. 1998, World Scientific, p.173, e-print: hep-ex/9805002.
28. G. Gromberg, private communication.
29. A.D. Watson, Z. Phys. C **12**, 123 (1982).

30. M. Glück, E. Reya, Z. Phys. C **39**, 569 (1988).
31. S. Keller, J.F. Owens, Phys. Rev. D **49**, 1199 (1994).
32. S.J. Brodsky, M. Burkardt, and I. Schmidt, Nucl. Phys. **B441**, 197 (1995).
33. R.G. Roberts, 'The Structure of the Proton', (Cambridge University Press 1990).
34. S. Frixione, M.L. Mangano, P. Nason, and G. Ridolfi, **hep-ph/9702287**.
35. NA59 Collaboration, CERN/SPSC 98-17, SPSC/P308, July 13, 1998.
36. P.J. Bussey et al., Nucl. Instr. Meth. **211**, 301-308 (1983).
37. N. Cabibbo, et. al., Phys. Rev. Lett. **9**, 435 (1962).
38. V.M. Strakhovenko, private communication.
39. A. Apyan , private communication.
40. V.M. Strakhovenko, April 14, 2000: e-print:hep-ph/0004131.
41. J. Schwinger, Phys. Rev. **75**, 1912 (1949).
42. K. Kirsebom et al., Nucl. Instr. Meth. B **119**, 79 (1996).
43. K. Kirsebom et al., Phys. Lett. B **459**, 347 (1999).
44. R. Milburn, NuMI-D-291, May, 1997.
45. D. Harris, private communication.

SEARCH FOR STRANGENESS AT NEW ULTRA-RELATIVISTIC HEAVY-ION COLLIDERS.

J.P. COFFIN FOR THE STAR AND THE ALICE COLLABORATIONS
*Institut de Recherches Subatomiques, IN2P3-CNRS/ULP, BP 28,
67037 STRASBOURG Cedex, France.*
E-mail : coffin@in2p3.fr

The Relativistic Heavy-Ion Collider (RHIC) is operating since a few months. The Large Hadron Collider (LHC) will be operational around 2005. They will be served by several experiments of considerable scale. Among them, the STAR and the ALICE experiments are dedicated to the study of fundamental aspects linked to an expected phase transition between the hadronic and a deconfined-quark state of nuclear matter. Numerous experimental approaches are offered by the versatility of these experiments, one of the most effective is the study of the strangeness production. An expected intriging form of strangeness is the H_0, the result of a *uuddss* quark state or of a $2\Lambda_0$ aggregation. The presentation will deal with the strangeness study and with possible experimental attempts to identify such a H_0 nuclear state, thus extending the promising general study of strangeness.

1 Introduction

The ultra relativistic heavy-ion colliders RHIC and LHC will be served by several detectors of large scale. Among them, they are STAR [8], in operation, in a not yet final configuration, since summer 2000, and ALICE [1], in preparation. These detectors are aimed at measuring different signals, strangeness among them, characteristic of the quark-gluon plasma formation, from observables analysed in the mid rapidity region. As such these detectors are not however devised, in first place, to study complex objects strongly charged in strangeness, i.e. extreme form of strangeness. Furthermore, although performed at much lower energy density, the search for this specific form of strange matter has not been successful up to now even with quite dedicated instruments [6], so expressing the difficulty of the task. Beside, the investigation of the mid rapidity region is not ideal for discovering this exotic matter because the chemical baryonic potential μ_B is expected to be close to zero and so hinders the distillation process governing the formation of droplets strongly charged in (anti)strangeness.

The new colliders will, however, provide much higher temperature and energy density, and longer lifetime for the expected plasma formation. Furthermore, theoretical speculations [10] suggest that μ_B should be different from zero and should allow the distillation process to take place. The reason is due to 1) fluctuations in the stopping power providing finite value of μ_B in a small fraction of all events 2) fluctuations of the net-baryon and net-strangeness content between different rapidity bins within any one event 3) strange (anti)baryon enhancement due to collective effects like chiral phase transition.

Hence, one may wonder to what extent some of these strange objects might be observed with the STAR and ALICE detectors. We have addressed this question by running some calculations and simulations to estimate the sensitivity of STAR and ALICE to the detection of strange objects of different lifetimes : stable or long-lived ($\tau \geq 10^{-7}$s), short-lived (10^{-11}s $\leq \tau \leq 10^{-7}$s) and unstable.

2 A brief description of the central part of the STAR and ALICE detectors

The STAR [8] and the ALICE [1] detectors are quite alike in the design of their main parts covering the mid rapidity region. The two measure in a pseudo rapidity domain of $|\eta| \leq 1$ for an angular acceptance of $45° \leq \theta \leq 135°$. Both include, from inner to outer, with respect to the beam axis, a Silicon tracker (ST) made of several layers of Silicon detectors, a large size Time Projection Chamber (TPC) and a time of flight (ToF). The TPC is quite relevant for measuring tracks over most of their length across the whole detector. The ST is essential to measure the initial part of the tracks from near the interaction point and to determine the secondary vertices

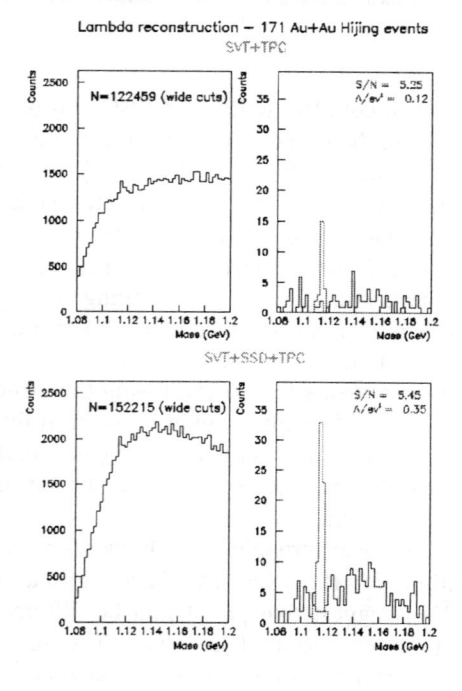

Figure 1. Λ^0 reconstruction efficiency obtained with ST+TPC (upper row) and ST+SSD+TPC (lower) with no cuts (left column) and appropriate cuts (right) on the background for selecting the signal. The plot is from reference [2].

resulting from particle decay. The more the layers of detection in the ST the more the measured space points, hence the best tracking. It is for this reason that an additional layer of Silicon strip detectors (SSD) has been proposed to be inserted between the ST and the TPC of STAR [7], it is presently under construction. Figure 1 illustrates that the gain in tracking efficiency, obtained by such an adding, is of about a factor 3.

The footprint of a stable or long-lived object will be a single track across the whole detector volume, as a result the TPC should be of primary importance. A short-lived object will decay in flight and should be recognised from track and secondary vertex reconstruction with the ST and the TPC. Secondary vertices from unstable-object decay being indiscernible from the primary vertex, the identification should based on the correlation of decay products. Here again the information extracted from the ST will be of decisive importance.

3 Detectability of strange objects in STAR and ALICE

3.1 Long-lived objects

Firstly, assumptions have to be made about the characteristics of the objects under investigation. For this, we rely primarily upon the work of Schaffner et al [9]. It is there suggested that such forms of strange matter should be preferentially extending from the triple magic strangelet (6u6d6s) to baryonic masses as high as 16 and with negative charge Z. The most promising candidates could be A_B= 10 (Z=-4), 12 (-6) and 16 (-6). However, simple coalescence estimate [9] yields production probabilities of strange clusters of the order of $10^{3-A-|S|}$. Then, large baryonic masses and high strangeness values S reduce rapidly the production probability. As a compromise we have considered, in this study, the case of an strange object characterised by A_B = 7 and Z = -1. This object being assumed to have a lifetime of $\tau \geq 10^{-7}$s, it will traverse the whole detector. As a result the TPC appears as the most important component to analyse, over most of its length, the track of a strange object. The relevant information is derived from the differential energy loss in the TPC gas $(1/\rho)(dE/dx)$ as a function of the linear momentum per charge unit p/Z, calculated from the Bethe and Bloch formula, for the A_B = 7, Z = -1 object as well as for other negatively charged particles or clusters. With a current energy resolution of about 7% [1], this object should be resolved from all other products over a large domain of p/Z. The ToF should, in principle, improve significantly the resolution [1]. However, in the case of very rare events, the background and the non gaussian-distribution tails make its use quite delicate to extract a signal [3, 4, 5].

Indeed, although 7% energy-resolution would normally be quite a respectable figure, in the present case, it does not guarantee that current "intruding" particles of multiplicity considerably larger than that of the searched strange objects may contribute and distort the distribution of these latter. Indeed, the number of intruders far in the tail of their distribution (several σ's) on which the searched signal is sitting may be appreciably altered. This effect has been discussed in some detail previously [1, 3, 4, 5].

Thus, it is necessary to evaluate the multiplicities of every possible species of intruders. They have been obtained from quantum statistical calculation including a phase transition [11] and from refrences. [1, 4]. From these multiplicities, one may calculate with 99% confidence level, by multiplying the particle multiplicity by the probability of intrusion [3, 4, 5], the maximum number (n_{max}) of intruders of each species into the searched strange-object distribution. The total number of intruders is taken as the sum N_{max} of all the contributing n_{max}. In order to identify an object characterised by $A_B = 7$ and $Z = -1$, a minimum number N_{min} of these latter has to be produced at such a rate that $N_{min} - N_{max}$ is of significant positive statistics. The

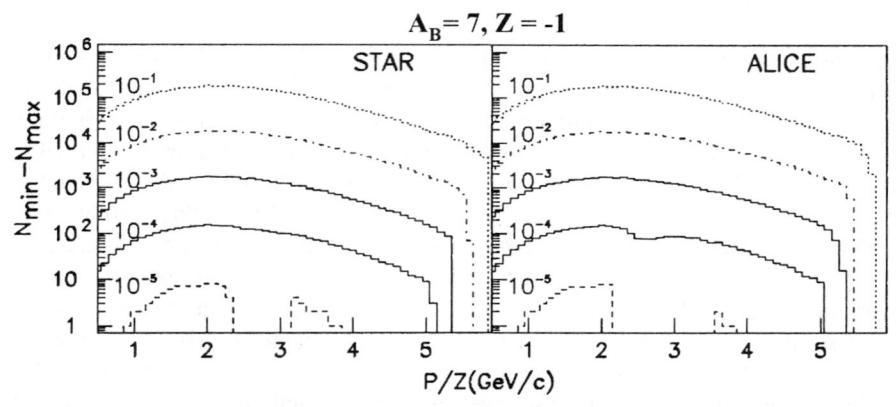

Figure 2. Difference between the minimum number of strange objects ($A_B = 7$, $Z = -1$) and the maximum number of contaminating particles as a function of p/Z for several assumptions (10^{-1} to 10^{-5}) on the production rate of this object within 5×10^7 Pb + Pb events.

difference $N_{min} - N_{max}$ is plotted in Fig. 2 as a function of p/Z for different multiplicity assumptions on the production rate of the strange object. One sees that the sensitivities of STAR and ALICE are quite comparable : A long-lived object like $A_B = 7$ and $Z = -1$, emitted at the rate of one particle every 10^5 events should be observable. It is recalled that a rate of about 10^7 and $5\ 10^7$ events per year is expected to be measured at STAR and ALICE, respectively. This sensitivity is not very dependent on the assumed characteristics of the object. It is however strongly dependent on the intruding particle multiplicities. Particle production cross-sections from first STAR results will be soon quite informative.

3.2 Short-lived objects

Short-lived exotic objects will decay in flight and their observation implies the identification of the decay pattern through track and vertex reconstruction. As a result, the ST and the TPC are both very useful. Furthermore, the optimal tracking conditions are established in the case where the secondary vertices are located in the so-called fiducial volume extending from a few millimeters away from the interaction point up to the first layer of the ST.

To illustrate the method, we shall consider the case of the simplest case of complex strange object, i.e. the H^0 which can be viewed as $2u2s2d$ or $\Lambda^0\Lambda^0$. In the case of metastability, where it decays via a weak interaction process, two main decay channels are dominant :

$$H^0 \Rightarrow \Sigma^- + p$$
$$\hookrightarrow n + \pi^-$$
$$H^0 \Rightarrow \Lambda^0 + p + \pi^-$$
$$\hookrightarrow p + \pi^-$$

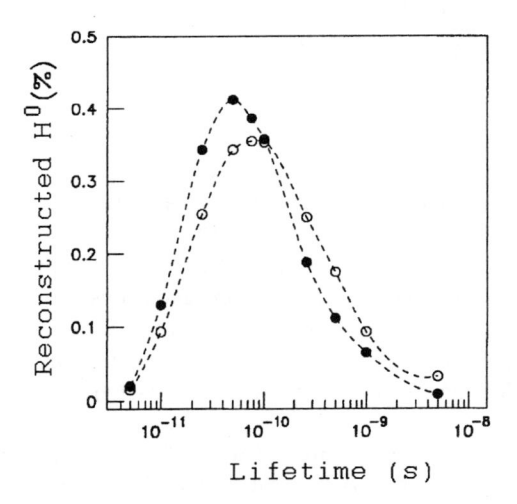

Figure 3. Percentage of reconstructed H^0 in STAR (open circles) and ALICE (dots) as a function of the H^0 lifetime.

The first mode is difficult to identify because it includes a very short Σ^- track plus a neutron which cannot be seen in the ST + TPC ensemble. Thus, we consider the second mode in which the sequential decay leads to the reconstruction of two vertices, each connecting the tracks of one proton and one π^-. As being not known the decay branching ratio has not been taken into account in the calculation.

In the evaluation presented here, a mass of 2215 MeV/c^2 has been assumed for the H^0. This latter has to be emitted within the $-1 \leq y \leq 1$ rapidity-domain while all decay products have to be kept into the $45° \leq \theta \leq 135°$ angular acceptance. As mentioned before, the secondary vertices have to be located in the fiducial volume, indeed it is the case where the highest reconstruction efficiency is obtained. As a result, the acceptance of the detector is dependent on the lifetime of the H^0. A lifetime of $\tau \approx 10^{-10}$ s turns out to be the most favorable lifetime value because it corresponds to the appropriate time for the secondary vertices to be well situated in the fiducial volume. Energy thresholds of 300 MeV for protons and 80 MeV for π^- 's have been taken. In this way, the integrated acceptance of H^0 in the STAR and ALICE detectors has been calculated.

The integrated acceptance multiplied by the squared (because $2\Lambda^0$) efficiency of track and vertex reconstruction yields the ratio of the number of reconstructed H^0 over the number of emitted in the acceptance, as shown in Fig. 3. The percentage of reconstructed H^0 is fairly comparable in STAR and ALICE. Assuming a H^0 lifetime of the order of 10^{-10}s, the percentage of reconstructed H^0 is in the domain of 0.40%-0.25%. In other words, one H^0 could be reconstructed among 250-400 H^0 emitted in the STAR and ALICE detector acceptance.

3.3 Unstable objects

This is the case where the strange objects decay via strong interaction, i.e. within less than about 10^{-20}s. The secondary vertices are thus indiscernible from the primary vertex and the only possibility to identify such an object is to correlate the decay products within an event. It is certainly the most difficult case.

Let us keep the case of the H^0, decaying instantaneously into two Λ^0's. Again, these two Λ^0's will decay in their turn into a proton and a π^-. They have to be correlated, event by event, via effective mass reconstruction among many other uncorrelated Λ^0's, these latter yielding quite a large background on top of which the H^0 signal is lying.

An effective mass ranging from 2300 up to 2500 MeV/c^2 has been assumed for the H^0. It has to be larger than twice that of the Λ^0, i.e. 2 x 1115 MeV/c^2. We suppose that the effective mass distribution is of Gaussian shape whose σ is considered ranging between 20 and 60 MeV/c^2. A combinatoric background due to all the uncorrelated Λ^0's within an event has been generated first. Then, it is considered that 0.5 and 8 Λ^0's are reconstructed in STAR and ALICE, respectively. These figures result from factors corresponding to the decay branching ratio of Λ^0 as well as to the fiducial volume, the angular acceptance, the energy thresholds and the track and vertex reconstruction as they are evaluated in the STAR and ALICE detectors. It is generally considered from simulations [1, 8] that the actual numbers of detected Λ^0's in a $Pb + Pb$ event, should be, respectively, of about 0.5 for 17 emitted in the acceptance of STAR and 8 for 300 emitted in the acceptance of ALICE. This allows for generating the background on top of which is added a Gaussian distribution simulating the H^0 signal for different Gaussian-distribution

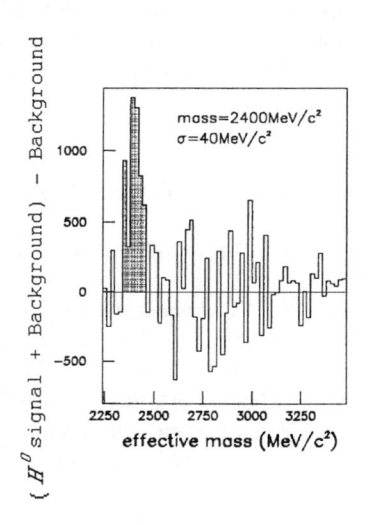

parameters.

Figure 4. H^0 effective mass distribution (grey zone) obtained after background subtraction from simulations performed with the parameter values indicated in the figure.

By subtracting the background fitted to the global distribution in the parts off the signal, one may extract the H^0 signal. Hence, one obtains a H^0 effective mass distribution, of which an example is given in Fig. 4 for 5×10^7 $Pb + Pb$ events, with the parameter choice indicated in the figure caption, and a H^0 multiplicity of 10^{-2} per Λ^0. This latter figure illustrates the minimum multiplicity of H^0 (relative to Λ^0) required for a reliable identification of the H^0 signal in both STAR and ALICE detectors.

Summary and Outlook

These preliminary calculations to evaluate the detectability of exotic strange objects in STAR and ALICE, show that the two detectors offer comparable and modest sensitivities to complex strange objects like the H^0. Of course, the expected large differences in the heavy-ion collision characteristics at the RHIC and LHC colliders, so far unknown, might be decisive in favour of one of them. Also, the sensitivities are moderately mass and momentum dependent but very dependent on the lifetime and the multiplicity of the searched objects. This suggests more accurate and reliable investigation. Some answer should come rapidly from forthcoming STAR results.

More work is needed to investigate further on a theoretical basis the best candidates for complex forms of strangeness as well as their characteristics. In the same time, simulations, including exact geometry and efficiency of the detectors, in which a few complex strange objects have been added to the realistic flux of measured current particles should yield more precise evaluation.

References

1. Alice Technical Design Report, Inner Tracking System, *CERN/LHCC 99-12* (1999)
2. Baudot J., Roy C. and Germain M., *Workshop on flow and strangeness, Obernai, France, (1999)*
3. Coffin J.P., *Proc. of Symp. on Fundamental Issues in Elementary Matter, Bad Honnef, Germany, Edited by Acta Phys. Hungarica-Heavy-Ion Phys.*, in press
4. Coffin J.P. et al., *Internal Note ALICE 95-49* (1995)
5. Coffin J.P. et al., J. Phys. G : Nucl. Part. Phys. **23** (1997) p. 2117
6. see for example Nagle J., *Proc. of Symp. on Fundamental Issues in Elementary Matter, Bad Honnef, Germany, Edited by Acta Phys. Hungarica-Heavy-Ion Phys.* in press
7. Proposal for a Silicon Strip Detector for STAR, *SN0400* (1999)
8. STAR Conceptual Design Report, *PUB-5347* (1992)
9. Schaffner-Bielich J. et al., *Phys. Rev.* **C55** (1997) p. 3038
10. Spieles C. et al, Phys. Rev. Lett. **76** (1996) p. 1776
11. Spieles C., private communication

CONFINEMENT IN THE BIG BANG AND DECONFINEMENT IN THE LITTLE BANGS AT CERN-SPS[*]

B. KÄMPFER, K. GALLMEISTER AND O.P. PAVLENKO

Forschungszentrum Rossendorf, PF 510119, 01314 Dresden, Germany
E-mail: kaempfer@fz-rossendorf.de

The evolution of strongly interacting matter during the cosmological confinement transition is reviewed. Despite of many proposed relics no specific signal from the rearrangement of quarks and gluons into hadrons has been identified by observations. In contrast to this, several observables in heavy-ion collisions at CERN-SPS energies point to the creation of a matter state near or slightly above deconfinement. We focus here on the analysis of dileptons and direct photons. Similarities and differences of the Big Bang and the Little Bang confinement dynamics are elaborated.

1 Introduction

The theory of strong interaction, QCD, points to a transition from a confined hadronic phase to the quark-gluon plasma at sufficiently high temperatures. The deconfinement temperature, T_c, is in the order of 170 MeV or slightly larger.[1] The very nature of the deconfinement transition depends on yet poorly constrained parameters, such as quark masses (cf. Ref.[2]). Above T_c the recent advanced QCD lattice calculations [3] deliver results on the equation of state of partonic matter, which can be understood within quasi-particle models.[4] Fig. 1 shows a few examples.

Hot deconfined matter must have existed in the early universe. According to the standard Big Bang cosmology the thermalized matter in the universe undergoes continuous cooling. That means that in the Big Bang the confinement transition at T_c happened, where quarks and gluons become strongly correlated, thus forming particles with large masses, such as protons and neutrons, and other hadrons as well. One intriguing question is whether the cosmic confinement transition left some specific imprint on the subsequent evolution of matter or a verifiable direct signal. Despite of many proposed relics, up to now no specific signal has been found.

One of the primary goals of the investigations of heavy-ion collisions at the CERN-SPS is the hunt for signals from deconfined matter. Indeed, there are indications that the quark-gluon plasma has already been encountered.[5] In Sec. 4 we shall consider in some detail the electromagnetic radiation from

---

[*] Work supported by BMBF 06DR921, WTZ UKR-008-98 and STCU-015.

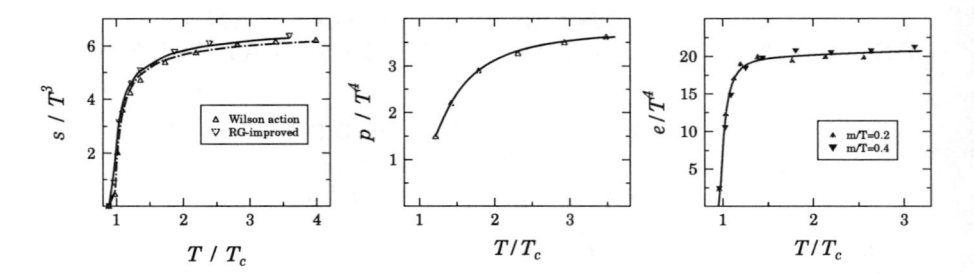

Figure 1. Equation of state of a gluon gas (left panel, entropy density s) and a two-flavor (middle panel, pressure p) and a four-flavor (right panel, energy density e) quark-gluon plasma as a function of the scaled temperature. Lattice QCD results (symbols) from the Bielefeld group; the curves represent an adjusted quasi-particle model (for details consult Ref.[4]).

the hot fireballs created in the collisions and deduce a temperature scale of $\mathcal{O}(T_c)$ from the data. Therefore, one can conclude that in the Little Bangs at CERN-SPS a matter state is created being near the borderline of confined and deconfined matter. This matter state resembles, to some extent as explained below, the matter in the universe at temperatures around T_c. The starting heavy-ion programme at RHIC and the future experiments at LHC are aimed at achieving matter states with temperatures clearly above T_c.

We are going to elaborate the similarities and differences of the confinement transition in the Big Bang and the Little Bangs (Sec. 2). In particular, we review possible relics of the cosmological confinement transition (Sec. 3). Then we present an analysis of dilepton and photon spectra observed in Little Bangs (Sec. 4). The conclusions can be found in Sec. 5.

2 Big Bang versus Little Bang dynamics

As starting point of describing the evolution of matter we chose relativistic hydrodynamics which is standard in cosmology [6] and which has been proven to be useful for heavy-ion collisions.[7] The dynamics of matter is governed by the local energy-momentum conservation,

$$T^{ij}{}_{;j} = 0, \qquad (1)$$

where we approximate the energy-momentum tensor by that of a perfect fluid, $T_{ij} = eu_iu_j + p(u_iu_j - g_{ij})$ with four-velocity u^i obeying $u^iu_i = +1$, g_{ij} is the metric tensor, e stands for the total energy density, and p denotes the thermodynamic pressure; the semicolon denotes the covariant derivative.

2.1 Friedmann's equation

To cast Eq. (1) into a tractable form one has to specify the space-time symmetry and the flow pattern. Basing on the cosmological principle[a] and Einstein's field equations for geometrodynamics one gets Friedmann's equation

$$\dot{e} = -\frac{3}{M_{\text{Pl}}}\sqrt{\frac{8\pi}{3}(e+p)}\sqrt{e} \qquad (2)$$

for the time evolution of the total energy density of matter; here M_{Pl} is the Plank mass determining even nowadays the cosmic dynamics.

The recent discovery of an accelerated expansion of the universe points to a substantial contribution of either a vacuum energy density and pressure, e^{vac} and $p^{\text{vac}} = -e^{\text{vac}}$, or a *quintessence* which dynamical behavior is not yet settled. Also the back extrapolation of the dark matter contribution meets uncertainties. With these caveats in mind we include in e and p only thermal excitations. The dynamical time scale is, from Eq. (2),

$$\frac{e+p}{\dot{e}} \sim \frac{M_{\text{Pl}}}{\sqrt{e}} \sim 10^{19}\,\text{fm/c} \quad \text{at} \quad T \sim 200\,\text{MeV}, \qquad (3)$$

showing that large energy densities drive fast evolution. This time scale is so large that quarks and gluons or, later on, hadrons and all leptons and the photons are in thermal and chemical equilibrium. These equilibrium conditions, however, cause a memory loss and, as we shall see below, little chances to find specific relics unless such ones which drop out of equilibrium.

2.2 Bjorken's equation

At sufficiently high energies, parton cascade and string models for describing heavy-ion collisions point to a dominant longitudinal motion of matter with four-velocity $u^i = \gamma(1,0,0,v_z)$ with $v_z = z/t$ and $\gamma = (1-v_z^2)^{-1/2}$. With this flow pattern, Eq. (1) becomes the celebrated Bjorken equation

$$\dot{e} = -\frac{1}{\tau}(e+p) \qquad (4)$$

being the Little Bang pendant to the Friedmann's Eq. (2) for Big Bang. Thermalization sets in at $\tau_1 = \mathcal{O}(1\,\text{fm/c})$, therefore the dynamical time scale

---

[a]This states homogeneity and isotropy in the 3D configuration space, which seem to be proven at early times by the tiny temperature fluctuations of $\Delta T/T < 10^{-4}$ of the present background radiation emerged from photon freeze-out at a world age of 300,000 years at temperature of 3,000 K.

is

$$\frac{e+p}{\dot{e}} = \tau \sim \tau_1 \sim 1\,\text{fm}/\text{c}. \tag{5}$$

Due to the shortness of this scale and the smallness of the considered systems, photons and leptons, once created, cannot come to equilibrium with the strongly interacting matter, rather they leave the fireballs nearly undisturbed and carry information on the early hot stages, where only strongly interacting matter can achieve local thermal equilibrium. The chemical equilibrium can also terminate early, thus opening another window to primordial stages.

Before discussing signals from hot matter in heavy-ion collisions let us consider possible relics from the confinement transition in Big Bang.

3 Relics of the cosmic confinement transition?

A discussion of this topic is hampered by the mentioned uncertainties of the nature of the deconfinement transition and the behavior of confined matter near T_c. One can use, for instance, various condensates, in particular the chiral condensate, as order parameters characterizing confinement. Within our phenomenological approach one has to resort to the behavior of the equation of state which may display a first-order phase transition (for light quarks) or a sharp cross over (for light u, d quarks and medium-heavy s quarks). Supposed ones describes with a bag model equation of state the deconfined matter, $p = \frac{1}{3}(e - 4B)$ (here B parameterizes the vacuum energy density), and with $p = \frac{1}{3}e$ the hadronic matter and adds appropriately the background contribution of photons and leptons, one finds the beginning of the transition at world age $t_1 \sim 6$ μsec and the end of the transition in case of a near to equilibrium transition with small surface tension $t_2 \sim 12$ μsec.[b]

Examples of possible cooling curves within the framework of classical nucleation theory are displayed in Fig. 2 for various values of the surface tension (see Ref.[8] for details). Since lattice QCD calculations point to small values of the surface tension at the boundary of confined and deconfined matter, a small supercooling is to be expected. Then frequent bubble nucleation sets in suddenly after some supercooling, and the resulting released latent heat reheates the matter to T_c, where nucleation ceases. Bubble growth determines the further evolution. This bubble growth is accompanied by shock

---

[b]General characteristic quantities at T_c are: horizon radius $R_H \sim 10$ km, Hubble time $t_H \sim 10^{-5}$ sec, energy density within the horizon M_H corresponding to $\sim 1 M_\odot$, and baryon charge within horizon $N_H^B \sim 10^{50}$.

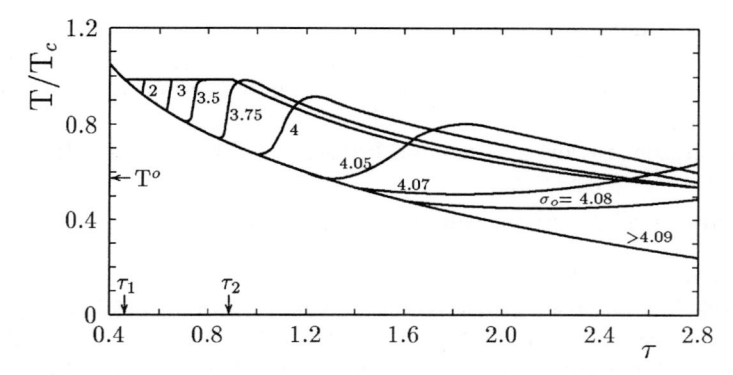

Figure 2. The temperature evolution during a first-order phase transition for various values of the surface tension parameter σ_0 as a function of the scaled dimensionless time $\tau = 2\mathcal{C}B^{1/2}t$. τ_1 and τ_2 denote the beginning and end of an equilibrium transition; T^o is the temperature of the maximum nucleation rate (cf. Ref.[8]), and $\mathcal{C} = \sqrt{8\pi/3}/M_{\text{Pl}}$.

waves [9] causing an irregular pattern of intersecting shock waves. It has been speculated that at these intersections matter is compressed and seeds for density enhancements are created. At sufficient density enhancement, matter can collapse to black holes.

Within the scenario of a first-order cosmic confinement transition various other possible imprints have been studied. Among them are:

– Isothermal baryon fluctuations: At given temperature the tiny, but nevertheless finite, baryo-chemical potential gives rise to different baryon densities in the confined and deconfined phases. If the baryon concentration in the region of shrinking quark matter at the end of the confinement era is not diffused away, then the nucleo-synthesis can be affected.[10]

– Vanishing sound velocity: The restoring pressure gradient in density-enhanced regions, created by fluctuations, vanishes and these regions can collapse in free fall. On scales being much larger than the typical bubbles, kinetically decoupled cold dark matter can be trapped in gravitational wells. Therefore, dark matter candidates can be distributed very inhomogeneously.[11] Also previously causally non-connected weak fluctuations on super-horizon scales can collapse to black holes.[12]

– Strangelets and quark nuggets: Weak processes establish equilibrium in the reactions $d \leftrightarrow u + l^- + \nu_l$ and $s \leftrightarrow u + l^- + \nu_l$ and the corresponding cross channels (l stands here for the electron or the muon). Thus, a substantial fraction of deconfined matter resides strange quarks. It has been speculated that strange quarks can stabilize quark matter.[13] Relics from the confinement transition can accordingly exist as stable quark nuggets or strangelets. The dedicated search for such exotic matter states in heavy-ion collisions, however, turned out negative.

For a recent review on such and further possible relics see Ref.[14]

4 Analysis of dilepton and photon spectra

By now a wealth of electromagnetic signals from the fireball in the Little Bangs at CERN-SPS has been registered. The spectra from the following collaborations are available: (i) CERES: Pb(158 AGeV) + Au → e^+e^-, (ii) NA50: Pb(158 AGeV) + Pb → $\mu^+\mu^-$, (iii) WA98: Pb(158 AGeV) + Pb → γ, (iv) CERES: S(200 AGeV) + Au → e^+e^-, (v) NA38: S(200 AGeV) + U → $\mu^+\mu^-$, (vi) HELIOS/3: S(200 AGeV) + W → $\mu^+\mu^-$, (vii) WA80: S(200 AGeV) + Au → γ (only upper bounds). In a schematic picture the electromagnetic signals can be considered as superposition of the following sources: (i) On very short time scales there are hard initial processes among the partons, being distributed according to primary nuclear parton distributions, such as the Drell-Yan process and charm production. (ii) On intermediate time scales there are the so-called secondary interactions among the constituents of the hot and dense, strongly interacting matter. This stage is often denoted as thermal era and the emitted dileptons as thermal dileptons. (iii) If the interactions among the hadrons in a late stage cease, there are hadronic decays into dileptons and other decay products.

We describe the hard processes by up-scaling the results of the event generator PYTHIA for pp collisions at appropriate energies (for details consult Ref.[15]). First we attempt a unifying description of the data by superpositioning the background (hadronic cocktail, Drell-Yan, correlated semileptonic decays of open charm-mesons, hard direct photons) and the thermal source parameterized by

$$\frac{dN_{l\bar{l}}}{d^4Q} = \frac{5\alpha^2}{36\pi^4} N_{\text{eff}} \exp\left\{-\frac{M_\perp \cosh(Y - Y_{\text{cms}})}{T_{\text{eff}}}\right\}, \qquad (6)$$

$$E\frac{dN_\gamma}{d^3p} = N_{\text{eff}} \frac{5\alpha\alpha_s T_{\text{eff}}^2}{12\pi^2} \int_0^1 ds\, s^2 \int_{-1}^{+1} d\xi\, e^{-A} \log\left[1 + \frac{\kappa}{\alpha_s}A\right], \qquad (7)$$

where $A = \frac{p_\perp \cosh y\, (1 - sv_0\xi)}{T_{\text{eff}}\sqrt{1-(sv_0)^2}}$, $\kappa = 2.912/(4\pi)$; $Q = (M_\perp \cosh Y, M_\perp \sinh Y, \vec{Q}_\perp)$ and $p = (p_\perp \cosh y, p_\perp \sinh y, \vec{p}_\perp)$ are the four-momenta of the the dileptons and photons with transverse mass $M_\perp = \sqrt{M^2 + Q_\perp^2}$ (here M is the invariant mass), transverse momenta $Q_\perp$ and $p_\perp$, and rapidities Y and y, respectively.

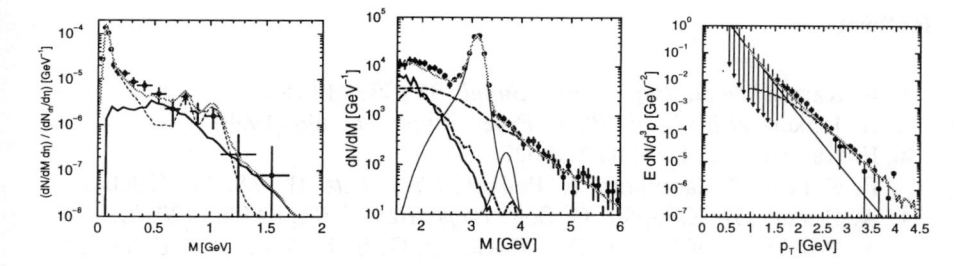

Figure 3. Comparison of our model with dilepton data (left panel for the CERES data,[17] dashed line: hadronic cocktail; middle panel for the NA50 data,[18] dashed line: Drell-Yan contribution, dot-dashed line: open charm contribution, thin lines: parameterizations of the J/ψ and ψ') and the photon data (right panel for WA98 data,[19] dashed line: hard direct photons, $v_0 = 0.3$). The thermal contribution (solid curves) is characterized by $T_{\text{eff}} = 170$ MeV and $N_{\text{eff}} = 3.3 \times 10^4$ fm^4. The sum of all contributions is depicted by the gray curves.

The two parameters T_{eff} and N_{eff} are to be adjusted to the experimental data. In Eqs. (6, 7) the time evolution of the volume of the fireball and the temperature have been replaced by averages.[15,16] Fig. 3 displays a few examples of the quality of our data description. Other examples, such as the transverse momentum spectra and an analysis of the sulfur beam induced reactions, can been found in Refs.[15,16]

The unique outcome of these studies is the value $T_{\text{eff}} = 160 \cdots 170$ MeV. Being aware of the schematic character of Eqs. (6, 7) one can implement a dynamical scenario. With parameters, partially fixed by hadronic observables, one finds a maximum temperature $\mathcal{O}(200)$ MeV or slightly above.[16,20] This is the most stringent proof that at CERN-SPS in heavy-ion collisions such temperatures are achieved which are in the deconfinement region. Of course, the effect of a finite baryon density in Eqs. (6, 7) and the assumption of thermalization need further consideration.

5 Summary

The analysis of electromagnetic signals emitted in the course of central heavy-ion collisions at CERN-SPS point to a state of strongly interacting matter with temperatures met also at confinement during the cosmic evolution. However, the dynamical time scales are vastly different. Equilibrium conditions mean memory loss, therefore, a specific imprint of the cosmic confinement transition has not yet been identified and seems unlikely.

References

1. F. Karsch, *Nucl. Phys. Proc. Suppl.* **83 - 84**, 14 (2000).
2. A. Peikert et al, *Nucl. Phys. Proc. Suppl.* **73**, 468 (1999).
3. F. Karsch et al, hep-lat/0010027,
   F. Karsch, E. Laermann, A. Peikert, *Phys. Lett.* B **478**, 447 (2000).
4. A. Peshier, B. Kämpfer, G. Soff, *Phys. Rev.* C **61**, 045203 (2000),
   A. Peshier, B. Kämpfer, O.P. Pavlenko, G. Soff, *Phys. Rev.* D **54**, 2399 (1996).
5. U. Heinz, hep-ph/0009170, *Nucl. Phys.* A in print.
6. P. Coles, F. Lucchin, *Cosmology*, (John Wiley & Sons, Chichester New York Brisbane Toronto Singapore 1995).
7. W. Greiner, H. Stöcker, *Phys. Rep.* **137**, 277 (1986).
8. B. Kämpfer, B. Lukács, G. Paál, *Cosmic phase transitions* (Teubner-Verlag, Stuttgart Leipzig 1994).
9. J. Ignatius et al, *Phys. Rev.* D **49**, 3854 (1994), **50**, 3738 (1994),
   M. Gyulassy et al, *Nucl. Phys.* B **237**, 477 (1984).
10. M.B. Christiansen, J. Madsen, *Phys. Rev.* D **53**, 5446 (1996),
    I.S. Suh, G.J. Mathews, *Phys. Rev.* D **58**, 123002 (1998).
11. C. Schmid, D.J. Schwarz, P. Widerin, *Phys. Rev.* D **59**, 043517 (1999),
    D.J. Schwarz, *Mod. Phys. Lett.* A **13**, 2771 (1998),
    C. Schmid, D.J. Schwarz, P. Widerin, *Phys. Rev. Lett.* **78**, 791 (1997).
12. K. Jedamzik, J.C Niemeyer, *Phys. Rev.* D **59**, 124014 (1999),
    K. Jedamzik, Phys. Rep. **307**, 155 (1998), *Phys. Rev.* D **55**, 5871(1997).
13. E. Farhi, R.L. Jaffe, *Phys. Rev.* D **30**, 2379, (1984),
    E. Witten, *Phys. Rev.* D **30**, 272, (1984),
    J. Schaffner-Bielich, *Nucl. Phys.* A **639**, 443c (1998).
14. B. Kämpfer, *Ann. Phys.* **9**, 605 (2000).
15. K. Gallmeister, B. Kämpfer, O.P. Pavlenko, C. Gale, hep-ph/0010332.
16. K. Gallmeister, B. Kämpfer, O.P. Pavlenko, *Phys. Lett.* B **473**, 20 (2000),
    K. Gallmeister, B. Kämpfer, O.P. Pavlenko, *Phys. Rev.* C **62**, 057901 (2000).
17. B. Lenkeit (CERES) *Nucl. Phys.* A **661**, 23c (1999).
18. E. Scomparin (NA50), *J. Phys.* G **25**, 235c (1999),
    P. Bordalo (NA50), *Nucl. Phys.* **A661**, 538c (1999).
19. M.M. Aggarwal et al (WA98), nucl-ex/0006007, nucl-ex/000608.
20. J. Alam, S. Sarkar, T. Hatsuda, T.K. Nayak, B. Sinha, hep-ph/0008074,
    R. Rapp, E.V. Shuryak, *Phys. Lett.* B **473**, 13 (2000),
    R. Rapp, J. Wambach, *Eur. Phys. J.* A **6**, 415 (1999),
    R.A. Schneider, W. Weise, hep-ph/0008083.

A CONFRONTATION WITH INFINITY *

GERARD 'T HOOFT

Institute for Theoretical Physics
University of Utrecht, Princetonplein 5
3584 CC Utrecht, the Netherlands.
e-mail: g.thooft@fys.ruu.nl

Early attempts at constructing realistic models for the weak interaction were offset by the emergence of infinite, hence meaningless, expressions when one tried to derive the radiative corrections. When models based on gauge theories with Higgs mechanism were discovered to be renormalizable, the bothersome infinities disappeared — they cancelled out. If this success seemed to be due to mathematical sorcery, it may be of interest to explain the physical insights on which it is actually based. This lecture contains much of the same material that was used in the Nobel Lecture, which was given in Stockholm, December 8, 1999.

1 Introduction

The efforts that were needed to tame the gauge theories, and our successes at this point, seem to imply that there are important lessons to be learned. I do realize the dangers of such a statement. Often in the past, progress was made precisely because lessons from the past were being ignored. Be that as it may, I nevertheless think these lessons are worth considering, and if researchers in the future should choose to ignore them, they must know what they are doing.

When I entered the field of elementary particle physics, no precise theory for the weak interactions existed[1]. It was said that any theory one attempted to write down was non-renormalizable. What was meant by that? In practice, what it meant was that when one tried to compute corrections to scattering amplitudes, physically impossible expressions were encountered. The result of the computations appeared to imply that these amplitudes should be infinite. Typically, integrals of the following form were found:

$$\int d^4k \, \frac{\text{Pol}(k_\mu)}{(k^2+m^2)\big((k+q)^2+m^2\big)} \;=\; \infty \;, \qquad (2.1)$$

where $\text{Pol}(k_\mu)$ stands for some polynomial in the integration variables k_μ. Physically, this must be nonsense. If, in whatever model calculation, the effects due to some obscure secondary phenomenon appear to be infinitely

---

*LECTURE DELIVERED AT LÜDERITZ, NAMIBIA, 17/11/2000, BASED ON THE NOBEL LECTURE 1999.

strong, one knows what this means: the so-called secondary effect is not as innocent as it might have appeared — it must have been represented incorrectly in the model to start with; one has to improve the model by paying special attention to the features that were at first thought to be negligible. The infinities in the weak-interaction theories were due to interactions from virtual particles at extremely high energies. High energy also means high momentum, and in quantum mechanics this means that the waves associated with these particles have very short wavelengths. One had to conclude that the short-distance structure of the existing theories was too poorly understood.

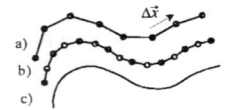

Figure 1. Differentiation

Short distance scales and short time intervals entered into theories of physics first when Newton and Leibniz introduced the notion of *differentiation*. In describing the motion of planets and moons, one had to consider some small time interval Δt and the displacement $\Delta \vec{x}$ of the object during this time interval (see Fig. 1a). The crucial observation was that, in the limit $\Delta t \to 0$, the ratio

$$\frac{\Delta \vec{x}}{\Delta t} = \vec{v} \qquad (2.2)$$

makes sense, and we call it "velocity". In fact, one may again take the ratio of the velocity *change* $\Delta \vec{v}$ during such a small time interval Δt, and again the ratio

$$\frac{\Delta \vec{v}}{\Delta t} = \vec{a} \qquad (2.3)$$

exists in the limit $\Delta t \to 0$; we call it "acceleration". Their big discovery was that it makes sense to write equations relating accelerations, velocities, and positions, and that in the limit where Δt goes to zero, you get good models describing the motion of celestial bodies (Fig. 1c).

Notice that this is not at all self-evident. According to the differential equations, we have to take the time intervals smaller than a microsecond, and the displacements of a planet during such time intervals ar just a few centimeters. Even though no astronomer in the times of Newton could resolve the motion of a planet with such accuracies, the equations only make logical sense if we take the intervals so short. Thus, our understanding of the planets required an idealized mathematical model, assuming things to happen at time and distance scales that are far beyond what we can actually observe.

The mathematics of differential equations grew out of this, and nowadays it is such a central element in theoretical physics that we often do not realize how important and how nontrivial these observations actually were. In modern theories of physics, we send distances and time intervals to zero all the time. Even in multidimensional field theories, we assume that the philosophy of differential equations applies. But occasionally it may happen that everything goes wrong. The limits that we thought we are familiar with, do not appear to exist. The behaviour of our model at the very tiniest time and distance scales then has to be completely reexamined.

Infinite integrals in particle theory were not new. They had been encountered many times before, and in some theories it was understood how to deal with them[2]. What had to be done was called "renormalization". Imagine a particle such as an electron to be something like a little sphere, of radius R and mass m_{bare}. Now attach an electric charge to this particle, of an amount Q. The electric-field energy would be

$$U = \frac{Q^2}{8\pi R}, \qquad (2.4)$$

and, according to Einstein's special theory of relativity, this would represent an extra amount of mass, U/c^2, where c is the speed of light. Particle plus field would carry a mass equal to

$$m_{\text{phys}} = m_{\text{bare}} + \frac{Q^2}{8\pi c^2 R}. \qquad (2.5)$$

It is this mass, called "physical mass", that an experimenter would measure if the particle were subject to Newton's law, $\vec{F} = m_{\text{phys}}\vec{a}$. What is alarming about this effect is that the mass correction diverges to infinity when the radius R of our particle is sent to zero. But we want R to be zero, because if R were finite it would be difficult to take into account that *forces* acting on the particle must be transmitted by a speed less than that of light, as is demanded by Einstein's theory of special relativity. If the particle were deformable, it would not be truly elementary. Therefore, finite-size particles cannot serve as a good basis for a theory of elementary objects.

In addition, there is an effect that alters the electric charge of the particle. This effect is called "vacuum polarization". During extremely short time intervals, quantum fluctuations cause the creation and subsequent annihilation of particle-antiparticle pairs. If these particles carry electric charges, the charges whose signs are opposite to our particle in question tend to move towards it, and this way they tend to neutralize it. Although this effect is usually quite small, there is a tendency of the vacuum to "screen" the charge of our particle. This screening effect implies that a particle whose charge is Q_{bare} looks like a particle with a smaller charge Q_{phys} when viewed at some distance. The relation between Q_{bare} and Q_{phys} again depends on R, and, as was the case for the mass of the particle, the charge renormalization also tends to infinity as the radius R is sent to zero — even though the effect is usually rather small at finite R.

It was already in the first half of the 20th century that physicists realized the following. The *only* properties of a particle such as an electron that we ever measure in an experiment are the physical mass m_{phys} and the physical charge Q_{phys}. So, the procedure we have to apply is that we should take the limit where R is sent to zero while m_{phys} and Q_{phys} are kept fixed. Whatever happens to the *bare* mass m_{bare} and the *bare* charge Q_{bare} in that limit is irrelevant, since these quantities can never be measured directly.

Of course, there is a danger in this argument. If, in Eq. (2.5), we send R to zero while keeping m_{phys} fixed, we notice that m_{bare} tends to *minus* infinity. Can theories in which particles have negative mass be nevertheless stable? The answer is no, but fortunately, Eq. (2.3) is replaced by a different equation in a quantized theory. m_{bare} tends to zero, not minus infinity.

2 The scale transformation

The modern way to discuss the relevance of the small-distance structure is by performing *scale transformations*, using the *renormalization group*[3], and we can illustrate this again by considering the equation of motion of the planets. Assume that we took definite time intervals Δt, finding equations for the displacements Δx. Imagine that we wish to take the limit $\Delta t \to 0$ very carefully. We may decide first to divide all Δt's and all Δx's by 2 (see Fig. 1b). We observe that, if the original intervals are already sufficiently small, the new results of a calculation will be very nearly the same as the old ones. This is because during small time intervals, planets and moons move along small sections of their orbits, which are *very nearly straight lines*. If they had been moving exactly along straight lines, the division by 2 would have made no difference at all. Planets move along straight lines if *no force acts on them*. The reason why differential equations were at all successful for planets is that *we may ignore the effects of the forces* (the "interactions") when time and space intervals are taken to be very small.

In quantized field theories for elementary particles, we have learned how to do the same thing. We reconsider the system of interacting particles at very short time and distance scales. If at sufficiently tiny scales the interactions among the particles may be ignored, then we can understand how to take the limits where these scales go all the way to zero. Since then the interactions may be ignored, all particles move undisturbedly at these scales, and so the physics is then understood. Such theories can be based on a sound mathematical footing; we understand how to do calculations by approximating space and time as being divided into finite sections and intervals and taking the limits in the end.

So, how is the situation here? Do the mutual interactions among elementary particles vanish at sufficiently tiny scales? Here is the surprise that physicists had to learn to cope with: they do not.

Many theories indeed show very bad behaviour at short distances. One of these is Fermi's theory for the weak interactions. In this theory, two fermions interact at a single point in space-time, both simultaneously transforming into other fermions. In addition, there may be related interactions where one fermion transforms into two others and an antiparticle. The strength of the coupling is expressed in terms of Fermi's fundamental coupling strength, G_F.

At large distance scales, the effects of this interaction are mild, as the associated cross sections are very small. At small distance scales, however, the cross sections appear to be relatively large, and the presence of this force is felt much more strongly there. As a consequence, such a theory has large interactions at small distance scales and vice versa. Therefore, at infinitesimally small distance scales, such a theory is ill-defined, and the model is unsuitable for an accurate description of elementary particles. Other examples of models with bad small-distance behaviour are certain 'chiral models' for scalar fields.

But some specially designed models are not so bad. Examples are: a model with spinless particles whose fields ϕ interact only through a term of the form $\lambda\phi^4$ in the Lagrangian, and a model in which charged particles interact through Maxwell's equations (quantum electrodynamics, QED). In general, we choose the distance scale Δx to be inversely proportional to a parameter called μ. A scale transformation by a factor of 2 amounts to adding $\ln 2$ to $\ln \mu$, and if the distance scale is Δx, then

$$\frac{\mu \mathrm{d}}{\mathrm{d}\mu}\Delta x = -\Delta x \ . \tag{3.1}$$

During the '60's, it was found that in *all* theories existing at the time, the interaction parameters, being either the coefficient λ for $\lambda\phi^4$ theory, or

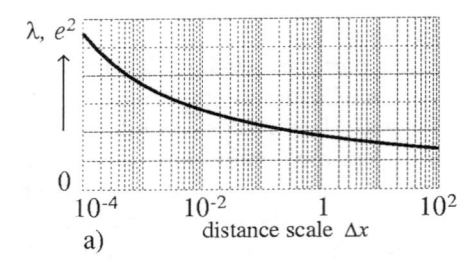

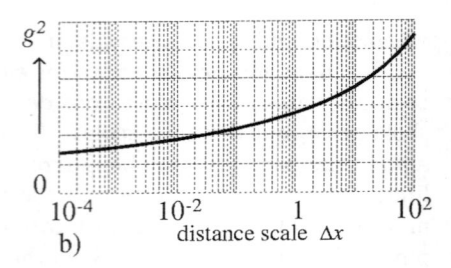

Figure 2. Scaling of the coupling strength as the distance scale varies, a) for $\lambda\phi^4$ theories and QED, b) for Yang-Mills theories.

the coefficient e^2 in quantum electrodynamics for electrons with charge e, the variation with μ is described by a positive function[4], called the β function:

$$\frac{\mu \mathrm{d}}{\mathrm{d}\mu}\lambda = \beta(\lambda) > 0 , \qquad (3.2)$$

consequently, comparing this with Eq. (3.2), λ is seen to *increase* if Δx decreases.

In the very special models that we just mentioned, the function $\beta(\lambda)$ behaves as λ^2 when λ is small, which is so small that the coupling only varies very slightly as we go from one scale to the next. This implies that, although there are still interactions, no matter how small the scales at which we look, these interactions are not very harmful, and a consequence of this is that these theories are "renormalizable". If we apply the perturbation expansion for small λ then, term by term, the expansion coefficients are uniquely defined, and we might be seduced into believing that there are no real problems with these theories.

However, many experts in these matters were worried indeed, and for good reason: If β is positive, then there will be a scale at which the coupling strength among particles diverges. The solution to Eq. (3.2) is (see Fig. 2a):

$$\lambda(\mu) = 1/(C - \beta_2 \ln \mu) , \quad \text{if} \quad \beta(\lambda) = \beta_2 \lambda^2 , \qquad (3.3)$$

where C is an integration constant, $C = 1/\lambda(1)$ if $\lambda(1)$ is λ measured at the scale $\mu = 1$. We see that at scales $\mu = \mathcal{O}\big[\exp\big(1/\beta_2\lambda(1)\big)\big]$, the coupling explodes. Since for small $\lambda(1)$ this is exponentially far away, the problem is not noticed in the perturbative formulation of the theory, but it was recognized that if, as in physically realistic theories, λ is taken to be not very small, there is real trouble at some definite scale. And so it was not so crazy to conclude that these quantum field theories were sick and that other methods should be sought for describing particle theories. In particular, this was thought to be inevitable in theories where, already at moderate scales, the interactions are strong, such as in all candidate theories for the strong force at that time.

However, some of the theories that were to be discovered later, such as the ones described below, would turn out not only to be renormalizable, but also *asymptotically free*. This means that the function $\beta(g)$ mentioned above has a negative sign. Consequently, the strength of the forces as a function of the distance scale behaves as in Fif. 2b. At small distance scales the coupling strength runs to zero logarithmically. These theories are the most convergent of all.

3 Yang-Mills theory

Fermi's theory for the weak force was not renormalizable at all. As stated earlier, the physical effects of fermi's interaction are felt at small distance scales much more strongly than at large distance scales.

In the 1970's, a whole new class of renormalizable quantum field theories was discovered. A simple prototype had been proposed by C.N. Yang and R.L. Mills in 1954. They introduced a new kind of field, which has the property that any particle of a certain type that traverses such a field, may transform into an other kind. Yang and Mills assumed that the *mass* of the particle is hardly affected. They discovered that *field equations* for such fields could be written down that are direct generalizations of electromagnetism.

Several physicists, among whom M. Veltman, insisted that the weak force should be described as the effect of such a Yang-Mills field. Yet their early models still appeared to generate uncontrollable infinities. The 'pure' Yang-Mills theory seemed to be renormalizable, but when it was dressed up to confront the weak interactions, renormalizability appeared to be lost. Veltman thought that the resolution of this problem could be found by subjecting the equations to a computer. But, to me it was clear that if we were to confront the infinities in our calculations for the weak-interaction processes, we had to disentangle the short-distance behaviour: we had to face the challenge of identifying a model for the weak interaction that shows the correct intertwining with the electromagnetic force at large distance scales but is sufficiently weakly interacting at small distances.

At *large* distance scales, the forces turn out to have a limited range, which is due to the mass terms in the equations for the force carriers, the W bosons.

Since the force is a vector force, the W bosons must have spin one, and such particles can come in *three* different polarization states: spin 1, 0, or −1. At *small* distance scales, we must insist on convergent behaviour, as in the pure Yang-Mills theories. Yet, these theories only feature massless vector bosons, like photons, which come in only two helicity states.

The number of possible modes at different scales has to match, and it is this way of reasoning that led to the resolution of our apparent problems: it is easy to add a degree of freedom at the short distance scale: we must have an additional scalar field, th Higgs field. How can we persuade a scalar field to combine with the vector field in such a way that, together, they form a massive vector field?

The answer to this question turned out to be spontaneous symmetry breaking [a]. We use a field with a quartic self-interaction but with a negative mass term, so that its energetically favored value is nonvanishing. The scalar field disappears from the physical spectrum, while, in its place a massless vector field becomes a massive one. The fact that scalar fields can thus be used to generate massive vector particles was known but not used extensively in the literature.[6] Also the fact that one could construct reasonable models for the weak interaction along these lines was known. These models, however, were thought to be inelegant, and the fact that they were the unique solution to our problems was not realized.

Not only did the newly revived models predict hitherto unknown channels for the weak interaction, they also predicted that at least one of the scalar field components would survive, to become a new, physically observable scalar particle, the Higgs boson[6]. The new weak interaction, the so-called neutral-current interaction, could be confirmed experimentally within a few years, but as of this writing, the Higgs boson is still fugitive. Some researchers suspect that it does not exist al all. Now if this were true then this would be tantamount to sending the Higgs mass all the way to infinity. An infinite-mass particle cannot be produced, so it can be declared to be absent. But such theories have bad small-distance behaviour. One can derive that the interaction strength at small distances is proportional to the Higgs mass; if that would be taken to be infinite then we would have landed in a situation where the small-distance behaviour was out of control. Such models simply do not work. Perhaps experimentalists will not succeed in producing and detecting Higgs particles, but this then would imply that entirely new theories

---

[a]The mass generation mechanism discussed here should, strictly speaking, not be regarded as spontaneous symmetry breaking, since in these theories the vacuum does not break the gauge symmetry. "Hidden symmetry" is a better phrase[5]. Let us simply refer to this mechanism as the "Higgs mechanism".

must be found to account for the small-distance structure. Candidates for such theories have been proposed. They seem to be inelegant at present, but of course that could be due to our present limited understanding, who knows? New theories would necessarily imply the existence of many presently unknown particle species, and experimenters would be delighted to detect and study such objects. We cannot lose here. Either the Higgs particle or other particles must be waiting there to be discovered, probably fairly soon.[7]

On short, what we achieved was a set of models that are mathematically precisely defined up to distance and time scales much smaller than those at which experiments can be done; we demand that at these very tiny length and time scales the scaled interactions should not be too strong.

Yet, there do exist forces which are already strong at moderate distance scales — the strong force. To these forces, the same philosophy applies even more: we must have a situation where the strentgh of the forces decrease at smaller distances. The outcome of our reasoning is here very remarkable. The good scaling behaviour of pure gauge theories (see Fig. 2b) allows us to construct a model in which the interactions at large distance scales is unboundedly strong, yet it decreases to zero (though only logarithmically) at small distances. Such a theory may describe the binding forces between quarks. It was found that these forces obtain a constant strength at arbitrarily large distances, where Coulomb forces would have decreased with an inverse square law. Quantum Chromodynamics, a Yang-Mills theory with gauge group $SU(3)$, could therefore serve as a theory for the strong interactions. It is the only allowed model in which the coupling strength is large but nevertheless the small-distance structure is under control.

The *Standard Model* is the most accurate model describing nature as it is known today. It is built exactly in accordance with the rules sketched above. Our philosophy is always that the experimentally obtained information about the elementary particles refers to their large distance behaviour. The small-distance structure of the theory is then postulated to be as regular as is possible without violating principles such as strict obedience of causality and Lorentz invariance. Not only do such models allow us to calculate their implications accurately, it appears that Nature really is built this way. In some sense, this result appears to be too good to be true. We shall shortly explain our reasons to suspect the existence of many kinds of particles and forces that could not yet be included in the Standard Model, and that the small-distance structure of the Standard Model does require modification.

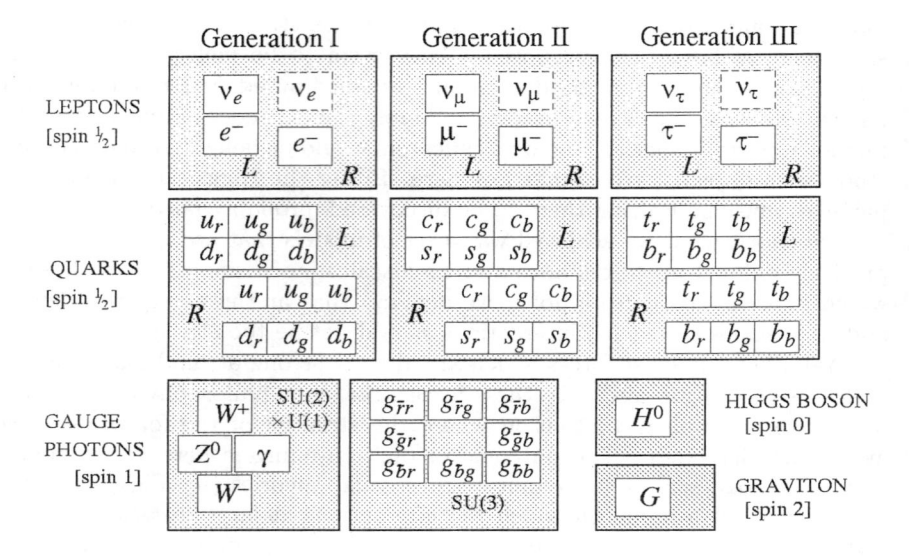

Figure 3. The Standard Model.

4 Future Colliders

Theoreticians are most eager to derive all they want to know about the structures at smaller distances using pure thought and fundamental principles. Unfortunately, our present insights are hopelessly insufficient, and all we have are some wild speculations. Surely, the future of this field still largely depends on the insights to be obtained from new experiments.

The present experiments at the Large Electron Positron Collider (LEP) at CERN are coming to a close. They have provided us with impressive precision measurements that not only gave a beautiful confirmation of the Standard Model, but also allowed us to extrapolate to higher energies, which means that we were allowed a glimpse of structures at the smallest distance ranges yet accessible. The most remarkable result is that the structures there appear to be smooth; new interactions could not be detected, which indicates that the mass of the Higgs particle is not so large, a welcome stimulus for further experimental efforts to detect it.

In the immediate future we may expect interesting new experimental results first from the Tevatron Collider at Fermilab, near Chicago, and then from the Large Hadron Collider (LHC) at CERN, both of which will devote much effort to finding the still elusive Higgs particle. Who will be first depends on what the Higgs mass will turn out to be, as well as other not yet precisely known properties of the Higgs. Detailed analysis of what we know at present indicates that Fermilab has a sizeable chance at detecting the Higgs first, and the LHC almost certainly will not only detect these particles, but also measure many of their properties, such as their masses, with high precision[8]. If supersymmetric particles exist, LHC will also be in a good position to be able to detect these, in measurements that are expected to begin shortly after 2005.

These machines, which will dissolve structures never seen before, however, also have their limits. They stop exactly at the point where our theories become highly interesting, and the need will be felt to proceed further. As before, the options are either to use hadrons such as protons colliding against antiprotons, which has the advantage that, due to their high mass, higher energies can be reached, or alternatively to use leptons, such as e^+ colliding against e^-, which has the advantage that these objects are much more pointlike, and their signals are more suitable for precision experiments.[8] Of course, one should do both. A more ambitious plan is to collide muons, μ^+ against μ^-, since these are leptons with high masses, but this will require numerous technical hurdles to be overcome. Boosting the energies to ever increasing values requires such machines to be very large. In particular the high-energy electrons will be hard to force into circular orbits, which is why design studies of the future accelerators tend to take the form of straight lines, not circles. These linear accelerators have the interesting feature that they could be extended to larger sizes in the more distant future.

My hope is that efforts and enthusiasm to design and construct such machines in the future will not diminish. As much international cooperation as possible is called for. A sympathetic proposal[9] is called ELOISATRON, a machine in which the highest conceivable energies should be reached in a gigantically large circular tunnel. It could lead to a hundredfold improvement of our spatial resolution. What worries me, however, is that in practice one group, one nation, takes an initiative and then asks other groups and nations to join, not so much in the planning, but rather in financing the whole thing. It is clear to me that the best international collaborations arises when all partners are made to be involved from the very earliest stages of the development onwards. The best successes will come from those institutions that are the closest approximations to what could be called "world machines". CERN claims to be a world machine, and indeed as such this laboratory has been, and hopefully will continue to be, extremely successful. Unfortunately, it still has an E in its name. This E should be made as meaningless as the N (after all, the physics studied at CERN has long ago ceased to be nuclear, it is subnuclear now). I would not propose to change the name, but to keep the name CERN only to commemorate its rich history.

5 Beyond the Standard Model

Other, equally interesting large scientific enterprises will be multinational by their very nature: plans are underway to construct neutrino beams that go right through the earth to be detected at the exit point, where it may be established how subtle oscillations due to their small mass values may have caused transitions from one type into another. Making world machines will not imply that competition will be eliminated; the competition, however, will not be between nations, but rather between the different collaborations who use different machines and different approaches towards physics questions.

The most interesting and important experiments are those of which we cannot guess the outcome reliably. This is exactly the case for the LHC experiments that are planned for the near future. What we do know is that the Standard Model, as it stands today, cannot be entirely correct, *in spite* of the fact that the interactions stay weak at ultra-short distance scales. Weakness of the interactions at short distances is not enough; we also insist that there be a certain amount of *stability*. Let us use the metaphor of the planets in their orbits once again. We insisted that, during extremely short time intervals, the effects of the forces acting on the planets have hardly any effect on their velocities, so that they move approximately in straight lines. In our present theories, it is as if at short time intervals several extremely strong forces act on the planets, but, for some reason, they all but balance out. The net force is so weak that only after long time intervals, days, weeks, months, the velocity change of the planets become apparent. In such a situation, however, a *reason* must be found as to why the forces at short time scales balance out. The way things are, at present, is that the forces balance out just by accident. It would be an inexplicable accident, and as no other examples of such accidents are known in Nature, at least not of this magnitude, it is reasonable to suspect that the true short-distance structure is not exactly as described in the Standard Model, but that there are more particles and forces involved, whose nature is as yet unclear. These particles and forces are arranged in a new symmetry pattern, and it is this symmetry that explains why the short-distance forces balance out.

It is generally agreed that the most attractive scenario is one involving "supersymmetry", a symmetry relating fermionic particles, whose spin is an integer plus one-half, and bosonic particles, which have integral spin.[10] It is the only symmetry that can be made to do the required job in the presence of the scalar fields that provide the Higgs mechanism, in an environment where all elementary particles interact weakly. However, when the interactions do eventually become strong then there are other scenarios. In that case, the objects playing the role of Higgs particles may be not elementary objects but composites, similar to the so-called Cooper pairs of bound electrons that perform a Higgs mechanism in ultracool solid substances, leading to superconductivity. Just because such phenomena are well known in physics, this is a scenario that cannot easily be dismissed. But, since there is no evidence at presence of a new strong interaction domain at the TeV scale, the bound-state Higgs theory is not welcome by most investigators.

One of the problems with the supersymmetry scenario is the supersymmetry breaking mechanism. Since at the distance scale where experiments are done at present no supersymmetry has been detected, the symmetry is broken. It is assumed that the breaking is "soft", which means that its effects are seen only at large distances, and only at the tiniest possible distance scales is the symmetry realized. Mathematically, this is a possibility, but there is as yet no plausible physical explanation of this situation. The only explanation can come from a theory at even smaller distance scales, where the gravitational force comes into play.

Until the early '80's, the most promising model for the gravitational force was a supersymmetric variety of gravity: supergravity.[11] It appeared that the infinities that were insurmountable in a plain gravity theory, would be overcome in supergravity. Curiously, however, the infinities appeared to be controlled by the enhanced symmetry and not by an improved small-distance structure of the theory. Newton's constant, even if controlled by a dilaton field, still is dimensionful in such theories, with consequently uncontrolled strong interactions in the small-distance domain. As the small-distance structure of the theory was not understood, it appeared to be almost impossible to draw conclusions from the theory that could shed further light on empirical features of our world.

An era followed with even wilder speculations concerning the nature of the gravitational force. By far the most popular and potentially powerful theory is that of the superstrings.[12] The theory started out by presenting particles as made up of (either closed or open) pieces of string. Fermions living on the string provide it with a supersymmetric pattern, which may be the origin of the approximate supersymmetry that we need in our theories. It is now understood that only in a perturbative formulation do particles look like strings. In a nonperturbative formalism there seems to be a need not only of strings but also of higher-dimensional substances such as membranes. But what exactly is the perturbation expansion in question? It is not the approximation that can be used at the shortest infinitesimal distances. Instead, the shortest distances seem to be linked to the largest distances by means of duality relations. Just because superstrings are also held responsible for the gravitational force, they cause curvature of space and time to such an extent that it appears to be futile to consider distances *short* compared to the Planck scale.

According to superstring theory, it is a natural and inevitable aspect of the theory that distance scales shorter than the Planck scale cannot be properly addressed, and we should not worry about it. When outsiders or sometimes colleagues from unrelated branches of physics attack superstring theory, I come to its defense. The ideas are very powerful and promising. But when among friends, I have this critical note. As string theory makes heavy use of differential equations it is clear that some sort of continuity is counted on. We should attempt to find an improved short-distance formulation of theories of this sort, if only to justify the use of differential equations or even functional integrals.

Rather than regarding the above as criticism against existing theories, one should take our observations as indications of where to search for further improvements. Emphasizing the flaws of the existing constructions is the best way to find new and improved procedures. Only in this way can we hope to achieve theories that allow us to explain the observed structures of the Standard Model and to arrive at more new predictions, so that we can tell our experimental friends where to search for new particles and forces.

References

1. For an account of the historical developments, see for instance: R.P. Crease and C.C. Mann, *The second creation: makers of the Revolution in Twentieth-century Physics* (New York, Macmillan, 1986), ISBN 0-02-521440-3.
2. See for instance A. Pais. *Inward bound: of matter and forces in the physical world,* Oxford University Press 1986, ISBN 0-19-851977-4, or
T.Y. Cao and S.S. Schweber, *The Conceptual Foundations and Philosophical Aspects of Renormalization Theory, Synthese* **97** (1993) 33, Kluwer Academic Publishers, The Netherlands.
3. K.G. Wilson and J. Kogut, *Phys. Reports* **12C** (1974) 75; H.D. Politzer,

Phys. Reports **14C** (1974) 129.
4. D.J. Gross, in *The Rise of the Standard Model*, Cambridge Univ. Press (1997), ISBN 0-521-57816-7 (pbk), p. 199.
5. S. Coleman, *Secret Symmetries*, in *Laws of Hadronic Matter*, A. Zichichi, ed., (Academic Press, New York and London, 1975).
6. P.W. Higgs, *Phys. Lett.* **12** (1964) 132; *Phys. Rev. Lett.* **13** (1964) 508; *Phys. Rev.* **145** (1966) 1156;
F. Englert and R. Brout, *Phys. Rev. Lett.* **13** (1964) 321.
7. See for instance: E. Accomando *et al*, *Phys. Reports* **299** (1998) 1, and P.M. Zerwas, *Physics with an e^+e^- Linear Collider at High Luminosity*, Cargèse lectures 1999, preprint DESY 99-178.
8. J. Ellis, *Possible Accelerators at CERN beyond the LHC*, preprint CERN-TH/99-350, hep-ph/9911440.
9. A. Zichichi, *Subnuclear Physics, the First Fifty Years, Highlights from Erice to ELN*, the Galvani Bicentenary Celebrations, Academia Delle Scienze and Bologna, 1998, pp. 117 - 135.
10. Supersymmetry has a vast literature. See for instance the collection of papers in: S. Ferrara, *Supersymmetry*, Vol. 1, North Holland, Amsterdam, etc., 1987.
11. S. Ferrara, *Supersymmetry*, Vol. 2, North defense, Amsterdam, etc., 1987.
12. See for instance J. Polchinski, *String Theory, Vol. 1, An introduction to the Bosonic String*, Cambridge Monographs on Mathematical Physics, P.V. Landshoff et al, eds. Cambridge Univ. Press 1998, ISBN 0-521-63303-6.

CURRENT STATUS OF QUARK GLUON PLASMA SIGNALS

L. GERLAND, S. SCHERER, D. ZSCHIESCHE, M. BLEICHER,
J. BRACHMANN, K. PAECH, C. SPIELES, H. WEBER, H. STÖCKER,
W. GREINER

Institut für Theoretische Physik, J.W. Goethe Universität
60054 Frankfurt a.M., Germany
E-mail: ziesche@th.physik.uni-frankfurt.de

S. BASS

Department of Physics, Duke University
27708-0305 Durham, NC, USA
and
RIKEN BNL Research Center, Brookhaven National Laboratory
Upton, NY 11973, USA

S. SOFF

Physics Department, Brookhaven National Laboratory
Upton, NY 11973, USA

Compelling evidence for a new form of matter has been claimed to be formed in Pb+Pb collisions at SPS. We discuss the uniqueness of often proposed experimental signatures for quark matter formation in relativistic heavy ion collisions. It is demonstrated that so far none of the proposed signals like J/Ψ meson production/suppression, strangeness enhancement, dileptons, and directed flow unambigiously show that a phase of deconfined matter has been formed in SPS Pb+Pb collisions. We emphasize the need for systematic future measurements to search for simultaneous irregularities in the excitation functions of several observables in order to come close to pinning the properties of hot, dense QCD matter from data.

1 Introduction

In the last few years researchers at Brookhaven and CERN have succeeded to measure a wide spectrum of observables with heavy ion beams, $Au + Au$ and $Pb + Pb$. While these programs continue to measure with greater precision the beam energy-, nuclear size-, and centrality dependence of those observables, it is important to recognize the major milestones passed thusfar in that work. The experiments have conclusively demonstrated the existence of strong nuclear A dependence of, among others, J/ψ and ψ' meson production and suppression, strangeness enhancement, hadronic resonance production, stopping and directed collective transverse and longitudinal flow of baryons and mesons – in and out of the impact plane, both at AGS and SPS energies –, and dilepton-enhancement below and above the ρ meson mass. These observations support that a novel form of "resonance matter" at high energy- and baryon density has been created in nuclear collisions. The global multiplicity and transverse energy measurements prove that substantially more entropy is produced in $A+A$ collisions at the SPS than simple superposition of $A \times pp$ would imply. Multiple initial and final state interactions play a critical role in all observables. The high midrapidity baryon density (stopping) and the observed collective transverse and directed flow patterns constitute one of the strongest evidence for the existence of an extended period ($\Delta \tau \approx 10$ fm/c) of high pressure and strong final state interactions. The enhanced ψ' suppression in $S+U$ relative to $p+A$ also attests to this fact. The anomalous low mass dilepton enhancement shows that substantial in-medium modifications of multiple collision dynamics exist, probably related to in-medium collisional broadening of vector mesons. The non-saturation of the strangeness (and anti-strangeness) production shows that novel non-equilibrium production processes arise in these reactions. Finally, the centrality dependence of J/ψ absorption in $Pb+Pb$ collisions presents further hints towards the nonequilibrium nature of such reactions. Is there evidence for the long sought-after quark-gluon plasma that thusfar has only existed as a binary array of predictions inside teraflop computers?

As we will discuss, it is too early to tell. Theoretically there are still too many "scenarios" and idealizations to provide a satisfactory answer. Recent results from microscopic transport models as well as macroscopic hydrodynamical calculations differ significantly from predictions of simple thermal models, e. g. in the flow pattern. Still, these nonequilibrium models provide reasonable predictions for the experimental data. We may therefore be forced to rethink our concept of what constitutes the deconfined phase in ultrarelativistic heavy-ion collisions. Most probably it is not a blob of thermalized quarks and gluons. Hence, a quark-gluon plasma can only be the source of *differences* to the predictions of these models for hadron ratios, the J/Ψ meson production, dilepton yields, or the excitation function of transverse flow. And there are experimental gaps such as the lack of intermediate mass $A \approx 100$ data and the limited number of beam energies studied thusfar, in particular between the AGS and SPS. Now the field is at the doorstep of the next milestone: $A+A$ at $\sqrt{s}=30-200$ AGeV which have started a few months ago.

2 J/Ψ production

The QCD factorization theorem is used to evaluate the PQCD cross sections of heavy quarkonium interactions with ordinary hadrons. However, the charmonium states (here denoted X) are not sufficiently small to ignore nonperturbative QCD physics. Thus, we evaluate the nonperturbative QCD contribution to the cross sections of charmonium-nucleon interaction by using an interpolation between known cross sections [3]. The J/Ψ-N cross section evaluated in this paper is in reasonable agreement with SLAC data [4].

Indeed, the A-dependence of the J/Ψ production studied at SLAC at $E_{inc} \sim 20$ GeV exhibits a significant absorption effect [4] leading to $\sigma_{abs}(J/\Psi$-$N) = 3.5 \pm 0.8$ mb. It was demonstrated [5] that, in the kinematic region at SLAC, the color coherence effects are still small on the internucleon scale for the formation of J/Ψ's. So, in contrast to the findings at higher energies, at intermediate energies this process measures the *genuine* J/Ψ-N interaction cross section at energies of ~ 15-20 GeV [5].

To evaluate the nonperturbative QCD contribution we use an interpolation formula [3] for the dependence of the cross section on the transverse size b of a quark-gluon configuration Three reference points are used to fix our parametrization of the cross sections (cf. Tab. 1). The X-N cross sections is calculated via: $\sigma = \int \sigma(b) \cdot |\Psi(x,y,z)|^2 dx\,dy\,dz$, where $\Psi(x,y,z)$ is the charmonium wave function. In our calculations we use the wave functions from a non-relativistic charmonium model (see [6]).

We follow the analysis of [7] to evaluate the fraction of J/Ψ's (in pp collisions) that come from the decays of the χ and Ψ'. The suppression factor S of J/Ψ's produced in the nuclear medium is calculated as: $S = 0.6 \cdot (0.92 \cdot S^{J/\Psi} + 0.08 \cdot S^{\Psi'}) + 0.4 \cdot S^{\chi}$. Here S^X are the respective suppression factors of the different pure charmonium states X in nuclear matter. The S^X are for minimum bias pA collisions within the semiclassical approximation (cf. [8]).

The charmonium states are produced as small configurations, then they evolve to their full size. Therefore, if the formation length of the charmonium states, l_f, becomes larger than the average internucleon distance, one has to take into account the evolution of the cross sections with the distance from the production point [5].

The formation length of the J/Ψ is given by $l_f \approx 2p/(m_{\Psi'}^2 - m_{J/\Psi}^2)$, where p is the momentum of the J/Ψ in the rest frame of the target. For a J/Ψ produced at midrapidity at SPS energies, this yields $l_f \approx 3$ fm. Due to the lack of better knowledge, we use the same $l_f \approx 3$ fm for the χ. For the Ψ' we use $l_f \approx 6$ fm, because it is not a small object, but has the size of a normal hadron, i.e. the pion. For $E_{lab} = 800$ AGeV we get a factor of two for the formation lengths due to the larger Lorentz factor.

However, this has a large impact on the Ψ' to J/Ψ-ratio depicted in Fig. 1, which shows the ratio $0.019 \cdot S_{\Psi'}/S_{J/\Psi}$ calculated with (squares (200 GeV) and triangles (800 GeV)) and without (crosses) expansion. The factor 0.019 is the measured value in pp collisions, because the experiments do not measure

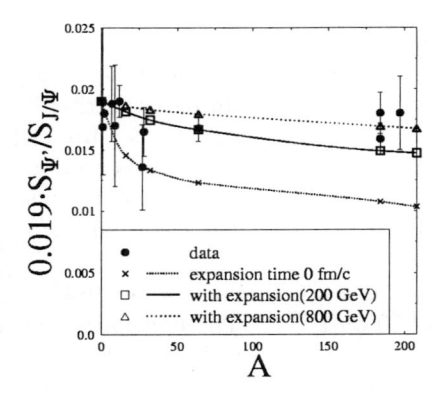

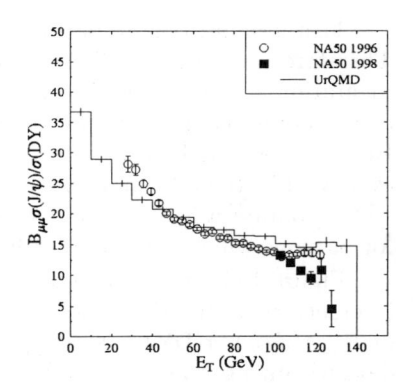

Figure 1: Left: The ratio $0.019 \cdot S_{\Psi'}/S_{J/\Psi}$ is shown in pA (crosses) in comparison to the data (circles). The squares and the triangles shows the ratio calculated with the expansion of small wave packages. Right: The ratio of J/ψ to Drell-Yan production as a function of E_T for Pb+Pb at 160 GeV.

the calculated value $S_{\Psi'}/S_{J/\Psi}$ but $(B_{\mu\mu}\sigma(\Psi'))/(B_{\mu\mu}\sigma(J/\Psi))$. $B_{\mu\mu}$ are the branching ratios for $J/\Psi, \Psi' \to \mu\mu$.

The calculations which take into account the expansion of small wave packages show better agreement with the data (circles) (taken from [9]) than the calculation without expansion time, i.e. with immediate J/Ψ formation, $l_f = 0$. We calculated this effect both at $E_{lab} = 200$ AGeV and 800 AGeV. The data have been measured at different energies ($E_{lab} = 200, 300, 400, 450, 800$ GeV and $\sqrt{s} = 63$ GeV). One can see that this ratio is nearly constant in the kinematical region of the data, but it decreases at smaller momentum (e.g. $E_{lab} = 200$ AGeV and $y < 0$) due to the larger cross section of the Ψ'.

However, the P-states yield two vastly different cross sections (see Tab. 1) for χ_{10} and χ_{11}, respectively. This leads to a higher absorption rate of the χ_{11} as compared to the χ_{10}. This new form of color filtering is predicted also for the corresponding states of other hadrons; e.g. for the bottomium states which are proposed as contrast signals to the J/Ψ's at RHIC and LHC!

$c\bar{c}$-state	J/Ψ	Ψ'	χ_{c10}	χ_{c11}
σ (mb)	3.62	20.0	6.82	15.9

Table 1: The total quarkonium-nucleon cross sections σ. For the χ two values arise, due to the spin dependent wave functions ($lm = 10, 11$).

Furthermore it is important to also take into account comoving mesons. Therefore we use the UrQMD model [10,11]. Particles produced by string fragmentation are not allowed to interact with other hadrons – in particular with a charmonium state – within their formation time (on average, $\tau_F \approx 1$ fm/c). However, leading hadrons are allowed to interact with a reduced cross section even within their formation time . The reduction factor is 1/2 for mesons which contain a leading constituent quark from an incident nucleon and 2/3 for baryons which contain a leading diquark.

Figure 1 shows the J/ψ to Drell-Yan ratio as a function of E_T for Pb+Pb interactions at 160 GeV compared to the NA50 data [12,13]. The normalization of $B_{\mu\mu}\sigma(J/\psi)/\sigma(\mathrm{DY}) = 46$ in pp interactions at 200 GeV has been fit to S+U data within a geometrical model [7]. The application of this value to our analysis is not arbitrary: the model of Ref. [7] renders the identical E_T-integrated J/ψ survival probability, $S = 0.49$, as the UrQMD calculation for this system. An additional factor of 1.25 [14] has been applied to the Pb+Pb calculation in order to account for the lower energy, 160 GeV, since the J/ψ and Drell-Yan cross sections have different energy and isospin dependencies.

The gross features of the E_T dependence of the J/ψ to Drell-Yan ratio are reasonably well described by the model calculation. No discontinuities in the shape of the ratio as a function of E_T are predicted by the simulation. The new high E_T data [13] decreases stronger than the calculation. This could be caused by underestimated fluctuations of the multiplicity of secondaries in the UrQMD model. This occurs, since high E_T-values are a trigger for very central events with a secondary multiplicity larger than in average [15].

3 Dilepton production

Beside results from hadronic probes, electromagnetic radiation – and in particular dileptons – offer an unique probe from the hot and dense reaction zone: here, hadronic matter is almost transparent. The observed enhancement of the dilepton yield at intermediate invariant masses ($M_{e^+e^-} > 0.3$ GeV) received great interest: it was prematurely thought that the lowering of vector meson masses is required by chiral symmetry restoration (see e.g.[16] for a review).

Fig. 2 shows a microscopic UrQMD calculation of the dilepton production in the kinematic acceptance region of the CERES detector for $Pb+Au$ collisions at 158 GeV. This is compared with the '95 CERES data[18]. Aside from the difference at $M \approx 0.4$ GeV there is a strong enhancement at higher invariant masses. It is expected that this discrepancy at $m > 1$ GeV could be filled up by direct dilepton production in meson-meson collisions[19] as well as by the mechanism of secondary Drell-Yan pair production proposed in[20].

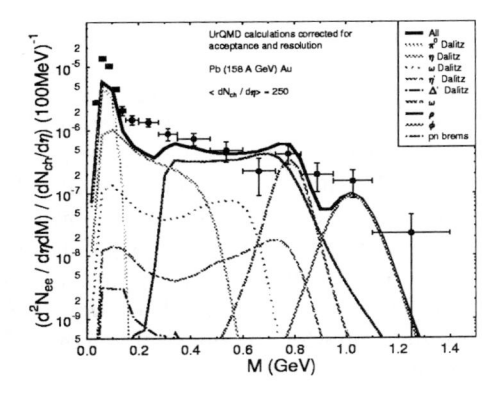

Figure 2: Microscopic calculation of the dilepton production in the kinematic acceptance region of the CERES detector for $Pb + Au$ collisions at 158 GeV. No in-medium effects are taken into account. Plotted data points are taken at CERES in '95.

4 Strangeness production

Strange particle yields are most interesting and useful probes to examine excited nuclear matter [21,22,23,24,25,26,27] and to detect the transition of (confined) hadronic matter to quark-gluon-matter. The relative enhancement of strange and especially multistrange particles in central heavy ion collisions with respect to peripheral or proton induced interactions have been suggested as a signature for the transient existence of a QGP-phase [21]. Here the main idea is that the strange (and antistrange) quarks are thought to be produced more easily and hence also more abundantly in such a deconfined state as compared to the production via highly threshold suppressed inelastic hadronic collisions. The relative enhanement of (anti)hyperons has clearly been measured by the WA97 an the NA49 collaboration in Pb-Pb collisions as compared to p-Pb collisions [25,26]. This data has been investigated within microscopic transport models (e.g. UrQMD [10]). In [22,28] it was shown that within such an approach strangeness enhancement is predicted for Pb-Pb due to rescattering. However, for central Pb-Pb collisions the experimentally observed hyperon yields are underestimated by the calculation in [22,28]. This result seems to confirm the conclusion that a deconfined QGP is formed in Pb-Pb collisions at SPS. But in [27,29] it was shown, that the antihyperon production by multi-mesonic reactions like $n_1\pi + n_2 K \to \bar{Y} + p$ could drive these rare particles towards local chemical quilibrium with pions, nucleons and kaons on a timescale of 1-3 fm/c. Accordingly this mechanism, which is a consequence of detailed balance could provide a convenient explanation for the antihyperon yields at CERN-SPS energies without any need of a deconfined quark-gluon-plasma phase. At the moment such back-reactions cannot be handled within the present transport codes. Therefore the aim for the future will be to find a way to include these processes in microscopic transport models.

5 Particle ratios

Ideal gas model calculations have been used for a long time to calculate particle production in relativistic heavy ion collisions, e.g. [30,31,32,33,34,35]. Fitting the particle ratios as obtained from those ideal gas calculations to the experimental measured ratios at SIS, AGS and SPS for different energies and different colliding systems yields a curve of chemical freeze-out in the $T-\mu$ plane. Now the question arises, how much the deduced temperature and chemical potentials depend on the model employed. Especially the influence of changing hadron masses and effective potentials should be investigated, as has been done for example in [36,37,38,39]. This is of special importance for the quest of a signal of the formation of a deconfined phase, i.e. the quark-gluon plasma. As deduced from lattice data [40], the critical temperature for the onset of a deconfined phase coincides with that of a chirally restored phase. Chiral effective models of QCD therefore can be utilized to give important insights on signals from a quark-gluon plasma formed in heavy-ion collisions.

Therefore we compare experimental measurements for Pb+Pb collisions at SPS with the ideal gas calculations and results obtained from a chiral SU(3) model [39,41]. This effective hadronic model predicts a chiral phase transition at $T \approx 150$MeV. Furthermore the model predicts changing hadronic masses and effective chemical potentials, due to strong scalar and vector fields in hot and dense hadronic matter, which are constrained by chiral symmetry from the QCD Lagrangean.

In [32] the ideal gas model was fitted to particle ratios measured in Pb+Pb collisions at SPS. The lowest χ^2 is obtained for $T = 168$MeV and $\mu_q = 88.67$MeV. Using these values as input for the chiral model leads to dramatic changes due to the changing hadronic masses in hot and dense matter [42] and therefore the freeze-out temperature and chemical potential have to be readjusted to account for the in-medium effects of the hadrons in the chiral model. We call the best fit the parameter set that gives a minimum in the value of χ^2, with $\chi^2 = \sum_i \frac{\left(r_i^{exp} - r_i^{model}\right)^2}{\sigma_i^2}$. Here r_i^{exp} is the experimental ratio, r_i^{model} is the ratio calculated in the model and σ_i represents the error in the experimental data points as quoted in [32]. The resulting values of χ^2 for different $T-\mu$ pairs are shown in figure 3.

In all calculations μ_s was chosen such that the overall net strangeness f_s is zero. The best values for the parameters are $T = 144$MeV and $\mu_q \approx 95$MeV. While the value of the chemical potential does not change much compared to the ideal gas calculation, the value of the temperature is lowered by more than 20 MeV. Furthermore Figure 3 shows, that the dropping effective masses and the reduction of the effective chemical potential make the reproduction

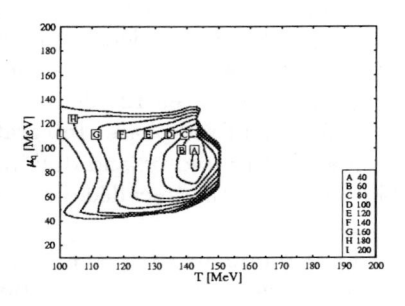

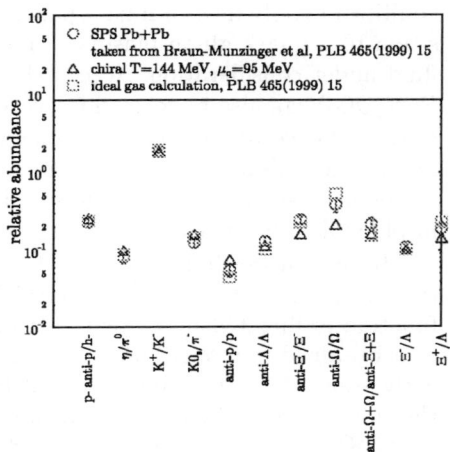

Figure 3: χ^2 (left) and resulting particle ratios compared to ideal gas calculation and data (right) for chiral model, data taken from [32]. The best fit parameters are $T = 144$ MeV and $\mu_q \approx 95$ MeV.

of experimentally measured particle ratios as seen at CERN's SPS within this model impossible for $T > T_c$. Using the best fit parameters a reasonable description of the particle ratios used in the fit procedure can be obtained (see fig.3, data from [32]).

We want to emphasize, that in spite of the strong assumption of thermal and chemical equilibrium the obtained values for T and μ differ significantly depending on the underlying model, i.e. whether and how effective masses and effective chemical potentials are accounted for. Note that we assume implicitly, that the particle ratios are determined by the medium effects and freeze out during the late stage expansion - no flavor changing collisions occur anymore, but the hadrons can take the necessary energy to get onto their mass shall by drawing energy from the fields. Rescattering effects will alter our conclusion but are presumably small when the chemical potentials are frozen.

6 Collective flow and the EOS

The in-plane flow has been proposed as a measure of the "softening" of the EoS[43], therefore we investigate the excitation function of directed in-plane flow. A three-fluid model with dynamical unification of kinetically equilibrated fluid elements is applied [44]. This model assumes that a projectile- and a target fluid interpenetrate upon impact of the two nuclei, creating a third fluid via new source terms in the continuity equations for energy- and momentum flux. Those source terms are taken from energy- and rapidity loss measurements in high energy pp-collisions. The equation of state (EoS) of this model assumes equilibrium only in each fluid separately and allows for a first order phase transition to a quark gluon plasma in fluid 1, 2 or 3, if the energy density in the fluid under consideration exceeds the critical value for two phase coexistence. Pure QGP can also be formed in every fluid separately, if the energy density in that fluid exceeds the maximum energy density for the mixed phase. Integrating up the collective momentum in x-direction at given rapidity, and dividing by the net baryon number in that rapidity bin, we obtain the so-called directed in-plane flow per nucleon.

Its excitation function (Fig. 4) shows a local minimum at 8 AGeV and rises until a maximum around 40 AGeV is reached. Fig. 4 shows the excitation function of directed flow calculated in the three-fluid model in comparison to that obtained in a one-fluid calculation. Due to non-equilibrium effects in the early stage of the reaction, which delay the build-up of transverse pressure[45], the flow shifts to higher bombarding energies. While measurements of flow at AGS[46] have found a decrease of directed flow with increasing bombarding energy, a minimum has so far not been observed.

In a recent investigation of the directed flow excitation functions [47] it has been shown, that the directed flow excitation functions are sensitive to the underlying EoS and that a different EoS can predict a slowly and smoothly decrease of the averaged directed flow as a function of bombarding energies. This different behaviour is due to the different phase transitions in the underlying equations of state. While in the two phase EoS based on a $\sigma - \omega$ model for the hadronic phase and a bag model for the deconfined phase a first-order phase transition occurs, the EoS in [47] provides a continues phase transition of the cross-over type.

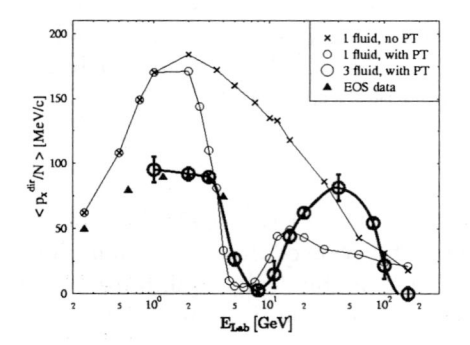

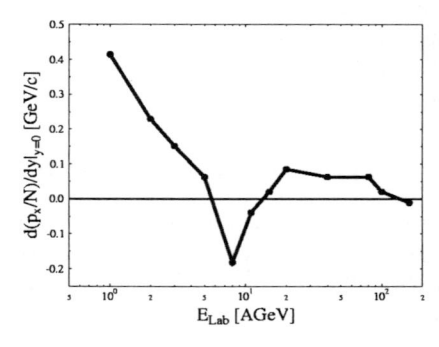

Figure 4: Excitation function of transverse flow as obtained from three fluid hydrodynamics with a first order phase transition and (Right) the slope of the directed in-plane momentum per nucleon at midrapidity.

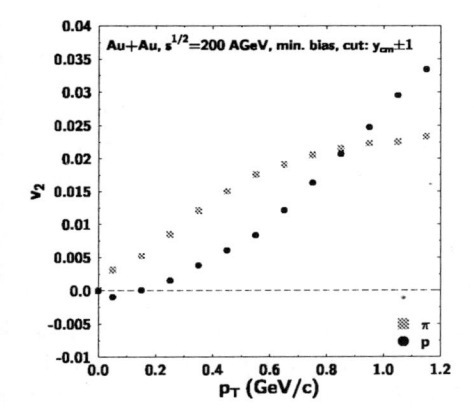

Figure 5: Elliptic flow parameter v_2 at midrapidity as a function of transverse momentum in minimum biased Au+Au reactions at $\sqrt{s} = 200A\text{GeV}$

The slope of the directed in-plane momentum per nucleon at midrapidity, $d(p_x/N)/dy$, is shown in Fig. 4 as a function of beam energy. We find a steady decrease of $d(p_x/N)/dy$ up to about top BNL-AGS energy, where the flow around midrapidity even becomes negative due to preferred expansion towards $p_x \cdot p_{long} < 0$. At higher energy, $E_{Lab} \simeq 40A$ GeV, the isentropic speed of sound becomes small and we encounter the following expansion pattern : flow towards $p_x \cdot p_{long} < 0$ can not build up ! Consequently, $d(p_x/N)/dy$ increases rapidly towards $E_{Lab} = 20 - 40A$ GeV, decreasing again at even higher energy because of the more forward-backward peaked kinematics which is unfavorable for directed flow.

Thus, the $Pb + Pb$ collisions (40 GeV) runs performed recently at the CERN-SPS may provide a crucial test of the picture of a quasi-adiabatic first-order hadronization phase transition at small isentropic velocity of sound.

7 Collective Flow at RHIC

Let us now compare the first results on elliptic flow (v_2) at $\sqrt{s_{NN}} = 130\text{GeV}$ as reported by the STAR-Collaboration[48] with a string hadronic model simulation: The experimental data indicates a strongly rising v_2 as a function of p_t with an average v_2 value of 6% at midrapidity and p_t approximately 600MeV. While the strong increase of v_2 with p_t has been predicted by the UrQMD model[49] the absolute magnitude of v_2 at $p_t = 600\text{MeV}$ is underpredicted by a factor 3 (cf. fig. 5).

When the formation time of hadrons in the initial strings is strongly reduced (to mimic short mean free paths in the early interaction region) the calculated flow values approach the hydrodynamic limit[50,49] and get in line with the measured elliptic flow values. This shows, that the pressure in the reaction zone is much higher than expected from simple stringlike models and supports the breakdown of pure string hadronic dynamics in the initial stage of Au-Au-collisions at RHIC energies. However, to get a consistent picture and to finally rule out the string hadronic approach the v_1 values and transverse momentum spectra [51] as given by the model calculation need to be exceeded by the experimental data.

8 Insights from quark molecular dynamics

Further insights about the possible formation of deconfined matter can be obtained from the Quark Molecular Dynamics Model (qMD)[52] which explicitly includes quark degrees of freedom. The qMD can provide us with detailed information about the dynamics of the quark system and the parton-hadron conversion. Correlations between the quarks clustering to build new hadrons can be studied [53].

Figure 6 shows (for S+Au collisions at SPS energies of 200 GeV/N) the number distribution for the mean path travelled by quarks forming a hadron (a) from the same initial hadron (solid line) and (b) from different initial hadrons (dotted line).

A measure of the relative mixing within the quark system and thus for thermalization is the relative number of hadrons formed by quarks from the same initial hadron correlation versus hadrons formed by quarks from different initial hadron correlations. This ratio is $r = 0.574 \pm 0.008$ for the S+Au collision. Since a value of $r = 1$ would indicate complete rearrangement of quarks and thus complete loss of correlations in the quark system, one would expect a much larger value of r, considering the presumed transition to the quark-gluon plasma in Pb+Pb collisions at 160 GeV/N,

Outlook

The latest data of CERN/SPS on flow, electro-magnetic probes, strange particle yields (most importantly multistrange (anti-)hyperons) and heavy quarkonia will be interesting to follow closely. Simple energy densities estimated from rapidity distributions and temperatures extracted from particle spectra indicate that initial conditions could be near or just above the domain of deconfinement and chiral symmetry restoration. Still the quest for an *unambiguous* signature remains open.

Directed flow has been discovered – now a flow excitation function, filling the gap between 10 AGeV (AGS) and 160 AGeV (SPS), will be extremely

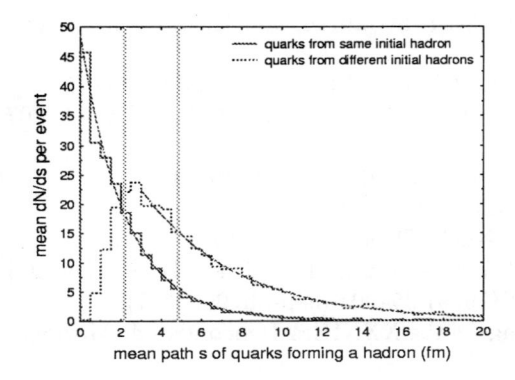

Figure 6: Hadronization in S+Au collisions at SPS (200 GeV/N): Number density distribution of mean diffusion path of quarks forming a hadron from the same initial hadron (solid line) and from different initial hadrons (dashed line) within qMD. Fitting the decay profiles yields diffusion lengths of 2.2 fm and 4.8 fm, respectively.

interesting: look for the softening of the QCD equation of state in the coexistence region. The investigation of the physics of high baryon density (e.g. partial restoration of chiral symmetry via properties of vector mesons) has been pushed forward by the 40 GeV run at SPS. Also the excitation function of particle yield ratios ($\pi/p, d/p, K/\pi$...) and, in particular, multistrange (anti-)hyperon yields, can be a sensitive probe of physics changes in the EoS. The search for novel, unexpected forms of matter, e.g. *hypermatter, strangelets* or even *charmlets* is intriguing. Such exotic QCD multi-meson and multi-baryon configurations would extend the present periodic table of elements into hitherto unexplored dimensions. A strong experimental effort should continue in that direction.

Now we have entered the exciting RHIC era, where the predicted deconfined and chirally restored phase should be formed and live long enough to produce clear and unambigious signals of it's existence. The LHC-program will top this scientific endeavour in 4 years.

Acknowledgments

This work was supported by DFG, GSI, BMBF, Graduiertenkolleg Theoretische und Experimentelle Schwerionenphysik, the A. v. Humboldt Foundation, and the J. Buchmann Foundation.

1. T. Matsui and H. Satz, Phys. Lett. **B178**, 416 (1986).
2. D. Kharzeev, Nucl. Phys. **A610**, 418c (1996).
3. L. Gerland, L. Frankfurt, M. Strikman, H. Stöcker, and W. Greiner, Phys. Rev. Lett. **81**, 762 (1998).
4. R. L. Anderson *et al.*, Phys. Rev. Lett. **38**, 263 (1977).

5. G. R. Farrar, L. L. Frankfurt, M. I. Strikman, and H. Liu, Phys. Rev. Lett. **64**, 2996 (1990).
6. L. Frankfurt, W. Koepf, and M. Strikman, Phys. Rev. **D54**, 3194 (1996).
7. D. Kharzeev, C. Lourenco, M. Nardi, and H. Satz, Z. Phys. **C74**, 307 (1997).
8. C. Gerschel and J. Hufner, Phys. Lett. **B207**, 253 (1988).
9. C. Lourenco, Nucl. Phys. **A610**, 552c (1996).
10. S. A. Bass et al., Prog. Part. Nucl. Phys. **41**, 225 (1998).
11. C. Spieles, R. Vogt, L. Gerland, S. A. Bass, M. Bleicher, H. Stocker and W. Greiner, Phys. Rev. **C60** (1999) 054901 [hep-ph/9902337].
12. A.Romana et al., in Proceedings of the XXXIIIrd Rencontres de Moriond, March 1998, Les Arcs, France.
13. M. C. Abreu et al., Phys. Lett. **B477**, 28 (2000).
14. R. Vogt, Phys. Rept. **310**, 197 (1999).
15. A. Capella, E. G. Ferreiro, and A. B. Kaidalov, hep-ph/0002300 (2000).
16. V. Koch, Int. Jour. Mod. Phys. **E6** (1997) 203.
17. W. Cassing, E. L. Bratkovskaya, R. Rapp, and J. Wambach, Phys. Rev. **C57** (1998) 916
18. G. Agakishiev et al., Phys. Lett. **B402** (1998) 405.
19. G. Q. Li and C. Gale, Phys. Rev. **C58** (1998) 2914.
20. C. Spieles et al., Eur. Phys. J. **C5** (1998) 349
21. J. Rafelski, B. Müller Phys. Rev. Lett.**48**, (1982) 1066; (E) **56** (1986) 2334; P. Koch, B. Müller, J. Rafelski *Phys. Rep.* **142**, (1986) 167; P. Koch, B. Müller, H. Stöcker, W. Greiner *Mod. Phys. Lett.* **A3**, (1988) 737
22. S. Soff, S. A. Bass, M. Bleicher, L. Bravina, M. Gorenstein, E. Zabrodin, H. Stöcker, W. Greiner Phys. Lett. **B471**, (1999) 89 and refs. therein
23. P. Senger, H. Ströbele J. Phys. **G25**, (1999) R59
24. R. Stock Phys. Lett.**B456**, (1999) 277
25. E. Andersen et al.(WA97 collaboration) Phys. Lett. **B433**, (1998) 209; S. Margetis et al.(NA49 collaboration) J. Phys. G **25**, (1999) 189
26. F. Sikler et al.(NA49 collaboration) Nucl. Phys. **A661**, (1999)
27. C. Greiner, S. Leupold nucl-th/0009036
28. S. Soff *et al.*, J. Phys. G in print, nucl-th/0010103.
29. C. Greiner, nucl-th/0011026.
30. D. Hahn and H. Stöcker, Nucl. Phys. **A452**, 723 (1986).
31. D. Hahn and H. Stöcker, Nucl. Phys. **A476**, 718 (1988).
32. P. Braun-Munzinger, J. Heppe, and J. Stachel, Phys. Lett. B **465**, 15 (1999).
33. J. Rafelski and J. Letessier, nucl-th/9903018 (1999).

34. F. Becattini, J. Cleymans, A. Keranen, E. Suhonen, and K. Redlich, hep-ph/0002267 (2000).
35. G. D. Yen and M. I. Gorenstein, Phys. Rev. **C59**, 2788 (1999).
36. H. Stöcker and W. Greiner, Z. Phys. A **286**, 121 (1978).
37. J. Theis, G. Graebner, G. Buchwald, J. A. Maruhn, W. Greiner, H. Stöcker and J. Polonyi, Phys. Rev. **D28** (1983) 2286.
38. J. Schaffner, I. N. Mishustin, L. M. Satarov, H. Stöcker, and W. Greiner, Z. Phys. **A341**, 47 (1991).
39. D. Zschiesche, P. Papazoglou, S. Schramm, C. Beckmann, J. Schaffner-Bielich, H. Stöcker, and W. Greiner, Springer Tracts in Modern Physics **163**, 129 (2000).
40. F. Karsch, hep-lat/9903031 (1998).
41. P. Papazoglou, D. Zschiesche, S. Schramm, J. Schaffner-Bielich, H. Stöcker, and W. Greiner, Phys. Rev. C **59**, 411 (1999).
42. D. Zschiesche, L. Gerland, S. Schramm, J. Schaffner-Bielich, H. Stöcker and W. Greiner, nucl-th/0007033.
43. D. H. Rischke, Y. Pürsün, J.A. Maruhn, H. Stöcker, W. Greiner, Heavy Ion Physics **1** (1995) 309.
44. J. Brachmann, A. Dumitru, J.A. Maruhn, H. Stöcker, W. Greiner, D.H. Rischke, Nucl. Phys. **A619** (1997) 391.
45. H. Sorge, Phys. Rev. Lett. **78**, 2309 (1997)
46. H. Liu et al. (E895 Collaboration), Nucl. Phys. **A638**, 451c (1998)
47. Y. B. Ivanov, E. G. Nikonov, W. Noerenberg, A. A. Shanenko and V. D. Toneev, nucl-th/0011004.
48. K. H. Ackermann *et al.* [STAR Collaboration], nucl-ex/0009011.
49. M. Bleicher and H. Stöcker, hep-ph/0006147.
50. P. Huovinen, priv. comm.;
    P. F. Kolb, J. Sollfrank and U. Heinz, Phys. Lett. **B459** (1999) 667 [nucl-th/9906003].
51. M. Bleicher *et al.*, Phys. Rev. **C62** (2000) 024904 [hep-ph/9911420].
52. M. Hofmann, S. Scherer, M. Bleicher, L. Neise, H. Stöcker, and W. Greiner, Phys. Lett. **B478** (200) 161
53. S. Scherer, M. Hofmann, M. Bleicher, L. Neise, H. Stöcker, and W. Greiner, N. Journ.Phys. *to be publ.*

Chiral Model Calculations of Nuclear Matter and Finite Nuclei

S. Schramm, C. Beckmann, H. Stöcker, D. Zschiesche, W. Greiner
Institut für Theoretische Physik
Johann Wolfgang Goethe-Universität
60054 Frankfurt, Germany

We discuss an approach modelling hadrons and nuclear matter on the basis of chiral symmetry. The masses of the hadrons are generated via spontaneous symmetry breaking. The equations are then used to calculate properties of nuclear matter and finite nuclei. The model includes the strangeness degree of freedom, enabling us to investigate hypernuclei. Some numerical results in comparison with experiment are presented.

1 Low-Energy QCD

Quantum chromodynamics (QCD) as the underlying theory of strong interactions is well established. Whereas high-energy experiments agree well with QCD calculations, there is no analogous success at low energies. Those theoretical difficulties originate from the momentum dependence of the QCD coupling strength. The QCD coupling $\alpha_{QCD}(q^2)$ shows the following behavior:

$$\alpha_{QCD}(q^2) \sim \frac{12\pi}{(33 - 2N_f)\log(|q^2|/\Lambda^2)} \tag{1}$$

where N_f is the number of quark flavors. The scale parameter Λ is about 200 MeV. For low momenta, the range of nuclear physics, the coupling increases and even diverges (an artifact of the approximation used in (1)). Thus in the region of low momenta and energies of $E < 1\,\text{GeV}$ the coupling becomes big and perturbation theory is not converging. Theoretical calculations have to either largely rely on phenomenological models or on brute-force numerical approaches, i.e. lattice gauge theory. With a lattice calculation, however, it is impossible to calculate properties of larger nuclei or even nuclear matter. In a QCD-inspired approach, which we will follow in this paper, one returns to a hadronic picture of strong interactions without referring to quarks or gluons directly, inmplementing symmetry properties found in the QCD Lagrangian.

2 Chiral Symmetry and QCD

There are many possible phenomenological descriptions of hadronic interactions. Following our line of reasoning to model strong interactions one identifies its basic symmetries and tries to preserve them in a model approach.

One essential symmetry is the so-called chiral symmetry. If we consider a fermion field (e.g. a quark field), we can write an operator $P_{L/R}$, projecting a masslos fermion spinor on its spin state pointing parallel or antiparallel to its momentum direction, respectively:

$$P_{\genfrac{}{}{0pt}{}{R}{L}} = \frac{1}{2}(\mathbb{1} \pm \gamma_5) \qquad (2)$$

Those operators are projection operators: the square of the projectors are the projectors themselves, i.e. $P_{L/R}P_{L/R} = P_{L/R}$ as can be directly inferred from Eq.(2). The mixed product vanishes: $P_L P_R = 0$.

To understand the importance of those projectors and projected fields let us consider some products of fermionic fields that occur in many Lagrangians, especially also in the Lagrangian of QCD. Starting with the vector current of a fermionic field ψ: $V_\mu = \bar\psi \gamma_\mu \psi$ we decompose the product into the left- and right-handed components by inserting the corresponding projection operators. Using $P_L + P_R = 1$, i.e. left and right-handed components form a complete set of states, we get

$$\overline{\psi}\gamma_\mu\psi \equiv \overline{(P_L + P_R)\psi}\gamma_\mu(P_L + P_R)\psi \qquad (3)$$

After some (anti-)commuting of γ matrices we arrive at the result

$$\overline{\psi}\gamma_\mu\psi = \overline{\psi_L}\gamma_\mu\psi_L + \overline{\psi_R}\gamma_\mu\psi_R \qquad (4)$$

Eq. (4) states that vector currents connect left-handed and right-handed fermion fields separately. All gauge theories, QED, QCD and the weak interactions, are based on the exchange of vector particles (the photon, the gluons, and the W and Z boson). Therefore all the fundamental interactions preserve the chirality of a fermion. A left-handed quark that emits or absorbs a gluon stays left-handed, and the same is true for electrons and photons. Thus we have two separate worlds, a left- and right-handed one.

In general other terms in the Lagrangian occur. Considering the scalar operator $S = \bar\psi\psi$ we repeat the same decomposition of the product into its chiral components:

$$S = \overline{\psi_R}\psi_L + \overline{\psi_L}\psi_R \qquad (5)$$

Scalar terms show the opposite behavior compared to the vector currents, they completely mix left- and right-handed components of the field. In a Lagrangian the scalar combination S shows up in fermionic mass terms: $\mathcal{L}_{mass} = -m\bar\psi\psi$. Therefore a mass term does not conserve chirality, it leads to chiral symmetry breaking. With the possible exception of neutrinos, fermions have masses and therefore break chiral symmetry. The practical question then arises, how strong the breaking is compared to the typical scales under consideration. This decides whether chiral symmetry is a useful concept to adopt or not. In the case of strong-interaction physics the chiral symmetry breaking terms are the mass terms of the quarks in the QCD Lagrangian

$$L_{mass}^{QCD} = -m_u \bar{u}u - m_d \bar{d}d - m_s \bar{s}s \tag{6}$$

Quark masses are not very well known experimentally with approximate values of $m_u \sim 5\,\text{MeV}$, $m_d \sim 9\,\text{MeV}$ and $m_s \sim 100\,\text{MeV}$ with large uncertainties. However, one can infer from those values that for nonstrange matter the chiral symmetry breaking is quite small compared to typical hadronic scales of roughly the proton mass or 1 GeV. For strange quarks one might still apply chiral symmetry concepts, keeping in mind that the violation could already be substantial.

3 Quark and Hadron Masses

A large amount of work has been done on the description of hadrons in terms of non-relativistic quark models. Here, quarks interact via a phenomenological potential, a harmonic oscillator in the simplest case:

$$H_{quark} = \sum_i^N \frac{\vec{p}_i^{\,2}}{2m_i} + \frac{1}{2} \sum_{i,j} V(|\vec{r}_i - \vec{r}_j|) \tag{7}$$

The spectra of most hadrons, with some failures, most notably the pions, can be reproduced rather successfully within such an approach. The masses of the quarks used in this approach are of the order of $m_u, m_d \sim 350$ MeV and $m_s \sim 500$ MeV. On the other hand, as discussede above, the quark masses entering the original QCD Lagrangian are much smaller. The assumption of approximate chiral symmetry relies on small masses in the theory. How can one resolve this mismatch?

We discussed that the coupling strength of QCD becomes large for low momenta. Assuming one has a quark-antiquark system with some kinetic and potential energy

$$E_{q\bar{q}} = E_{kin} + E_{pot} \quad . \tag{8}$$

If the quarks interact very strongly with each other it might happen that in an attractive channel the potential energy becomes so big (and negative) that it compensates for and exceeds the kinetic energy term $|E_{pot}| > E_{kin}$. In such a case the total energy of the system is negative $E_{q\bar{q}} = E_{kin} + E_{pot} < 0$. This signals an *instability* of the ground state. The system gains energy by putting a quark-antiquark pair into the vacuum. Therefore, the vacuum fills up with quark-antiquark pairs. If those pairs overlap one has also to take into account the interaction between pairs (including Pauli blocking), and a repulsive force will develop that prevents the system from generating an infinite density of pairs. Thus the true groundstate of the strongly interacting system is filled with pairs of quarks that fulfill basic known symmetries of the vacuum which is a scalar. This sea of quarks is called the *quark condensate* or also *scalar condensate*, which can also be observed in lattice calculations. As said before the pairs form scalars, where left- and right-handed particles are coupled together. Therefore the ground state of QCD breaks the chiral symmetry although the theory itself can be completely (and in practice nearly is) chirally symmetric. One might think of the propagation of a quark in this medium. Assume, for instance, a left-handed quark scatters from one of the pairs in the vacuum. Those include combinations left-handed antiquark/right-handed quark $\overline{q_L}q_R$. The incoming quark annihilates with the left-handed antiquark and a right-handed quark emerges, thus chirality is not conserved. This mode of symmetry breaking is called dynamical or spontaneous symmetry breaking, in contrast to explicit symmetry breaking where one puts a symmetry breaking term like Eq.(6) directly into the Lagrangian. As a result, one can have small explicit quark masses, and therefore a nearly chirally symmetric Lagrangian, but nevertheless end up with effectively heavy quarks and therefore hadrons due to the non-trivial vacuum structure of the theory.

4 Hadronic Models based on Chirality

Let us start with a simple point-like nucleonic interaction term, which is just the square of the scalar product (5):

$$S^2 = \left(\bar{N}N\right)^2 \tag{9}$$

A decomposition into left- and right-handed components yields (we forget about any operator-ordering subtleties for the moment):

$$S^2 = \left(\bar{N}_L N_R\right)^2 + \left(\bar{N}_R N_L\right)^2 + 2\left(\bar{N}_L N_R\right)\left(\bar{N}_R N_L\right) \tag{10}$$

The first two terms break chiral symmetry (2 right-handed particles in, 2 left-handed out and vice versa) whereas the last term does not change the number of left- and right-handed particles. Therefore we introduce a similar combination $P = \bar{N}i\gamma_5 N$, which can be shown to have pseudoscalar quantum numbers. In this case we get

$$P^2 = -\left(\bar{N}_L N_R\right)^2 - \left(\bar{N}_R N_L\right)^2 + 2\left(\bar{N}_L N_R\right)\left(\bar{N}_R N_L\right) \tag{11}$$

Thus in the sum
$$\mathcal{L}_{int} = G(S^2 + P^2) \tag{12}$$
the chiral symmetry violating terms cancel and we end up with a nucleon (or quark) interaction that is chirally symmetric. This term is very well known, originally introduced by Nambu and Jona-Lasinio[1]. From (12) we can directly derive a chirally symmetric model of nucleons and mesons. We identify the two-nucleon object $S = \overline{N}N$ with a hadronic object of the same quantum numbers, the σ meson. We can do the same with the pseudoscalar combination. P has just the quantum numbers of a pion (in a one-flavor theory). Thus replacing one scalar and pseudoscalar product in (12) with the corresponding hadronic fields we obtain
$$\mathcal{L}_{int} = 2G\overline{N}\left[\sigma + i\gamma_5\pi\right]N \tag{13}$$
which is the interaction of the linear σ model[2]. Note, that the chirally symmetric construction yields the same coupling strength G between nucleons and the sigma meson as between nucleons and pions, respectively. Effectively this procedure is also called bosonization, since we neglect the fermionic substructure of the composite operators and replace them with bosonic fields. In the next step we can replace all bilinear operators in (102 with bosonic fields, which yields
$$\mathcal{L}_{int} = 2G\left[\sigma^2 + \pi^2\right] \tag{14}$$
Thus, a chirally symmetric hadronic model includes couplings as in Eq. (13) and purely mesonic terms in the combination $\sigma^2 + \pi^2$. Following this general approach in our specific model we investigate a flavor-SU(3) extension of the chiral model discussed above, including hyperons and strange mesons. The exact procedure is rather involved (see [7]) but in effect one replaces the fields in (13) and (14) by the corresponding SU(3) multiplets. The Lagrangian that we consider has the general structure:
$$\mathcal{L} = \mathcal{L}_{\text{kin}} + \sum_{W=X,Y,V,\mathcal{A},u} \mathcal{L}_{\text{BW}} + \mathcal{L}_{\text{VP}} + \mathcal{L}_{\text{vec}} + \mathcal{L}_0 + \mathcal{L}_{\text{SB}}. \tag{15}$$

$\mathcal{L}_{\text{kin}}$ is the kinetic energy term, $\mathcal{L}_{\text{BW}}$ includes the interaction terms of the different baryons with the various spin-0 and spin-1 mesons. The baryon masses are generated by the nonstrange ($<q\bar{q}>$) scalar condensate σ and the strange ($<s\bar{s}>$) scalar condensate ζ. In the medium the scalar fields are reduced, leading to a lowering of the baryonic masses. $\mathcal{L}_{\text{VP}}$ contains the interaction terms of vector mesons with pseudoscalar mesons. $\mathcal{L}_{\text{vec}}$ generates the masses of the spin-1 mesons through interactions with spin-0 mesons, and $\mathcal{L}_0$ gives the meson-meson interaction terms which induce the spontaneous breaking of chiral symmetry. It also includes the scale breaking logarithmic potential. Finally, $\mathcal{L}_{\text{SB}}$ introduces an explicit symmetry breaking of the $U(1)_A$, the $SU(3)_V$, and the chiral symmetry. All these terms have been discussed in detail in [4].

5 Numerical Results

Finally, let us look at some results of the chiral model we discussed so far. The first step is to fix the coupling constants so that the model can reproduce the masses of the various SU(3) multiplets of hadrons. The resulting fit, which we will use for the other results discussed, can reproduce the masses of the hadronic multiplets very well (see [7]). With vacuum properties in control we can switch to nuclear matter properties. There are two independent parameters in the model that can be tuned to reproduce saturation properties of nuclear matter. We obtain nuclear saturation at the canonical values of $\rho = 0.15\,\text{fm}^{-3}$ with a binding energy $E/A = -16$ MeV. The compressibility of nuclear matter has the value of $\kappa = 276$ MeV. Such a low and reasonable compressibility is usually not obtained in $\sigma - \omega$-type chiral models. Here, especially the SU(3) nature of the model with the additional strange condensate softens nuclear matter.

Studying high temperatures one can observe first-order phase transitions within this model, where the scalar fields drop rapidly to small values. Since the baryon masses are generated by their coupling to the scalar fields the masses are reduced significantly, indicating a phase transition to a (nearly) chirally symmetric phase. Note that beyond this transition the validity of the model is very questionable, as one would expect quark degrees of freedom, which have explicitly been excluded in the model, to become dominant. Further investigations have either to include colored fields or one has to match quark models to the hadronic model at high temperatures and densities. In Fig. 1 one can see that the masses drop in two steps. This behavior originates from jumps in the values of the non-strange (σ) and strange (ζ) scalar fields, which take place at slightly different temperatures.

There have been various studies of measured hadronic particle ratios in ultrarelativistic heavy-ion collisions as measured at CERN. Thermal fits of the ratios yield a temperature of beyond 160 MeV strongly indicating that before chemical freeze-out the system was in a quark-gluon phase [5]. We repeat the same type of fit using our hadronic model. As can be seen in Fig. 2 the resulting quality of the fit is similar, the temperature, however, is only T = 144 MeV. This clearly shows - even assuming a thermal fit of the data is a sensible approach - that the extracted temperature values are quite uncertain and conclusions regarding the state of matter in the collision are still on shaky

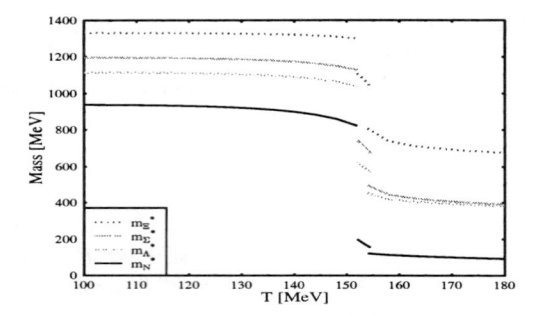

Figure 1: Baryonic masses as function of temperature. Two neighboring first-order phase transitions can be observed.

grounds.

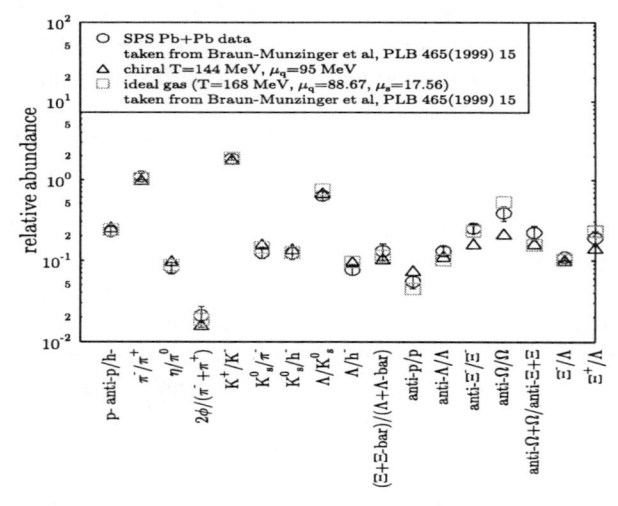

Figure 2: Hadronic particle ratios as measured in 160 AGeV ^{208}Pb on ^{208}Pb collisions. The values of the experiment and different theoretical fits are shown.

Using the same equations as for nuclear matter, but taking into account a finite number of nucleons, we calculated nuclei and their properties. The results are presented in [4]. It is encouraging that the general properties and magic numbers of the nuclei can be reproduced without refitting parameters.

In addition, exploiting the fact that we started out with a SU(3) description, we investigated hypernuclear properties. Fig. 3 shows the binding energies of a Λ particle in different nuclei and for various energy levels. A very good

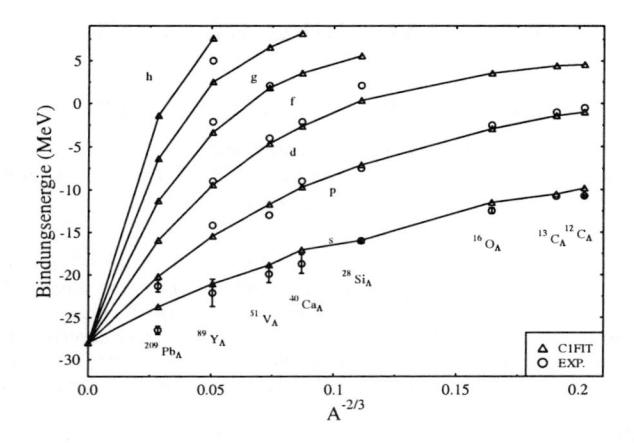

Figure 3: Binding energies of the Λ particle in various energy states of different hypernuclei. Experimental results are shown as circles, theoretical values are depicted as triangles.

agreement with experiment can be achieved.

1. Y. Nambu and G. Jona-Lasinio, Phys. Rev. **122**, 345 (1961); **124**, 246 (1961)
2. M. Gell-Mann, M. Levy, Nuovo Cim. **16**, 705 (1960)
3. D. Zschiesche, P. Papazoglou, S. Schramm, J. Schaffner-Bielich, H. Stöcker, W. Greiner, Phys. Rev. C, to be published.
4. P. Papazoglou, D. Zschiesche, S. Schramm, J. Schaffner-Bielich, H. Stöcker and W. Greiner, Phys.Rev. **C59**, 411 (1999).
5. P. Braun-Munzinger, I. Heppe, and J. Stachel, Phys.Lett. **B465**, 15 (1999).
6. G. Mao, P. Papazoglou, S. Hofmann, S. Schramm, H. Stöcker and W. Greiner, Phys.Rev. **C59**, 3381 (1999).
7. P. Papazoglou, S. Schramm, J. Schaffner, H. Stöcker and W. Greiner, Phys. Rev. **C57**, 2576 (1998).

PARTON SHOWERS AND MULTIJET EVENTS *

R. KUHN, A. SCHÄLICKE, F. KRAUSS AND G. SOFF

Institut für Theoretische Physik, TU Dresden, 01062 Dresden, Germany
Max Planck Institut für Physik komplexer Systeme, 01187 Dresden, Germany
E-mail: kuhn@theory.phy.tu-dresden.de

A Monte-Carlo event–generator has been developed which is dedicated to simulate electron–positron annihilations. Especially a new approach for the combination of matrix elements and parton showers ensures the independence of the hadronization parameters from the CMS energy. This enables for the first time the description of multijet–topologies, e.g. four jet angles, over a wide range of energy, without changing any parameter of the model. Covering all processes of the standard model our simulator is capable to describe experiments at present and future accelerators, i.e. the LEP collider and a possible Next Linear Collider(NLC).

1 Introduction

Multijet events play a crucial role in present and future high energy particle physics. Already in past experiments multijet observables have lead to proofs of the theory of strong interaction, e.g. the underlying symmetry group has been established. With rising energies the production of multijet events via the electroweak interaction becomes more important, e.g. the creation of ZH and W-pairs involves at least four jets and dominates the QCD background. In addition, the majority of signals for new physics, e.g. supersymmetry is related to multijet topologies.

Our Monte-Carlo generator APACIC++ (A PArton Cascade In C++)[1] in combination with our matrix element generator AMEGIC++ (A Matrix Element Generator in C++)[2] was developed with the aim to describe these multijet events in a correct manner over a wide range of energy. This was achieved with a new approach for combining the advantages of matrix elements and parton showers, which leads to a good describtion of experiments at present accelerators, i.e. the LEP collider. Including extensions of the standard model, primarily supersymmetry, APACIC++/AMEGIC++ will be dedicated for the search of new physics at a possible Linear collider.

The paper is outlined as follows. Tracking the physics features related to event generation in the subsequent sections we describe briefly the treatment of initial state radiation, matrix element generation and evaluation, the

---

*DEDICATED TO PROF. F. SELLSCHOP ON THE OCCASION OF HIS 70TH BIRTHDAY.

combination with the parton shower and the parton shower itself.

2 Initial state radiation

At the beginning of every event generation the initial state has to be defined. At present our package supports e^+e^- as colliding particles, but due to initial state radiation of photons the energy as well as the momentum are not fixed yet. Different approaches describing the subsequent radiation of photons are the structure function ansatz[3], the electromagnetic shower[4] and the Yennie-Frautschi-Suura (YFS) scheme[5]. Within AMEGIC++ the first and the last version are implemented. The structure function ansatz considers the effect of diminishing the electron energies by initial state photons without generating them explicitly. However, the YFS-approach allows a direct generation of photons in a theoretical well defined way up to an arbitrary order of α_{QED}. In the present state we have implemented this scheme in the soft photon limit, i.e. an exponantiation of all effects to leading logarithmic order. In Fig. 1 one can see, that AMEGIC++ agrees with KoralZ[7] to the considered order. Further more we display the effect of higher order corrections, which lead to significant changes. Hence, we will extend our treatment of initial state radiation accordingly.

3 Matrix elements

Now, the inital state is set and the event generation proceeds with the determination of the jet structure. Jets are defined within different schemes, e.g. the DURHAM[8] and the JADE[9] cluster algorithm. Utilizing these schemes the matrix elements are regularized, i.e. the soft and collinear divergencies are avoided. Within APACIC++ different matrix element generators are supplied, namely Excalibur, Debrecen[10] and AMEGIC++ . They all differ in their field of application. Our prefered choice is AMEGIC++ , which is applicable for all standard model tree level processes up to 6 massive outgoing particles. Its fully automatic procedure for the determination and integration of the amplitudes can be divided into three major steps:

1. The Feynman diagrams are achieved through the mapping of the eligible vertices onto tree topologies.

2. The diagrams are translated into helicity amplitudes and stored into word-strings for easy evaluation.

3. Integrating the amplitudes with Rambo[11], Sarge[12] or a multi-channel

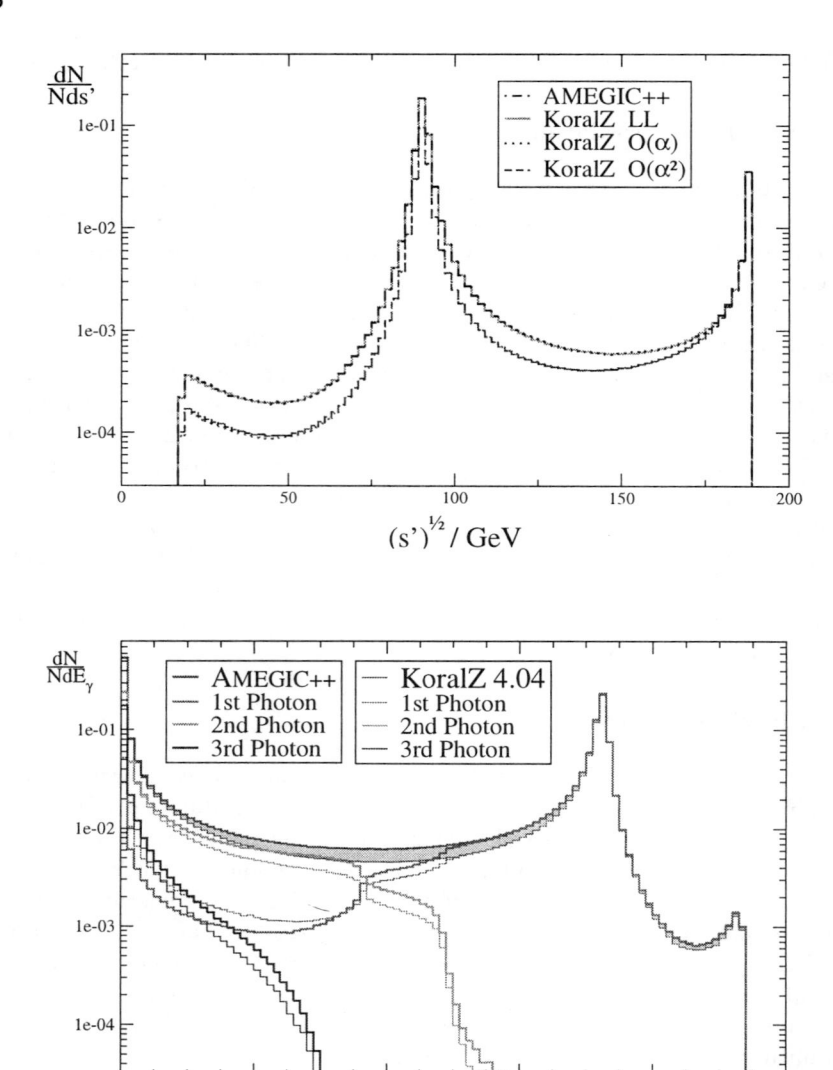

Figure 1. **AMEGIC++** and **KoralZ** are compared in the s'-distribution at 189 GeV CMS-energy on the upper panel. On the lower panel the corresponding energy distribution of the ISR-photons is displayed.

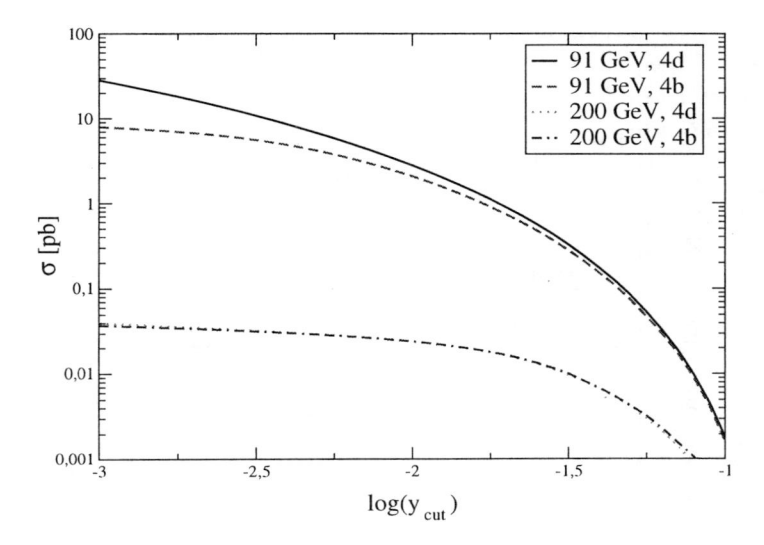

Figure 2. The 4 jet rate for four massless and massive quarks at 91 GeV and 200 GeV CMS-energy are displayed.

approach[13] (which can include the former ones) the total cross section is derived.

For further details we refer the reader to a more concise description of our program[2]. A comparison between massless and massive four jet cross sections as evaluated with AMEGIC++ is depicted in Fig. 2.

4 Combining matrix elements and parton shower

Once the jet structure is established by the matrix elements a parton shower should evolve these different jets. Since particles calculated via a hard matrix element are naturally on their mass-shell and a parton shower can handle off-shell particles only, it is obvious that a scheme for combining these two steps is indispensable. Moreover such a procedure should take advantage of the virtues of matrix elements, i.e. the description of jet correlations, and parton shower, i.e. the evolution of jets. This can be achieved following four steps:

1. The number and the flavour of the outgoing jets are determined utilizing the different matrix elements.

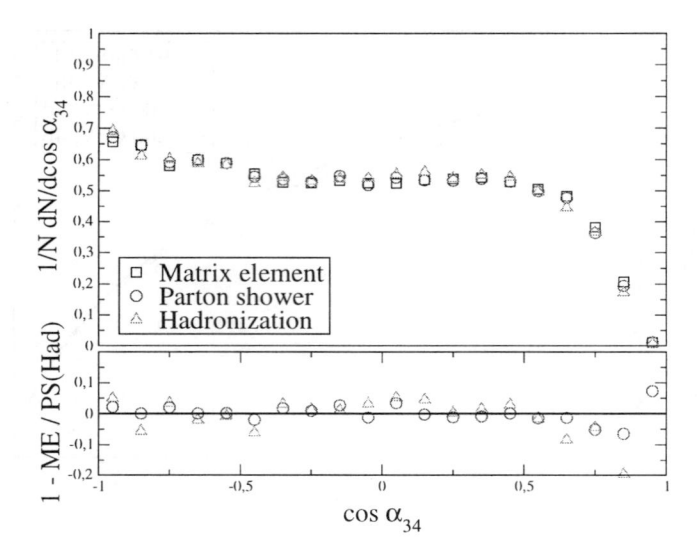

Figure 3. The α_{34} (the angle between the lowest energy jets) angle for QCD four jet events at $\sqrt{s} = 206\,\text{GeV}$.

2. The kinematical configuration is chosen according to the matrix element. An extra weight appears, when higher order corrections are taken into account, i.e. a combination of rescaled coupling constants $\alpha_S(y_{\text{cut}}s)$ and Sudakov form factors care for an exact treatment of leading logarithms[14]. This guaranties a smooth transition of the kinematics from the matrix element to the parton shower regime.

3. One of the contributing Feynman diagrams is chosen in order to gain the colour configuration of the event. The probability for the selections can be obtained for instance in a parton shower like manner[15,16].

4. A history of the parton branchings is deduced from the chosen Feynman diagram. Now the partons can be provided with virtual mass utilizing the Sudakov form factor originating from the parton shower.

The success of this combination scheme is especially reflected in four jet events, see Fig. 3. Needless to say, that this total agreement between the different phases of event generation could not be achieved using a parton shower starting from two partons only. A detailed comparison of the four jet angles between the different event generators is presented in[17].

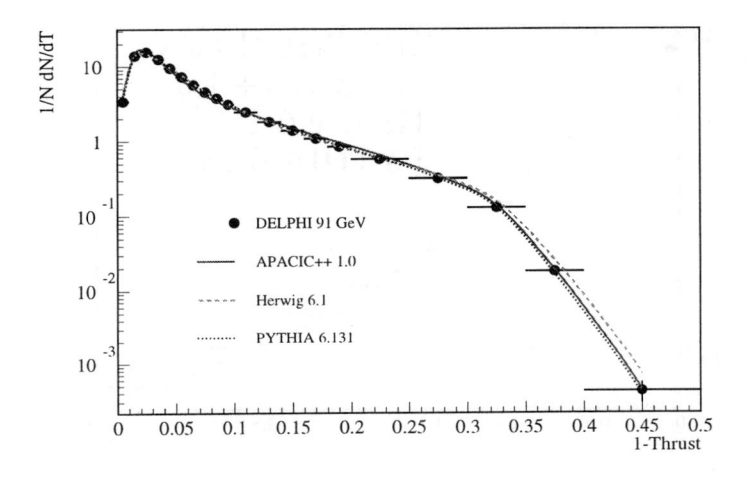

Figure 4. The eventshape variable thrust at $\sqrt{s} = 91$ GeV.

5 Parton shower and fragmentation

After all partons gained a virtual mass the evolution of jets can proceed. Different schemes according to different approximations are implemented in APACIC++ , i.e. the ordering by virtualities (LLA) and angles (MLLA). Further details can be found in many textbooks, see for instance[18]. In addition, azimuthal correlations between the different planes of parton branchings are taken into account.

Subsequently the outgoing partons have to be hadronized. This is performed with the help of the Lund-string model provided by Pythia[4].

6 Results

We performed a comparison between Ariadne[19], Herwig[20], Pythia, our event generator APACIC++/AMEGIC++ and the data of the DELPHI collaboration at different CMS-energies. In Fig. 4 the thrust distribution (an event shape variable) at 91 GeV is displayed and shows a good overall agreement with the data. The transversal momentum $p_\perp^{out}$ could not be described correctly in

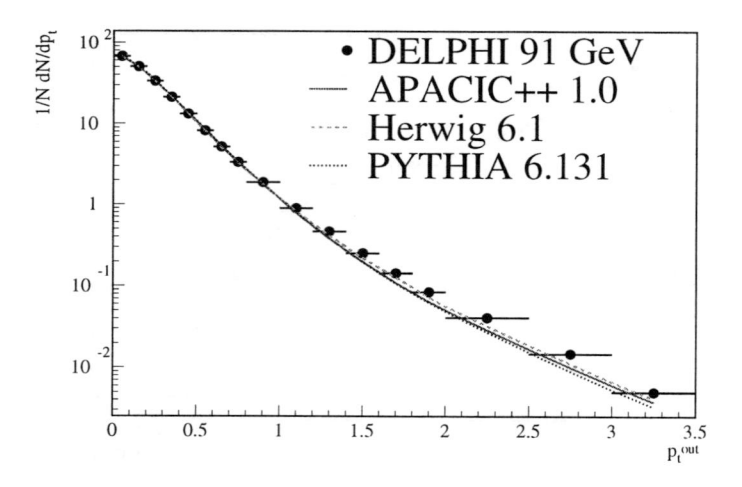

Figure 5. The transversal momentum $p_\perp^{out}$ at $\sqrt{s} = 91\,\text{GeV}$.

the high momentum region. This is seemingly a common feature of all event generators, see Fig. 5. Even though our program includes the full information of matrix elements an energy extrapolation has been achieved, see Fig. 6. This is an important feature of APACIC++/AMEGIC++ , since a pure matrix element generator does not have this property.

We conclude, that we reached the aim of providing an event generator, which is able to describe multijet topologies with a proper energy scaling.

Acknowledgments

F.K., R.K. and A.S. would like to thank J. Drees, K. Hamacher and U. Flagmeyer for helpful discussions. During the process of tuning the APACIC++– parameters to experimental data by U. Flagmeyer we were able to identify and cure some shortcomings and bugs of the program.

For R.K. and F.K. it is a pleasure to thank L. Lonnblad and T. Sjostrand for valuable comments and S. Catani and B. Webber for the pleasant collaboration on the combination of matrix elements and parton showers.

This work is supported by BMBF.

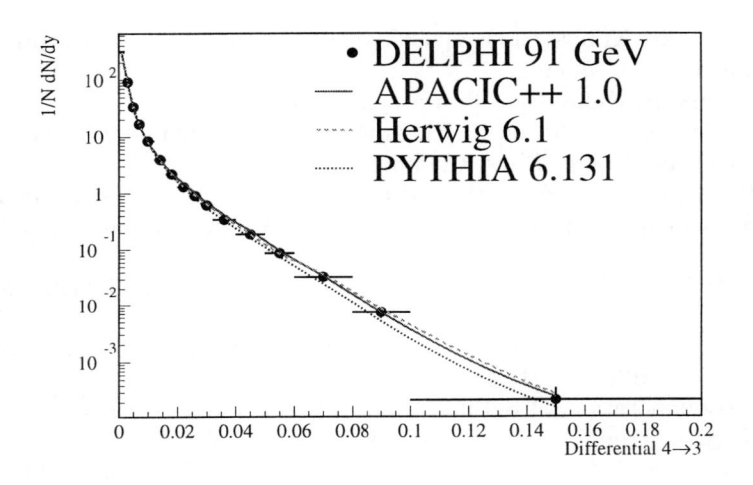

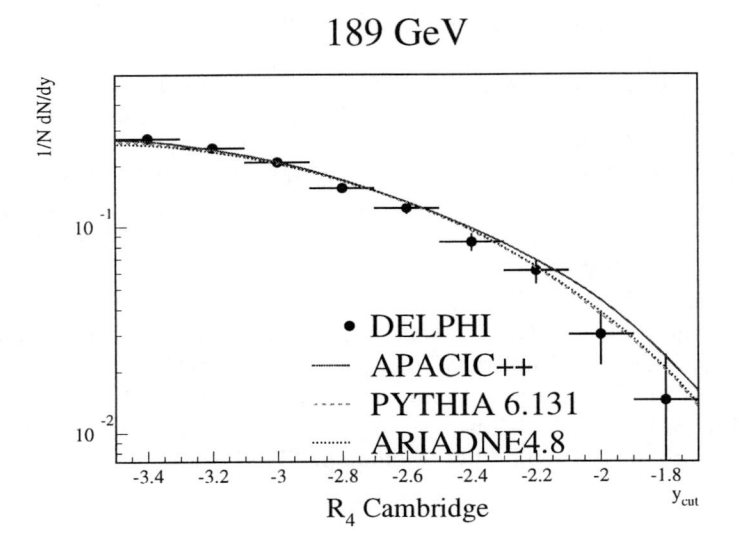

Figure 6. The energy extrapolation from the differential three jet rate from 91 GeV (upper panel) to 189 GeV (lower panel).

References

1. R. Kuhn, F. Krauss, B. Ivanyi, G. Soff; hep-ph/0004270, accepted by Comp. Phys. Commun..
2. F. Krauss, R. Kuhn, G. Soff in preparation.
3. F. A. Berends, R. Pittau, R. Kleiss, Nucl. Phys. B426 (1994) 344.
4. T. Sjostrand, Comp. Phys. Commun. 82 (1994) 74.
5. D. R. Yennie, S. C. Frautschi and H. Suura, Ann. Phys. 13 (1961) 379.
6. F. A. Berends, R. Pittau, R. Kleiss, Comp. Phys. Commun. 85 (1995) 437.
7. S. Jadach, B.F.L. Ward, Z. Was, Comp. Phys. Commun. 124 (2000) 233.
8. S. Catani, Yu. L. Dokshitzer, M. Olsson, G. Turnock, B. R. Webber, Phys. Lett. B269 (1991) 432.
9. Jade–Collaboration, S. Bethke et al., Phys. Lett. B213 (1988) 235.
10. Z. Nagy, Z. Trocsanyi, Nucl. Phys. B, Proc. Suppl. 64 (1998) 63.
11. R. Kleiss, W. J. Stirling, S. D. Ellis, Comp. Phys. Commun. 40 (1986) 359; R. Kleiss, W. J. Stirling, Nucl. Phys. B385 (1992) 413.
12. P. D. Draggiotis, A. van Hameren and R. Kleiss, Phys. Lett. B483 (2000) 124.
13. R. Kleiss, R. Pittau, Comp. Phys. Commun. 83 (1994) 141.
14. S. Catani, F. Krauss, R. Kuhn, B. R. Webber, in preparation
15. F. Krauss, R. Kuhn, G. Soff, J. Phys. G 26 (2000) L11.
16. F. Krauss, R. Kuhn, G. Soff, Acta Phys. Pol. B30 (1999) 3875.
17. A. Ballestrero et al., hep-ph/0006259, appeared in *Reports of the working groups on precision calculations for LEP2 physics*, CERN 2000-009, p. 137, ISBN 92-9083-171-5
18. R. K. Ellis, W. J. Stirling, B. R. Webber, *QCD and Collider Physics*, Cambridge Monographs on Particle Physics, Nuclear Physics and Cosmology, Cambridge University Press, 1. Edition (1996).
19. L. Lonnblad, Comp. Phys. Commun. 71 (1992) 15.
20. G. Marchesini, B. R. Webber, G. Abbiendi, I. G. Knowles, M. H. Seymour, L. Stanco, Comp. Phys. Commun. 67 (1992) 465.

SIGNATURES OF THE QUARK GLUON PLASMA: A PERSONAL OVERVIEW

C. GREINER

Institut für Theoretische Physik, Universität Giessen, D-35392 Giessen, Germany
E-mail: carsten.greiner@theo.physik.uni-giessen.de

In this talk 'Signatures of the Quark-Gluon Plasma' are being reviewed. We first discuss, on a no-QGP basis, the two prominent indications of (a) enhanced strangeness production and of (b) anomalous J/ψ-suppression: We elaborate in particular on a recent idea of antihyperon production solely by multi-mesonic reactions. As a possible source for an enhanced dissociation of $c\bar{c}$ pairs we summarize the findings within the 'early'-comover absorption scenario of prehadronic string excitations. As an exotic candidate we then finally adress the stochastic formation of so called disoriented chiral condensates: An experimentally feasible DCC, if it does exist, has to be a rare event following an unusual and nontrivial distribution on an event by event basis.

1 Motivation and Summary

The prime intention for present ultrarelativistic heavy ion collisions at CERN and at Brookhaven lies in the possible experimental identification of the quark gluon plasma (QGP), a theoretically hypothesized new phase of matter, where quarks and gluons are deliberated and move freely over an extended, macroscopically large region. Recently, referring to several different experimental findings within the Lead Beam Programme at the CERN-SPS, strong 'circumstantial evidence' for the temporal formation of the QGP has been conjectured[1]. As a first and principle objection, however, any theoretical predictions in favor for the QGP can, strictly speaking, only be regarded as qualitative or as semi-quantitative: A satisfactory theoretical understanding of either the microscopic dynamics or of the hadronisation of a hypothetical deconfined phase is at present not really given. It is also of scientific importance still to confront the excitement with further possible criticism. We therefore want to review, on a *no-QGP basis*, where quantitaive predictions are possible, the two most prominent indications for the QGP: enhanced strangeness production[2] and anomalous J/ψ-suppression[3], both being proposed already a long time ago.

The main idea behind the collective enhancement of strangeness is that the strange (and antistrange) quarks are thought to be produced more easily and hence also more abundantly in such a deconfined state as compared to the production via highly threshold suppressed inelastic hadronic collisions.

The analysis of measured abundancies of hadronic particles within thermal models[4] strongly supports the idea of having established an equilibrated fireball in some late stage of the reaction. In this respect, especially a nearly fully chemically equilibrated yield of strange antibaryons, the antihyperons, had originally been advocated as the appropriate QGP candidate[2]. Although intriguing, after all this may not be the correct interpretation of the observed antihyperon yields: In the following section 2 we will elaborate in brief on our very recent, yet conservative idea[5] of rapid antihyperon production solely by multi-mesonic reactions like $n_1\pi + n_2 K \to \bar{Y} + p$. This might indeed well explain the observed excess of antihyperons.

A suppression of the J/Ψ yield in ultra-relativistic heavy-ion collisions (in comparison to Drell-Yan pairs) is seen as the other plausible signature because the strongly bound J/Ψ should dissolve in the QGP due to color screening. Indeed, a significant reduction of the J/Ψ yield when going from proton-nucleus to nucleus-nucleus collisions has been observed, especially for Pb + Pb at 160 GeV/A. Besides the QGP as a possible explanation more conservative views of possible J/Ψ-absorption like on the still incoming nucleons and also on the produced secondaries, the 'comovers', have been envisaged with rather good success in explaining the data. In these hadronic scenarios, however, the to be assumed and unknown annihilation crosssections of the $c\bar{c}$-states on the mesons are highly debated. As a further and intuitive appealing alternative we will consider in section 3 within a microscopic simulation the effect of $c\bar{c}$ dissociation on the individual, highly excited hadronic strings in the prehadronic phase of the heavy-ion collision. Such a picture can in fact be regarded as an 'early'-comover absorption scenario[6]. A satisfactorary agreement with the various data can be achieved by choosing one phenomenological parameter within a rather plausible range.

As a last and exotic candidate for a direct signature stemming from the QGP phase transition (and which can be adressed in RHIC experiments at Brookhaven) we then summarize our ideas of stochastic formation of so called disoriented chiral condensates (DCC). The idea of DCC[7] first appeared in a work of Anselm but it was made widely known due to Bjorken, and Rajagopal and Wilczek. The spontaneous growth and subsequent decay of these configurations emerging after a rapid chiral phase transition from the QGP to the hadronic world would give rise to large collective fluctuations in the number of produced low momentum neutral pions compared to charged pions. In section 4 we briefly summarize our recent findings on the important question on the the likeliness of an instability leading potentially to a large DCC yield of low momentum pions[8]. Our investigations show that an experimentally feasible DCC, if it does exist in nature, has to be a rare event with some finite prob-

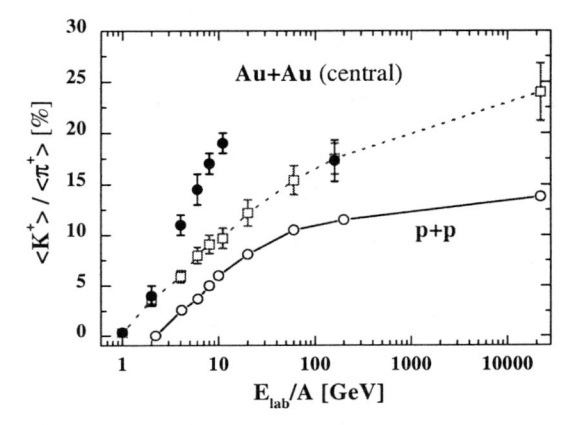

Figure 1. Calculated K^+/π^+-ratio[10] around midrapidity for central Au+Au reactions (open squares) from SIS to RHIC energies in comparison to experimental data. For visualization of the collective strangeness enhancement the corresponding ratio for elementary p+p collisions (open circles) is also depicted.

ability following a nontrivial and nonpoissonian distribution on an event by event basis. DCCs could then (only) be revealed experimentally by inspecting higher order factorial cumulants θ_m ($m \geq 3$) in the sampled distribution of low momentum pions.

2 Strangeness and Antihyperon Production

Since a relative enhancement of strangeness is observed already in hadron-hadron collisions for increasing energy (see fig. 1), which is certainly not due to any macroscopic or bulk effects, the to be measured strangeness should be compared relative to p+p collisions at the same energy. The arguments for enhanced strangeness production via the QGP should generally apply already for the most dominant strange particles, the kaons[2]. However, nonequilibrium inelastic hadronic reactions can explain to a very good extent the overall strangeness production seen experimentally[9]: Within a microscopic transport simulation an enhancement of the scaled kaon yield due to hadronic rescattering both with increasing system size and energy was found. The outcome for the most dominant strange particles, the K^+-mesons, is summarized in

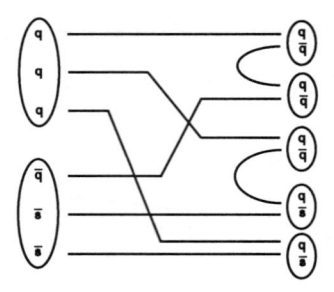

Figure 2. Schematic picture for $\bar{\bar{\Xi}} + N \leftrightarrow 3\pi + 2K$.

fig. 1. After the *primary* string fragmentation of intrinsic p-p–collisions the hadronic fireball starts with a K^+/π^+ ratio still far below chemical equilibrium with $\approx 6-8\%$ at AGS to SPS energies before the hadronic rescattering starts. The major amount of produced strange particles (kaons, antikaons and Λs) at CERN SPS-energies can then be understood in terms of early and still energetic, secondary and ternary non-equilibrium interactions. (At the lower AGS energies, the relative enhancement factor of ≈ 3 can not be fully explained within the pure cascade type calculations [9] without any possible in-medium modifications of the kaons.)

Still, applying the usual concept of binary collisions within the transport approaches, the experimentally observed enhancement of antihyperons can by far not be explained by succesive binary (strangeness exchange) reactions[2]. This fact then gives the strong support for some new exotic mechanism like, most plausible, the temporary formation of a deconfined and strangeness saturated new state of matter[1]. As outlined in the introduction, this might not be the full story[5]. Multimesonic 'back-reactions' (see eg fig. 2 for a particular illustration) involving n pions and n_Y kaons of the type

$$n\pi + n_Y K \leftrightarrow \bar{Y} + N \qquad (1)$$

corresponding to the inverse of the strong binary baryon-antibaryon annihilation process can easily account for a fast production of the antihyperon species. It is the latter annihilation process which dictates the timescale of how fast the antihyperon densities do approach local chemical equilibrium with the pions, nucleons and kaons. A simplified master equation for the number of antihyperons as a function of time can be written in the most direct form[5]

$$\frac{d}{dt}\rho_{\bar{Y}} = -\Gamma_{\bar{Y}}\left\{\rho_{\bar{Y}} - \rho_{\bar{Y}}^{eq}\right\}, \qquad (2)$$

where production due to the multi-mesonic 'back-reactions' is hidden in the second term $\Gamma_{\bar{Y}}\rho_{\bar{Y}}^{eq}$. It is further plausible to assume that the annihilation crosssections are approximately the same like for $N\bar{p}$ at the same relative momenta, i.e. $\sigma_{p\bar{Y}\to n\pi+n_YK} \approx 50$ mb. The equilibration timescale $(\Gamma_{\bar{Y}})^{(-1)} \sim 1/(\sigma_{N\bar{Y}}v_{\bar{Y}N}\rho_B)$ is thus to a good approximation proportional to the inverse of the baryon density. Adopting an initial baryon density of approximately 1–2 times normal nuclear matter density ρ_0 for the initial and thermalized hadronic fireball, the antihyperons will equilibrate on a timescale of 1–3 fm/c! This timescale competes with the expansion timescale of the late hadronic fireball, which is in the same range or larger. In any case it becomes clear that these multimesonic, hadronic reactions, contrary to binary reactions, can explain most conveniently a sufficiently fast equilibration before the (so called) chemical freeze-out occurs at the parameters given by the thermal model analyses[4].

To be more quantitative some explicit coupled master equations for an expanding system have to be considered and are presently pursued. In addition, one has to invent some clever strategy to handle such multi-particle 'back-reactions' within the sophisticated transport codes.

3 J/Ψ-suppression via Dissociation by Strings

Within a microscopic hadronic transport calculation one can exploit various assumptions (models) for the $c\bar{c}$ formation and propagation and also take into account the Drell-Yan process explicitly. As one particular scenario we now report on the effect of $c\bar{c}$ dissociation in the prehadronic phase of the heavy-ion collision[6]. This is motivated by the fact, that the very early collision phase is not described by hadrons but by highly excited strings. As each individual string carries a lot of internal energy (to produce the later secondaries) in a small and localized space-time volume the quarkonia state might get completely dissociated by the intense color electric field inside a single string [11].

In the transport treatment one explicitely follow the motion of the $c\bar{c}$ pair in the (pre-)hadronic matter throughout the collision dynamics. The $c\bar{c}$ pair may now be either destroyed in collisions with nucleons with a dissociation crosssection of $3-6$ mb (see the discussion in[6]) or by dissociation on the very energetic prehadronic excitations. Several hundred of these strings are temporarily formed during a central Pb-Pb collision at SPS energies in the early collision stage. The dynamical evolution of the strings is now included explicitly [6]. The fragmentation of the strings into hadrons starts after some phenomenologically accepted formation time $\tau_f \approx 0.8$ fm/c. For the

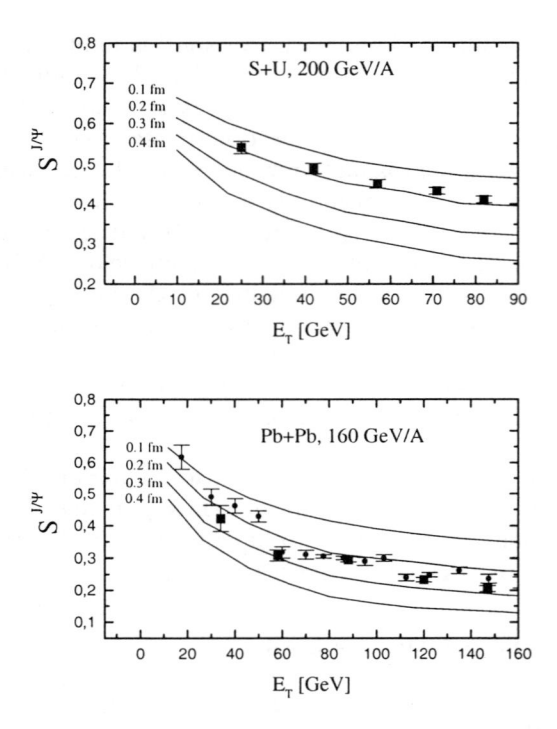

Figure 3. The J/Ψ survival probability $S^{J/\Psi}$ for S + U at 200 A·GeV (upper part) and Pb + Pb at 160 A·GeV (lower part) as a function of the transverse energy E_T in comparison to experimental data (full squares and circles). The calculated results are shown for the string radii R_s =0.1, 0.2, 0.3 and 0.4 fm.

dissociation we assume further that a $c\bar{c}$ state immediately gets broken apart whenever it moves into the region of the color electric field of a string. In this sense strings are completely 'black' for $c\bar{c}$ states[11]. The field energy density contained in a string is given by $\sigma/(\pi R_S^2)$, where $\sigma \approx 1 \text{GeV/fm}$ denotes the QCD string tension. For a string radius $R_S \approx 0.3\,fm$ one accordingly has a local high color electric energy density of $\approx 4\,GeV/fm^3$, which substantially screens the binding potential of the charmonium state [3].

The comparison to J/Ψ suppression in nucleus-nucleus collisions is now performed on an event-by-event basis using the neutral transverse energy E_T as a trigger as in the experiments. For p + U and a string radius of $R_S = 0.4$ fm only 2% of the J/Ψ's are absorbed by strings. The absorption is thus dominated, as expected intuitively, by the $c\bar{c}$-baryon dissociation on nucleons.

This turns out to be completely different for heavy-ion collisions, where the absorption on strings becomes a much more important effect. In Fig. 3 our results are shown for S + U and Pb + Pb as a function of the transverse energy and for different string radii $R_s = 0.1,...,0.4$ fm. The later serves as the one phenomenological parameter to be adressed. A moderate to strong dependence on the string radius R_s is observed with $R_s \approx 0.2$-0.3 fm giving the best fit to the experimental data. For this string radius 40% of the absorbed J/Ψ's are dissociated by strings in central collisions of Pb + Pb.

To summarize, various (hadronic) models can at present achieve a rather good agreement to all data. The underlying ideas are to some extent different. It is, of course, also possible that all of them might attribute to the suppression of $c\bar{c}$ states, so that the obliged requirement of a QGP to understand the results of most central Pb+Pb collisions is not really given.

4 Stochastic Disoriented Chiral Condensates

In this last section we will give a brief report on our recent findings on the stochastic nature of DCC formation and how to possibly identify their existence experimentally[8]. This work resulted from an earlier investigation[12], where we adressed for the first time the potential likeliness of an instability leading to a sufficiently large DCC event during the evolution of a fireball undergoing a phase transition within the linear σ-model.

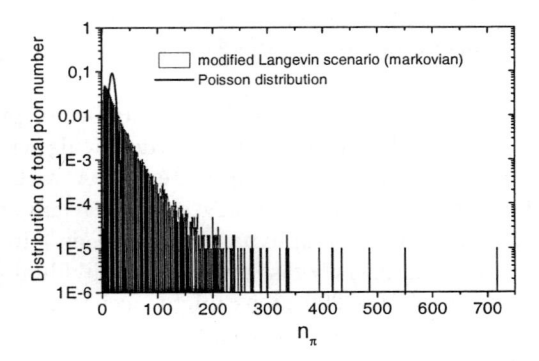

Figure 4. Statistical distribution $P(n_\pi)$ of the final yield n_π in low momentum pion number of a single DCC for a rapidly expanding situation (see ref.[8] for details) compared with a corresponding simple poissonian distribution.

The main idea is that the final fluctuations depend critically on the initial conditions chosen for the evolving chiral order parameter, thus deciding to some extent whether the system enters temporarily the instable region during the 'roll-down' period of the order parameter[12,8]. In fact, a semi-classical and dissipative dynamics of the order parameter and the pionic fields can be obtained by an effective and complex action, where the interaction with the thermal pions has been integrated out. (One of the most prominent topics in modern statistical quantum field theory is to describe the evolution and behavior of the long wavelength modes at or near thermal equilibrium and also to understand the non-equilibrium evolution of a phase transition.) To the end, we have utilized the following stochastic Langevin equations of motion for the order parameters $\Phi_a = \frac{1}{V} \int d^3x \, \phi_a(\mathbf{x}, t)$ in a D-dimensional ('Hubble') expanding volume $V(\tau)$ to describe the evolution of collective pion and sigma fields[12,8]:

$$\ddot{\Phi}_0 + \left(\frac{D}{\tau} + \eta\right) \dot{\Phi}_0 + m_T^2 \Phi_0 = f_\pi m_\pi^2 + \xi_0,$$

$$\ddot{\Phi}_i + \left(\frac{D}{\tau} + \eta\right) \dot{\Phi}_i + m_T^2 \Phi_i = \xi_i, \qquad (3)$$

with $\Phi_0 = \sigma$ and $\Phi_i = (\pi_1, \pi_2, \pi_3)$ being the chiral meson fields and $m_T^2 = \lambda \left(\Phi_0^2 + \sum_i \Phi_i^2 + \frac{1}{2}T^2 - f_\pi^2\right) + m_\pi^2$ denotes the effective transversal ('pionic') masses. These coupled Langevin equations resemble in its structure a phenomenological Ginzburg-Landau description of phase transition. Aside from a theoretical justification one can regard the Langevin equation as a practical tool to study the effect of thermalization on a subsystem, to sample a large set of possible trajectories in the evolution, and to address also the question of all thermodynamically possible initial configurations in a systematic manner.

In fig. 4 we show the statistical distribution in the number of produced long wavelength pions N_π out of the evolving chiral order fields within the DCC domain $V(\tau)$ for one particular set of parameters[8]. A rather rapid and ($D =$)3-dimensional expansion has been employed. (The results majorly depend on how fast the assumed cooling and expansion proceeds.) In general one finds that only for D=3 and sufficiently fast expansion individual unusual strong fluctuations of the order of 50 - 200 pions might occur, although the average number $\langle n_\pi \rangle$ of the emerging long wavelength pions only posesses a moderate and *undetectable* value of 5 -20.

In these interesting cases the final distribution does *not* follow a usual Poissonian distribution (comp. fig. 4), which represents a very important outcome of our investigation. (Critical, dynamical) Fluctuations with a large

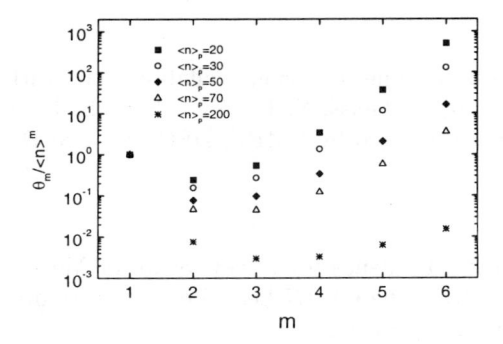

Figure 5. The reduced factorial cumulants for $m = 1$ to 6 for the pion number distribution of low momentum stemming from a single emerging DCC (of the previous figure) and an additional poissonian distributed background pion source with different mean values $\langle n \rangle_P = 20 - 200$.

number of produced pions are still likely with some small but finite probability! Unusual events out of sample contain a multiple in the number of pions compared to the average. One should indeed interpret those particular events as semi-classical 'pion bursts' similar to the mystique Centauro candidates. This result suggests a very important conclusion: If DCCs are being produced, an experimental finding will be a rare event following a strikingly, nontrivial and nonpoissonian distribution. A dedicated event-by-event analysis for the experimental programs (e.g. the STAR TPC at RHIC) is then unalterable.

The further analysis of this unusual distribution by means of the cumulant expansion shows that the reduced higher order factorial cumulants $\theta_m / <n_\pi>^m$ for $m \geq 3$ exhibit an abnormal, exponentially increasing tendency, as illustrated in fig. 5. There an additional incoherent Poissonian background of (low momentum) pions stemming from other possible sources has been added. We advocate that an analysis by means of the higher order cumulants serves as a new and powerful signature. In conclusion, the occurence of a rapid chiral phase transition (and thus DCCs) might then probably only be identified experimentally by inspecting higher order facorial cumulants θ_m ($m \geq 3$) for taken distributions of low momentum pions.

Acknowledgments

The topics reviewed have been done in various collaborations with T. Biro, E. Bratkovskaya, W. Cassing, J. Geiss, S. Leupold, S. Loh, U. Mosel and Z. Xu. This work has been supported by BMBF, DFG and GSI Darmstadt.

References

1. U. Heinz and M. Jacob, 'Evidence for a New State of Matter: An Assessment of the Result from the CERN Lead Beam Programme', CERN Press Office (2000), nucl-th/0002042.
2. P. Koch, B. Müller and J. Rafelski, *Phys. Rep.* **142**, 167 (1986).
3. T. Matsui and H. Satz, *Phys. Lett.* B **178**, 416 (1986).
4. J. Cleymans, contribution to this conference.
5. C. Greiner and S. Leupold, nucl-th/0009036; C. Greiner, nucl-th/0011026.
6. J. Geiss, C. Greiner, E. Bratkovskaya, W. Cassing and U. Mosel, *Phys. Lett.* B **447**, 31 (1999).
7. D. Anselm, *Phys. Lett.* B **217**, 169 (1989); J.D. Bjorken, *Int. J. Mod. Phys.* A **7**, 4819 (1992); K. Rajagopal and F. Wilczek, *Nucl. Phys.* B **404**, 577 (1993).
8. Z. Xu and C. Greiner, *Phys. Rev.* D **62**, 036012 (2000).
9. J. Geiss, W. Cassing and C. Greiner, *Nucl. Phys.* A **644**, 107 (1998).
10. J. Geiss, PHD thesis, Universität Giessen (1998); W. Cassing, *Nucl. Phys.* A **661**, 468c (1999).
11. S. Loh, C. Greiner and U. Mosel, *Phys. Lett.* B **404**, 238 (1997).
12. T.S. Biró and C. Greiner, *Phys. Rev. Lett.* **79**, 3138 (1997).

4. Neutrino Physics and Nuclear Astrophysics

PERSPECTIVES OF NUCLEAR PHYSICS:
From Superheavies via Hypermatter to Antimatter and the Structure of a Highly Correlated Vacuum

Walter Greiner

Abstract

The extension of the periodic system into various new areas is investigated. Experiments for the synthesis of superheavy elements and the predictions of magic numbers are reviewed. Different ways of nuclear decay are discussed like cluster radioactivity, cold fission and cold multifragmentation, including the recent discovery of the tripple fission of ^{252}Cf. Furtheron, investigations on hypernuclei and the possible production of antimatter-clusters in heavy-ion collisions are reported. Various versions of the meson field theory serve as effective field theories at the basis of modern nuclear structure and suggest structure in the vacuum which might be important for the production of hyper- and antimatter. A perspective for future research is given.

There are fundamental questions in science, like e. g. "how did life emerge" or "how does our brain work" and others. However, the most fundamental of those questions is "how did the world originate?". The material world has to exist before life and thinking can develop. Of particular importance are the substances themselves, i. e. the particles the elements are made of (baryons, mesons, quarks, gluons), i. e. elementary matter. The vacuum and its structure is closely related to that. On this I want to report today. I begin with the discussion of modern issues in nuclear physics.

The elements existing in nature are ordered according to their atomic (chemical) properties in the **periodic system** which was developed by Mendeleev and Lothar Meyer. The heaviest element of natural origin is Uranium. Its nucleus is composed of $Z = 92$ protons and a certain number of neutrons ($N = 128-150$). They are called the different Uranium isotopes. The transuranium elements reach from Neptunium ($Z = 93$) via Californium ($Z = 98$) and Fermium ($Z = 100$) up to Lawrencium ($Z = 103$). The heavier the elements are, the larger are their radii and their number of protons. Thus, the Coulomb repulsion in their interior increases, and they undergo fission. In other words: the transuranium elements become more instable as they get bigger.

In the late sixties the dream of the superheavy elements arose. Theoretical nuclear physicists around S.G. Nilsson (Lund)[1] and from the Frankfurt school[2,3,4] predicted that so-called closed proton and neutron shells should counteract the repelling Coulomb forces. Atomic nuclei with these special **"magic" proton and neutron numbers** and their neighbours could again be rather stable. These magic proton (Z) and neutron (N) numbers were thought to be $Z = 114$ and $N = 184$ or 196. Typical predictions of their life times varied between seconds and many thousand years. Fig.1 summarizes the expectations at the time. One can see the islands of superheavy elements around $Z = 114$, $N = 184$ and 196, respectively, and the one around $Z = 164$, $N = 318$.

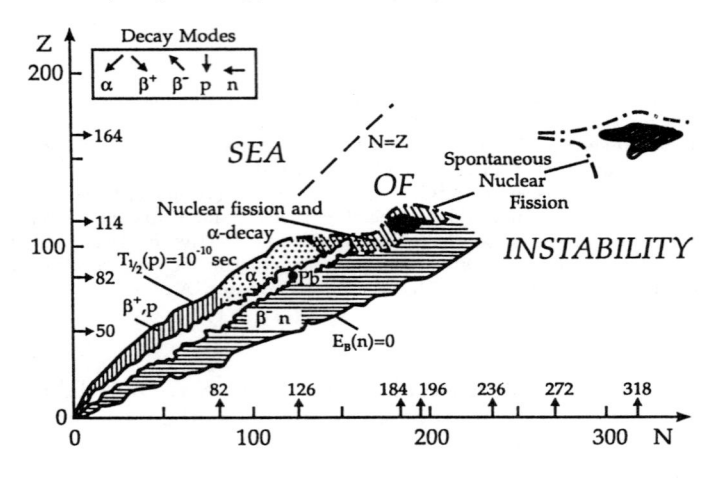

Figure 1: The periodic system of elements as conceived by the Frankfurt school in the late sixties. The islands of superheavy elements ($Z = 114$, $N = 184$, 196 and $Z = 164$, $N = 318$) are shown as dark hatched areas.

The important question was how to produce these superheavy nuclei. There were many attempts, but only little progress was made. It was not until the middle of the seventies that the Frankfurt school of theoretical physics together with foreign guests (R.K. Gupta (India), A. Sandulescu (Romania))[5] theoretically understood and substantiated the concept of bombarding of double magic lead nuclei with suitable projectiles, which had been proposed intuitively by the russian nuclear physicist Y. Oganessian[6]. The two-center shell model, which is essential for the description of fission, fusion and nuclear molecules, was developped in 1969-1972 together with my then students U. Mosel and J. Maruhn[7]. It showed that the shell structure of the two final fragments was visible far beyond the barrier into the fusioning nucleus. The collective potential energy surfaces of heavy nuclei, as they were calculated in the framework of the two-center shell model, exhibit pronounced valleys, such that these valleys provide promising doorways to the fusion of superheavy nuclei for certain projectile-target combinations (Fig. 4). If projectile and target approach each other through those "cold" valleys, they get only minimally excited and the barrier

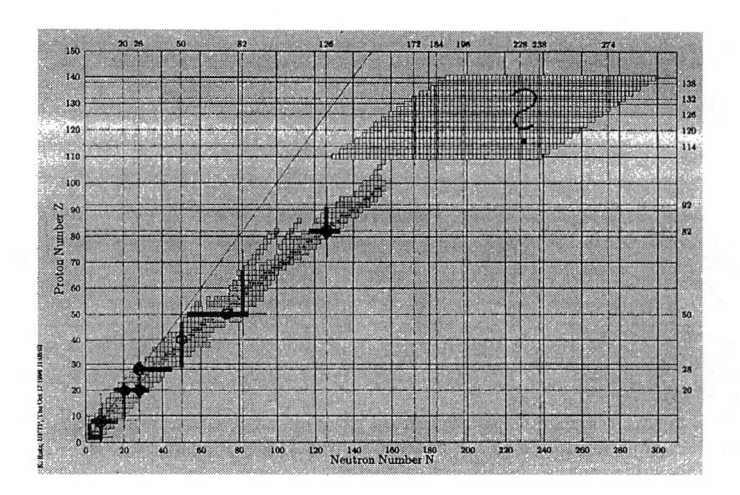

Figure 2: The shell structure in the superheavy region around $Z = 114$ is an open question. As will be discussed later, meson field theories suggest that $Z = 120, N = 172, 184$ are the magic numbers in this region.

which has to be overcome (fusion barrier) is lowest (as compared to neighbouring projectile-target combinations). In this way the correct projectile- and target-combinations for fusion were predicted. Indeed, Gottfried Münzenberg and Sigurd Hofmann and their group at GSI [8] have followed this approach. With the help of the SHIP mass-separator and the position sensitive detectors, which were especially developped by them, they produced the pre-superheavy elements $Z = 106, 107, \ldots 112$, each of them with the theoretically predicted projectile-target combinations, and only with these. Everything else failed. This is an impressing success, which crowned the laborious construction work of many years. The before last example of this success, the discovery of element 112 and its long α-decay chain, is shown in Fig. 6. Very recently the Dubna–Livermore–group produced two isotopes of $Z = 114$ element by bombarding ^{244}Pu with ^{48}Ca (Fig. 3). Also this is a cold–valley reaction (in this case due to the combination of a spherical and a deformed nucleus), as predicted by Gupta, Sandulescu and Greiner [9] in 1977. There exist also cold valleys for which both fragments are deformed [10], but these have yet not been verified experimentally. The very recently reported $Z = 118$ isotope fused with the cold valley reaction [12] ^{58}Kr + ^{208}Pb by Ninov et al. [13] yields the latest support of the cold valley idea.

Studies of the shell structure of superheavy elements in the framework of the

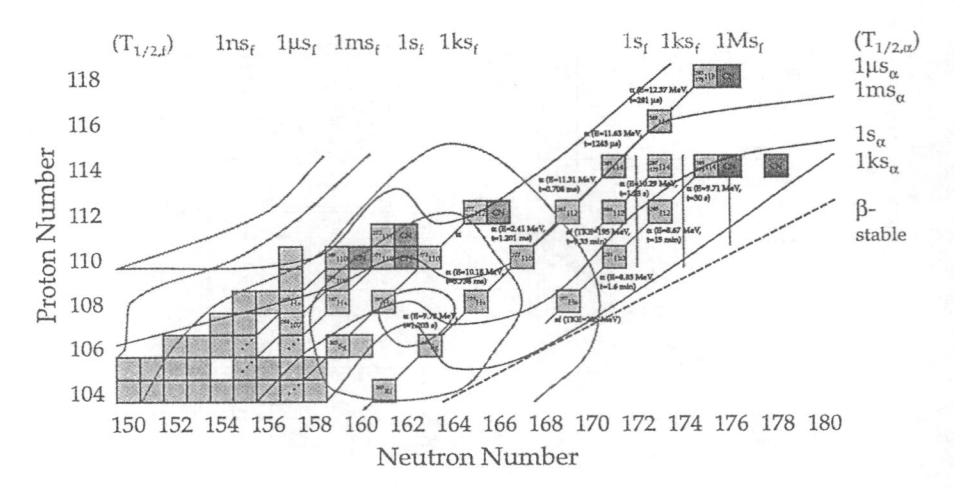

Figure 3: The $Z = 106 - 112$ isotopes were fused by the Hofmann–Münzenberg (GSI)–group. The two $Z = 114$ isotopes were produced by the Dubna–Livermore group. It is claimed that three neutrons are evaporated. Obviously the lifetimes of the various decay products are rather long (because they are closer to the stable valley), in crude agreement with early predictions [3,4] and in excellent agreement with the recent calculations of the Sobicevsky-group [11]. The recently fused $Z = 118$ isotope by V. Ninov et al. at Berkeley is the heaviest one so far.

meson field theory and the Skyrme–Hartree–Fock approach have recently shown that the magic shells in the superheavy region are very isotope dependent [14] (see Fig. 7). **According to these investigations $Z = 120$ being a magic proton number seems to be as probable as $Z = 114$.** Additionally, recent investigations in a chirally symmetric mean–field theory (see also below) result also in the prediction of these two magic numbers[38,40]. The corresponding magic neutron numbers are predicted to be $N = 172$ and - as it seems to a lesser extend - $N = 184$. Thus, this region provides an open field of research. R.A. Gherghescu et al. have calculated the potential energy surface of the $Z = 120$ nucleus. It utilizes interesting isomeric and valley structures (Fig. 8). The charge distribution of the $Z = 120, N = 184$ nucleus indicates a hollow inside. This leads us to suggest that it might be essentially a fullerene consisting of 60 α-particles and one additional binding neutron per alpha. This is illustrated in Fig 5. The protons and neutrons of such a superheavy nucleus are distributed over 60 α particles and 60 neutrons (forgetting the last 4 neutrons).

The determination of the chemistry of superheavy elements, i. e. the calculation of the atomic structure — which is in the case of element 112 the shell

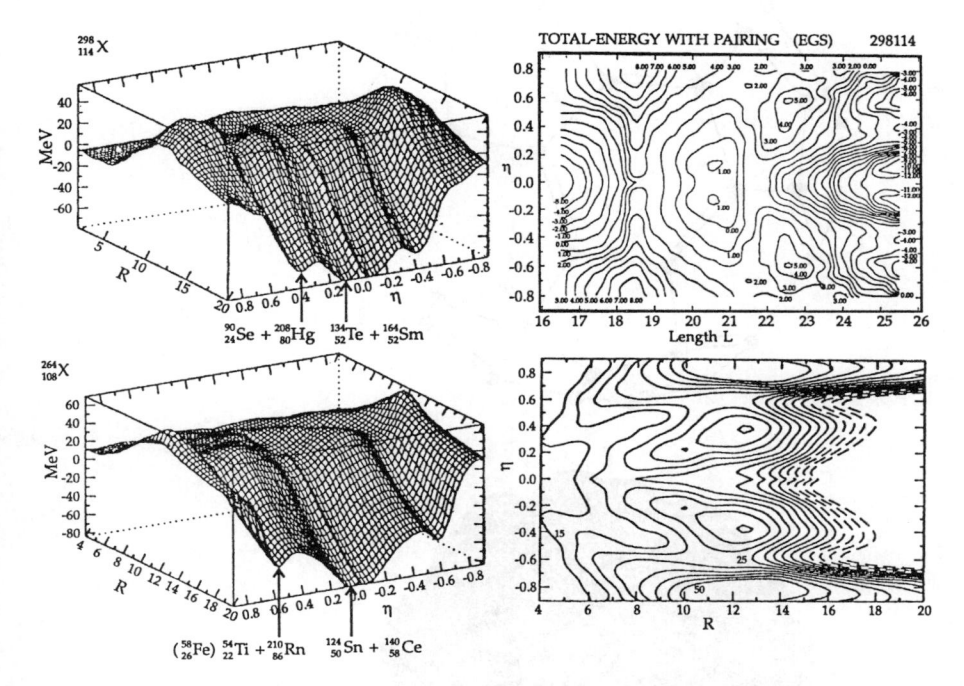

Figure 4: The collective potential energy surface of $^{264}108$ and $^{184}114$, calculated within the two center shell model by J. Maruhn et al., shows clearly the cold valleys which reach up to the barrier and beyond. Here R is the distance between the fragments and $\eta = \dfrac{A_1 - A_2}{A_1 + A_2}$ denotes the mass asymmetry: $\eta = 0$ corresponds to a symmetric, $\eta = \pm 1$ to an extremely asymmetric division of the nucleus into projectile and target. If projectile and target approach through a cold valley, they do not "constantly slide off" as it would be the case if they approach along the slopes at the sides of the valley. Constant sliding causes heating, so that the compound nucleus heats up and gets unstable. In the cold valley, on the other hand, the created heat is minimized. The colleagues from Freiburg should be familiar with that: they approach Titisee (in the Black Forest) most elegantly through the Höllental and not by climbing its slopes along the sides.

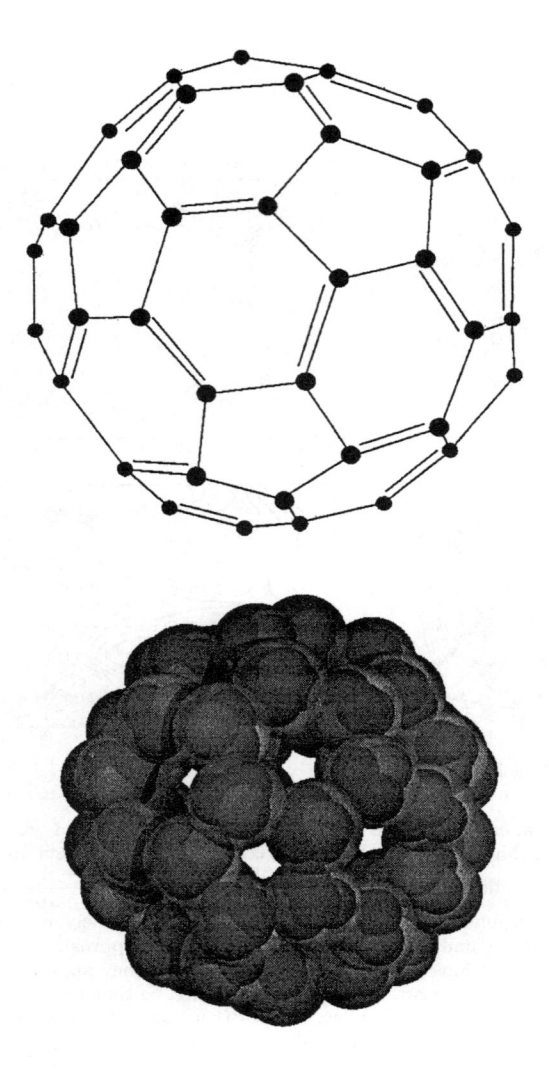

Figure 5: Typical structure of the fullerene ^{60}C. The double bindings are illutsrated by double lines. In the nuclear case the Carbon atoms are replaced by α particles and the double bindings by the additional neutrons. Such a structure would immediately explain the semi–hollowness of that superheavy nucleus, which is revealed in the mean–field calculations within meson–field theories. (Lower picture by H. Weber.)

$$^{70}\text{Zn} + {}^{208}\text{Pb} \rightarrow {}^{277}112 + 1n$$

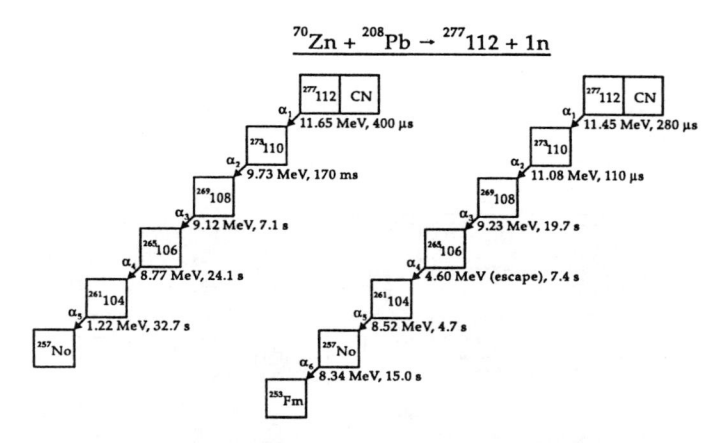

Figure 6: The fusion of element 112 with ^{70}Zn as projectile and ^{208}Pb as target nucleus has been accomplished for the first time in 1995/96 by S. Hofmann, G. Münzenberg and their collaborators. The colliding nuclei determine an entrance to a "cold valley" as predicted as early as 1976 by Gupta, Sandulescu and Greiner. The fused nucleus 112 decays successively via α emission until finally the quasi-stable nucleus ^{253}Fm is reached. The α particles as well as the final nucleus have been observed. Combined, this renders the definite proof of the existence of a $Z = 112$ nucleus.

structure of 112 electrons due to the Coulomb interaction of the electrons and in particular the calculation of the orbitals of the outer (valence) electrons — has been carried out as early as 1970 by B. Fricke and W. Greiner[15]. Hartree-Fock-Dirac calculations yield rather precise results.

The potential energy surfaces, which are shown prototypically for $Z = 114$ in Fig 4, contain even more remarkable information that I want to mention cursorily: if a given nucleus, e. g. Uranium, undergoes fission, it moves in its potential mountains from the interior to the outside. Of course, this happens quantum mechanically. The wave function of such a nucleus, which decays by tunneling through the barrier, has maxima where the potential is minimal and minima where it has maxima. This is depicted in Fig. 9.

The probability for finding a certain mass asymmetry $\eta = \dfrac{A_1 - A_2}{A_1 + A_2}$ of the fission is proportional to $\psi^*(\eta)\psi(\eta)d\eta$. Generally, this is complemented by a coordinate dependent scale factor for the volume element in this (curved) space, which I omit for the sake of clarity. Now it becomes clear how the so-called **asymmetric** and **superasymmetric** fission processes come into being. They result from the enhancement of the collective wave function in the cold valleys. And that is indeed, what one observes. Fig. 10 gives an impression of it.

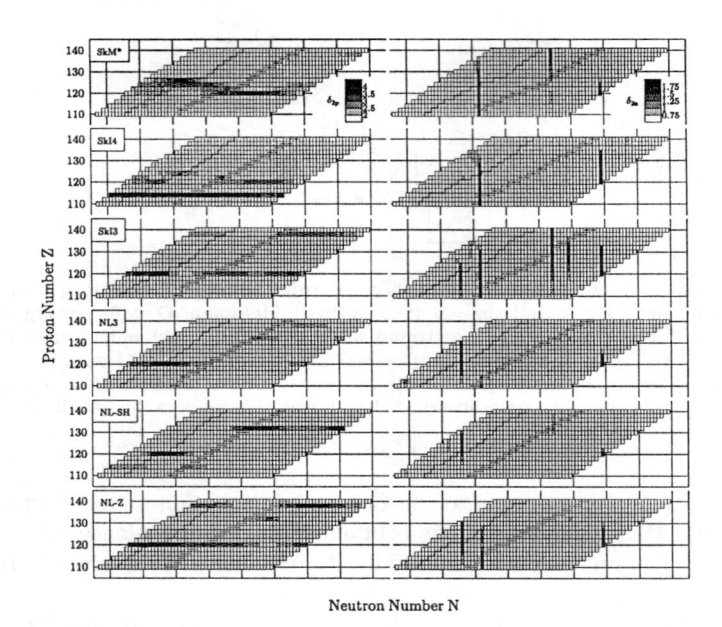

Figure 7: Grey scale plots of proton gaps (left column) and neutron gaps (right column) in the N-Z plane for spherical calculations with the forces as indicated. The assignment of scales differs for protons and neutrons, see the uppermost boxes where the scales are indicated in units of MeV. Nuclei that are stable with respect to β decay and the two-proton dripline are emphasized. The forces with parameter sets SkI4 and NL-Z reproduce the binding energy of $^{264}_{156}108$ (Hassium) best, i.e. $|\delta E/E| < 0.0024$. Thus one might assume that these parameter sets could give the best predictions for the superheavies. Nevertheless, it is noticed that NL-Z predicts only $Z = 120$ as a magic number while SkI4 predicts both $Z = 114$ and $Z = 120$ as magic numbers. The magicity depends — sometimes quite strongly — on the neutron number. These studies are due to Bender, Rutz, Bürvenich, Maruhn, P.G. Reinhard et al. [14].

For a large mass asymmetry ($\eta \approx 0.8, 0.9$) there exist very narrow valleys. They are not as clearly visible in Fig. 4, but they have interesting consequences. Through these narrow valleys nuclei can emit spontaneously not only α-particles (Helium nuclei) but also ^{14}C, ^{20}O, ^{24}Ne, ^{28}Mg, and other nuclei. Thus, we are lead to the **cluster radioactivity** (Poenaru, Sandulescu, Greiner [16]).

By now this process has been verified experimentally by research groups in Oxford, Moscow, Berkeley, Milan and other places. Accordingly, one has to revise what is learned in school: there are not only 3 types of radioactivity (α-, β-, γ-radioactivity), but many more. Atomic nuclei can also decay through spontaneous cluster emission (that is the "spitting out" of smaller nuclei like carbon, oxygen,...). Fig. 11 depicts some nice examples of these processes.

The knowledge of the collective potential energy surface and the collective masses $B_{ij}(R, \eta)$, all calculated within the Two-Center-Shell-Modell (TCSM), allowed H. Klein, D. Schnabel and J. A. Maruhn to calculate lifetimes against fission in an "ab initio" way [17].

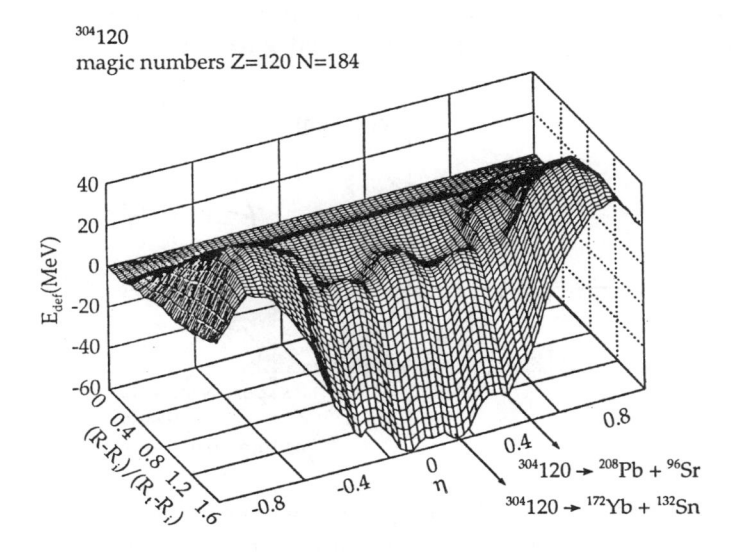

Figure 8: Potential energy surface as a function of reduced elongation $(R - R_i)/(R_t - R_i)$ and mass asymmetry η for the double magic nucleus ^{304}120. ^{304}120$_{184}$.

Utilizing a WKB-minimization for the penetrability integral

$$\mathcal{P} = e^{-I}, \quad I = \min_{\forall\,\text{paths}} \tfrac{2}{\hbar} \int_S \sqrt{2m(V(R,\eta) - E)}\, ds$$

$$= \min_{\forall\,\text{paths}} \tfrac{2}{\hbar} \int_0^1 \sqrt{\underbrace{2m g_{ij}}_{B_{ij}}(V(x_i(t)) - E)\tfrac{dx_i}{dt}\tfrac{dx_j}{dt}}\, dt \qquad (1)$$

where $ds^2 = g_{ij}\,dx_i\,dx_j$ and g_{ij} – the metric tensor – is in the well-known fashion related to the collective masses $B_{ij} = 2m g_{ij}$, one explores the minimal paths from the nuclear ground state configuration through the multidimensional fission barrier (see Fig. 12).

The thus obtained fission half lives are depicted in the lower part of figure 12. Their distribution as a function of the fragment mass A_2 resembles quite well the asymmetric mass distribution. Cluster radioactive decays correspond to the broad peaks around $A_2 = 20, 30\ (200, 210)$. The confrontation of the calculated fission half lives with experiments is depicted in Fig. 13. One notices "nearly quantitative" agreement over 20 orders of magnitude, which is – for an ab-initio calculation – remarkable!

Finally, in Fig. 14, we compare the lifetime calculation discussed above with one based on the Preformation Cluster Model by D. Poenaru et al. [18] and recognize an amazing degree of similarity and agreement.

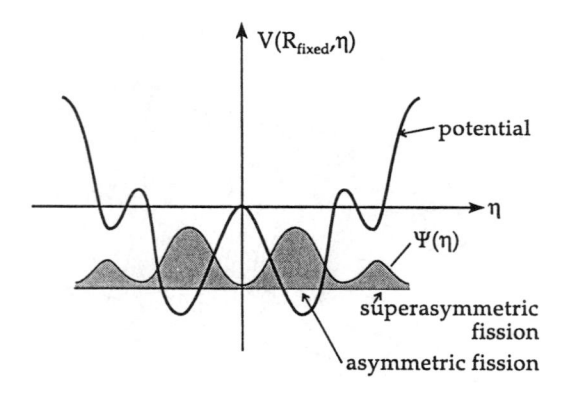

Figure 9: The collective potential as a function of the mass asymmetry $\eta = \dfrac{A_1 - A_2}{A_1 + A_2}$. A_i denotes the nucleon number in fragment i. This qualitative potential $V(R_{\text{fixed}}, \eta)$ corresponds to a cut through the potential landscape at $R = R_{\text{fixed}}$ close to the scission configuration. The wave function is drawn schematically. It has maxima where the potential is minimal and vice versa.

The systematics for the average total kinetic energy release for spontaneously fissioning isotopes of Cm and No is following the Viola trend, but ^{258}Fm and ^{259}Fm are clearly outside. The situation is similar also for ^{260}Md, where two components of fission products (one with lower and one with higher kinetic energy) were observed by Hulet et al.[19]. The explanation of these interesting observations lies in two different paths through the collective potential. One reaches the scission point in a stretched neck position (i.e. at a lower point of the Coulomb barrier - thus lower kinetic energy for the fragments) while the other one reaches the scission point practically in a touching-spheres-position (i.e. higher up on the Coulomb barrier and therefore highly energetic fragments)[20]. The latter process is cold fission; i.e., the fission fragments are in or close to their ground state (cold fragments) and all the available energy is released as kinetic energy. Cold fission is, in fact, typically a cluster decay. The side-by-side occurence of cold and normal (hot) fission has been named

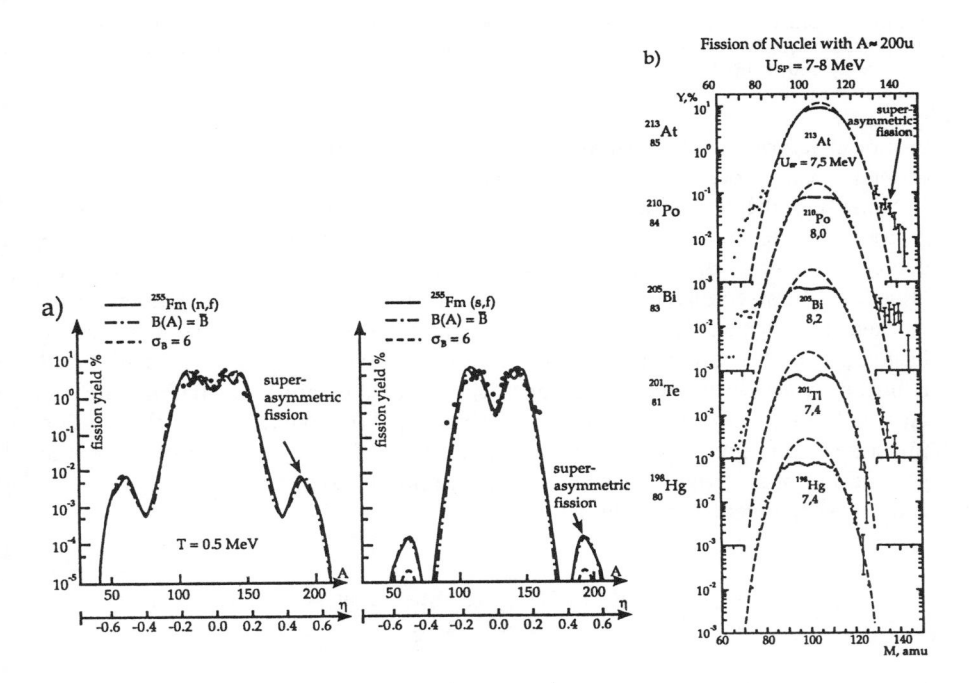

Figure 10: Asymmetric (a) and symmetric (b) fission. For the latter, also superasymmetric fission is recognizable, as it has been observed only a few years ago by the russian physicist Itkis — just as expected theoretically.

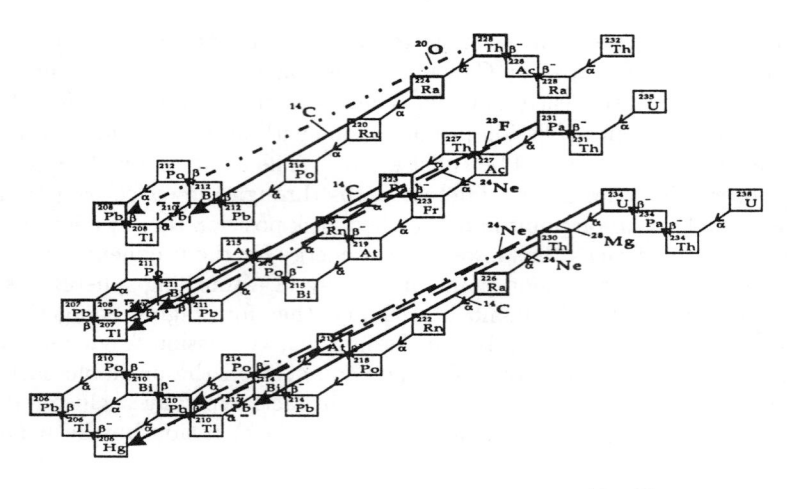

Figure 11: Cluster radioactivity of actinide nuclei. By emission of ^{14}C, ^{20}O,... "big leaps" in the periodic system can occur, just contrary to the known α, β, γ radioactivities, which are also partly shown in the figure.

bi-modal fission [20]. There has now been put forward a phantastic idea [21] in order to study cold fission (Cluster decays) and other exotic fission processes (ternary-, multiple fission in general) very elegantly: By measuring with e. g. the Gamma–sphere characteristic γ-transitions of individual fragments in coincidence, one can identify all these processes in a direct and simple way (Fig. 15). First confirmation of this method by J. Hamilton, V. Ramaya et al. worked out excellently [22]. This method has high potential for revolutionizing fission physics! With some physical intuition one can imagine that **triple -** and **quadriple** fission processes and even the process of **cold multifragmentation** will be discovered - absolutely fascinating! We have thus seen that fission physics (cold fission, cluster radioactivity, ...) and fusion physics (especially the production of superheavy elements) are intimately connected.

Indeed, very recently, tripple fission of

$$^{252}\text{Cf} \rightarrow {}^{146}\text{Ba} + {}^{96}\text{Sr} + {}^{10}\text{Be}$$
$$\rightarrow {}^{112}\text{Ry} + {}^{130}\text{Sn} + {}^{10}\text{Be}$$
$$\rightarrow ...$$

has been identified by measuring the various γ-transitions of these nuclei in coincidence (see Fig. 16). Even though the statistical evidence for the ^{10}Be line is small (≈ 50 events) the various coincidences seem to proof that spontaneous

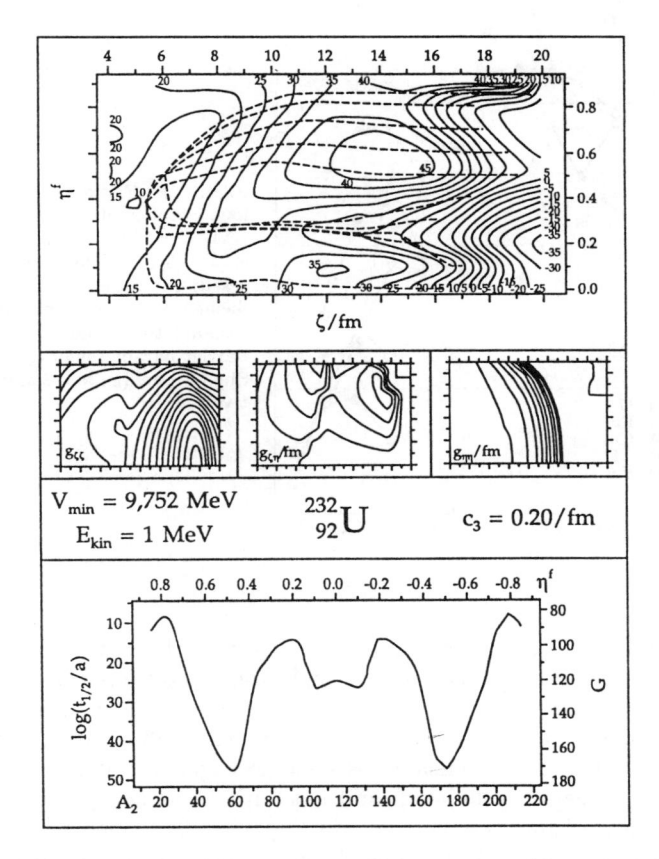

Figure 12: The upper part of the figure shows the collective potential energy surface for $^{232}_{92}$U with the groundstate position and various fission paths through the barrier. The middle part shows various collective masses, all calculated in the TCSM. In the lower part the calculated fission half lives are depicted.

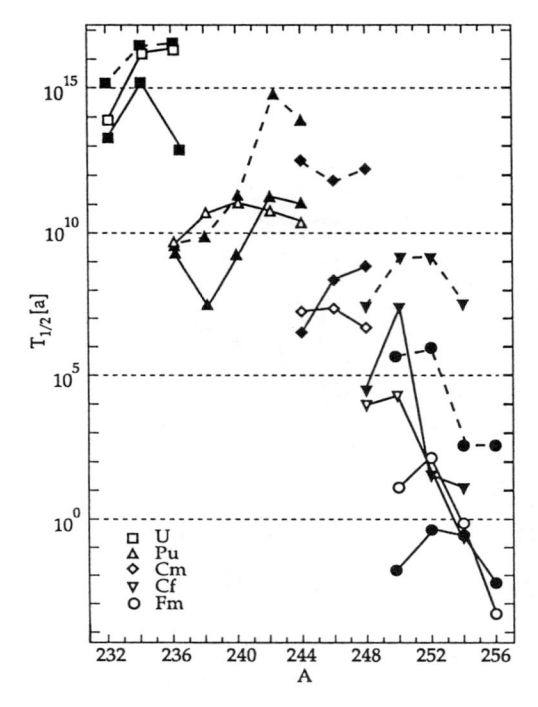

Figure 13: Fission half lives for various isotopes of $Z = 92$ (□), $Z = 94$ (△), $Z = 96$ (◇), $Z = 98$ (▽) and $Z = 100$ (○). The black curves represent the experimental values. The dashed and dotted calculations correspond to a different choice of the barrier parameter in the Two Center Shell Model ($c_3 \approx 0.2$ and 0.1 respectively).

tripple fission out of the ground state of ^{252}Cf with the heavy cluster ^{10}Be as a third fragment exists. Also other tripple fragmentations can be expected. One of those is also denoted above. In fact, there are first indications, that this break-up is also observed. The most amazing observation is, however, the following: The cross coincidences seem to suggest that one deals with a simultaneous three-body breakup and not with a cascade process. For that one expects a configuration as shown in Fig. 17. Consequently the ^{10}Be will obtain kinetic energy while running down the combined Coulomb barrier of ^{146}Ba and ^{96}Sr and, therefore, the 3368 keV line of ^{10}Be should be Doppler-broadened. Amazingly, however, it is not and, moreover, it seems to be about 6 keV smaller than the free ^{10}Be γ-transition. If this turns out to be true, the only explanation will be that the Gamma is emitted while the nuclear molecule of the type shown in Fig. 17 holds. The molecule has to live longer than about 10^{-12} sec. The nuclear forces from the ^{146}Ba and ^{96}Sn cluster to the left and right from ^{10}Be lead to a softening of its potential and therefore to a somewhat smaller transition energy. Thus, if experimental results hold, one has discovered

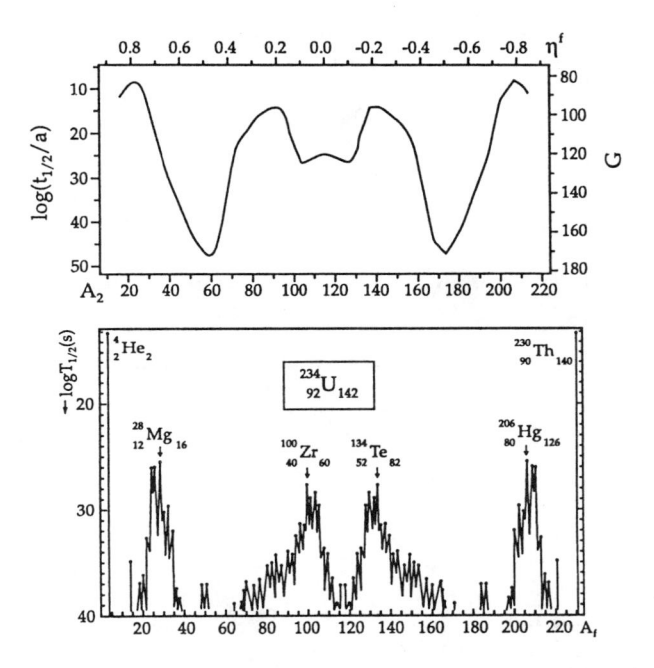

Figure 14: Comparison of the fission half lives calculated in the fission model (upper figure – see also Fig. 12) and in the Preformation Cluster Model [18]. In both models the deformation of the fission fragments is not included completely.

long living ($\approx 10^{-12}$ sec) complex nuclear molecules. This is phantastic! Of course, I do immediately wonder whether such configurations do also exist in e.g. U + Cm soft encounters directly at the Coulomb barrier. This would have tremendous importance for the observation of the spontaneous vacuum decay [26], for which "sticking giant molecules" with a lifetime of the order of 10^{-19} sec are needed. The nuclear physics of such heavy ion collisions at the Coulomb barrier (giant nuclear molecules) should indeed be investigated!

As mentioned before there are other tri-molecular structures possible; some with ^{10}Be in the middle and both spherical or deformed clusters on both sides of ^{10}Be. The energy shift of the ^{10}Be-line should be smaller, if the outside clusters are deformed (smaller attraction ⇔ smaller softening of the potential) and bigger, if they are spherical. Also other than ^{10}Be-clusters are expected to be in the middle. One is lead to the molecular doorway picture. Fig. 18 gives a schematic impression where within the potential landscape cluster-molecules

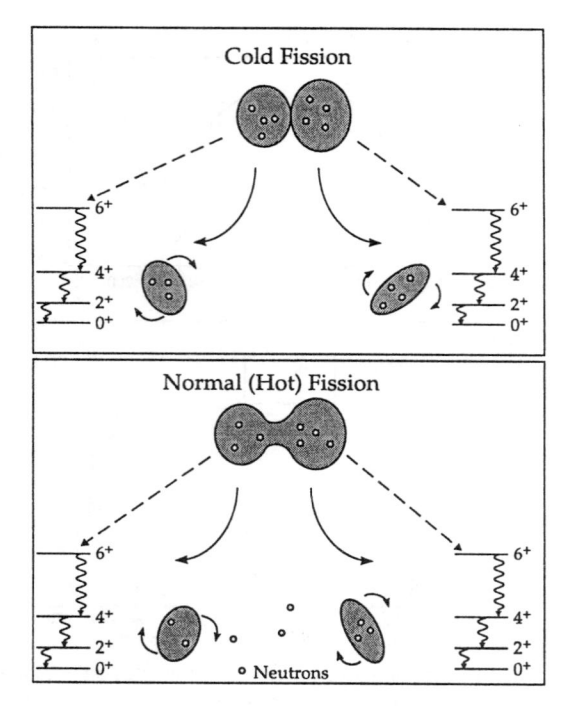

Figure 15: Illustration of cold and hot (normal) fission identification through multiple γ-coincidencs of photons from the fragments. The photons serve to identify the fragments.

are expected to appear, i.e. close to the scission configuration. Clearly, there will not be a single tri-molecular configuration, but a variety of three-body fragmentations leading to a spreading width of the tri-molecular state. This is schematically shown in Fig. 19.

Finally, these tri-body nuclear molecules are expected to perform themselves rotational and vibrational (butterfly, whiggler, β-, γ-type) modes. The energies were estimated by P. Hess et al[24]; for example rotational energies typically of the order of a few keV (4 keV, 9 keV, ...). A new molecular spectroscopy seems possible!

The "cold valleys" in the collective potential energy surface are basic for understanding this exciting area of nuclear physics! It is a master example for understanding the **structure of elementary matter**, which is so important for other fields, especially astrophysics, but even more so for enriching our

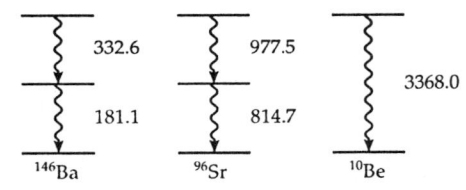

Figure 16: The γ-transitions of the three fission products of ^{252}Cf measured in coincidence. Various combinations of the coincidences were studied. The free 3368 keV line in ^{10}Be has recently been remeasured by Burggraf et al.[23], confirming the value of the transition energy within 100 eV.

"Weltbild", i.e. the status of our understanding of the world around us.

Nuclei that are found in nature consist of nucleons (protons and neutrons) which themselves are made of u (up) and d (down) quarks. However, there also exist s (strange) quarks and even heavier flavors, called charm, bottom, top. The latter has just recently been discovered. Let us stick to the s quarks. They are found in the 'strange' relatives of the nucleons, the so-called hyperons (Λ, Σ, Ξ, Ω). The Λ-particle, e. g., consists of one u, d and s quark, the Ξ-particle even of an u and two s quarks, while the Ω (sss) contains strange quarks only. Fig. 20 gives an overview of the baryons, which are of interest here, and their quark content.

If such a hyperon is taken up by a nucleus, a **hyper-nucleus** is created. Hyper-nuclei with one hyperon have been known for 20 years now, and were extensively studied by B. Povh (Heidelberg)[27]. Several years ago, Carsten Greiner, Jürgen Schaffner and Horst Stöcker[28] theoretically investigated nuclei with many hyperons, **hypermatter**, and found that the binding energy per baryon of strange matter is in many cases even higher than that of ordinary matter (composed only of u and d quarks). This leads to the idea of extending the periodic system of elements in the direction of strangeness.

One can also ask for the possibility of building atomic nuclei out of **anti-**

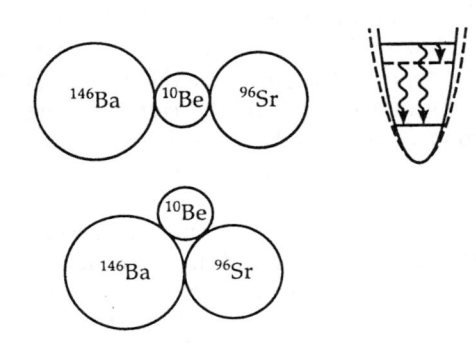

Figure 17: Typical linear cluster configuration leading to tripple fission of ^{252}Cf. The influence of both clusters leads to a softening of the ^{10}Be potential and thus to a somewhat smaller transition energy. Some theoretical investigations indicate that the axial symmetry of this configuration might be broken (lower lefthand figure).

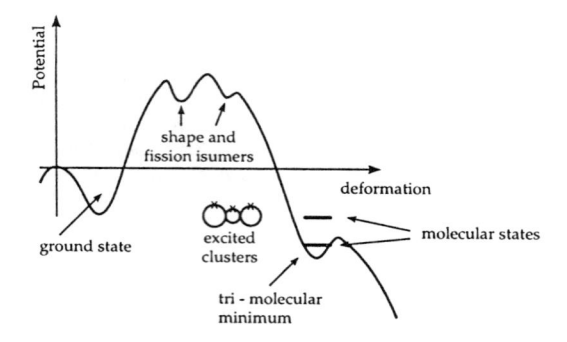

Figure 18: Cluster molecules: Potential energy curve of a heavy nucleus showing schematically the location of groundstate, shape- and fission-isomeric states and of tri-molecular states.

matter, that means searching e. g. for anti-helium, anti-carbon, anti-oxygen. Fig. 21 depicts this idea. Due to the charge conjugation symmetry antinuclei should have the same magic numbers and the same spectra as ordinary nuclei. However, as soon as they get in touch with ordinary matter, they annihilate with it and the system explodes.

Now the important question arises how these strange matter and antimatter clusters can be produced. First, one thinks of collisions of heavy nuclei, e. g. lead on lead, at high energies (energy per nucleon $\geq$ 200 GeV). Calculations with the URQMD-model of the Frankfurt school show that through **nuclear shock waves**[29,30,31] nuclear matter gets compressed to 5–10 times of its usual value, $\rho_0 \approx 0.17$ fm^3, and heated up to temperatures of $kT \approx 200$ MeV. As a consequence about 10000 pions, 100 Λ's, 40 Σ's and Ξ's and about as many antiprotons and many other particles are created in a single collision. It seems conceivable that it is possible in such a scenario for some Λ's to get captured by a nuclear cluster. This happens indeed rather frequently for one or two Λ-particles; however, more of them get built into nuclei with rapidly decreasing probability only. This is due to the low probability for finding the right conditions for such a capture in the phase space of the particles: the numerous particles travel with every possible momenta (velocities) in all directions. The chances for hyperons and antibaryons to meet gets rapidly worse with increasing number. In order to produce multi-Λ-nuclei and antimatter nuclei, one has to look for a different source.

In the framework of meson field theory the energy spectrum of baryons has a peculiar structure, depicted in Fig. 22. It consists of an upper and a lower continuum, as it is known from the electrons (see e. g. [26]). Of special interest in the case of the baryon spectrum is the potential well, built of the

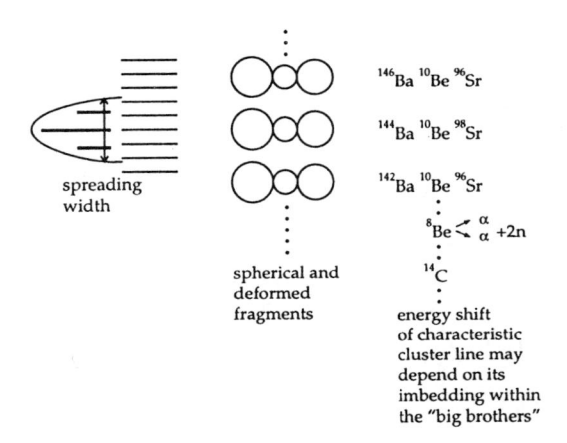

Figure 19: Microstructure of tri-molecular states: Various tri-cluster configurations are spread out and mix with background states. Thus the tri-molecular state obtains a spreading width.

scalar and the vector potential, which rises from the lower continuum. It is known since P.A.M. Dirac (1930) that the negative energy states of the lower continuum have to be occupied by particles (electrons or, in our case, baryons). Otherwise our world would be unstable, because the "ordinary" particles are found in the upper states which can decay through the emission of photons into lower lying states. However, if the "underworld" is occupied, the Pauli-principle will prevent this decay. Holes in the occupied "underworld" (Dirac sea) are antiparticles.

The occupied states of this underworld including up to 40000 occupied bound states of the lower potential well represent the **vacuum**. The peculiarity of this strongly correlated vacuum structure in the region of atomic nuclei is that — depending on the size of the nucleus — more than 20000 up to 40000 (occupied) bound nucleon states contribute to this polarization effect. Obviously, we are dealing here with a **highly correlated vacuum**. A pronounced shell structure can be recognized [32,33,34]. Holes in these states have to be interpreted as bound antinucleons (antiprotons, antineutrons). If the primary nuclear density rises due to compression, the lower well increases while the upper decreases and soon is converted into a repulsive barrier (Fig. 23). This compression of nuclear matter can only be carried out in relativistic nucleus-nucleus collision with the help of shock waves, which have been proposed by the Frankfurt school[29,30] and which have since then been confirmed extensively (for references see e. g. [35]). These **nuclear shock waves** are accompanied by

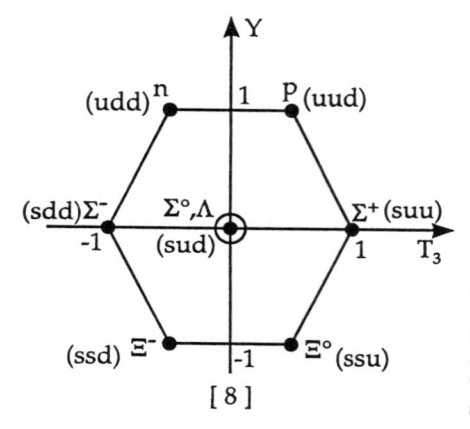

Figure 20: Important baryons are ordered in this octet. The quark content is depicted. Protons (p) and neutrons (n), most important for our known world, contain only u and d quarks. Hyperons contain also an s quark. The number of s quarks is a measure for the strangeness.

heating of the nuclear matter. Indeed, density and temperature are intimately coupled in terms of the hydrodynamic Rankine-Hugoniot-equations. Heating as well as the violent dynamics cause the creation of many holes in the very deep (measured from $-M_B c^2$) vacuum well. These numerous bound holes resemble antimatter clusters which are bound in the medium; their wave functions have large overlap with antimatter clusters. When the primary matter density decreases during the expansion stage of the heavy ion collision, the potential wells, in particular the lower one, disappear.

The bound antinucleons are then pulled down into the (lower) continuum. In this way antimatter clusters may be set free. Of course, a large part of the antimatter will annihilate on ordinary matter present in the course of the expansion. However, it is important that this mechanism for the production of antimatter clusters out of the highly correlated vacuum does not proceed via the phase space. The required coalescence of many particles in phase space suppresses the production of clusters, while it is favoured by the direct production out of the highly correlated vacuum. In a certain sense, the highly correlated vacuum is a kind of cluster vacuum (vacuum with cluster structure). The shell structure of the vacuum levels (see Fig. 22) supports this latter suggestion. Fig. 24 illustrates this idea.

The mechanism is similar for the production of multi-hyper nuclei (Λ, Σ, Ξ, Ω). Meson field theory predicts also for the Λ energy spectrum at finite primary nucleon density the existence of upper and lower wells. The lower well belongs to the vacuum and is fully occupied by Λ's.

Dynamics and temperature then induce transitions ($\Lambda\bar{\Lambda}$ creation) and de-

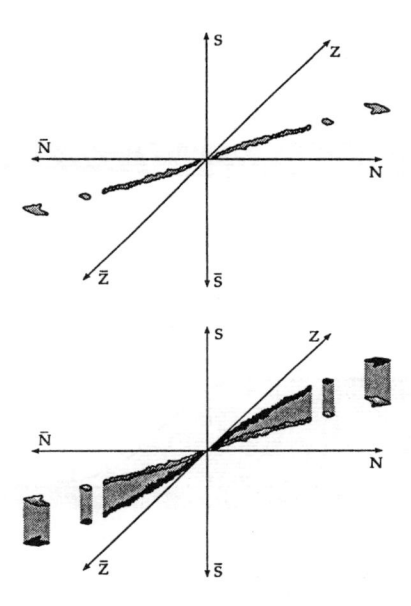

Figure 21: The extension of the periodic system into the sectors of strangeness $(S, \bar{S})$ and antimatter $(\bar{Z}, \bar{N})$. The stable valley winds out of the known proton (Z) and neutron (N) plane into the S and $\bar{S}$ sector, respectively. The same can be observed for the antimatter sector. In the upper part of the figure only the stable valley in the usual proton (Z) and neutron (N) plane is plotted, however, extended into the sector of antiprotons and antineutrons. In the second part of the figure it has been indicated, how the stable valley winds out of the Z-N-plane into the strangeness sector.

posit many Λ's in the upper well. These numerous bound Λ's are sitting close to the primary baryons: in a certain sense a giant multi-Λ hypernucleus has been created. When the system disintegrates (expansion stage) the Λ's distribute over the nucleon clusters (which are most abundant in peripheral collisions). In this way multi-Λ hypernuclei can be formed.

Of course this vision has to be worked out and probably refined in many respects. This means much more and thorough investigation in the future. It is particularly important to gain more experimental information on the properties of the lower well by (e, e' p) or (e, e' p p') and also ($\bar{p}_c p_b$, $p_c \bar{p}_b$) reactions at high energy ($\bar{p}_c$ denotes an incident antiproton from the continuum, p_b is a proton in a bound state; for the reaction products the situation is just the opposite)[36]. Also the reaction (p, p' d), (p, p' ^{3}He), (p, p' ^{4}He) and others of similar type need to be investigated in this context. The systematic scattering of antiprotons

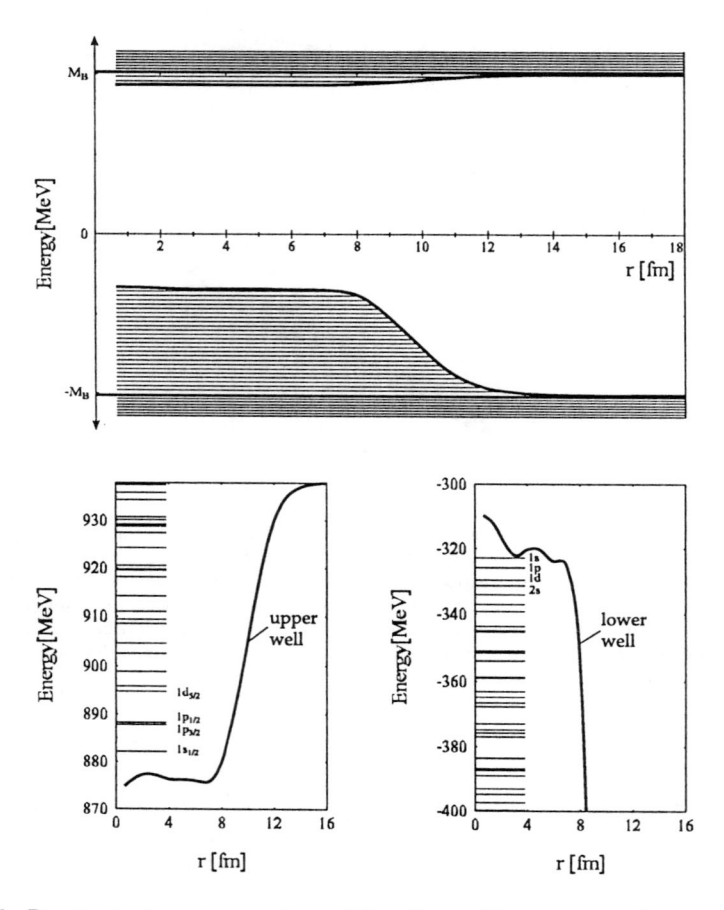

Figure 22: Baryon spectrum in a nucleus. Below the positive energy continuum exists the potential well of real nucleons. It has a depth of 50-60 MeV and shows the correct shell structure. The shell model of nuclei is realized here. However, from the negative continuum another potential well arises, in which about 40000 bound particles are found, belonging to the vacuum. A part of the shell structure of the upper well and the lower (vacuum) well is depicted in the lower figures.

on nuclei can contribute to clarify these questions. Problems of the meson field theory (e. g. Landau poles) can then be reconsidered. An effective meson field theory has to be constructed. Various effective theories, e. g. of Walecka-type on the one side and theories with chiral invariance on the other side, seem to give different strengths of the potential wells and also different dependence on the baryon density [37]. The Lagrangians of the Dürr-Teller-Walecka-type and of the chirally symmetric mean field theories look quite differently. We exhibit them — without further discussion — in the following equations:

$$\mathcal{L} = \mathcal{L}_{\text{kin}} + \mathcal{L}_{\text{BM}} + \mathcal{L}_{\text{vec}} + \mathcal{L}_I + \mathcal{L}_{\text{SB}}$$

Non-chiral Lagrangian:

$$\mathcal{L}_{\text{kin}} = \frac{1}{2}\partial_\mu s \partial^\mu s + \frac{1}{2}\partial_\mu z \partial^\mu z - \frac{1}{4}B_{\mu\nu}B^{\mu\nu} - \frac{1}{4}G_{\mu\nu}G^{\mu\nu} - \frac{1}{4}F_{\mu\nu}F^{\mu\nu}$$

$$\mathcal{L}_{\text{BM}} = \sum_B \overline{\psi}_B \left[i\gamma_\mu \partial^\mu - g_{\omega B}\gamma_\mu \omega^\mu - g_{\phi B}\gamma_\mu \phi^\mu - g_{\rho B}\gamma_\mu \tau_B \rho^\mu \right.$$
$$\left. - e\gamma_\mu \frac{1}{2}(1+\tau_B)A^\mu - m_B^* \right]\psi_B$$

$$\mathcal{L}_{\text{vec}} = \frac{1}{2}m_\omega^2 \omega_\mu \omega^\mu + \frac{1}{2}m_\rho^2 \rho_\mu \rho^\mu + \frac{1}{2}m_\phi^2 \phi_\mu \phi^\mu$$

$$\mathcal{L}_I = -\frac{1}{2}m_s^2 s^2 - \frac{1}{2}m_z^2 z^2 - \frac{1}{3}bs^3 - \frac{1}{4}cs^4$$

Chiral Lagrangian:

$$\mathcal{L}_{\text{kin}} = \frac{1}{2}\partial_\mu \sigma \partial^\mu \sigma + \frac{1}{2}\partial_\mu \zeta \partial^\mu \zeta + \frac{1}{2}\partial_\mu \chi \partial^\mu \chi - \frac{1}{4}B_{\mu\nu}B^{\mu\nu} - \frac{1}{4}G_{\mu\nu}G^{\mu\nu} - \frac{1}{4}F_{\mu\nu}F^{\mu\nu}$$

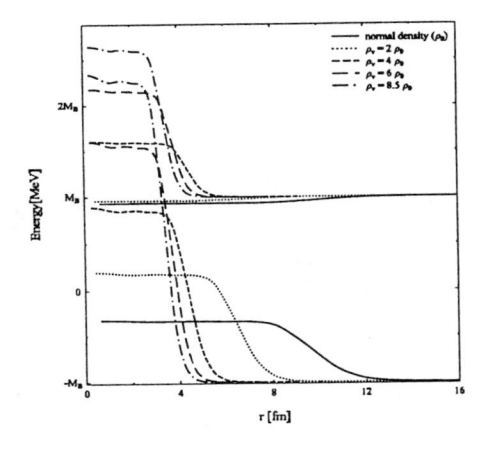

Figure 23: The lower well rises strongly with increasing primary nucleon density, and even gets supercritical (spontaneous nucleon emission and creation of bound antinucleons). Supercriticality denotes the situation, when the lower well enters the upper continuum.

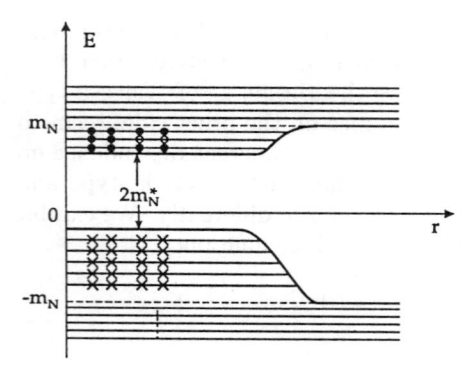

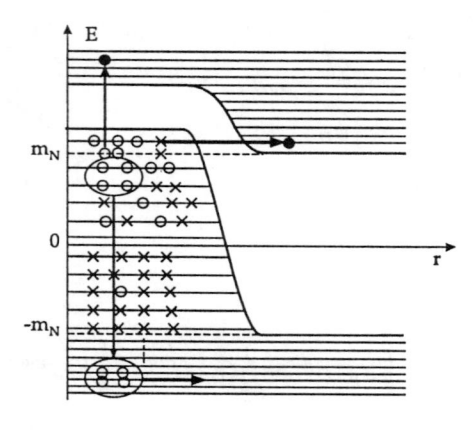

Figure 24:
Due to the high temperature and the violent dynamics, many bound holes (antinucleon clusters) are created in the highly correlated vacuum, which can be set free during the expansion stage into the lower continuum. In this way, antimatter clusters can be produced directly from the vacuum. The horizontal arrow in the lower part of the figure denotes the spontaneous creation of baryon-antibaryon pairs, while the antibaryons occupy bound states in the lower potential well. Such a situation, where the lower potential well reaches into the upper continuum, is called supercritical. Four of the bound holes states (bound antinucleons) are encircled to illustrate a "quasi-antihelium"
formed. It may be set free (driven into the lower continuum) by the violent nuclear dynamics.

$$\mathcal{L}_{\text{BM}} = \sum_B \overline{\psi}_B \left[i\gamma_\mu \partial^\mu - g_{\omega B}\gamma_\mu \omega^\mu - g_{\phi B}\gamma_\mu \phi^\mu - g_{\rho B}\gamma_\mu \tau_B \rho^\mu \right.$$
$$\left. - e\gamma_\mu \frac{1}{2}(1+\tau_B)A^\mu - m_B^* \right] \psi_B$$

$$\mathcal{L}_{\text{vec}} = \frac{1}{2} m_\omega^2 \frac{\chi^2}{\chi_0^2} \omega_\mu \omega^\mu + \frac{1}{2} m_\rho^2 \frac{\chi^2}{\chi_0^2} \rho_\mu \rho^\mu + \frac{1}{2} m_\phi^2 \frac{\chi^2}{\chi_0^2} \phi_\mu \phi^\mu + g_4^4(\omega^4 + 6\omega^2 \rho^2 + \rho^4)$$

$$\mathcal{L}_I = -\frac{1}{2} k_0 \chi^2 (\sigma^2 + \zeta^2) + k_1 (\sigma^2 + \zeta^2)^2 + k_2 \left(\frac{\sigma^4}{2} + \zeta^4 \right) + k_3 \chi \sigma^2 \zeta - k_4 \chi^4$$
$$+ \frac{1}{4} \chi^4 \ln \frac{\chi^4}{\chi_0^4} + \frac{\delta}{3} \ln \frac{\sigma^2 \zeta}{\sigma_0^2 \zeta_0^2}$$

$$\mathcal{L}_{\text{SB}} = -\left(\frac{\chi}{\chi_0} \right)^2 \left[m_\pi^2 f_\pi \sigma + \left(\sqrt{2} m_K^2 f_K - \frac{1}{\sqrt{2}} m_\pi^2 f_\pi \right) \zeta \right]$$

The non-chiral model contains the scalar-isoscalar field s and its strange counterpart z, the vector-isoscalar fields ω_μ and ϕ_μ, and the the ρ-meson ρ_μ as well as the photon A_μ. For more details see [37]. In contrast to the non-chiral model, the $SU(3)_L \times SU(3)_R$ Lagrangian contains the dilaton field χ introduced to mimic the trace anomaly of QCD in an effective Lagrangian at tree level (For an explanation of the chiral model see [37,38]).

The connection of the chiral Lagrangian with the Walecka-type can be established by the substitution $\sigma = \sigma_0 - s$ (and similarly for the strange condensate ζ). Then, e.g. the difference in the definition of the effective nucleon mass in both models (non-chiral:$m_N^* = m_N - g_s s$, chiral:$m_N^* = g_s\sigma$) can be removed, yielding:

$$m_N^* = g_s\sigma_0 - g_s s \equiv m_N - g_s s \tag{2}$$

for the nucleon mass in the chiral model.

Nevertheless, if the parameters in both cases are adjusted such that ordinary nuclei (binding energies, radii, shell structure,...) and properties of infinite nuclear matter (equilibrium density, compression constant K, binding energy) are well reproduced, the prediction of both effective Lagrangians for the dependence of the properties of the correlated vacuum on density and temperature is remarkably different. This is illustrated to some extend in Fig. 25. Accordingly, the chirally symmetric meson field theory predicts much higher primary densities (and temperatures) until the effects of the correlated vacuum are strong enough so that the mechanisms described here become effective. In other words, according to chirally symmetric meson field theories the antimatter-cluster-production and multi-hypermatter-cluster production out of the highly correlated vacuum takes place at considerably higher heavy ion energies as compared to the predictions of the Dürr-Teller-Walecka-type meson field theoories. This in itself is a most interesting, quasi-fundamental question to be clarified. Moreover, the question of the nucleonic substructure (form factors, quarks, gluons) and its influence on the highly correlated vacuum structure has to be studied. The nucleons are possibly strongly polarized in the correlated vacuum: the Δ resonance correlations in the vacuum are probably important. Is this highly correlated vacuum state, especially during the compression, a preliminary stage to the quark-gluon cluster plasma? To which extent is it similar or perhaps even identical with it? It is well known for more than 10 years that meson field theories predict a phase transition qualitatively and quantitatively similar to that of the quark-gluon plasma [39] — see Fig. 26.

The extension of the periodic system into the sectors hypermatter (strangeness) and antimatter is of general and astrophysical importance. Indeed, microseconds after the big bang the new dimensions of the periodic system, we have touched upon, certainly have been populated in the course of the baryo-

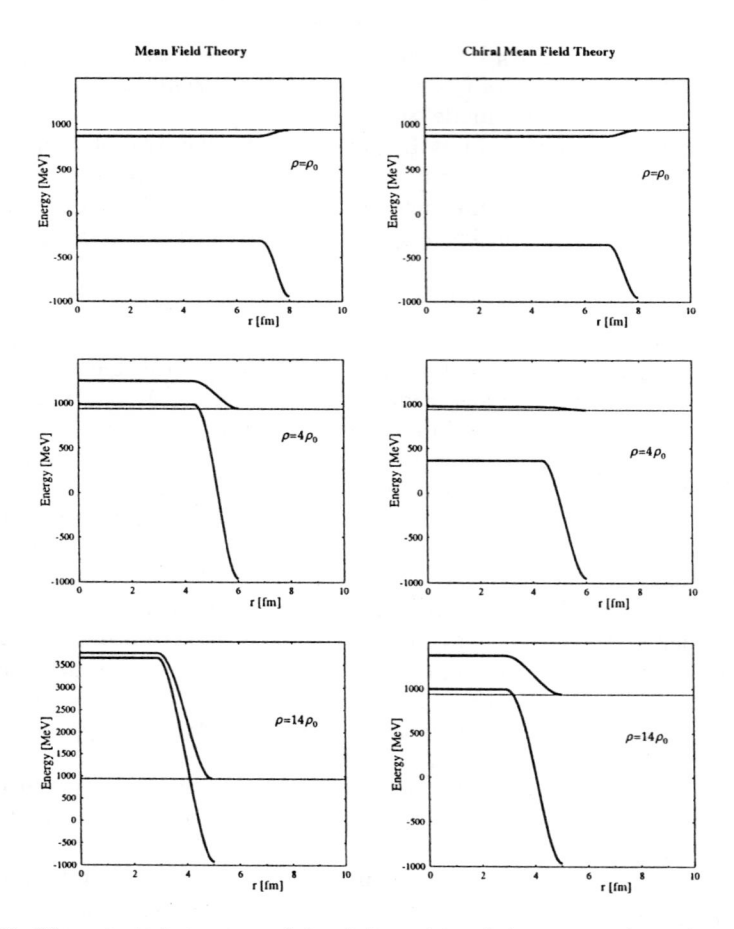

Figure 25: The potential structure of the shell model and the vacuum for various primary densities $\rho = \rho_0$, $4\rho_0$, $14\rho_0$. At left the predictions of ordinary Dürr-Teller-Walecka-type theories are shown; at right those for a chirally symmetric meson field theory as develloped by P. Papazoglu, S. Schramm et al. [37,38]. Note however, that this particular chiral mean-field theory does contain ω^4 terms. If introduced in both effective models, they seem to predict quantitatively similar results.

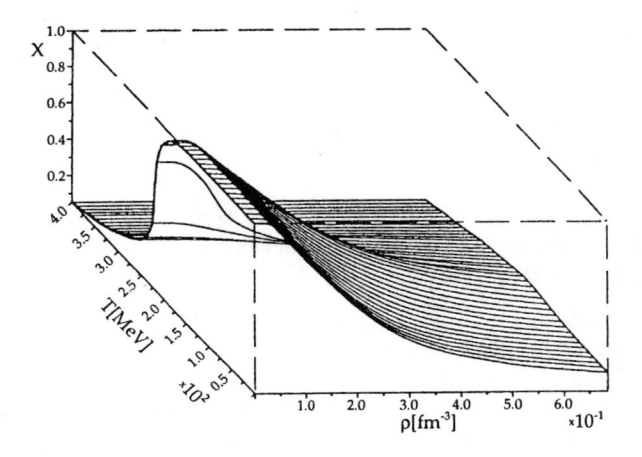

Figure 26: The strong phase transition inherent in Dürr-Teller-Walecka-type meson field theories, as predicted by J. Theis et al. [39]. Note that there is a first order transition along the ρ-axis (i.e. with density), but a simple transition along the temperature T-axis. Note also that this is very similar to the phase transition obtained recently from the Nambu-Jona-Lasinio-approximation of QCD [41].

and nucleo-genesis. Of course, for the creation of the universe, even higher dimensional extensions (charm, bottom, top) come into play, which we did not pursue here. It is an open question, how the depopulation (the decay) of these sectors influences the distribution of elements of our world today. Our conception of the world will certainly gain a lot through the clarification of these questions.

For the Gesellschaft für Schwerionenforschung (GSI), which I helped initiating in the sixties, the questions raised here could point to the way ahead. Working groups have been instructed by the board of directors of GSI, to think about the future of the laboratory. On that occasion, very concrete (almost too concrete) suggestions are discussed — as far as it has been presented to the public. What is missing, as it seems, is a **vision on a long term basis**. The ideas proposed here, the verification of which will need the **commitment for 2–4 decades of research**, could be such a vision with considerable attraction for the best young physicists. The new dimensions of the periodic system made of hyper- and antimatter cannot be examined in the "stand-by" mode at CERN (Geneva); a dedicated facility is necessary for this field of research, which can in future serve as a home for the universities. The GSI — which has unfortunately become much too self-sufficient — could be such a home for new generations of physicists, who are interested in the **structure of elementary matter**. GSI would then not develop just into a detector laboratory for CERN, and as such become obsolete. I can already see the enthusiasm in the eyes of young scientists, when I unfold these ideas to them — similarly as it was 30 years ago, when the nuclear physicists in the state of Hessen initiated the construction of GSI.

I am grateful to Dipl.-Phys. Thomas Bürvenich for helping me in the technical production of these proceedings.

1. S.G: Nilsson et al. Phys. Lett. 28 B (1969) 458
   Nucl. Phys. A 131 (1969) 1
   Nucl. Phys. A 115 (1968) 545
2. U. Mosel, B. Fink and W. Greiner, Contribution to "Memorandum Hessischer Kernphysiker" Darmstadt, Frankfurt, Marburg (1966).
3. U. Mosel and W. Greiner, Z. f. Physik 217 (1968) 256, 222 (1968) 261
4. a) J. Grumann, U. Mosel, B. Fink and W. Greiner, Z. f. Physik 228 (1969) 371
   b) J. Grumann, Th. Morovic, W. Greiner, Z. f. Naturforschung *26a* (1971) 643
5. A. Sandulescu, R.K. Gupta, W. Scheid, W. Greiner, Phys. Lett. *60*B (1976) 225
   R.K. Gupta, A. Sandulescu, W. Greiner, Z. f. Naturforschung *32*a (1977) 704
   R.K. Gupta, A.Sandulescu and W. Greiner, Phys. Lett. *64*B (1977) 257
   R.K. Gupta, C. Parrulescu, A. Sandulescu, W. Greiner Z. f. Physik A283 (1977) 217
6. G. M. Ter-Akopian et al., Nucl. Phys. A*255* (1975) 509
   Yu.Ts. Oganessian et al., Nucl. Phys. A*239* (1975) 353 and 157
7. D. Scharnweber, U. Mosel and W. Greiner, Phys. Rev. Lett *24* (1970) 601
   U. Mosel, J. Maruhn and W. Greiner, Phys. Lett. *34*B (1971) 587
8. G. Münzenberg et al. Z. Physik A309 (1992) 89
   S.Hofmann et al. Z. Phys A*350* (1995) 277 and 288
9. R. K. Gupta, A. Sandulescu and Walter Greiner, Z. für Naturforschung *32*a (1977) 704
10. A. Sandulescu and Walter Greiner, Rep. Prog. Phys 55. 1423 (1992); A. Sandulescu, R. K. Gupta, W. Greiner, F. Carstoin and H. Horoi, Int. J. Mod. Phys. E1, 379 (1992)
11. A. Sobiczewski, Phys. of Part. and Nucl. 25, 295 (1994)
12. R. K. Gupta, G. Münzenberg and W. Greiner, J. Phys. G: Nucl. Part.

Phys. 23 (1997) L13
13. V. Ninov, K. E. Gregorich, W. Loveland, A. Ghiorso, D. C. Hoffman, D. M. Lee, H. Nitsche, W. J. Swiatecki, U. W. Kirbach, C. A. Laue, J. L. Adams, J. B. Patin, D. A. Shaughnessy, D. A. Strellis and P. A. Wilk, preprint
14. K. Rutz, M. Bender, T. Bürvenich, T. Schilling, P.-G. Reinhard, J.A. Maruhn, W. Greiner, Phys. Rev. C *56* (1997) 238.
15. B. Fricke and W. Greiner, Physics Lett *30*B (1969) 317
    B. Fricke, W. Greiner, J.T. Waber, Theor. Chim. Acta (Berlin) *21* (1971) 235
16. A. Sandulescu, D.N. Poenaru, W. Greiner, Sov. J. Part. Nucl. 11(6) (1980) 528
17. Harold Klein, thesis, Inst. für Theoret. Physik, J.W. Goethe-Univ. Frankfurt a. M. (1992)
    Dietmar Schnabel, thesis, Inst. für Theoret. Physik, J.W. Goethe-Univ. Frankfurt a.M. (1992)
18. D. Poenaru, J.A. Maruhn, W. Greiner, M. Ivascu, D. Mazilu and R. Gherghescu, Z. Physik A*328* (1987) 309, Z. Physik A*332* (1989) 291
19. E. K. Hulet, J. F. Wild, R. J. Dougan, R. W.Longheed, J. H. Landrum, A. D. Dougan, M. Schädel, R. L. Hahn, P. A. Baisden, C. M. Henderson, R. J. Dupzyk, K. Sümmerer, G. R. Bethune, Phys. Rev. Lett. *56* (1986) 313
20. K. Depta, W. Greiner, J. Maruhn, H.J. Wang, A. Sandulescu and R. Hermann, Intern. Journal of Modern Phys. A*5*, No. 20, (1990) 3901
    K. Depta, R. Hermann, J.A. Maruhn and W. Greiner, in "Dynamics of Collective Phenomena", ed. P. David, World Scientific, Singapore (1987) 29
    S. Cwiok, P. Rozmej, A. Sobiczewski, Z. Patyk, Nucl. Phys. A491 (1989) 281
21. A. Sandulescu and W. Greiner in discussions at Frankfurt with J. Hamilton (1992/1993)
22. J.H. Hamilton, A.V. Ramaya et al. Journ. Phys. G **20** (1994) L85 - L89
23. B. Burggraf, K. Farzin, J. Grabis, Th. Last, E. Manthey, H. P. Trautvetter, C. Rolfs, *Energy Shift of first excited state in* ^{10}Be *?*, accepted for publication in Journ. of. Phys. G
24. P. Hess et al., *Butterfly and Belly Dancer Modes in* $^{96}Sr + {}^{10}Be + {}^{146}Ba$, in preparation
25. E.K. Hulet et al. Phys Rev C *40* (1989) 770.
26. W. Greiner, B. Müller, J. Rafelski, QED of Strong Fields, Springer Verlag, Heidelberg (1985). For a more recent review see W. Greiner, J.

Reinhardt, *Supercritical Fields in Heavy-Ion Physics*, Proceedings of the 15th Advanced ICFA Beam Dynamics Workshop on Quantum Aspects of Beam Physics, World Scientific (1998)
27. B. Povh, Rep. Progr. Phys. *39* (1976) 823; Ann. Rev. Nucl. Part. Sci. *28* (1978) 1; Nucl. Phys. A*335* (1980) 233; Progr. Part. Nucl. Phys. *5* (1981) 245; Phys. Blätter *40* (1984) 315
28. J. Schaffner, Carsten Greiner and H. Stöcker Phys. Rev. C*45* (1992) 322; Nucl. Phys. B*24B* (1991) 246; J. Schaffner, C.B. Dover, A. Gal, D.J. Millener, C. Greiner, H. Stöcker: Annals of Physics*235* (1994) 35
29. W. Scheid and W. Greiner, Ann. Phys. *48* (1968) 493; Z. Phys. *226* (1969) 364
30. W. Scheid, H. Müller and W. Greiner Phys. Rev. Lett. 13 (1974) 741
31. H. Stöcker, W. Greiner and W. Scheid Z. Phys. A 286 (1978) 121
32. I. Mishustin, L.M. Satarov, J. Schaffner, H. Stöcker and W.Greiner Journal of Physics G (Nuclear and Particle Physics) *19* (1993) 1303
33. P.K. Panda, S.K. Patra, J. Reinhardt, J. Maruhn, H. Stöcker, W. Greiner, Int. J. Mod. Phys. E 6 (1997) 307
34. N. Auerbach, A. S. Goldhaber, M. B. Johnson, L. D. Miller and A. Picklesimer, Phys. Lett. B182 (1986) 221
35. H. Stöcker and W. Greiner, Phys. Rep. 137 (1986) 279.
36. J. Reinhardt and W. Greiner, to be published.
37. P. Papazoglou, D. Zschiesche, S. Schramm, H. Stöcker, W. Greiner, J. Phys. G 23 (1997) 2081; P. Papazoglou, S. Schramm, J. Schaffner-Bielich, H. Stöcker, W. Greiner, Phys. Rev. C 57 (1998) 2576.
38. P. Papazoglou, D. Zschiesche, S. Schramm, J. Schaffner-Bielich, H. Stöcker, W. Greiner, nucl-th/9806087, accepted for publication in Phys. Rev. C.
39. J. Theis, G. Graebner, G. Buchwald, J. Maruhn, W. Greiner, H. Stöcker and J. Polonyi, Phys. Rev. D 28 (1983) 2286
40. P. Papazoglou, PhD thesis, University of Frankfurt, 1998; C. Beckmann et al., in preparation
41. S. Klimt, M. Lutz, W. Weise, Phys. Lett. B*249* (1990) 386.

H.E.S.S. — AN ARRAY OF STEREOSCOPIC IMAGING ATMOSPHERIC CHERENKOV TELESCOPES CURRENTLY UNDER CONSTRUCTION IN NAMIBIA

R. STEENKAMP

Department of Physics, University of Namibia, Private Bag 13301, 340 Mandume Ndemufayo Avenue, Pionierspark, Windhoek, Namibia
E-mail: rsteenkamp@unam.na

The H.E.S.S. experiment is an array of sophisticated Imaging Atmospheric Cherenkov Telescopes (IACTs) that is currently under construction on the farm Göllschau in the Khomas Highland of Namibia. The site is located about 100 km from the capital of Namibia and about 20 km from the Gamsberg itself. This experiment is planned to succeed the older HEGRA experiment and is planned to be an order of magnitude more sensitive than HEGRA.

1 Introduction

The acronym H.E.S.S., which stands for High Energy Stereoscopic System, was chosen in honour of Victor F. Hess, who discovered the existence of cosmic rays in 1911/2.

Like its predecessor, HEGRA (High Energy Gamma Ray Astronomy), the H.E.S.S. experiment will use a stereoscopic Atmospheric Cherenkov Technique (ACT) to observe γ-ray induced air showers in the earth's atmosphere. By measuring the spatial extent, height and most importantly the orientation of these air showers, one can successfully do Very High Energy (VHE) γ-ray astronomy.

The term 'VHE' refers to γ rays with photon energies larger than about 100 GeV. Photon energies this high can only be produced by non-thermal particle populations (particle populations with a power-law energy spectra as opposed by 'thermal' particle populations with Maxwellian energy distributions) and particle decay processes. This means that H.E.S.S. will be an ideal instrument with which to observe the so-called non-thermal universe.

One of H.E.S.S.'s immediate predecessors, HEGRA, is an array of 5 (one prototype and 4 research telescopes) 3 m diameter IACTs currently situated at La Palma in the Canary Islands. HEGRA successfully demonstrated the so-called *stereoscopic* concept[1,2] that pertains to the simultaneous observation of gamma-ray induced air showers with multiple IACTs. Stereoscopy enables not only greater resolution of sources, but also allows the measurement of their gamma-ray energy spectra[5].

H.E.S.S. is being planned as a next generation full-scale instrument dedicated to long-term Astrophysical research[3]. It was designed to have an order of magnitude better flux sensitivity and the ability to detect gamma-ray photons above a lower energy threshold than HEGRA. To this effect the light collection surface (reflector) was significantly increased and an advanced light detector ('camera') was designed. Naturally H.E.S.S. must also be able to do high quality stereoscopy at a similar or better level than HEGRA. This implies a large separation between individual IACTs.

Lastly, H.E.S.S. is designed with almost total automation and ease of maintenance in order to minimise human intervention on maintenance and observation procedures. This is necessary because the experiment is situated quite some distance from the experts that designed it.

The HEGRA IACT array has a lower photon energy detection threshold of 500 GeV, an angular resolution capacity per photon of 0.1° and the minimum detectable energy flux per photon is 10^{-12} erg/cm^{2}s in 100 h. By contrast, the H.E.S.S. IACT array is projected to have a lower photon energy detection threshold of 40 GeV. For stereoscopy resulting in acceptable angular resolution of the incident photon this lower threshold has to be limited to 100 GeV. The angular resolution per photon is $\lesssim 0.1°$ and the minimum detectable energy flux at 1 TeV is projected to be 10^{-13} erg/cm^{2}s in 50 h.

Figure 1[3,8] shows the energy flux spectrum of the Crab Nebula, the 'candlestick-standard' for gamma-ray astronomers. The solid lines represent the component synchrotron (S) and inverse Compton (IC) components as predicted by theory. Data points are shown as measured by the COMPTEL and EGRET experiments on the GRO satellite. To the far right of the graph are some data-points produced by Extensive Air Shower (EAS) experiments. The data gap between 10 GeV to 20 TeV is clearly seen. A stereoscopic ACT like H.E.S.S. can therefore successfully provide measurements in the region in question. Also, the low flux sensitivity of H.E.S.S., as indicated, will enable the detection of new source populations at the 10 'milli-Crab' level.

The chosen site for the H.E.S.S. experiment is on the farm Göllschau located about 100 km south-west of Windhoek and about 20 km from the Gamsberg itself. The coordinates of the site is 23°16'18.4"S, 16°30'00.8"E and is about 1800 ± 20 m above sea level.

2 Specification of the H.E.S.S. IACTs

Each IACT consists of a light collector and a light detector. The light collector, henceforward the reflector, consists of a steel structure supporting a spherical segmented reflector with an overall hexagonal shape on an Altitude-

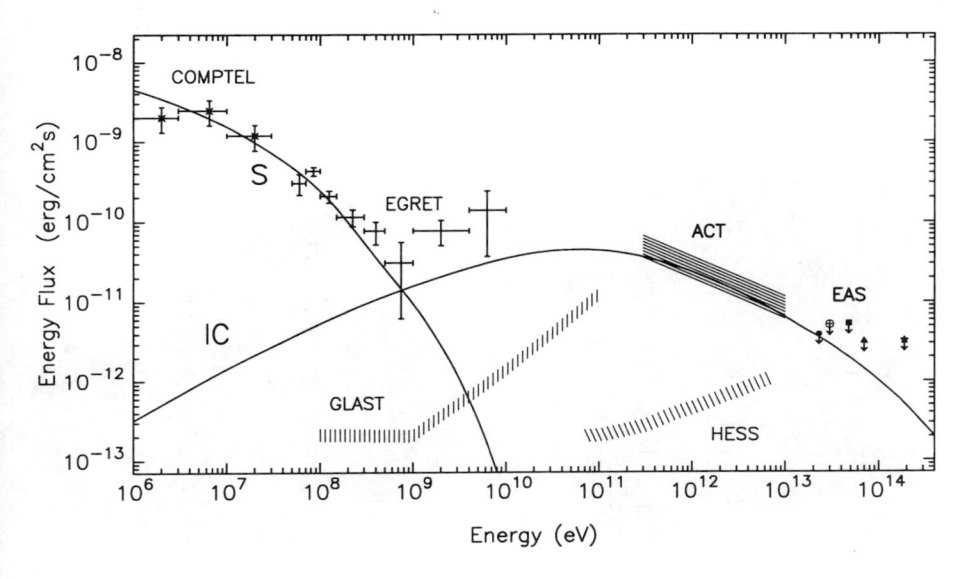

Figure 1. The energy flux spectrum of the Crab Nebula as measured by the COMPTEL and EGRET experiments on the GRO satellite. EAS data points are also shown. The projected flux sensitivity of HESS is shown as well as the energy range in which the inverse Compton component of the spectrum can be investigated by an ACT like H.E.S.S.

Azimuth mounting (see Figure 2a). The reflector itself will consist of 382 small circular front-aluminised glass mirrors, each 60 cm in diameter with a focal length of 15 m. A thin quartz layer protects the aluminisation from the elements.

The resulting reflector has a total reflective area of 108 m^2 and is 12 m in diameter. To avoid progressive misalignment of the mirror tiles during normal observations, each individual mirror is designed to be not only separately adjustable, but automatically adjustable with computer control. As shown in Figure 2(b), each mirror is connected to the reflector space frame with a tripod mount with servo motors on 2 of the 3 legs of the tripod to facilitate mirror adjustment. The plan is to periodically calibrate the mirror tiles on each IACT by focusing starlight onto a CCD camera on the main IACT camera's lid.

The IACT's light detector, henceforward the camera, is based on both the cameras used by HEGRA and the French CAT (Cherenkov Array at Thémis). While the overall telescope design and software systems are being designed in Germany, the camera is being designed and built in France.

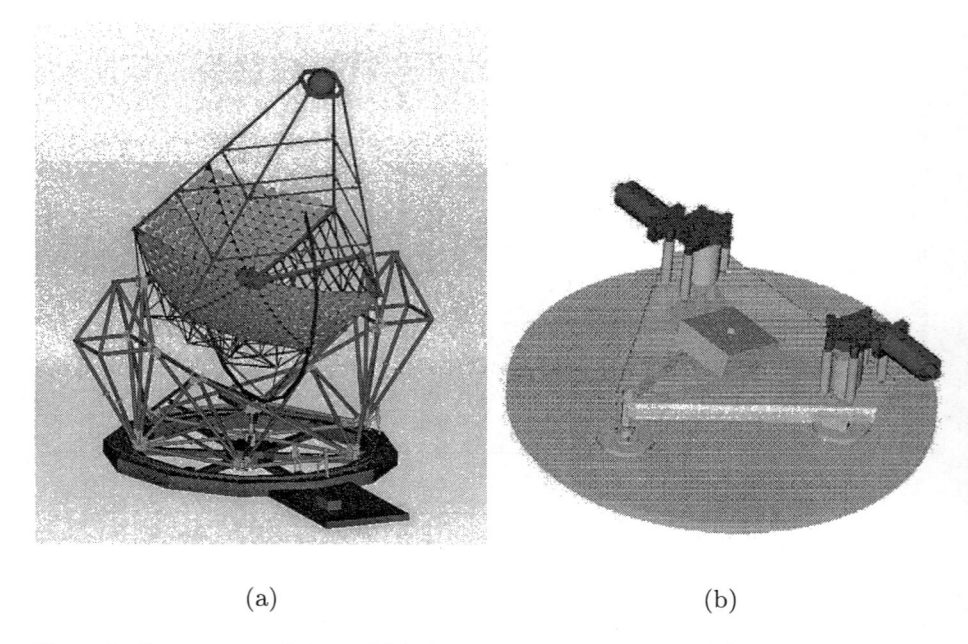

Figure 2. Computer renderings of (a) the telescope structure and (b) the mounting of a mirror tile.

The camera is made up of 960 PMTs arranged in 60 removable drawers of 16 PMTs each. Each PMT covers 0.16° of the sky, giving the camera a total field of view of 5°. As Cherenkov flashes are detected by individual PMTs, a triggering signal is produced by single pixel triggers. These are then fed into a topological trigger in the camera housing itself to pick out air showers that are of interest. The idea is to do the triggering as soon as possible to avoid data losses that would have been the case if the analogue signals had to be transmitted to a central facility. See Figure 3(a) for an exploded diagrammatic view of the camera housing.

For now (Phase I) conventional PMTs (Photonis XP2960) will be used, but alternative detectors are being investigated for use in the the second generation (Phase II) telescopes.

Initially 4 IACTs in a square formation, 120 m apart, are planned for Phase I. For Phase II it is planned that the array will be expanded to a total of 12–16 IACTs. This will make H.E.S.S. the largest experiment of its kind. Several different spatial arrangements are currently under study.

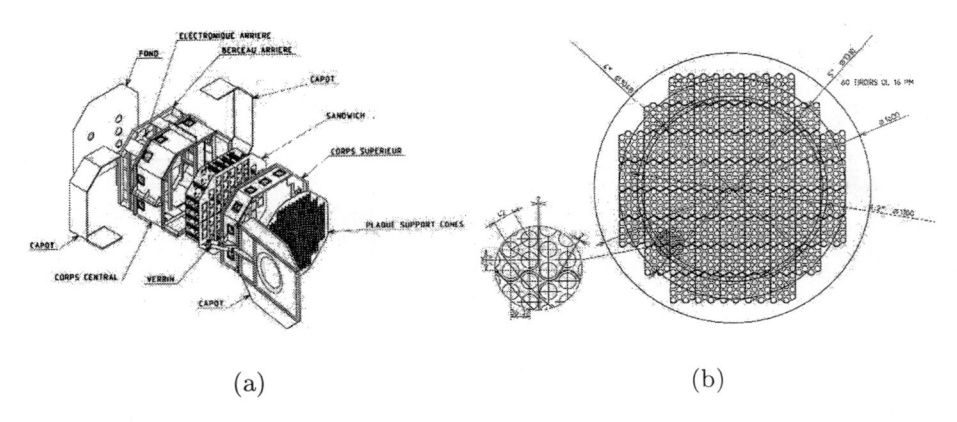

Figure 3. (a) An exploded view of the camera housing. (b) A front view of the PMT array, showing the drawer arrangement.

Since the ACT is concerned with gamma-ray-initiated air showers in the earth's atmosphere and the propagation of the resulting Cherenkov radiation through the atmosphere, it follows that the atmosphere above the Gamsberg region must be carefully monitored. Especially of interest are aerosols and thin high altitude clouds that will influence observation conditions. To this effect the use of a multi-frequency LIDAR (LIght Detection And Ranging) and also the possible use of smaller secondary ACD has been proposed by the Durham group.

A small optical telescope, called ATOM (Automatic Telescope for Optical Monitoring), has been proposed by the Landessternwarte in Heidelberg. This telescope is intended to be used in multi-wavelength observations of AGNs (Active Galactic Nuclei).

Recently a proposal has been received from the ROTSE (Robotic Optical Transient Search Experiment) group to become part of the H.E.S.S. collaboration and to locate the next generation ROTSE-II instrument with which GRBs (Gamma-Ray Bursts) will be studied on the H.E.S.S. site. So far a decision on this matter has not been reached.

3 Stereoscopy

The concept of stereoscopy is mentioned several times in the preceding text. The details are well documented in literature[1,2,6]. Now, though not an expert on this technique, I shall briefly explain the basic principles. Figure 4 shows

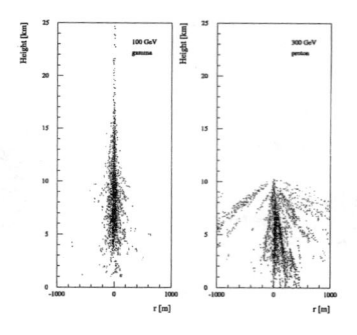

Figure 4. Simulation results of air showers.

simulation results form a gamma-ray- and a proton-induced air shower[6].

Now the Cherenkov radiation from a gamma-ray-induced air shower will form images on the 4 different cameras as shown in Figure 5(a)[9] and as shown in Figure 5(b)[6] this information can be used to determine, among other things, the shower axis and thus pinpointing a gamma-ray source with great accuracy ($\lesssim 0.1°$).

4 Astrophysical Objectives of H.E.S.S.

Within our galaxy H.E.S.S. will observe and/or search for SuperNova Remnants (SNRs), pulsars, pulsar nebulae, accreting neutron stars, stellar black holes, microquasars, Giant Molecular Clouds (GMCs) and diffuse VHE radiation from both the Galactic disk and Galactic halo. Outside our galaxy H.E.S.S. will observe Active Galactic Nuclei or AGNs (radio galaxies, quasars, blazars), rich clusters of galaxies and nearby normal and starburst galaxies.

Some of the astrophysical problems that will be addressed are the search for and identification of cosmic-ray accelerators, furthering the understanding of AGNs (Physics of AGN jets in particular) and to probe the high-energy history of galaxy and cluster formation by detection of VHE radiation from galactic clusters. Matters of observational cosmology will also be addressed by studying pair halos, formed by pair production processes due to interactions of VHE ($\gg$ 10 TeV) photons with the Diffuse Extragalactic BAckground (DEBRA) fields, around VHE sources. This would lead to answers about the density of the DEBRA fields and would provide an accurate measure of distances up to redshifts $z \lesssim 1$.

Last, but not least will be the search for VHE γ-rays from dark matter

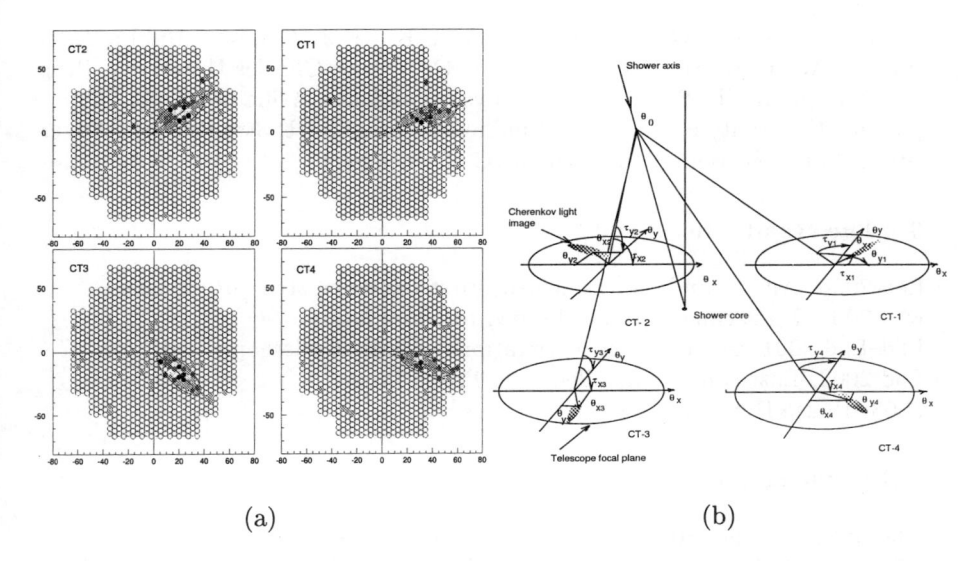

Figure 5. (a) Simulation results of the images that a gamma-ray-induced air shower will produce on the 4 cameras of the H.E.S.S. array. (b) Geometry of the calculation of the shower axis.

(WIMPS[a], superstrings and other left-overs — or topological defect, if you will — from the Big Bang).

5 List of Collaborating Institutions and their Representatives

From Germany: Max-Plank-Institut für Kernphysik, Heidelberg, Astrophysics (H. Völk) & Particle Physics (W. Hofmann) Divisions; Humbold Universität Berlin (T. Lohse); Ruhr-Universität Bochum (R. Schlickeiser); Universität Hamburg (G. Heinzelmann); Landessternwarte Heidelberg (S. Wagner); Universität Kiel (W. Stamm).

From France: LPNHE Ecole Polytechnique, Palaiseau (B. Degrange); LPC College de France, Paris (M. Punch); Universités Paris VI–VII, LPNHE (M. Rivoal); Université de Grenoble (H. Henri?); CERS, Toulouse (A.R. Bazer-Bachi); CEA Saclay (P. Goret).

From Italy: Universitá della Basilicata, Potenza & INFN sezione di Roma, Rome (G. Auriemma); IAS-CNR, Rome (F. Giovannelli).

---

[a] Weakly Interacting Massive particles

From abroad: Durham University, U.K. (K.J. Orford); Dublin Institute for Advanced Studies, Ireland (L.O'C. Drury); Charles University, Prag, Czech Republic (L. Rob); Yerevan Physics Institute, Armenia (A.G. Akhperjanian); University of Namibia, Windhoek, Namibia (R. Steenkamp); University of Potchefstroom, South Africa (O.C. de Jager).

6 The Road Ahead

End-2000: All 4 foundations finished; First telescope structure finished
Mid-2001: First camera to be shipped to Namibia
Mid–End 2001: Start of first observations with first telescope
End 2002: Last camera installed (end Phase I)
2003: Phase II commences

Acknowledgements

The author wishes to thank the organisers of this conference, especially Simon Connell, for arranging for financial assistance enabling me to attend this conference. The author also wishes to acknowledge and thank Heinrich Völk, Werner Hofmann and other members of the H.E.S.S. collaboration for many insightful discussions about the experiment.

References

1. F.A. Aharonian et al., *Towards a Major Atmospheric Cherenkov Detector II* (Calgary), ed. R.C. Lamb, p. 81, (1993)
2. P.M. Chadwick et al., *Space. Sci. Rev.* **75**, 153 (1995)
3. F.A. Aharonian et al. *HESS Letter of Intent* MPIK H-V11, (1997), [http://www-hfm.mpi-hd.mpg.de/HESS/public/hessloi3.ps.gz].
4. F.A. Aharonian et al. *HESS Letter of Intent: Appendix A*, (1997), [http://www-hfm.mpi-hd.mpg.de/HESS/public/PhJ.ps.gz].
5. W. Hofmann, *Towards a Major Atmospheric Cherenkov Detector V* (Durban), ed. O.C. de Jager, p. 284, (1997),
6. F.A. Aharonian et al., *Astropart. Phys.* **6**, 343 (1997)
7. A. Köhnle, *Proc. 26th ICRC* (Salt Lake City) **5**, 239 (1999)
8. A. Köhnle, *Proc. 26th ICRC* (Salt Lake City) **5**, 271 (1999)
9. A. Konopelko, Heidelberg Conference, (2000).

COSMIC PARTICLE ACCELERATION - ELECTRONS VS. NUCLEI.

O. C. DE JAGER

Unit for Space Physics, Potchefstroom University for CHE, Potchefstroom 2520, South Africa
E-mail: okkie@fskocdj.puk.ac.za

Although cosmic radiation inlcudes accelerated nuclei from H through Fe in certain abundances, the observed high energy electron component is much less intense as a result of energy loss processes in our galaxy. The situation appears to be reversed when we observe the gamma-radiation from some of the well-known gamma-ray sources. Since electrons, being lighter particles, radiate more efficiently than nuclei of the same energy, we expect to see the radiation from electrons to be dominant at the sources of cosmic ray acceleration. We have selected galactic pulsars and their plerions for a case study, and we will also review the role which ions play in the acceleration process.

1 Introduction

It is believed that Supernova Remnants are responsible for the spectrum and composition of high energy to ultrahigh energy galactic cosmic rays [1]. In this case Fermi shock acceleration with proper geometrical considerations can reproduce the detailed spectra of H, He, C, O, Si and Fe up to the so-called "Knee" near 10^{15} eV. It is thus clear from the outset that nuclear acceleration (at least in our galaxy) is taking place. The origin of cosmic rays above the knee is still uncertain, but it appears as if the composition becomes heavier above the "Knee", with elements from He through Fe, but lacking H [2]. One possibility is that these heavier elements are accelerated in the progenitor stellar winds after preacceleration has taken place in the supernova shell of the exploded star [3].

Furthermore, high energy to very high energy electrons/positrons (in a ratio of 10 to 1) are also observed near Earth, which cannot be explained by a solar origin (see e.g. refs. [4] and [6]), but the accelerated cosmic electrons constitute only a percent of the total cosmic radiation. The reason for this is the severe energy losses experienced by these light particles as a result of propagation effects in the galaxy (see the review by Stephens [5]).

It is not sufficient to only observe the accelerated particles at Earth, since the charged particles are deflected by the galactic magnetic field and arrive isotropically at our solar system. We therefore have to find another approach to pinpoint the origin of cosmic rays, and this done by means of the radiation produced at the source - even during the acceleration process. There are several radiation mechanisms, such as synchrotron radiation from energetic electrons on magnetic fields, which are always present. However, these electrons can also produce observable relativistic bremstrahlung if the ambient gas density is large enough. Inverse Compton scattering will also always be present in our galaxy, since the Cosmic Microwave Background Radiation (CMBR), galactic infrared emission from dust and optical starlight provide target photons for the production of gamma radiation via this process. Nuclei also produce gamma radiation via a few processes: Collisions between very high energy nuclei and target gas result in spallation products such as pions, of which the π^o component decays to 2γ's, whereas the $\pi^{+/-}$ component decays to $e^{+/-}$, which can in turn result in the abovementioned electron radiation. A high target density of soft photons can also result in the so-called photo-meson process, if the primary hadron's energy is above the threshold for this process.

The following question therefore arises: do we see both electron and nuclear components at the sources, where the acceleration is taking place? We will first review particle acceleration in general and then consider pulsars and their plerions as a specific laboratory for the study of more than one acceleration mechanism.

2 Particle Acceleration in Astrophysical Sources

The main acceleration mechanisms for electrons and nuclei can be reduced to electric field acceleration, even though they manifest in totally different forms. This section gives a general review of two of the most important mechanisms for particle acceleration to very high energies:

2.1 Dynamo Electric Field Acceleration

Rotating magnetised compact objects such as pulsars generate regions of field-aligned electric fields, and accelerate primary charged particles from the conductive surface [7]. The primary gamma-radiation radiation from the electron component converts to secondary and tertiary electron pairs in the superstrong magnetic fields, since the radiation energy per photon is typically well above the rest mass of an electron. The multiplicity for the cascade is typically large, and is needed to explain the intensity and spectra of pulsed gamma-ray emission emergent from pulsar magnetospheres [8]. Thus, pulsed gamma-ray emission from pulsars should reflect high energy emission from electrons and positrons in equal numbers.

2.2 Shock Acceleration

A Supernova explosion typcially results in a shock front, where Fermi acceleration can take place if the shock is collissionless. The particles scatter across the shock and gain energy with each encounter, with the energy gain per encounter depending on the velocity of the shock and the shock compression ratio. In this case the finite probability for particle escape from the shock versus ongoing acceleration results in a typical power law spectrum [9].

This mechanism is very successful to explain the composition of the nuclear component of cosmic rays as discussed in the previous section, and the detection of high energy gamma-rays from Supernova shells associated with dense molecular clouds was considered to be the first direct evidence of protons and nuclei accelerated by Supernova shells, since such nuclei would produce gamma-rays through spallation products in the target molcular gas. This popular belief was challenged by de Jager & Mastichiadis [10], who have shown that the electrons responsible for the typical bright radio shells will automatically radiate a similar intensity of high energy gamma-rays through relativistic bremstrahlung on the molecular gas. Thus, no need to invoke a cosmic nuclear component. Furthermore, if this nuclear component was produced up to the "Knee" of the cosmic ray spectrum, ground-based TeV gamma-ray telescopes should have seen the spectrum extending from the high energy gamma-ray region (100 MeV to 10 GeV) to at least TeV energies. This was not seen, which strengthened the claim of de Jager & Mastichiadis that the cosmic ray nuclear production is probably smaller than considered previously.

Even particle acceleration in the exotic Gamma Ray Bursts can be explained in terms of relativistic shock acceleration [11], and in the case of relativistic shocks, it can be shown that the acceleration timescale is comparable to the gyroperiod of the particle. This theoretical prediction will be exploited in the next section.

In the case of (relativistic) pulsar wind shocks, Harding & Gaisser [12] have shown that the maximum particle energy (in the absence of severe synchrotron losses) is equal to the full polar cap potential associated with the host neutron star. We will investigate this claim further in the next section when we attempt to close the pulsar current.

A combination of dynamo- and shock acceleration can take place near the central engines of supermassive objects, where the gas around the center of a galaxy forms an accretion disk, which should result in dynamo generated electric fields. Such dynamo processes lead to the well-known jets seen from Active Galactic Nuclei. The jets are also known to produce knots, which are in turn interpreted in terms of shocks, where impulsive injection can take place, and hence particle acceleration to give the time variable emission from optical to TeV energies. The time variability is then a result of the build-up process resulting from the finite time to accelerate a particle to its maximum energy (while radiating), whereas the decay of a typical flare can be interpreted in terms of energy losses, until the energetic population is depleted (see e.g. ref [13]).

The rest of this paper is devoted to another type of system, where we also see both acceleration mechanisms taking place, and we will point to specific features associated with these acceleration mechanisms.

3 The Pulsar-Plerion Connection

Without discussing the details of pulsars, we can review the general features: Consider a magnetised neutron star with radius R (~ 10 km), surface magnetic field strength $B_o \sim 10^{12}$ G and spin angular frequency $\Omega = 2\pi/P$ (with spin period $P = 0.001$ s to a few seconds). The strong dipolar magnetic field results in a corotating magnetosphere out to the light cylinder radius $r_L = c/\Omega$, which defines a volume of "closed- and open" field lines, with the latter closing outside the light cylinder and defining an "open" field line vacuum potential drop of

$$\Delta V_{\rm pc} = \frac{B_o R^3 \Omega^2}{2c^2} = 3 \times 10^{12} B_{12} R_6^3 / P^2 \text{ volt.} \quad (1)$$

For the last expression we have normalised the surface field strength in units of 10^{12} G and R in units of 10^6 cm (or 10 km). It is this potential drop over the "polar cap" (corresponding to the "open" field line region), which can in principle accelerate charged particles to energies well above a TeV. We will first discuss the current through the polar cap before returning to the discussion on the fate of these accelerated particles.

From Maxwell's equations we can derive the current of primary particles extracted from the surface of the neutron star. This is called the "Goldreich-Julian" current [7]:

$$I_{\rm GJ} \sim \frac{B_o R^3 \Omega^2}{2c}, \quad (2)$$

which is flowing from the polar cap along the open field line region to $r > r_L$. It is also interesting to note that the product of $\Delta V_{\rm pc}$ and $I_{\rm GJ}$ is close to the total energy loss rate due to magnetic dipole radiation and also the spindown power $-I\Omega\dot{\Omega}$ (with $I \sim 10^{45}$ g.cm^2 the moment of inertia). This spindown power is usually more than the total solar output for pulsars younger than about 100,000 years.

Particles of equal charges must be extracted from the neutron star surface, otherwise the surface will develop a charge imbalance. Typically electrons and iron nuclei will be extracted, with 56 electrons for each iron nucleus. Both species will be accelerated along the open field lines towards the light cylinder, but a screening effect will take place: Electrons will radiate high energy γ-rays through curvature radiation, but it should be noted that the maximum photon energy emergent from the magnetosphere is well defined by the shape of the cross section for the conversion of gamma-rays to $e^{+/-}$ pairs in the superstrong magnetic fields. A pair-photon cascade results above the polar cap until the gamma-rays can escape without further pair creation, in which case the observable photon energies (Figure 1) will be below threshold for further pair creation [8]. It was shown that this maximum energy is typically in the 5 to 30 GeV range, and a hard photon spectrum will be seen below this energy. The spectral shape will also depend on the cascade multiplicity, and the hardest spectra will be associated with the smallest multiplicities. Whereas young pulsars are associated with large multiplicities, the mulitplicity associated with older pulsars will be less. Nel & de Jager found that the spectra of pulsars harden with increasing age, consistent with this interpretation. They also found that the observed cutoffs are in the 5 to 30 GeV range [14].

Figure 1 shows a scatterplot of pulsed gamma-rays from the Vela pulsar (PSR B0833-45) as observed by the EGRET instrument on the Compton Gamma Ray Observatory [15]. This pulsar is one of the youngest in the Southern Hemisphere, and will be a prime target for the H.E.S.S. telescope in Namibia (see Steenkamp [16] for a review of the H.E.S.S. experiment). Two sharp clusters is seen per period with a phase separation of $\sim$ 150 degrees, and the events from these peaks have much higher energies compared to the uniform background events seen towards phase 1.0. The spectrum cuts off near 10 GeV due to magnetic pair creation, and pulsed events should be marginally detectable for the H.E.S.S. experiment for which triggers above 10 GeV is possible.

The net effect is that the pair plasma will result in a screening effect, so that only some fraction of $\Delta V_{sc} < \Delta V_{pc}$ is available for acceleration to energies $q\Delta V_{sc} < q\Delta V_{pc}$ above the polar cap (the subscripts "sc" and "pc" refer respectively to the space-charge limited- and full vacuum polar cap potentials).

The pair plasma (after losing a fraction of their energy due to gamma-radiation as seen in Figure 1), together with the iron nuclei, will escape from the open field line region to $r > r_L$, forming a pair plasma out to a parsec scale. This is what is called a "plerion" (the Greek for "wind bag"), but the pulsar current must eventually close outside the light cylinder in this plerion.

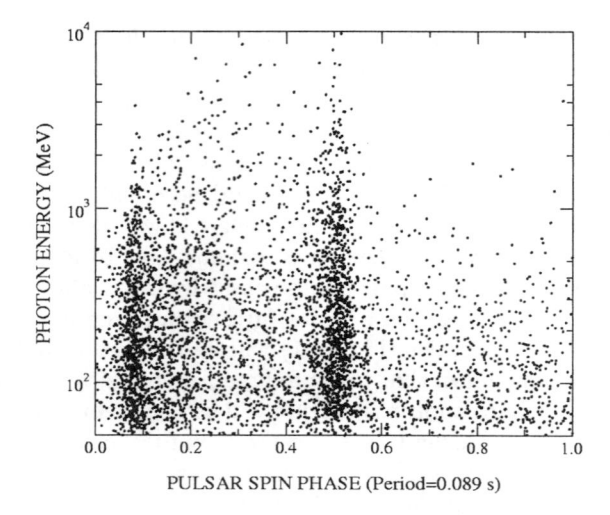

Figure 1. A scatterplot of the energy per photon vs the arrival phases of gamma-rays as measured by the EGRET instrument on the Compton Gamma-Ray Observatory. The phases were obtained by folding the arrival times at the solar system barycentre with the correct timing parameters.

The electrodynamics of this pulsar/plerion system is extremely difficult. In fact, nobody has been able to close the pulsar current self-consistently, and the best way I can explain the pulsar circuit is to use an electronic circuit analogy: $\Delta V_{\rm pc}$ is equivalent to the total available vacuum EMF (electromotive force), which is short-circuited to some degree by the "internal resistance" $R_{\rm m}$ resulting from the $e^{-/+}$-photon cascade above the polar cap. The subscript "m" refers to "magnetospheric". The actual space charge potential drop available for acceleration in the closed field line region is then $\Delta V_{\rm sc} = \Delta V_{\rm pc} - I_{\rm GJ} R_{\rm m}$. This, in turn, is equal to $I_{\rm GJ} R_{\rm w}$, where $R_{\rm w}$ is the resistance corresponding to the parsec-scale plerion, since the pulsar current must close in the plerion. Here the subscript "w" refers to the "wind" resistance. Figure 2 shows a schematic diagram of the pulsar current circuit.

This discussion brings to mind the claim of Harding & Gaisser [12], which states that the maximum energy which a charged particle can attain in the pulsar wind shock is equal to the charge times the full polar cap potential. The authors are actually referring to the potential drop $\Delta V_w = I_{\rm GJ} R_w = \Delta V_{\rm sc}$ over the plerion, and for typical Crab-like pulsars, it can be shown that the space-charge potential is comparable to the vacuum potential, whereas older pulsars should give smaller plerionic potentials given the more significant effect of screening, which is consistent with the observation that almost the total spindown power is converted to particles and radiation inside the light cylinder in the case of the older pulsars.

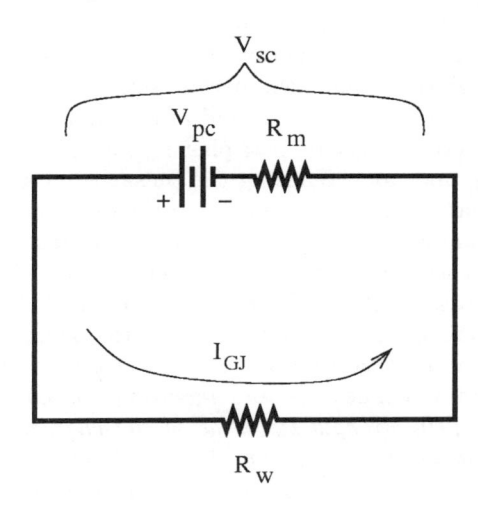

Figure 2. An equivalent circuit diagram of the pulsar current as explained in the text.

The total spindown power of a pulsar can then be written as the sum of two terms:

$$-I\Omega\dot{\Omega} = I_{GJ}^2 R_m + I_{GJ}^2 R_w, \qquad (3)$$

where the first term corresponds to the pulsed component (Figure 1) from a region the size of a few hundred kilometers (at $r < r_L$), whereas the second component corresponds to the unpulsed plerionic component from a parsec scale size, which is typically resolved by high resolution optical/X-ray instruments. In the case of young pulsars, the screening effect is less severe, in which case the pulsar converts almost all its spindown power into plerionic emission. This is the case for the 950-year old Crab Nebula, which is so bright that it acts as the standard candle for X-ray and Gamma Ray Astronomy. Older pulsars, on the other hand, convert almost all their spindown power into pulsed emission, so that the magnetospheric screening effect is more severe, with the consequence that the plerionic contribution is less dominant.

There is growing evidence that the iron nuclei accelerated from the neutron star surface, and entering the plerion, is responsible for the generation of waves in the pulsar-wind shock interface between the pulsar and the parsec-scale plerion. These waves are then responsible for the further gyroresonant acceleration of electron/positron pairs from the pulsar [17]. (At the pulsar wind shock we find that the ram pressure from the pulsar wind balances the ambient pressure in the rest of the plerion, resulting in a standing shock). It is believed that through some unknown mechanism, a significant fraction of the electromagnetic energy density in the wind zone (outside the light cylinder) is converted to the electron/positron pair plasma, thus raising the mean energy of the pair plasma, and upon reaching the pulsar-wind shock, the "heated" electrons can resonate with the ion-mediated waves to much higher energies. The acceleration timescale for such electrons in the relativistic pulsar-wind shock can be as fast as the gyroperiod of the electron. If the system is young enough such as the case for the Crab Nebula, the magnetic field strength at the shock will be large enough for synchrotron losses to dominate other loss mechanisms during the acceleration phase. Balancing the acceleration timescale and the synchrotron loss timescale, and *rewriting the maximum electron energy and magnetic field strength in terms of the characteristic synchrotron energy, we arrive at an expression containing only fundamental constants of nature, which equals 25 MeV.* [18] Furthermore, the acceleration process at the pulsar-wind shock is stochastic, resulting in a broad band spectrum of particles and synchrotron radiation, with the maximum radiation energy extending up to 25 MeV as mentioned above.

Figure 3 shows the synchrotron surface brightness distribution of the accelerated component of the pair plasma in the plerion beyond the pulsar wind shock in the case of the Crab Nebula. These electrons are convected away from the pulsar-wind shock and move towards larger radii while radiating and losing energy. The consequence is that the extended source assumes a finite size, which shrinks towards higher energies due to the quadratic energy dependence of synchrotron losses on the electron energy. The particle-in-cell simulation [19] procedure followed in the calculation of Figure 3 matches the observed X-ray surface brightness, and the synchrotron cutoff seen at 25 MeV for the Crab Nebula [18] is consistent with the maximum acceleration expected for gyroresonant acceleration. *The detection of the spectral feature at 25 MeV in the Crab Nebula is perfectly consistent with the acceleration of electrons and positrons on a timescale of the gyroperiod as expected for relativistic shocks, and we have an electron feature, even though the acceleration process may have been mediated by a nuclear process.*

Figure 4 is similar to Figure 3, except that the pairs also inverse Comp-

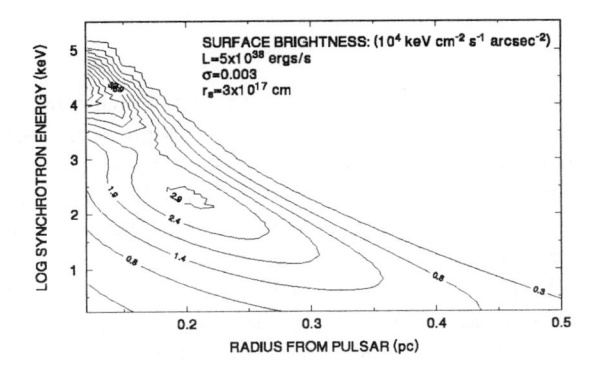

Figure 3. The synchrotron surface brightness distribution of the Crab Nebula calculated from a MHD flow solution (field strength and velocity vs. radius) which matches the boundary conditions of the Crab Nebula's expansion velocity and an input power law spectrum which matches the surface integrated X-ray spectrum.

ton scatter soft photons to very high gamma-ray energies. Note that one requires arcsecond resolution to confirm Figure 4, and the 0.1 degree resolution gamma-ray telescopes will only see a point source, given the source in Figure 4, placed at a distance of 2 kpc. However, it was shown that the spatially integrated surface brightness distribution from Figure 4 matches the total observed spectrum of gamma-ray emission out to 50 TeV [19].

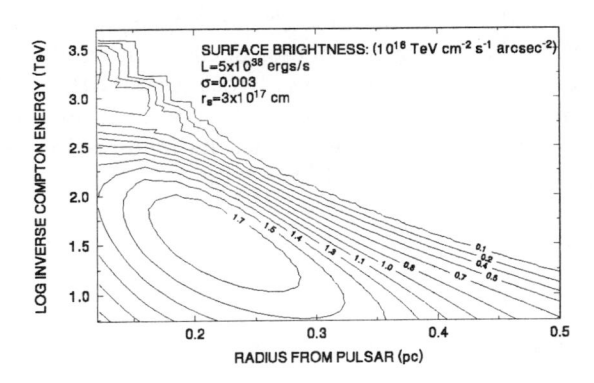

Figure 4. The directly calculated surface brightness distribution for inverse Compton scattered very high energy gamma-rays using the normalisation parameters for Fig. 3. See text.

4 Conclusion

We have discussed the composition of cosmic rays, which requires acceleration of both electrons/positrons and nuclei to ultra high energies. This field of study is extremely wide, requiring a discussion of several types of sources of cosmic rays and the physics involved. I have chosen to discuss only the basic features of particle acceleration, after which we have concentrated on pulsar-plerion systems, which show evidence of both electric field- and shock acceleration. Whereas direct electric field acceleration is confined to a small dimension of a few hundred kilometer, it is clear that the detection of pulsed gamma-rays from pulsars allows us to probe such a small volume and the physics associated with it.

We have also shown how one can (schematically) close the pulsar current, and the rest of the rotational energy loss of the pulsar is spent in an extended region outside the pulsar on a scale size of a few light years. In this way we have discussed the roll which iron nuclei may play in the acceleration process of electrons and positrons in a shock, even though only the electron pairs are directly seen via their radiation. The latter is seen in both syncrotron- (given a magnetic field) and inverse Compton radiation (given soft target photons), and by combining this multiwavelength spectral/spatial information, we are able to test the validity of magneto-hydrodynamic models for the expansion of the relativistic plasma in the plerionic environment [19].

Finally, future ground-based experiments such as MAGIC and H.E.S.S. may be able to detect the tail of the pulsed spectrum (terminated by magnetic pair production) under ideal conditions [20], since space-borne observations and theory have shown that the pulsed photon spectra cut off above 5 to 30 GeV.

It should be mentioned that the study of pulsars with H.E.S.S. will constitute only a minor fraction of its total observational programme, and the full power of H.E.S.S. will be realised when its high-resolution imaging and spectroscopic capability is exploited [16].

Acknowledgments

The author and the Unit for Space Physics of the Potchefstroom University for CHE would like to thank the South African Department for Arts, Culture and Science for their contribution towards the H.E.S.S. project in Namibia.

References

1. E.G. Berezhko and L.T. Ksenofontov, Proc. 26th Int. Cosmic Ray Conf. **3**, 381 (1999).
2. M.A.K. Glasmacher, et al., Astroparticle Physics **10**, 291 (1999).
3. S. Markoff, et al., Proc. 26th Int. Cosmic Ray Conf. **4**, 411 (1999).
4. K. Tang, The Astrophysical Journal **278**, 851 (1984).
5. S.A. Stephens, Proc. 26th Int. Cosmic Ray Conf. **3**, 241 (1999).
6. R.J. Protheroe, The Astrophysical Journal **254**, 391 (1982).
7. P. Goldreich and W.H. Julian, The Astrophysical Journal **157**, 869 (1969).
8. A.K. Harding, The Astrophysical Journal **245**, 267 (1981).
9. H. Moraal and W.I. Axford, Astronomy & Astrophysics **125**, 540 (1983).
10. O.C. de Jager and A. Mastichiadis, The Astrophysical Journal **482**, 874 (1997).
11. M. Baring, Proc. 26th Int. Cosmic Ray Conf. **4**, 5 (1999).
12. A.K. Harding and T.K. Gaisser, The Astrophysical Journal **358**, 561 (1990).
13. L. Maraschi, G. Ghisellini, and A. Celotti, The Astrophysical Journal **397**, L5 (1992).
14. H.I. Nel & O.C. de Jager, Asrtophysics & Space Science **230**, 299 (1995).
15. G. Kanbach et al., Astronomy & Astrophysics **289**, 855 (1994).
16. R. Steenkamp, *These Proceedings - Luderitz2000*, ed. R. Tegen & S. Connell, 2001.
17. Y.A. Gallant and J. Arons, The Astrophysical Journal **435**, 230 (1994).
18. O.C. de Jager et al., The Astrophysical Journal **457**, 253 (1996).
19. O.C. de Jager et al., *in Proc. of the International Symposium on Gamma Ray Astronomy*, ed. F. Aharonian & H. Völk, AIP, 2001a.
20. O.C. de Jager et al., *in Proc. of the International Symposium on Gamma Ray Astronomy*, ed. F. Aharonian & H. Völk, AIP, 2001b.

5. Atomic and Nuclear Physics in the Study of Diamond

HYDROGEN MOBILITY IN DIAMOND

S. KALBITZER

Ion Beam Technology – Consulting

Bahofweg 2, D-69121 Heidelberg, Germany
Fone/fax: +49 6221 474396
E-mail: SKalbitzer@aol.com

Due to the high Debye temperature of diamond, C_D, the hydrogen isotope ^1H is expected to undergo quantum-mechanical transport up to elevated temperatures. Based on detailed experimental and theoretical information on the $^1H/Si$ system the transport parameters for $^1H/C_D$ are predicted in both the high-temperature limit and the few-phonon regime.

1. Introduction

Hydrogen is known as a fast diffusor in many solid matrices, in particular the lightest isotope ^1H. A critical temperature of a fraction of the Debye temperature, $T_c \sim \theta_D/3$, may be chosen to separate (quasi)classical from quantum-mechanical transport. The scheme of fig.1 is a qualitative visualization of this behavior.[1] While in the many-phonon regime at $T > T_c$ data can be analyzed in an Arrhenius plot, where $\log D \propto 1/T$, a double logarithmic plot is appropriate for the few-phonon regime at $T < T_c$, where $D \propto T^n$.

Because the system $^1H/Si$ has been explored in considerable detail by both experiment and theory, it appears feasible to employ similarity relations in order to predict ^1H diffusion in the homologue system of diamond, C_D, where much less information is available. Apparently, around room temperature H solubility in good material is low and its mobility high, as it is in Si. While bulk concentrations appear negligible, H in surface positions has been detected, where it acts as a dangling bond terminator.[2]

2. Information background from $^1H/Si$

A detailed experimental study of $D(^1H)$ in Si has shown $T_c \sim 200$ K, which is in astonishingly good agreement with the above demarcation estimate, since $\theta_D \sim 640$ K.[3,4] Figs.2 and 3 show these two transport regimes, a steep high-temperature section and a flat low-temperature one, in their respective plots.

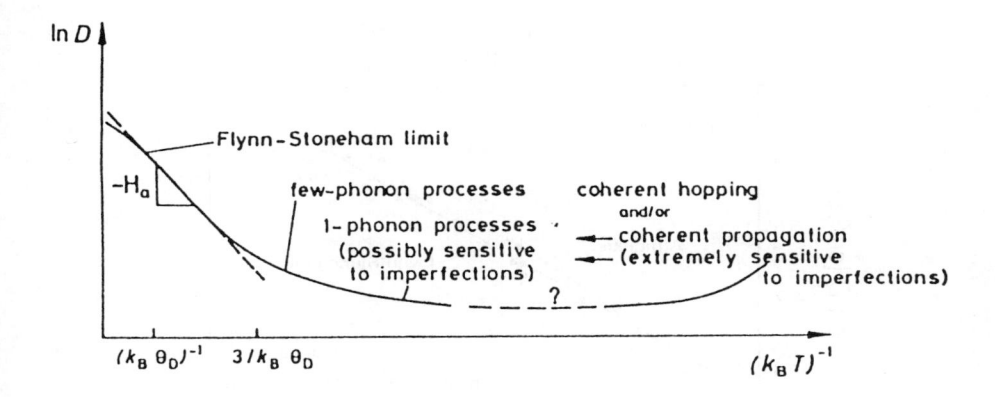

Fig. 1: Temperature dependence of the "small polaron" diffusion.[1]

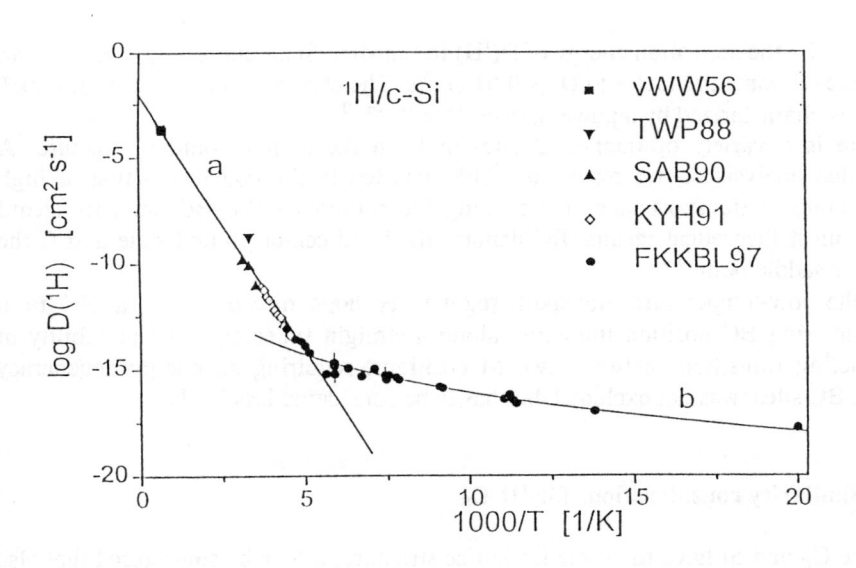

Fig.2: Arrhenius plot of D(T) of ^{1}H/Si.[3] Line a: classical regime; curve b: tunneling transport. Transition at $T_c = \theta_D/3$.

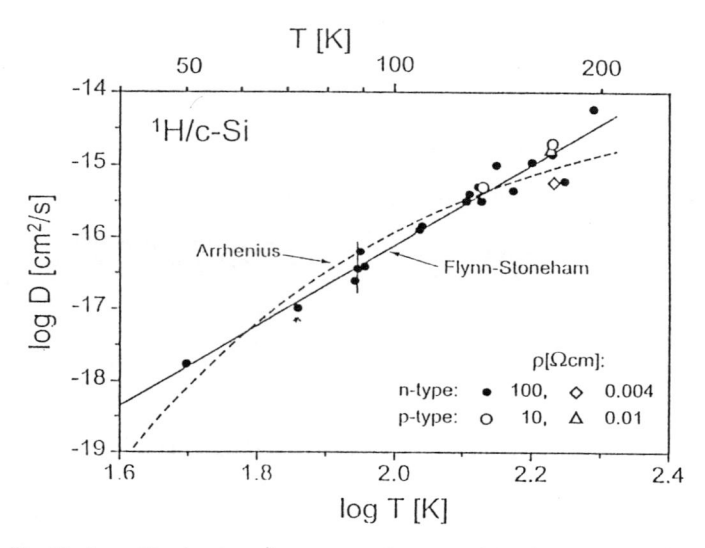

Fig.3: Double logarithmic plot of curve b of fig.2. Straight line: $D(T) \propto T^n$, $n = 5.6 \pm 0.3$; dashed curve: fit to curve b of fig.2 by a straight line.

At $T > T_c$, the activation energy of $D(^1H)$ in trap-free Si amounts to $Q \approx 0.5$ eV and the pre- exponential factor to $D_0 \sim 0.01$ cm^2/s. The observed hopping transport at $T < T_c$ is characterized by a power law of $D \propto T^{5.6 \pm 0.3}$.

There is a variety of interstitial sites of H in the diamond-lattice structure. A detailed analysis of our results on ^{1}H/Si has led to the conclusion that at high temperatures transport takes place along the cornered BC-C-BC path, in accord with most theoretical results. BC denotes the bond-center ground state and C the corner saddle point.[3,4]

In the low-temperature transport regime ^{1}H hops directly from a BC to a neighboring BC position tunneling along a straight trajectory. The possibility of tunneling transitions between two M positions, requiring an energy degeneracy with BC sites, was not excluded, but has to be considered less likely.[4]

3. Similarity considerations for ^{1}H/C$_D$

Since C_D and Si have the same fcc lattice structure, it may be conjectured that also the transport mechanisms of ^{1}H/C$_D$ follow this similarity. In particular, it is reasonable to assume that the energetically lowest position of ^{1}H is again the BC lattice site, where the electron density is highest.

In C_D, the transition temperature, however, is expected to be located at elevated temperatures, around $T_c \sim 700$ K, since $\theta_D \sim 2200$ K. As a consequence, phonon assisted tunneling should be observable in C_D already at room temperature. Again, a temperature dependence of $D \propto T^n$, with $n \sim 5 - 7$, is expected according to the Flynn-Stoneham formalism for two-phonon processes, as presented in the appendix section.[5] The assumption of $n = 5.6$ for ^{1}H/C_D is not very critical, since a deviation by $\Delta n \sim 1$ would lead to tolerable errors in $D(T)$, e.g. at $T_c = \theta_D/3$ to a factor of 3.

Assuming the same transport paths, the corresponding transport parameters of ^{1}H/C_D should scale with the respective characteristic lattice parameters, such as distance, potential and frequency. In the appendix some detailed considerations are given. Tab.I is a list of the relevant parameters for both systems.

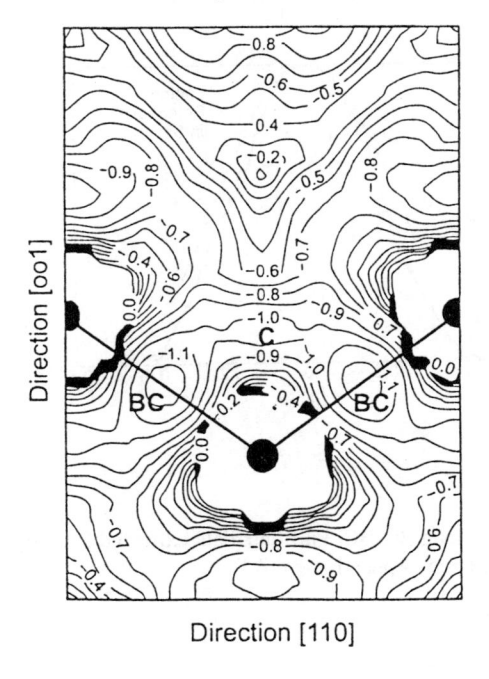

Fig.4: Equi-potential contours for ^{1}H/Si around the interstitial site at the bond-center site BC.[6] Classical jump over the barrier along BC-C-BC; inelastic tunneling along BC-BC.

The next step is to estimate the form factor γ for the tunnel barrier in ^{1}H/C_D. In ^{1}H/Si we derived $\gamma = 0.52$ from the experimental data on the tunneling transport. This value is distinctly lower than the expected value of $\gamma = 0.637$ for a simple sinusoidal shape as given by eq. A1 in the appendix section. There we derive $\gamma \sim 0.5$ with consideration of the zero point energy for the coincidence state.

Fig.5 shows the barrier profile of the BC-BC tunnel transition in ^{1}H/Si as extracted from the theoretical work of Herrero and Ramirez.[6] A good fit by the generalized sinusoidal function of eq. A4 is obtained with an exponent n ≅ 1.5. The corresponding form factor is γ = 0.56 without zero-point energy correction. Assuming a zero-point energy of about 10% of the barrier height we find a correction to γ ~ 0.50, in excellent agreement with the experimental figure of γ = 0.52. Fig.6 shows how important the consideration of the zero-point energy is.

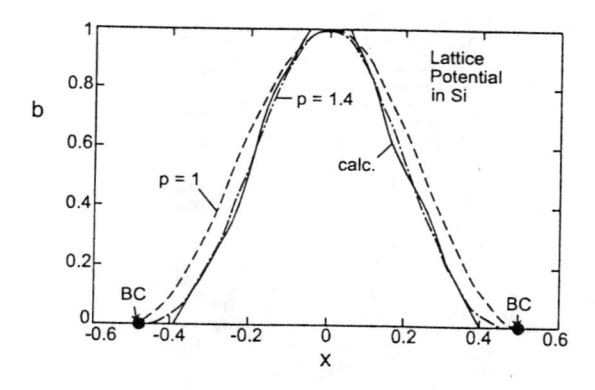

Fig.5: Potential profile along BC-BC. Full line: data extracted from fig.4; dashed line: fit with a sinusoidal function (Appendix: eq. A4).

It is obvious that the potential barrier for high-temperature diffusion cannot be higher than the tunnel barrier plus the polarization energy, as described in the appendix by eq. A8.

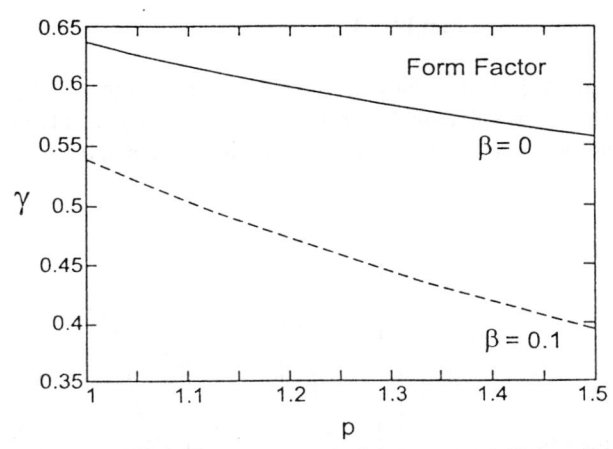

Fig.6: Tunneling form factor γ vs potential exponent p (Appendix: eq. A5).

With the further conditions that the Arrhenius prefactor amounts to $D_0 \sim 0.015$ cm^2/s, that classical diffusion D matches hopping transport H at $T_c = \theta_D/3$, that hopping at room temperature is as high as $H \sim 10^{-13}$ cm^2/s we obtain further constraints on the choice of the parameters Q, B, U, and γ. Thus, the predicted parameter set for ^{1}H/C$_D$ is Q = 1.3 eV, B = 1.2 eV, U = 0.2 eV and γ = 0.45.

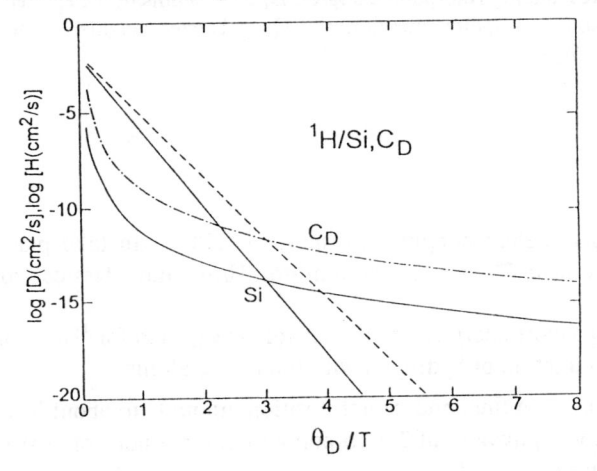

Fig. 7: Comparison of the diffusion systems ^{1}H/Si and ^{1}H/C$_D$ in an Arrhenius plot. At room temperature, equivalent to reciprocal reduced temperatures of $1/\tau = \theta_D/T = 2.13$ and 7.33 in Si and C$_D$, respectively, diffusion is over-the-barrier in Si and tunneling in C$_D$.

4. Comparison of diffusion in ^{1}H/Si and ^{1}H/C$_D$

Fig.7 compares the measured data of ^{1}H/Si with the calculated data for ^{1}H/C$_D$. The tunneling diffusion of ^{1}H/C$_D$ was calculated according to the theory presented in the Appendix.[6,7] A complete list of the used parameters is given by tab.I.

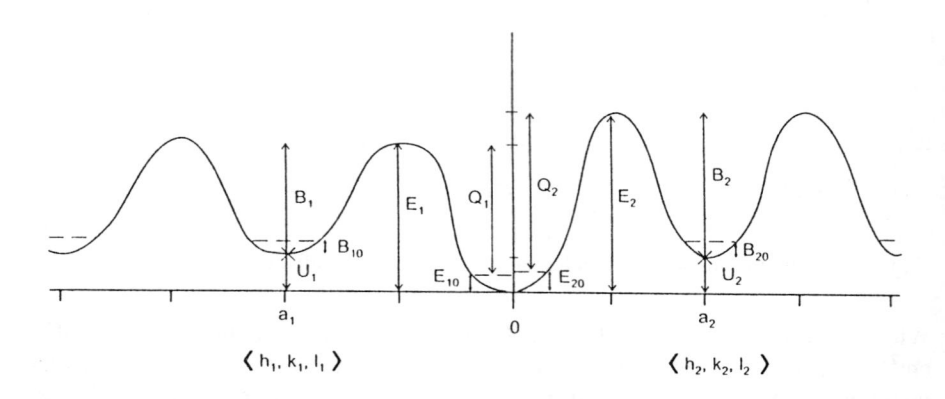

Fig.8: Energy levels of ^{1}H in different lattice positions: E_1, E_2 = depths of the relaxed states, E_{10}, E_{20} = corresponding zero-point energies, B_1, B_2 = depths of the excited states, B_{01}, B_{02} = corresponding zero-point energies, Q_1, Q_2 = barrier heights in the lattice directions <h_1,k_1,l_1>, <h_2,k_2,l_2>.

5. Conclusions

With the proposed parameters hopping transport of ^{1}H/C$_D$ can take place up to reduced temperatures near $T_c = \theta_D/3$, similar to ^{1}H/Si, and classical diffusion above.

The corresponding transport energies are higher for ^{1}H/C$_D$ than for ^{1}H/Si, probably due to the stronger interaction of hydrogen with the lattice atoms.

The binding energy of ^{1}H in the bond-center position amounts to about $\hbar\omega_D$ in both C$_D$ and Si, which is the equivalent of 2 zero-point energy phonons of ½·0.19 eV in C$_D$ and ½·0.055 eV in Si.

Appendix: Transport theory

The estimation of quantum mechanical transition rates start with the assumption of a simple lattice potential. For a sinusoidal potential of the form

$$f(x) = \tfrac{1}{2}B\,[1+\cos(2\pi x/a)], \qquad -a/2 \leq x \leq a/2, \tag{A1}$$

Kehr quotes for the tunneling matrix element:[7]

$$J = 8B\chi^{-1/4}\exp(-\gamma\chi^{1/2}), \quad \chi = 2ma^2B/\hbar^2, \tag{A2}$$

where B = barrier height, a = tunneling distance, and m = mass of the diffusor; the form factor amounts to $\gamma = 2/\pi$ for the above potential, but varies noticeably with its shape.
The zero-point energy of the particle in the above potential well is obtained from eq. A1 as:

$$B_0 = \tfrac{1}{2}\hbar\omega = \hbar\pi B^{1/2}/a(2m)^{1/2} = 0.142 B^{1/2}/a \tag{A3}$$

with units of energy in eV and of distance in Å. The concomitant effect of $B_0 > 0$ is to enhance the tunneling probability by reducing the form factor γ in the exponent of eq. A2.
For fitting realistic lattice potentials, as the one shown in fig. 5, the necessary modification of eq. A1 is simply achieved by a generalized sinusoidal function:

$$b(x,p) = \{\tfrac{1}{2}[1+\cos(2\pi x/a)]\}^p, \tag{A4}$$

where $b(x,p) = f(x,p)/B$.

Its effect on the form factor γ of the tunneling integral is:

$$\gamma(p) = \int [b(x,p)^{1/2} - b(x_0,p)^{1/2}]dx, \tag{A5}$$

with $x_0(p) = \pm a\,\cos^{-1}[2(B_0/B)^{1/p} - 1]/2\pi$ as upper and lower integration limits, respectively.
The effect of these zero-point energy corrections on γ are displayed by fig.6.
For the tunneling path BC->BC of ^{1}H/Si, with $p \approx 1.5$, we obtain $\gamma \sim 0.64$ without and $\gamma \sim 0.50$ with a zero-point energy ratio of $B_0/B \sim 0.10$. This reduction of the

tunneling exponent by about 20% produces an increase of J by orders of magnitude, and even much more of the transport coefficient since $D \propto J^2$. Thus, this parameter is also of crucial importance for predicting the hopping transport properties of $^1H/C_D$. With our actual choice of $p = 2$, corresponding to a steeper lattice potential of $1H/CD$, the form factor reduces to $\gamma \sim 0.45$.

The temperature dependence of the hopping transport coefficient H in the tunneling regime is by a power law reflecting the operating phonon process. For the low temperature regime of $T < \Theta_D/3$ Flynn and Stoneham specified:[5]

$$H = g\, a^2\, \Gamma(T),$$
$$\Gamma(T) = AT/\Theta_D)^n, \qquad (A6)$$
$$A = 57600\pi\omega_D J^2 U^2 \exp(-5U/\hbar\omega_D)/(\hbar\omega_D)^4.$$

The symbols mean: g = geometrical factor, a = tunneling distance, Γ = tunneling frequency, $\omega_D = k\Theta_D/h$ = Debye frequency, Θ_D = Debye temperature, U = coincidence energy, J = tunnel transition matrix element, n = order of phonon process, e.g. n = 5 - 7 for a two-phonon process.

For the high temperature branch, usually presentable by an Arrhenius relation, we use the traditional equation:

$$D(T) = ga^2 v_D \exp(-Q/kT), \quad v_D = \omega_D/2\pi, \qquad (A7)$$

Q is the activation energy obtained from an Arrhenius plot of D(T); the prefactor $D_0 = ga^2 v_D$ scales with distance and frequency from 0.01 in Si to 0.015 in C_D.

The next step is to match the two transport regimes at the transition temperature, $T_c = \Theta_D/3$. The following relation between the activation energies Q_1 and Q_2 along the lattice directions 1 and 2 is useful for estimating limits for these quantities for different transport paths:

$$Q_1 \leq B_2 + U - B_{20}, \qquad (A8)$$

where the zero point energy level is at $B_{20} \sim 0.1 B_2$ and the energy difference is $U = U_1 = U_2$ of the relaxed and excited particle state, respectively. Fig.8 is a scheme of these relations. Thus, depending on the value of U, Q_1 can be somewhat larger, but also smaller than B_2, by up to the estimated order of 100 meV.

With the proposed values of tab.I, i.e. $B_2 \sim 1.2$ eV and $U \sim 0.2$ eV with an uncertainty of about 10%, there is reasonable agreement with the upper limit on $Q_1 \sim 1.3$ eV set by eq. A8.

Tab. I: Transport parameters for ^{1}H in Si and C_D

Parameter	Si	C_D
a (Å)	1.92	1.26
θ_D (K)	640	2200
ω_D (s^{-1})	$0.84 \cdot 10^{14}$	$2.88 \cdot 10^{14}$
ε (eV)	0.055	0.190
U (eV)	0.055	0.2
B (eV)	0.57	1.2
γ	0.52	0.45
J (eV)	$\sim 4 \cdot 10^{-8}$	$\sim 1 \cdot 10^{-5}$
N	5.6	5.6
D_0 (cm^2/s)	0.01	0.015
Q (eV)	0.57	1.30
D_{Tc} (cm^2/s)	$\sim \cdot 10^{-14}$	$\sim 10^{-12}$
D_{300K} (cm^2/s)	$\sim 10^{-11}$ *	$\sim \cdot 10^{-14}$ **

* over-the-barrier diffusion

** tunnel diffusion

References

1. A. Seeger, in *Proc. Int. Conf. On Diffusion in Metals and Alloys*, edited by F. J.Kedves and D. L. Beke (Trans Tech Publications, Aedermannsdorf, 1983), p. 39
2. S. Jans, S. Kalbitzer, P. Oberschachtsiek, and J. P. F. Sellschop, Nucl. Instrum. Methods Phys. Res. B **85**, 321 (1994)
3. Ch. Langpape, S. Fabian, Ch. Klatt, and S. Kalbitzer, Appl. Phys. A: Mater. Sci. Process. **64**, 207 (1996)
4. S. Fabian, S. Kalbitzer, Ch. Klatt, M. Behar, and Ch. Langpape, Phys. Rev. B **58**, 16144 (1998)
5. C. P. Flynn and A. M. Stoneham, Phys. Rev. B **1**, 3966 (1970)
6. C. P. Herrero and R. Ramirez, Phys. Rev. B **51**, 16761 (1995)
7. K. Kehr, in *Hydrogen in Metals I*, edited by G. Alefeld and J. Völkl (Springer Verlag, Berlin, 1978), p. 203

6. Applications of Pure and Applied Physics in Technology

TUMOR THERAPY WITH HIGH-ENERGY HEAVY-ION BEAMS

D. SCHARDT

for the Heavy-ion Therapy Collaboration*

Gesellschaft für Schwerionenforschung, Biophysics Division,
Planckstr.1, D-64291 Darmstadt, Germany
E-mail: d.schardt@gsi.de
*GSI Darmstadt, FZ Rossendorf, Radiological Clinic and DKFZ Heidelberg

Heavy-ion beams offer favourable conditions for the treatment of deep-seated local tumors. The well defined range and the small lateral beam spread make it possible to deliver the dose with millimeter precision. In addition, heavy ions have an enhanced biological efficiency in the Bragg peak region which is caused by the dense ionization and the resulting reduced cellular repair rate. Furthermore, heavy ions offer the unique possibility of in-vivo range monitoring by applying Positron-Emission-Tomography (PET) techniques. Taking advantage of these clinically relevant properties, a therapy unit using ^{12}C beams with energies of 80-430 MeV/u was constructed at GSI. The fully active beam delivery system includes a magnetic raster scan device providing a high degree of dose conformation to the target volume while healthy tissue and radiosensitive structures are spared to a maximum extent. In the framework of a clinical study 68 patients have been treated since December 1997 with promising results so far. Plans for a dedicated heavy-ion treatment center at the Radiological Clinic Heidelberg will be further pursued.

1 Introduction

The application of high-energy beams of heavy charged particles to radiotherapy was first considered in 1946 when Robert R. Wilson († 2000) described the potential benefits of proton beams and predicted "...that precision exposures of well defined small volumes within the body will soon be feasible" [1]. Two years later the 184 inch synchrocyclotron at LBL Berkeley became available for experiments and the physical and radiobiological properties of proton beams were thoroughly investigated. In contrast to photon beams protons have a favourable depth-dose profile (Bragg curve) and the narrow Bragg peak can be precisely adjusted to the desired depth by the beam energy (200 MeV are needed to penetrate 25 cm of water). Ions heavier than protons like carbon in addition offer an enhanced biological efficiency in the Bragg peak region (Fig.1) which can be explained by the very dense ionization towards the end of the particle track.

Patient treatments started in 1954 at LBL Berkeley, first with protons and later with helium beams. Radiotherapy with heavier ions was initiated by Tobias et al. [2,3] at the BEVALAC facility at LBL. There most of the patient treatments (1975-1992) were performed with beams of ^{20}Ne (670 MeV/u) which at that time appeared to be most attractive because of their high relative biological efficiency (RBE) combined with a low oxygen enhancement ratio (OER) in the treatment target volume. The beams were delivered to the patient by passive beam shaping systems including a scatterer or wobbler magnets for broadening the beam and a number of passive elements like ridgefilter, range modulator, collimator and bolus [4]. Until its closure in 1992 the BEVALAC was the only facility worldwide using heavy ions for the treatment of localized deep-seated tumors.

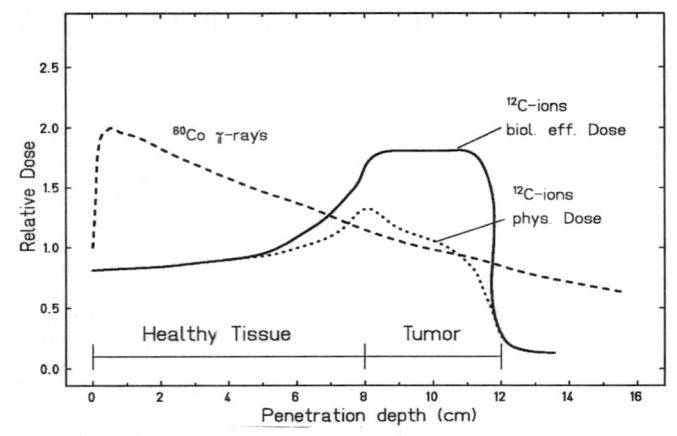

Figure 1. Depth-dose profiles of photon and carbon beams. The extended Bragg peak of carbon ions is generated by superposition of several Bragg curves with different particle energies. The increased biological effectiveness in the Bragg peak region is an advantage for the treatment of deep-seated tumors, especially in the vicinity of radiosensitive structures.

While the medical programme at the BEVALAC had to be integrated into the environment of a nuclear physics research facility, the first heavy-ion medical accelerator HIMAC [5] dedicated to radiotherapy started in 1994 at NIRS Chiba (Japan), using similar technical concepts as those pioneered at Berkeley.

At GSI Darmstadt (Germany) a new concept [6] was developed, differing significantly from the previous designs at the BEVALAC and HIMAC: moving a narrow pencil beam over the target volume (raster scan) a tumor conform treatment can be achieved to a high degree, restricting the biologically most effective ions to the target volume and minimizing the dose to the surrounding normal tissue. In spite of the demanding technical concept the fully active rasterscan system has proven to operate reliably since the first patient treatment in December 1997. A similar method of target conform proton irradiation has been developed at PSI (Switzerland) [7]. In the following, some basic properties of heavy-ion therapy beams are discussed and an overview of the heavy-ion therapy program at GSI is given.

2   Physical and radiobiological properties of heavy-ion therapy beams

The slowing-down process of heavy charged particles is governed by continuous interactions with the atomic shell of the absorber nuclei (tissue) and well described by the Bethe-formula. The resulting depth-dose profile (Bragg curve) which exhibits a flat plateau region and a distinct peak near to the end of range of the particles represents the major physical advantage of heavy charged particle beams in radiotherapy. For heavy ions, however, nuclear reactions along the penetration path may cause a significant alteration of the radiation field. At energies of several hundred MeV/u which are required for radiotherapy applications the most frequent nuclear interactions are peripheral collisions where the beam particles may loose one or several nucleons. The fragments continue travelling with nearly the same velocity and direction. These nuclear reactions lead to an attenuation of the primary beam flux and a build-up of lower-Z fragments with increasing penetration depth. As the range of the particles scales with A/Z^2 the depth-dose profile of heavy-ion beams shows a characteristic fragment tail beyond the Bragg peak (Fig.2).

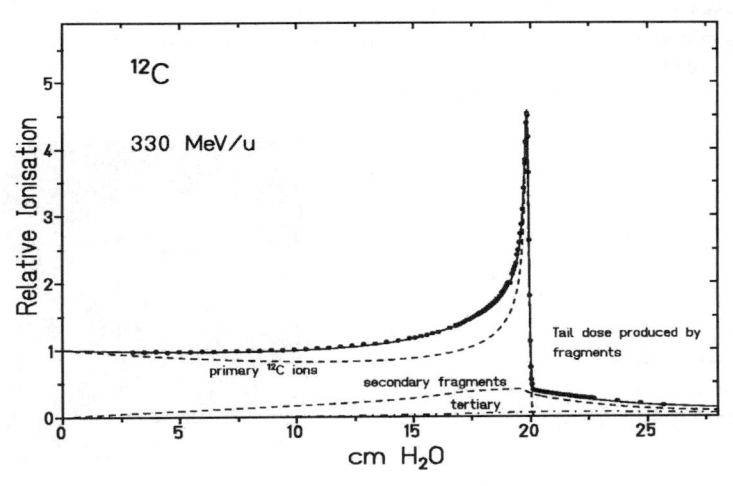

Figure 2. Bragg curve for a carbon beam in water measured at GSI with large parallel-plate ionization chambers. The data points are compared to a model calculation [8] (solid line) and the dashed lines indicate the calculated contributions from the primary particles and from nuclear fragments.

The production of nuclear fragments as a function of penetration depths in water and their angular and velocity distributions were experimentally investigated for ^{20}Ne beams at LBL Berkeley [9] and for various primary beams (^{10}B, ^{12}C, ^{14}N, ^{16}O, ^{20}Ne) [10,11] at GSI. Similar studies were performed at HIMAC/Chiba [12].

The radiobiological efficieny of charged particles is mainly characterized by their local ionization density which can directly be correlated to the local density of DNA damage. As a result of extensive irradiation experiments with cell cultures at LBL, NIRS and GSI it was found that carbon beams meet the therapy requirements best possible [13,14]. At high energies in the entrance region carbon ions have a sufficiently low ionization density and act like photons, producing mostly repairable DNA damage. Towards the Bragg peak the ionization density increases significantly, resulting in irrepairable damages and high cell killing power. These findings were confirmed by measuring molecular DNA damage and repair along therapeutic beams [15].

The treatment planning program [16] developed at GSI and DKFZ Heidelberg is based on a biological model [17] which takes into account the variance of RBE in a mixed particle field (incl. fragmentation). The physical dose is adjusted to the RBE by an iterative procedure in order to achieve a homogeneous biological effect.

3 In-situ range verification with PET-techniques

An interesting positive aspect of the nuclear fragmentation effect discussed above is the formation of shortlived positron-emitting isotopes which can be used for an in-situ monitoring of their stopping points by Positron-Emission-Tomograpy (PET)-techniques [18,19]. This yields an experimental verification of correct treatment planning and beam delivery, especially the monitoring of the penetration depths which is invaluable for treating tumors near critical structures. This feature is unique to heavy-ion beams as here the induced β^+-activity in tissue mainly stems from positron-emitting projectile fragments which have nearly the same end-of-range as the primary particles.

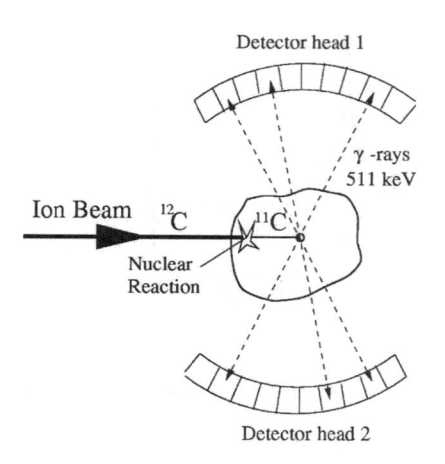

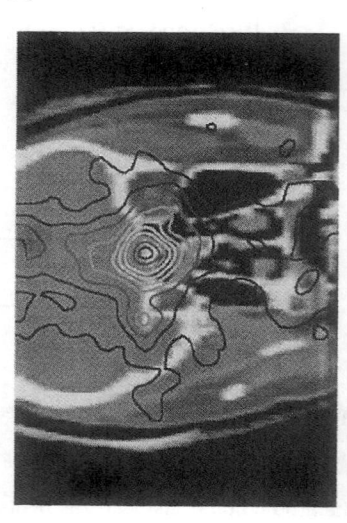

Figure 3. Left: Principle of the in-situ range verification by PET techniques. Right: Superposition of a measured β^+-activity distribution (contour lines) and the corresponding frontal slice of patient CT data. The carbon beam entered from the left side, the PET-detector heads were located vertically above and below the patient table.

The principle of the measurement is sketched in Fig.3. Along the penetration path in tissue the primary ^{12}C ions may undergo a nuclear reaction and continue travelling as ^{11}C fragments with about the same velocity. As they have the same nuclear charge they reach almost the same depth as the primary ions (slightly less because of the lower mass number). The spatial distribution of the β^+-activity of the ^{11}C ions can be obtained by coincident recording of the annihilation radiation in two opposite detector heads and applying tomographic reconstruction algorithms. The β^+-activity distribution is then compared to the expected distribution which is calculated based on the patient CT-data, the treatment plan and the actual irradiation conditions. Superposition of the measured and calculated β^+-activity distributions then reveals possible differences with an accuaracy of about 2.5 mm. This method has proven to be a valuable tool for the quality assurance of heavy-ion therapy.

4 The GSI treatment unit

In order to have a maximum benefit of the physical and biological advantages of heavy-ion beams for the treatment of deep-seated tumors the concept of the GSI treatment unit includes a fully active beam delivery system [20]. A pencil beam (typically 5 mm half-width) with well-defined energy and corresponding depth of the Bragg peak is moved by fast horizontal and vertical scanning magnets slice-by-slice over the target volume. By choosing appropriate steps in the beam energy the whole target volume is uniformly irradiated (Fig.4). In order to ensure that each pixel of the target slices receives the desired dose, the scanning system has to be intensity controlled and needs a feedback from a fast beam monitor. This implies great demands on the control and safety systems as well as strong requirements on the accelerator performance such as stability and reproducibility of the absolute beam position. In comparison with passive beam delivery, the active system has a clear advantage as there are no restrictions in shaping the target volume and any prescribed 3-dimensional dose distribution can in principle be generated. Furthermore, beam losses and contamination by nuclear fragmentation in passive beam shaping elements in front of the patient are minimized in active systems.

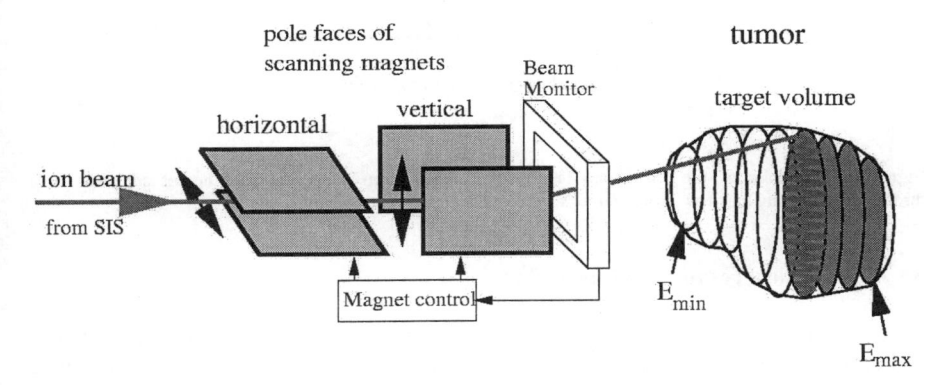

Figure 4. Principle of the magnetic scanning system at GSI. The target volume is irradiated by moving the ion beam (80-430 MeV/u ^{12}C) with fast magnets over each slice. The required beam energies - corresponding to the depth of the Bragg peak for each slice - are supplied on a pulse-to-pulse operation by the synchrotron (SIS) control system.

The construction of the therapy unit at GSI started in 1993 and included the treatment area (Cave M) with a new beamline, a control room, and a medical annex building (Fig.5). Within the Heavy-Ion Therapy Collaboration three other institutes take care of various parts of the project: The Radiological Clinic Heidelberg (all clinical aspects such as patient selection, diagnostic, dose calculation), the German Cancer Research Center DKFZ Heidelberg (patient immobilisation, treatment planning, dosimetry), and the Research Center FZ Rossendorf near Dresden (PET camera). The first patient was treated in December 1997. Since then 68 patients were treated, most of them with inoperable malignant tumors in the skull base.

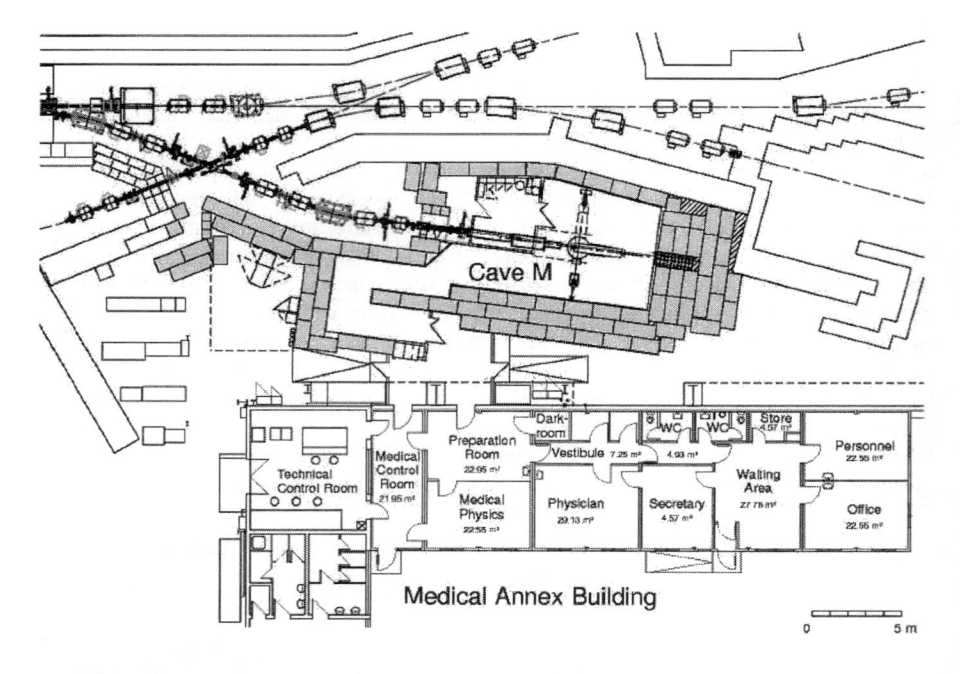

Figure 5. Groundplan of the GSI therapy facility. The carbon beam is entering from the left side and passes through the scanning magnets located 6m upstream of the target position.

5 First clinical results and future plans

First clinical results [21] were reported for 45 patients with chordomas, chondrosarcomas and other skull base tumors treated at the GSI therapy unit. Most of these patients received a fractionated carbon ion irradiation in 20 consecutive days with a median total dose of 60 GyE which was well tolerated. The one-year local control rate was 94%. As can be seen from a typical treatment plan shown in Fig. 6, the target volume often is close to the brain stem or the optical nerves, which can be spared very well by the precision irradiation.

Encouraged by the positive clinical results obtained in this pilot project the plans for a dedicated hospital-based ion treatment facility [22] are further pursued. The proposed facility to be built near the Radiological Clinics and DKFZ in Heidelberg is planned for the treatment of 1000 patients per year and includes three treatment rooms, one of which is equipped with a rotating gantry. With such a device the ion beam can be delivered from any angle to the patient table. Simlilar facilities are planned in Austria (Med-Austron), France (Lyon) and Italy (TERA foundation). Most efforts in the field of particle therapy are presently made by Japan. Following the dedicated carbon-ion treatment facility HIMAC, in operation since 1994, the new combined proton/ion facility at Hyogo will probably start operation in 2001, and other facilities are under construction or being planned.

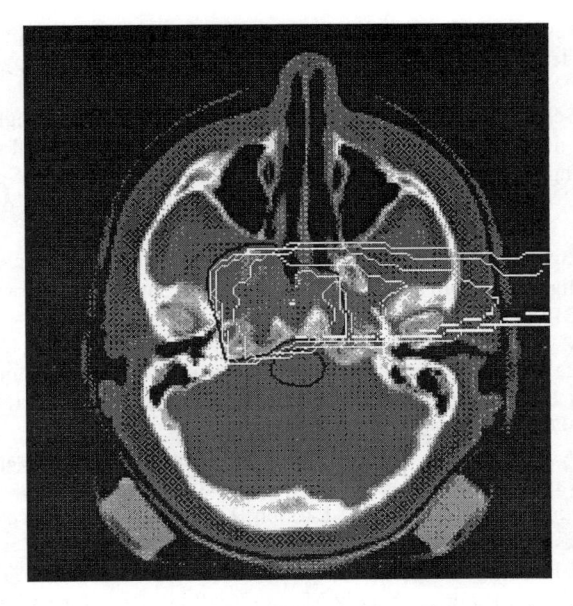

Figure 6. Carbon-ion treatment plan for a large tumor in the skull base. The target volume is close to the brain stem and optical nerve which can be spared very well by the carbon ion irradiation.

6 Acknowledgements

I would like to thank all colleagues of the heavy-ion therapy project for the very fruitful collaboration and in particular Gerhard Kraft for his enthusiasm and many stimulating discussions during all phases of the project.

References

1. Wilson R. R., Radiological use of fast protons. *Radiology* **47** (1946) pp.487-491.
2. Tobias C. A. and Todd P.W., Heavy charged particles in cancer therapy, In *Radiobiology and Radiotherapy*. Institute Monograph No.24 (1967).
3. Petti P. L. and Lennox A. J., Hadronic radiotherapy, *Ann. Rev. Nucl. Part. Sci.* **44** (1994) pp.155-197.
4. Chu W.T., Ludewigt B. A. and Renner, T. R., Instrumentation for treatment of cancer using proton and light-ion beams, *Rev. Sci. Instrum.* **64** (1993) pp. 2055-2122.
5. Hirao Y., Ogawa H., Yamada S. et al., Heavy-ion synchrotron for medical use - HIMAC project at NIRS Japan, *Nucl. Phys.* **A538** (1992) pp. 541c-550c.
6. Kraft G., Gademann G. (Eds.), Einrichtung einer experimentellen Strahlen-therapie bei der Gesellschaft für Schwerionenforschung Darmstadt, *GSI-Report* **93-23** (1993).

7. Blattman H., Munkel G., Pedroni E. et al., Conformal radiotherapy with a dynamic application technique at PSI, *Progress in Radio-Oncology V*, ed. by Kogelnik H. D., (Monduzzi Editore, Bologna 1995), pp. 347-352.
8. Sihver L., Schardt D. and Kanai T., Depth-dose distributions of high-energy carbon, oxygen and neon beams in water, *Jpn. J. Med. Phys.* **18** (1998) pp. 1 - 21.
9. Schimmerling W., Miller J., Wong M., Rapkin M., Howard J., Spieler H. G. and Jarret B. V., Fragmentation of 670 A MeV Neon 20 as a function of depth in water, *Radiat. Res.* **120** (1989) pp. 36-71.
10. Schall I., Schardt D., Geissel H. et al., Charge-changing nuclear reactions of relativistic light-ion beams ($5 \leq Z \leq 10$) passing through thick absorbers, *Nucl. Instr. and Meth.* **B11** (1996) pp.221-234.
11. Golovkov M., Aleksandrov D., Chulkov L., Kraus G., and Schardt D., Fragmentation of 270 A MeV carbon ions in water, In *Advances in Hadrontherapy*, ed. by Amaldi U., Larsson B., Lemoigne Y., (Excerpta Medica, Int. Congr. Series 1144, Elsevier Science, 1997), pp. 316-324.
12. Fukumura A., Hiraoka T., Tomitani T. et al, Attenuation of therapeutic heavy-ion beams in various thick targets due to projectile fragmentation, In *Advances in Hadrontherapy*, ed. by Amaldi U., Larsson B., Lemoigne Y., (Excerpta Medica, Int. Congr. Series 1144, Elsevier Science, 1997), pp. 325-330.
13. Kraft G., RBE and its interpretation, *Strahlenther. Onkol.* **175** (1999) pp. 44-47.
14. Weyrather W. K., Ritter S., Scholz M. and Kraft G., RBE for carbon track-segment irradiation in cell lines of differing repair capacity, *Int. J. Radiat. Biol.* **75** (1999) pp. 1357 - 1364.
15. Heilmann J., Taucher-Scholz G., Haberer T., Scholz M., Kraft G., Measurement of intracellular DNA double strand breaks, induction and rejoining along the tracks of carbon and neon particles in water, *Int. J. Radiat. Oncol. Biol. Phys.* **34** (1996) pp. 599-608.
16. Krämer M., Jäkel O., Haberer T., Kraft G., Schardt D. and Weber U., Treatment planning for heavy-ion radiotherapy, submitted to *Phys. Med. Biol.*
17. Scholz M. and Kraft G., Calculation of heavy-ion inactivation probabilities based on track structure, X-ray sensitivity and target size, *Rad. Prot. Dosim.* **52** (1994) pp. 29-33.
18. Enghardt W., Fromm W. D., Geissel H. et al., The spatial distribution of positron-emitting nuclei generated by relativistic light-ion beams in organic matter, *Phys. Med. Biol.* **37** (1992) pp. 2127-2131.
19. Enghardt W., Debus J., Haberer T. et al., The application of PET to quality assurance of heavy-ion tumor therapy, *Strahlenther. Onkol.* **175** (1999) Suppl. II, pp. 33-36.
20. Haberer T., Becher W., Schardt D. and Kraft G., Magnetic scanning system for heavy-ion therapy, *Nucl. Instr. and Meth. in Phys. Res.* **A330** (1993) pp. 296-305.
21. Debus J., Haberer T., Schultz-Ertner D. et al., Fractionated carbon ion irradiation of skull base tumors at GSI. First clinical results and future perspectives, *Strahlenther. Onkol.* **176** (2000) pp. 211-216.
22. Groß K. D. and Pavlovic M. (Eds.), Proposal for a dedicated ion beam facility for cancer therapy, GSI Darmstadt 1998.

IBA TECHNIQUES TO STUDY RENAISSANCE POTTERY TECHNIQUES

A.BOUQUILLON, J.CASTAING, J.SALOMON, A.ZUCCHIATTI

Centre de Récherche et Restaurations des Musées de France, UMR 171 du CNRS,
6 rue des Pyramides, 75041 Paris, France
E-mail: Anne.Bouquillon@ culture.fr   zucc@ge.infn.it

F.LUCARELLI, P.A.MANDO'

Dipartimento di Fisica and Istituto Nazionale di Fisica Nucleare,
largo E.Fermi 2, 50125 Firenze, Italy

P.PRATI

Dipartimento di Fisica and Istituto Nazionale di Fisica Nucleare,
via Dodecaneso 33, 16146 Genova, Italy

G.LANTERNA

Opificio delle Pietre Dure,
viale B.Strozzi 1, 50100 Firenze, Italy

M.G.VACCARI

Museo Nazionale del Bargello,
via del Proconsolo 4, 50122 Firenze, Italy

The application of Ion Beam Analysis, associated to Scanning Electron Microscopy is examined in connection with an extensive program on structural and chemical analyses of glazed terracotta's from the Italian Renaissance, launched by a French-Italian collaboration in the framework of the European COST-G1 scientific action. The objectives of the collaboration are reviewed. The compatibility of data from different specimen and various laboratories are discussed. Examples of the PIXE and statistical analyses on some artefacts of the *"Robbiesche"* type, supplied by the Louvre Museum of Paris and the Opificio delle Pietre Dure of Florence, are given to illustrate the performances of IBA in this particular field.

1   Introduction

The collections of art museums all over the world, host glazed, coloured terracotta sculptures of the Italian Renaissance, attributed to the della Robbia family, a dynasty of sculptors and ceramists active and respected in Florence for more than a century. They created monumental pieces (altarpieces, tabernacles...) as well as smaller images, for private devotional practice. All their production is characterized by an apparent continuity in style and by a similar treatment of the materials. The appeal of such "*Robbiesche*" productions was so strong, that these artefacts were abundantly imitated and copied from the Renaissance to modern times. Physical and chemical analyses could well extend the knowledge we have on such artefacts and give the curators and historians, besides valuable information for conservation and restoration, the opportunity to add new arguments and data to stylistic considerations, so to strengthen the reasoning that leads to author attribution. The particular nature of the artefacts requires techniques that are non-destructive or at least non-invasive: sampling must be reduced to minimum amounts and limited to hidden areas. In this framework, an international collaboration, encouraged by the European scientific action COST-G1, has been launched with the aim of improving the technical procedures required in the IBA analysis of glazes and of creating a database of analytical information for the "*Robbiesche*". The investigation has so far concerned 42 artefacts supplied by the Louvre museum or by the restoration laboratory of the Opificio delle Pietre Dure of Florence. This paper will emphasize the glaze structural and chemical composition studies by Particle Induced X-ray Emission (PIXE) and, to a lesser extent by Scanning Electron Microscopy (SEM).

2   The structure and composition of a glaze

The Renaissance glazes are heterogeneous vitreous covers, similarly to modern ones. Their main constituent is silica that forms the glass network and was extracted from river sand. Silica was mixed with potassium-rich wine dregs acting as a flux, to reduce the fusion temperature, and then fired. The frit obtained was powdered and mixed with "*calcina*" (lead and tin burned powder) used to add lead-oxide that enhances the flux effect and tin-oxide that makes the glaze opaque, to hide imperfections of the underlying clay body. Also added were colouring agents in the form of metallic oxides and some more river sand. The glaze powder was stirred with water, well homogenised and painted on a fired biscuit (clay body of the sculpture), then dried in air and fired at a temperature close to 950°C. The result of this complex preparation is illustrated in figure 1 .We can observe the heterogeneity of the glaze: inside a grey glassy matrix, un-melted quartz or feldspars grains are present, as well as bubbles. The small white grains are cassiterite (SnO_2) crystals

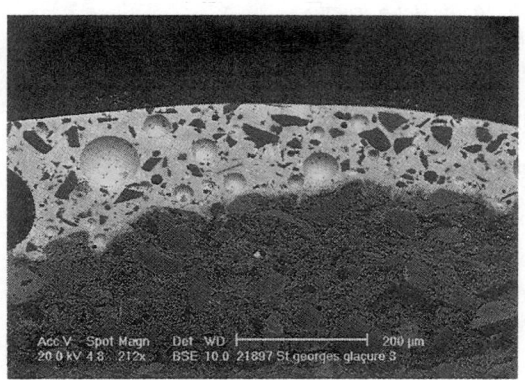

Figure 1. The cross section of a white glaze from: "St. George kills the dragon" Louvre museum

responsible of the opacity. The glaze has a mean thickness of 200µm and the limit between paste and glaze is marked by a thin interface (10µm), that has special mechanical properties and is chemically different from the glaze because it corresponds to the mixing of glaze and clay. This can be a limitation for some of our analyses.

3 Technical aspects of the analysis of glazes by PIXE and SEM

A broad range of chemical and physical analyses, are planned for glaze and clay studies. Sampling has already been performed on the 42 examined artefacts. At present the structural and chemical composition of the glazes is deduced from the combination of results obtained, from only two techniques:

- Scanning electron microscopy (SEM) coupled with an EDS system for structural characterization and fast chemical analysis of polished glaze sections. SEM has been performed both at the Louvre and the Opificio.
- PIXE measurements, for more precise chemical analyses, both at the INFN KN3000 accelerator in Florence and at the AGLAE particle accelerator in the Louvre museum.

SEM measurements, whenever available, give indications on the microscopic structure of the glaze which is in turn related to its preparation recipes and firing conditions. This kind of information is generally categorical (e.g. large or small bubbles, coarse or fine grinding, good or poor fusion, thin or thick interface,...). It has allowed, to estimate to what extent PIXE and µ-PIXE elemental concentrations can be representative of the glaze composition, averaged by the beam size over the mineral microstructures and defects, but was not used systematically at this early stage of our research campaign.

Specimen of very different nature have been made available: whole artefacts, polished sections, fragments. They have required the optimisation of procedures for PIXE analysis, all aimed at obtaining an oxide composition well representative of the examined object. While a detailed account on the analytical procedures has been given elsewhere [1][2], we would like to recall their guidelines:

- In the direct surface analysis of the glaze of a whole sculpture or on potsherds, we have scanned, whenever possible, two different areas (usually 500µm*500µm) in zones that do not appear weathered and should show the original chemical composition.
- With micro samples we have performed a fast PIXE mapping based on two representative elements, Pb as a marker of the glaze and Ca typical of the

- marly paste, to locate the glaze area and, whenever possible, averaged the composition by performing a linear scan along the glazed layer.
- For artefacts where different kinds of samples were available we have compared the results to identify possible interferences of elements from the marly paste (in polished sections) or from surface dirt (in fragments or objects).
- As many points as possible have been measured for each artefact to understand glaze uniformity.

PIXE data sorting is performed, at both accelerator laboratories, by the GUPIX program, in its thick target configuration; 23 element oxides are searched from Na to Pb. The calculation procedures have been optimised and regularly controlled by repeating systematically (over 350 measurements performed), in alternation with glaze measurements, PIXE analyses of a set of glass standards from the Corning museum and the British Glass Industry Research Association. As shown in figure 2, several series of oxide concentrations, extracted at both AGLAE and INFN-KN3000 at different times, are consistent to one another and reproduce the certified values. In only two cases K_2O certified concentration is not reproduced, but the PIXE value, in defect at both laboratories, is confirmed by SEM-EDS. The compatibility of results from the two laboratories is therefore very good and encouraging. All data can be confidently included in the same data base for statistical considerations.

4 Some selected results

The most important goal of our collaboration is the constitution of the largest possible data base of analytical information to assist curators and historians in the conservation, restoration and attribution of the Renaissance masterpieces. We hope to achieve this goal in the long term, after extension of the present PIXE and SEM-EDS measurements to many more objects and implementation of the data base with results from other techniques. However it is already possible to extract some interesting results on groups of objects that are similar in shape, style and use.

4.1 Two pairs of kneeling Angels

The sculpture department of the Louvre museum keeps in its reserves two pairs of kneeling angels stylistically different but all attributed to the della Robbia "*bottega*". The two pairs entered in the Louvre collections in 1862 and 1911. Both pairs are included in the « *robbiesche* » but are stylistically different. The first one is typical of the Buglioni's or some other florentine workshop drawing its inspiration from della Robbia. The second could be representative of the "*bottega*" but executed by an apprentice rather than by the masters (Luca ageing or Andrea).

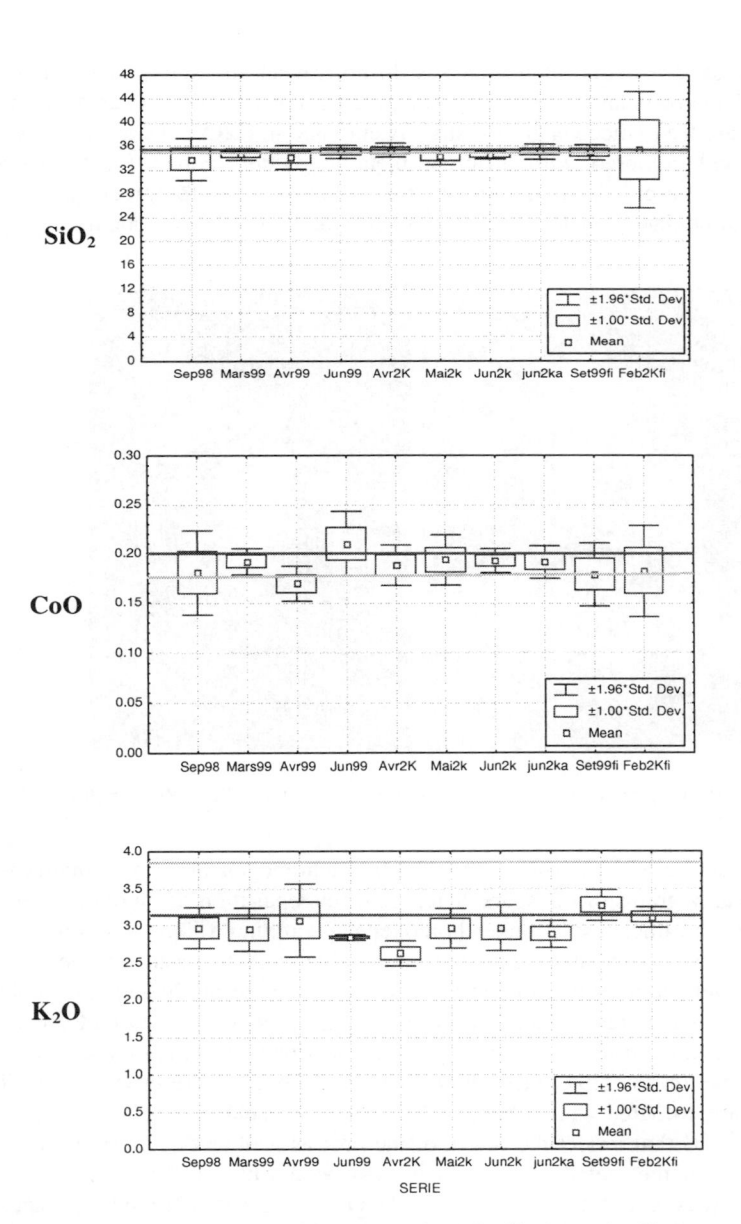

Figure 2. The certified (gray line) oxide concentration (%) of both a major (Si, top panel) and a minor (Co, middle panel) glass component, is well reproduced by series of measurements taken at AGLAE and INFN-KN3000 (boxes) as well as by SEM-EDS(dark line). K is not reproduced in only two cases but PIXE (in both laboratories) and SEM give similar results.

The clay pastes are marls: their analyses do not reveal any significant difference. The glazes have been analysed directly on the artefacts as well as on cross sections. All of them are lead-siliceous glazes, slightly enriched in K2O and CaO opacified by tin-oxides, and coloured with different metallic oxides: cobalt for blue glaze, manganese for purple, lead-antimoniate for yellow, copper for green (associated with lead-antimoniate to make the hue vivid or with manganese to fade it).

Figure 3. Kneeling Angels from the pair C52-C53 (left) and RF33-RF34 (right)

The microstructure of the glazes is very similar : a homogeneous glassy matrix with un-melted quartz grains or feldspars, heterogeneous repartition of tin oxides crystals and bubbles. The thickness is rather homogeneous, around 150µm for both pairs. All the chemical composition data are compatible with the Renaissance recipes published in old ceramics treatises in particular that written by Picolpasso [3] about the traditions of the Italian majolica. The homogeneity of each pair of angels is evident. Noticeable differences can be highlighted when we compare the two pairs. But if we focus on the blue glaze, the difference becomes more pronounced. The blue glazes are coloured with cobalt-oxides. Cobalt is usually associated with other metallic elements that reflect the origin of the ores as well as the different methods of purification of the metal. The cobalt is associated with Ni, Cu, Fe for the pair C52/53, whereas noticeable amounts of As_2O_3 are measured in the blue glazes of the other pair. According to the data published by Gratuze et al [4], these two assemblages could reveal different supplying trades: the one without As should corresponds to the Schneeberg mining district and is commonly found for ceramics from XIV-XVth century, the As enriched one is found frequently for more recent artefacts(XV-XVIIIth century). These results suggest already a possible chronology of the four artefacts but have of course to be confirmed and complemented by the analysis more data.

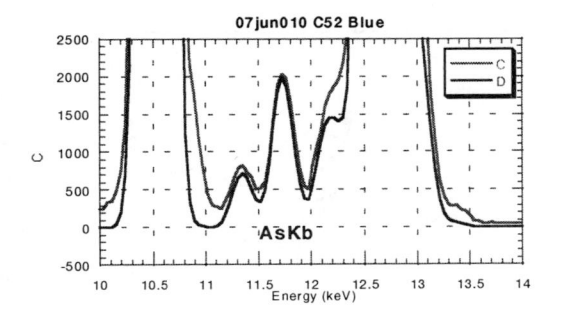

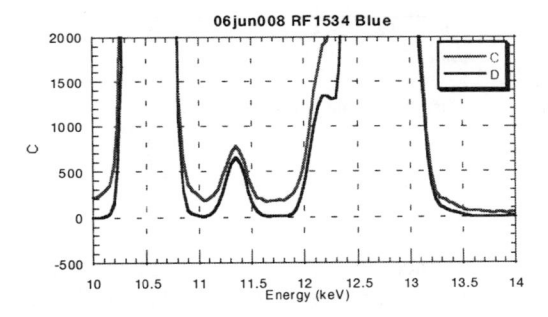

Figure 4. The PIXE spectrum of an angel from the pair C52-C53 (top) and RF33-RF3 (bottom) showing the marked difference in As content. The grey line (C) in the experimental spectrum; the dark line (D) is the GUPIX fit.

4.2 Comparison of some attributed artefacts from OPD

The glaze composition of artefacts coming from the Opificio have been analysed by PIXE: Altarpiece of Montalcino (OPD), Virgin and Child of Arezzo (MAR), Virgin and Child of Palermo (MPA) all attributed to Andrea della Robbia; Portrait of boy of Naples (TFN) of Luca della Robbia, Baptismal Font of Prato (PRA) of Benedetto Buglioni, Head of young man of Florence (TGF). The characterisation of any attributed artefact is important for comparison with non attributed objects. We have preliminarily investigated the reproducibility of PIXE results within the same artefact and the possibility of distinguishing one from another. We have performed a discriminant analysis for white glazes, on which data were most abundant in this group of objects. Part of the specimen were SEM polished sections irradiated at AGLAE, most were whole objects or fragments analysed at the INFN-KN3000. In the space defined by the first two discriminant functions (Root1, Root2 in figure 5), each artefact constitutes an individual group, with no exception. The points of the baptismal font of Prato and the head of young man from Florence (probably a more recent object) are significantly distant from the della Robbia artefacts. Those attributed to Andrea sit very close to one another, while the portrait of boy, an undisputed masterpiece of Luca, tends to separate from the other objects of the *"bottega"*.

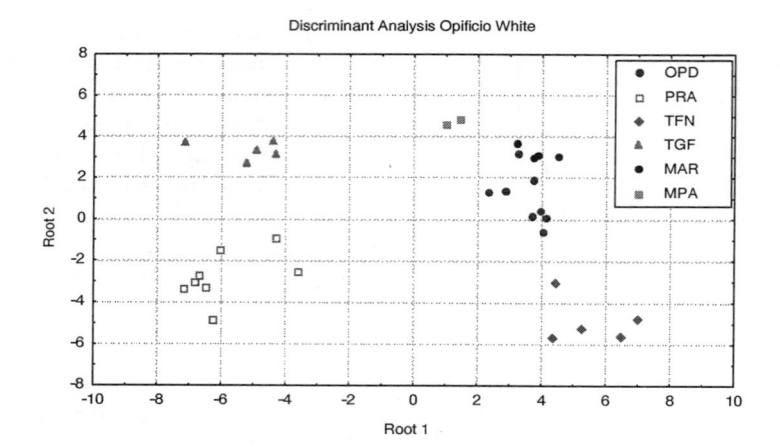

Figure 5. Discriminant analysis of white glazes from attributed objects.

5 Conclusions

PIXE, μ-PIXE and SEM have been combined in an analysis of Renaissance terracotta sculptures, made complex by the variety of specimens (whole objects, fragments, sections) by the need of combining data from different laboratories, by the number of artefacts (42) examined. Reproducibility of results among different laboratories has been achieved and maintained throughout the analysis campaign. The large database available allows since now to address questions on groups of similar related artefacts. Increasing the database and pushing further the statistical analysis should hopefully allow to improve workshop and author attribution.

References

1. A.Zucchiatti, A.Bouquillon, J.Salomon, J.R.Gaborit: Study of Italian Renaissance Sculptures using an external beam μ–probe – NIM B161-163(2000)699-707
2. A.Bouquillon, J.R.Gaborit, G.Lanterna, J.Salomon , A. Zucchiatti : Analysis of glazes with μ-beams of charged particles: examples from the study of some Della Robbia terracotta sculptures. NIM B in press
3. Picolpasso C.V.(1548) : Les trois libvres de l'art du potier – traduit par M.Claudius Popelyn –eds Librairie internationale Paris 1861
4. Gratuze B. Soulier I., Blet M., Vallauri L. (1996) : De l'origine du cobalt : du verre à la céramique – Revue d'Archéométrie, 20 p.77-94

THE BIAS IN THICKNESS CALIBRATION EMPLOYING PENETRATING RADIATION

J. A. OYEDELE

Department of Physics, University of Namibia, Private Bag 13301, Windhoek, Namibia.
E-mail: oyedelej@unam.na

Penetrating radiations are suitable for sheet thickness calibration and should lead to very accurate results. However, in the case of on-line thickness calibration, there may be an error or bias in the measurement which is attributable to thickness and source fluctuations. This bias is shown to increase with the degree of thickness irregularity and with the measurement time-interval in some systems. In contrast, the variance of the probability density function describing the bias decreases with increasing measurement time-interval. Conditions under which the bias is minimized are enumerated.

1 Introduction

The use of penetrating radiation for thickness calibration is well-known and accepted [1,2]. Aside from the simplicity of the measurement technique, the material being examined is not disturbed during the process of measurement thereby making the technique suitable for both static and dynamic systems. Furthermore, the error in the technique has been studied so that accurate results should be obtained from the measurements [3,4,5]. Since most of the early thickness measurements were done on static systems, the error in a given measurement is usually attributed mainly to statistical source fluctuation and instrumentation characteristics. However, in the case of dynamic system or on-line thickness calibration, the time variation of the material thickness may also contribute to the error in the measurement. The combined effect of source and thickness fluctuations during a measurement can be complicated as the measurement procedure may reduce the effect of one of the fluctuations while enhancing that of the other.

In this work, the error or bias attributable to source and thickness fluctuations in on-line thickness calibration by radiation transmission technique is presented. The investigation is extended to different and distinct source and thickness fluctuation.

2 Radiation Transmission and Error

Consider a sheet of thickness $z(v,t)$ being moved with speed v across the path of a radiation beam of intensity $I_o(t)$ as shown in figure 1. The detector response during a time-interval τ is

$$R = \frac{R_o}{<I_o>\tau} \int_0^\tau I_o(t)\exp[-\mu_m \rho z(v,t)]dt \qquad (1)$$

where μ_m and ρ are respectively the mass absorption coefficient and mass density of the sheet, and $<I_o>$ and R_o are respectively the mean incident intensity and the detector response in the absence of the sheet.

When, as usual, the calibration procedure assumes that the sheet's thickness is uniform and that the intensity of the radiation remains constant during the measurement time-interval, the detector response reduces to

$$R_c = R_o \exp[-\mu_m \rho z_c] \qquad (2)$$

where subscript c indicates that the sheet's thickness remains constant. The corresponding transmittance is

$$T_c = R_c / R_o = \exp[-\mu_m \rho z_c] \qquad (3)$$

and the corresponding average thickness of the sheet is

$$<z>_c = \frac{1}{\tau} \int_0^\tau z_c dt = z_c \qquad (4)$$

For the case of fluctuating source and thickness, the transmittance is, from equation (1),

$$T = R/R_o = \frac{1}{<I_o>\tau} \int_0^\tau I_o(t)\exp[-\mu_m \rho z(v,t)]dt \qquad (5)$$

and an average thickness

$$<z(v,t)> = \frac{1}{\tau} \int_0^\tau z(v,t)dt \qquad (6)$$

When the incident radiation and the sheet's thickness is not constant but, as usual and for convenience, equation (4) is used instead of equation (6), then there is an error in the calibration process which is given by

$$\Delta z = <z(v,t)> - <z>_c \tag{7}$$

A relationship between $<z(v,t)>$ and $<z>_c$ can be obtained by noting that different thicknesses and source fluctuations can be associated with a given transmittance so that equation (3) can be equated to equation (5) to get

$$z_c = -\frac{1}{\mu_m \rho} \ln[\frac{1}{<I_o>\tau} \int_0^\tau I_o(t)\exp[-\mu_m \rho z(v,t)]dt] \tag{8}$$

Substituting equation (8) in equation (7):

$$\Delta z = \frac{1}{\tau}\int_0^\tau z(v,t)dt + \frac{1}{\mu_m \rho}\ln[\frac{1}{<I_o>\tau}\int_0^\tau I_o(t)\exp[-\mu_m \rho z(v,t)]dt] \tag{9}$$

Equation (9) gives the error or bias in on-line thickness calibration by radiation transmission technique.

3 Probability Density Function

The bias or error, Eq. (9), is easily evaluated for simple variations of thickness and intensity with time. However, if the incident intensity is specified by a probability density function (PDF), the error will also be specified by a PDF as follows.
Let

$$a(t) = [I_o(t)/(<I_o>\tau)]\exp[-\mu_m \rho z(v,t)] \tag{10}$$

and

$$A(\tau) = \int_0^\tau a(t)dt \tag{11}$$

Equation (9) becomes

$$\Delta z = <z(v,t)> + \frac{1}{\mu_m \rho} \ln A(\tau) \qquad (12)$$

Consider a Gaussian distribution for the PDF of the incident intensity. For a given thickness z(v,t), a(t) and A(τ) will also be Gaussian. The PDF for A(τ) can therefore be written as

$$B(A) = (1/\sqrt{2\pi}\sigma_A)\exp[-(A-\mu_A)^2/2\sigma_A^2] \qquad (13)$$

where σ_A^2 and μ_A are respectively the variance and mean of the PDF.

The combination of equations (12) and (13) gives the probability density function of the error Δz :

$$R(\Delta z) = \frac{\mu_m \rho}{\sqrt{2\pi}\sigma_A} \exp[-\frac{[\exp\mu_m\rho(\Delta z - <z(v,t)>) - \mu_A]^2}{2\sigma_A^2}]$$
$$x \exp[\mu_m\rho(\Delta z - <z(v,t)>)] \qquad (14)$$

4 Applications and Discussions

It is informative to evaluate the error, equations (9) and (14), for four different idealized sheets and radiation source fluctuations. Such sheets and fluctuations are (i) regular sheet and steady source (ii) saw-tooth shaped sheet and steady source (iii) regular sheet and source whose PDF is Gaussian and (iv) saw-tooth shaped sheet and source whose PDF is Gaussian. Figure 2 shows such regular and saw-tooth shaped sheets having thicknesses z(v,t) = z_c and z(v,t) = z_o − γvt, t_{n-1} < t < t_n , where z_c and z_o are constants and γ/2 is the slope of the non-uniform sheet.

4.1 Regular sheet and steady source

When the sheet is regular or uniform (z(v,t) = z_c) and the source is steady so that I(t) = I_o, then the error, Eq. (9), reduces to

$$\Delta z_{co} = 0 \qquad (15)$$

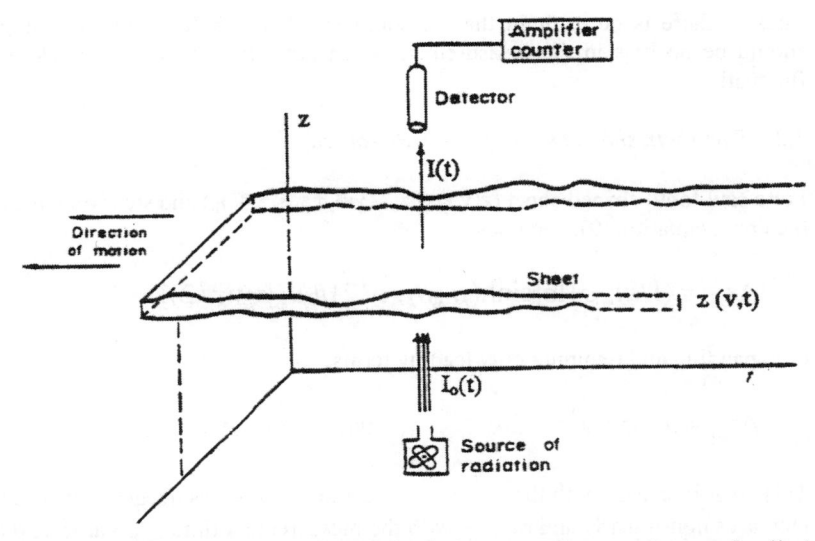

Fig. 1. Schematic diagram of a non-uniform sheet being moved over a beam of radiation.

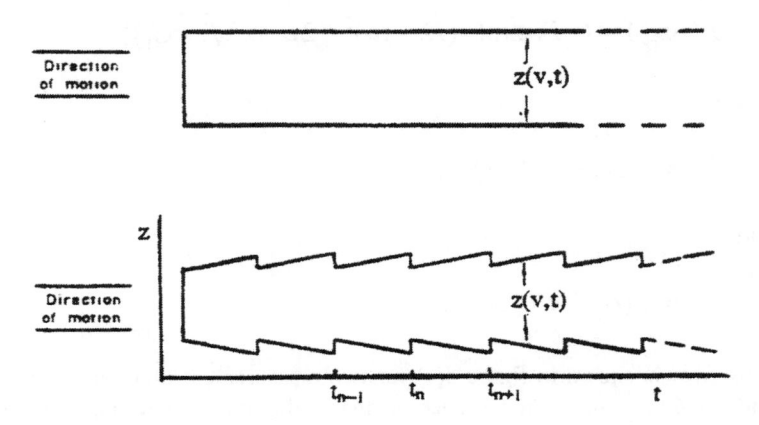

Fig. 2. Cross-sectional view of (a) a uniform sheet and (b) a saw-tooth shaped sheet.

That is, there is no error in the measurement. This result is not surprising as there should be no bias in the measurement when both the source and thickness are not fluctuating.

4.2 Saw-tooth shaped sheet and steady source

For a saw-tooth shaped sheet ($z(v,t) = z_o - \gamma v t$, $t_{n-1} < t < t_n$) and steady source ($I(t) = I_o$), the error, equation (9), becomes

$$\Delta z_{so} = (1/\mu_m \rho) \ln[\sinh(\mu_m \rho \gamma v \tau / 2)/(\mu_m \rho \gamma v \tau / 2)]$$

or expanding and retaining only leading terms,

$$\Delta z_{so} \approx \mu_m \rho \gamma^2 v^2 \tau^2 / 24 \qquad \text{with} \quad [(\mu_m \rho \gamma v \tau)^2 / 24] \leq 1 \qquad (16)$$

This error increases with the slope, γ, or degree of thickness irregularity. The error also increases non-linearly and rapidly with the measurement time interval, τ, and the speed of motion, v, as shown in Fig. 3. Consequently, the use of short measurement time-intervals and low speeds will reduce the error or bias in this case.

4.3 Regular sheet and source having Gaussian distribution

When the sheet is regular and the source has a PDF which is Gaussian, the error, Eq. (14), reduces to

$$R(\Delta z_{cg}) = G_1 D_1 \exp[-(D_1 - \mu_1)^2 /(2\mu_m \rho / \sqrt{2\pi} G_1)] \qquad (17)$$

where

$$D_1 = f(\Delta z - < z(v,t) >)$$

and

$$G_1 = f(\sigma_o^2 / < I_o >^2 \tau)$$

This error is shown in figure 4(a) for two measurement time-intervals $\tau = 5$ seconds and $\tau = 60$ seconds. The variance or the readily measurable full width at half height (FWHH) of the PDF is large at shorter time-interval (Fig. 4a(i)) but small at longer time-interval (Fig. 4a(ii)). This result shows that there is a wider range of errors when

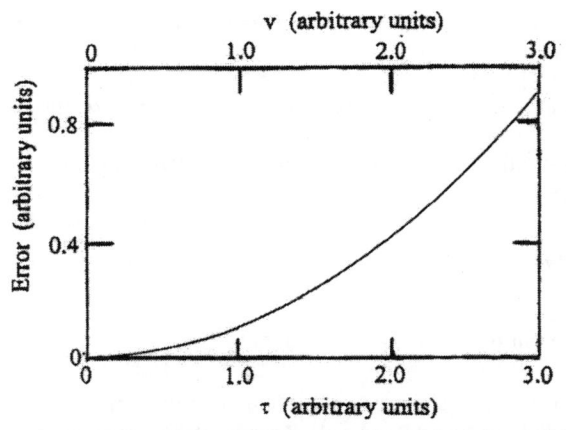

Fig. 3. The variation of error with measurement time-interval, τ, and speed, v, when the sheet is saw-tooth shaped and the radiation source is steady.

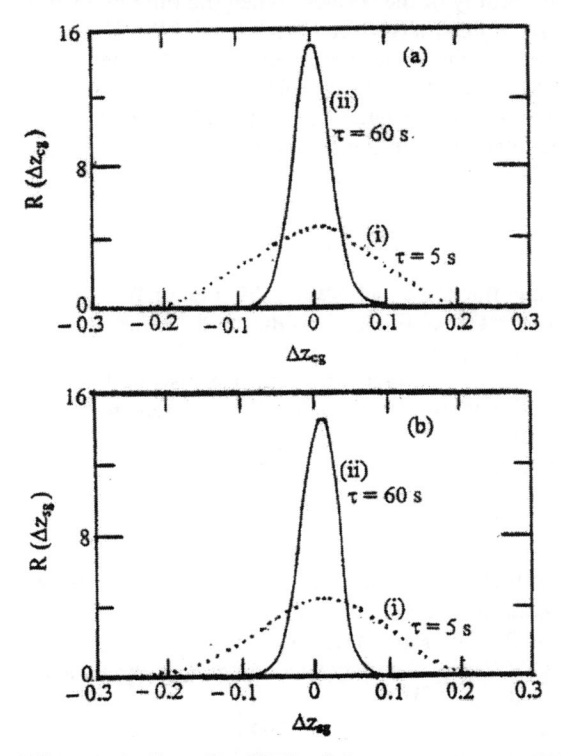

Fig. 4. The probability density function (PDF) of the error when the PDF of the incident intensity is Gaussian and the sheet is (a) uniform and (b) saw-tooth shaped. The time-interval is (i) 5 s and (ii) 60 s.

measurements are taken at short time-intervals compared with those taken at longer time-intervals.

4.4 Saw-tooth shaped sheet and source having Gaussian distribution

If the sheet is saw-tooth shaped and the source has a PDF which is Gaussian, the error, Eq. (14), becomes

$$R(\Delta z_{sg}) = G_2 D_2 \exp[-(D_2 - \mu_2)^2 /(2\mu_m \rho / \sqrt{2\pi} G_2)] \tag{18}$$

with G_2 and D_2 similar to G_1 and D_1 respectively.

The error, Eq. (18), is shown in Fig. 4(b) for two different measurement time-intervals $\tau = 5$ seconds and $\tau = 60$ seconds. This result is similar to that obtained for the uniform sheet (Fig. 4a) although there is now a shift in peak from $\Delta z_{sg} = 0$ which reflects the non-uniformity of the surface. When the time interval is small, the FWHH is large (Fig. 4b(i)), while at large time-interval, the FWHH is small (Fig. 4b(ii)). The result therefore confirms the earlier observation (Fig. 4a) that there is a wider range of errors when measurements are taken at short time-intervals compared with those taken at longer time intervals. However, this is in contrast to the increase of error with measurement time-interval when the source is non-fluctuating and the sheet is saw-tooth shaped (Fig. 3(b)).

5 Conclusion

The bias in on-line thickness calibration by radiation transmission results from thickness and radiation source fluctuations during the measurement time-interval. This bias increases with the speed of motion and with the measurement time interval in some systems. However, the range of errors decreases with the measurement time-interval. The use of low speeds and suitable measurement time-interval as well as radiation source with small variance will reduce the error in the measurement and improve the results obtained.

References

1. Hsu H. H., Pratt J. C. and Shunk E. R. *Nucl. Instrum. Methods* **193**, 235 (1982).
2. Knoll G. F. *Radiation detection and Measurement* (Wiley, New York, 1992).
3. Haggmark L. G. *J. Appl. Phys.* **45**, 3196 (1974).
4. Tsoulfanidis N. *Measurement and Detection of Radiation* (Hemisphere Publishing Corporation, Washington, 1983).
5. Bland C. J. *Arab J. Nucl. Sci. Appl.* **32**, 204 (1999).

THE MEASUREMENT OF VERY OLD RADIOCARBON AGES BY ACCELERATOR MASS SPECTROMETRY

A.E. LITHERLAND AND K.H. PURSER*
IsoTrace Laboratory, University of Toronto, Canada

In 1992 it was reported from the IsoTrace Laboratory that carbon monoxide derived from natural gas had a $^{14}C/C$ ratio less than 1.6×10^{-18} (equivalent radiocarbon age over 110,000 years). This test measurement was done to demonstrate that the radiocarbon content of carbon obtained from deep underground sources was much smaller than had previously been established. The measurement was done prior to the construction of the Borexino 5 Mg test scintillator, the material for which was derived from a deep oil well. When operated underground the scintillator was reported in 1998 to give a low counting rate of ^{14}C beta particles corresponding to a $^{14}C/C$ ratio of $1.94 \pm 0.09 \times 10^{-18}$. While such levels would obscure the recoil electrons from the solar pp neutrinos, neutrinos from Be^7 should be observable. To find oil or natural gas having lower ratios of $^{14}C/C$, or older carbon, points to the desirability of developing methods for measuring the isotope ratio in samples much smaller than 5Mg. Accelerator Mass Spectrometry (AMS) is an obvious choice for the study of old carbon samples but has so far received little attention for this purpose. This is partly because of the possibility of atmospheric CO_2 contamination and partly because of the lack of a compelling reason for undertaking such a study. A group is now planning to explore this problem at the IsoTrace Laboratory. Here we will confine our comments to (1) some early measurements of background rates, (2) AMS tandem backgrounds in general and (3) possible high-intensity low-background ion sources that will make possible the achievement of lower AMS isotope ratios.

* Also: Krytek Corporation, Peabody, MA 01960

Dedication

It is a pleasure to dedicate this paper to Friedl Sellschop on the occasion of his seventieth birthday. Among his many other achievements, Friedl was an early pioneer in the field of neutrino physics. Shortly following the publication by Reines and Cowan[1] of the free antineutrino absorption cross section using reactor-produced antineutrinos, Friedl organised a collaboration between the University of Witwatersrand, Johannesburg, South Africa, and the Case Institute of Technology, Cleveland, Ohio. The purpose was to study solar and cosmic ray produced neutrino interactions using a deep-mine detector. By 1963 Friedl had obtained substantial funding and a mining collaboration that allowed construction, 3200 meters below ground level, of a special underground laboratory and the fabrication of the largest neutrino detector at that time. The detector consisted of 5,000 gallons (16Mg) of light-oil scintillator fluid monitored by photomultipliers. Using this facility Reines and Sellschop made the first detection of naturally occurring muon neutrinos during February 1965[2,3].

Introduction

In 1992 it was reported by Beukens[4], using Accelerator Mass Spectrometry, that the carbon in CO derived from natural gas methane had a $^{14}C/C$ ratio less than 1.6×10^{-18}. This experiment was suggested by Raghavan[5], who was considering the use of liquid scintillators for solar neutrino measurements. Later it was reported by Alimoneti et al. in 1998[6] that underground experiments at the deep Gran Sasso underground laboratory showed that very low levels of radiocarbon contamination within scintillators could be achieved by using hydrocarbon derivatives from oil wells. The purpose of the Gran Sasso measurements was to study the counting rates of ^{14}C beta particles, and other background events, to determine the feasibility of studying low energy (<1MeV) recoil electrons produced by interactions of solar neutrinos within a much larger scintillator [5,7,8]. The 5 Mg test detector gave a very low counting rate of beta particles that corresponded to a $^{14}C/C$ ratio of $1.94 \pm 0.09 \times 10^{-18}$ or a ^{14}C age of about 110,000 years. Even at these very low levels of contamination, radiocarbon beta rays would obscure the recoil electrons from the solar pp neutrinos but those from Be^7 should be observable.

The C^{14} activity in a given sample of gas or oil will depend upon the natural radioactivity of the surrounding rocks, all of which will contain low-levels of the neutron and alpha-particle emitting isotopes of uranium and thorium. Alimoneti et al. [6] have provided an estimate for such a background and showed that one could expect the ratio of $^{14}C/C$ in limestone to be about 5×10^{-21}. As the levels of U and Th fluctuate widely in minerals it may be possible that natural gas or oil having $^{14}C/C$ as low as 10^{-22} (180,000 years) exists. To find such materials, which may be somewhat randomly distributed, requires a measuring technique that will accept much smaller samples than 5 Mg. The technique must also measure their $^{14}C/C$ ratios in a few days when the ratio is as low as ten orders of magnitude below the $^{14}C/C$ level in the atmosphere and at the surface of the earth ($\sim 10^{-12}$). Accelerator Mass Spectrometry (AMS) is the obvious choice as the procedure is inherently free from cosmic-ray and heavy element backgrounds and can count the ^{14}C atoms present with high efficiency ($\sim 5\%$). AMS, however, has backgrounds peculiar to the use of ion sources, mass spectrometers and accelerators and these will be discussed with particular reference to the IsoTrace Laboratory. To illustrate the problems further, additional comments will be made on two other accelerator systems, which are being used for AMS. The contamination due to sample preparation chemistry will not be discussed.

The IsoTrace Facility

Figure 1 shows a schematic diagram of the IsoTrace facility at the University of Toronto. The system consists of seven major sections, which are as follows; (1) a source of negative carbon ions (Negative Ion Source); (2) an electric and magnetic analysis section for selecting the ion species to be injected (EA1 and MA1);

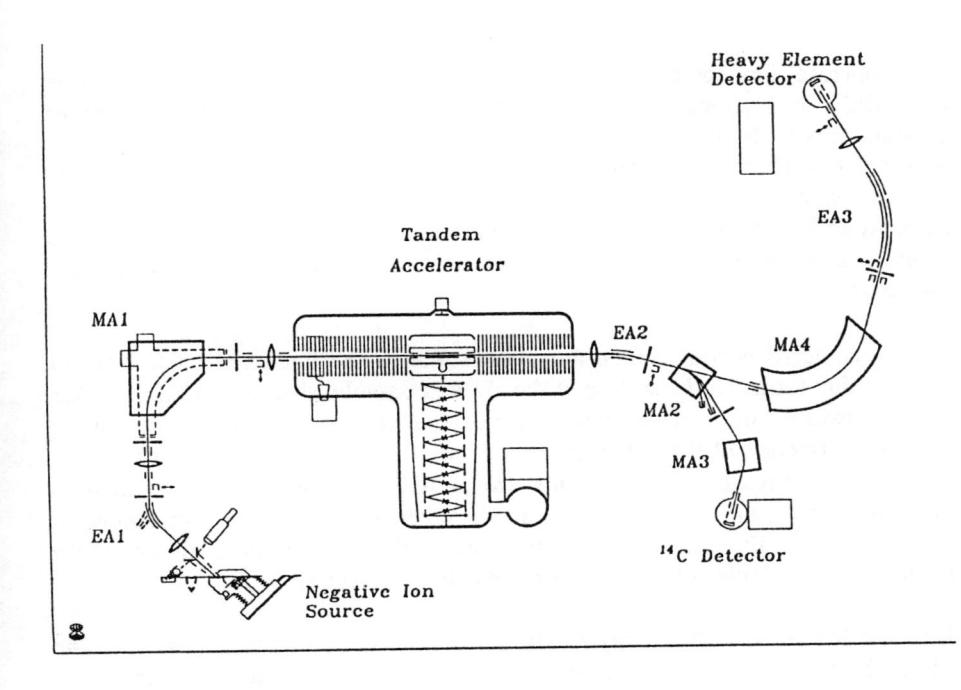

Figure 1

(3) an acceleration stage for negative ions between ground and the high voltage terminal; (4) a gas stripper that converts the negative ions into positive ions; (5) a positive ion acceleration stage; (6) high-energy electric and magnetic analysis, (EA2, MA2, MA3); and (7) a gaseous ionisation detector that can measure both the energy and rate of energy loss for each incoming particle (^{14}C). It can be seen that in addition there is a dedicated analysis line (MA4 and EA3) that is used for heavy element AMS. This program will not be discussed here.

The Negative Ion Source: The IsoTrace ion source is a variant of the caesium sputter source design first described by Hortig [9]. Its characteristics are that the efficiency for converting graphite to useful C^- ions is high (about 5%), with the C^- ions leaving the source being dominantly in the ground state. There has never been confirmation that the interfering $^{14}N^-$ isobar is produced. However, at a $^{14}C/C$ level of 10^{-20} this could be a problem as it is 4 orders of magnitude lower than has been tested. The possibility that a very rare meta-stable [10] excited state of N^- may exist will have to be investigated.

The IsoTrace ion source consists of a reservoir of molten caesium that is maintained at a temperature between 120 and 200°C. At these temperatures caesium vapour from the reservoir is directed through a hot (1,200°C) porous tungsten frit where individual atoms are converted at the exit into Cs^+ ions by surface ionisation. In practice, a beam of positive caesium ions, having an intensity of 50-100µA, can be extracted and these ions are directed, without mass analysis, directly onto the graphite sample from which neutral carbon atoms are sputtered.

As the sputtered neutral carbon atoms pass though the fractional mono-layer of caesium that forms on the surface of the graphite sample, electron tunnelling takes place between this surface film and the sputtered carbon atoms causing a substantial fraction of the outgoing carbon atoms to transform into negative ions. At IsoTrace the resulting $^{12}C^-$ currents have an intensity of about 10µA. A modern graphite target generates $^{14}C^-$ ions at a rate of about 75/second. Unfortunately, as described below, the process also produces molecular ions and these are one of the significant contributions to the background at low ^{14}C levels.

Injected Ion Mass and Energy Analysis: The two deflection elements, EA1 and MA1, form a mass spectrometer that, for singly charged negative ions, transmits only a specific ion mass to the injection point of the tandem accelerator. The electric analyser, EA1, which specifies ion energy, eliminates low-energy tails from the sputter ion source. These could introduce backgrounds. The magnetic analyser, MA1, transmits negative ions of the required momentum. It should be emphasised, however, that single analysers could also transmit other ions due to scattering and fragmentation of the negative ions between the two analysers. This is discussed in greater detail below, in the section on backgrounds for the high-energy analysers.

An insulated magnet box allows the energy of the ions to be adjusted within the magnetic deflection stage. Such an arrangement makes possible rapid switching between masses 12, 13 and 14 permitting normalisation of ^{14}C counting rates to the primary intensities of ^{12}C and ^{13}C.

Both $^{13}CH^-$ (0.74 eV binding energy) and $^{12}CH_2^-$ (0.21 eV binding energy) molecules are also injected into the accelerator as mass-14 ions. These unwanted molecules contribute seriously to the high-energy backgrounds, as described below.

Care must be taken to minimise the injection of $^{14}NH^-$ anion, because the resulting continuum of ^{14}N positive ions can generate background. Low-energy tails of the $^{14}NH^-$ beam are avoided by use of the electrostatic analyser, EA1, and by good momentum resolution in the magnetic element, MA1. However some of these anions are injected by fragmentation processes as described below.

Terminal Stripping: After acceleration through the first tandem acceleration region the negative ions have an energy of approximately 2.0 MeV. At this energy approximately 35% of the incoming ^{14}C ions are stripped by differentially pumped argon gas within the terminal to a positive charge state of +3. It is believed that all molecules will then be disintegrated by Coulomb explosions to $^{13}C^{+Q}$ and $^{12}C^{+Q}$ plus H^+ ions as no triply charged CH or CH_2 molecules have ever been seen. These mass 13 and 12 carbon ions can be particularly troublesome, as the beams are intense and are injected with high efficiency into the positive ion acceleration tube. A continuum background of both ^{12}C and ^{13}C ions is then generated when charge-changing collisions with residual gas molecules occur at random locations along the positive ion acceleration tube. The final energy spectrum, shown in figure 2 below,

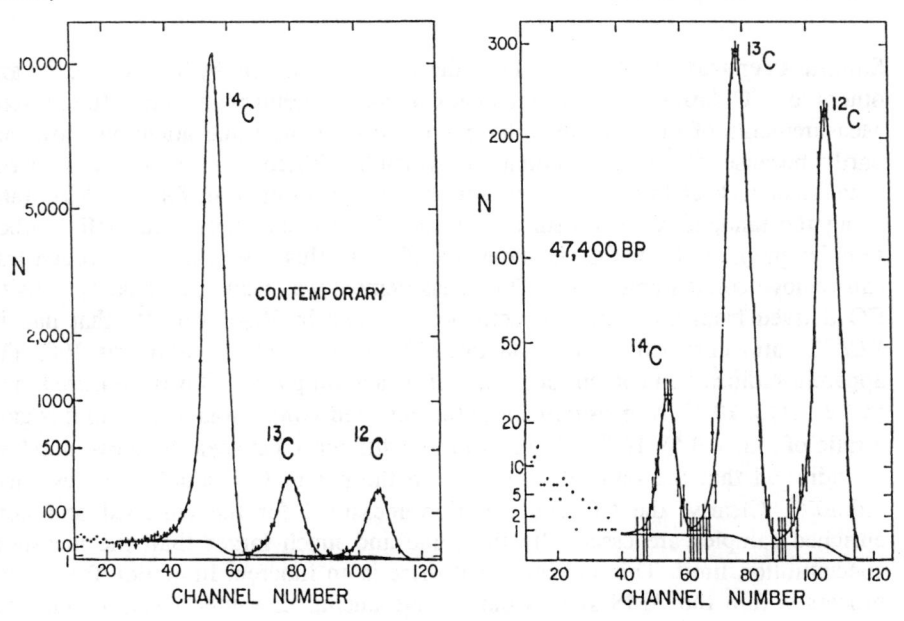

Figure 2

is generated when a small slice of this continuum, having the momentum necessary to pass unimpeded through the magnets MA2 and MA3, is scattered in the residual gas of the electric analyser, EA2. The ^{12}C and the ^{13}C ions are from molecular break-up.

Post Acceleration Analysis: Basically, the IsoTrace high-energy analysis system consists of a single electric analysis followed by a pair of identical magnetic analysers with a defining slit at a beam crossover between the magnetic elements. The combination of post acceleration elements, EA2, MA2 and MA3, act to define for each charge state a specific mass and energy that can be transmitted without attenuation. The dual magnet system, with an intermediate aperture, is needed for the reduction of gas and wall scattering.

Final Detector: The ions transmitted through the post-acceleration analysis system are directed into a gaseous ion chamber that can measure the total energy and the rate of energy loss for each incoming particle. This device permits discrimination against the ^{12}C and ^{13}C interference events introduced from the dissociation of the ^{13}CH$^-$ and ^{12}CH$_2^-$ molecules as they have a different energy. This is shown in figure 2. If any rare nitrogen events are introduced by residual ^{14}NH$^-$ ions from the ion source, they can be identified by dE/dx measurements.

IsoTrace Backgrounds

Sample Preparation: AMS has been discussed extensively by Tunis et al [11] and others for ^{14}C dating measurements back to about 50,000 years before the present. Measurements of older samples have not received the same attention, however, partly because of ^{14}C contamination, which is difficult to remove, and partly because of lack of interest. The present measured lower limit for the ^{14}C/C ratio using the tandem AMS system at IsoTrace Laboratory facility, for solid carbon samples prepared from CO, is about 6×10^{-16}. In this case a liquid nitrogen trap can remove condensable gases. This measurement was made by Beukens [4] using CO derived from underground methane, provided by Raghavan [12], that had its ^{14}C/^{13}C ratio enriched [13] by a factor 200 and the ^{13}C/^{12}C ratio was 100. The apparent radiocarbon content of an non-enriched sample of CO was measured to be $(5.7 \pm 1.4) \times 10^{-16}$. A measurement of the enriched sample was made and this gave a ratio of $(5.6 \pm 0.5) \times 10^{-16}$. The error on the difference between these two numbers [4] indicated that the ratio of ^{14}C/C within the parent CO sample was less than 1.6×10^{-18}. Clearly, the ^{14}C concentration measured for the enriched and non-enriched samples are essentially the same and much larger than the measured concentration limit. This indicates that there is an inherent limitation from some process within the AMS system that is introducing unwanted backgrounds near

$^{14}C/C \sim 6\times10^{-16}$. Contamination during sample preparation [4] is most likely. An additional small contribution from contamination of ^{14}C within the ion source is possible. The data in Table I show a comparison of some old carbon sample measurements made at IsoTrace [4].

Table I

IsoTrace Radiocarbon Backgrounds

Material	$^{14}C/C$ ($\times 10^{-16}$)
CO_2 from Acid Hydrolysis (Carrera Marble)	8.9±0.1
CO_2 from Natural Gas	9.0±0.1
CO_2 from Combustion of Anthracite	9.5±0.2
Acetylene from Calcium Carbide	1.6 ±1.2

A comparison with the data given earlier for the samples prepared from CO and those prepared from CO_2 indicate that, when the liquid nitrogen trap was used to remove condensable gases such as CO_2, the background ratio is lowered from about 9×10^{-16} to about 6×10^{-16}. This discovery indicates clearly that condensable gases are present as a result of the sample preparation. However, their complete removal seems to be difficult so that the use of CO_2 for studying the oldest carbon is problematical. At the IsoTrace Laboratory both the CO and CO_2 are routinely converted to acetylene, with the help of lithium metal. The acetylene is then electrically cracked onto aluminum backings. While the latter process does not generate a measurable contamination, some contamination of the lithium is suspected [4]. This is because the creation of cracked carbon from acetylene (produced from CaC_2), without the use of lithium, indicates a much lower value of $^{14}C/C$, as shown in table I. The contamination problem is still under investigation

The Ion Source: In all existing ^{14}C AMS systems the ion sources used have been variants of the caesium sputter source design, popularised by Middleton[14] and others. Its features are that the efficiency for converting graphite to C^- ions is high (about 5%), with the C^- ions leaving the source being dominantly in the ground state. There has never been confirmation that the interfering $^{14}N^-$ isobar is produced. However, at a $^{14}C/C$ level of 10^{-22} this may be a problem as it that 6 orders of magnitude lower than has been tested to date. An excited state of N^- may be meta-stable [10] so this problem will have to be revisited.

The IsoTrace ion source consists of a reservoir of molten caesium that is maintained at a temperature between 120 and 200°C. At these temperatures caesium vapour from the reservoir is directed through a hot (1,200°C) porous tungsten frit where it is converted into Cs^+ ions at the exit by surface ionisation. In practice, a beam of positive caesium ions, having an intensity of 50-100μA, can be extracted from the frit and is directed, without mass analysis, to impact the graphite sample from which neutral carbon atoms are sputtered. The contribution of C^+ ions to the Cs^+ beam is at present unknown and must be measured. For example 1ppb of contemporary carbon could give an apparent $^{14}C/C$ ratio of 10^{-21}.

As the sputtered neutral carbon atoms pass though the fractional mono-layer of caesium on the surface of the graphite sample, electron tunnelling takes place between the surface and the sputtered carbon atoms causing a significant fraction of the outgoing carbon atoms to transform into negative ions. At IsoTrace the $^{12}C^-$ currents used have an intensity of about 10μA and are generated from a modern graphite target together with $^{14}C^-$ ions at a rate of about 60/second. Allowing for inevitable charge state distribution effects at the stripper leads to final counting rates for a modern sample of ~20/sec. The materials of the support for the solid carbon target in the ion source must also be free of ^{14}C to the same extent as the C^+ content of the Cs^+ beam.

An alternative Middleton-type sputter ion source [14] used at the Woods Hole Oceanographic Institution (WHOI) AMS tandem uses about 3 times higher primary caesium beam current and proportionately greater C^- output. This source will reliably produce 70 μA of C^- beam and with a modern carbon sample ~350 $^{14}C^-$ ions per second. The machine transmission efficiency between source and detector is ~40%, causing the particle detector system at that current to register ^{14}C events at a rate of ~140/sec (~5.0×10^5/hour). The rates for other isotope ratios are shown in table II

Table II
Counting ^{14}C at very low levels:

Radiocarbon Age	$^{14}C/C$ Concentration	^{14}C Detector Counting Rate	^{14}C Detector Rate with 200 x enrichment
Modern	1.2×10^{-12}	142/sec	-------
59,000 years	1.0×10^{-15}	430/hour	-------
106,000 years	1.0×10^{-18}	0.43/hour	2064/day
144,000 years	1.0×10^{-20}	0.72 /week	21/day

Using existing source technology (70 microamperes of $^{14}C^-$) the counting rates expected are given above.

Molecular Fragmentation at Low Energies: The molecular ions generated by sputtering show a tendency to fragment in flight spontaneously [15,16] and by any collisions with residual gas. These fragmented molecules can cause a variety of problems.

The $^{14}NH^-$ ion is stable (0.38 eV binding energy) and beams of this species are observed readily from negative ion sources. These ions, having mass 15, should not, in principle, be injected into the accelerator. However, a small fraction may be injected at mass-14 from the tail of the NH^- mass-15 peak unless an electric analyser is used in addition to a magnet. At the terminal these molecules dissociate to produce a potential ^{14}N background if a second charge exchange takes place during the acceleration of the positive ions. Even if an electric analyser is used after the ion source the NH_2^- ions that decay to or are induced to become, by collisions with residual gas, NH^- before the magnetic analyser will be injected into the tandem as mass-14 ions. Fortunately low enough levels of ^{14}N can be separated from the ^{14}C ions in the final ion counter. It is not obvious that the level will be low enough for very rare ^{14}C ions to be measured and so this problem will have to be studied.

At IsoTrace the ME/Q^2 interference is large, see figure 2, but the ion energies are 14/13 and 14/12 times those of the ^{14}C and so are easily resolved from them during normal radiocarbon dating. They are between 5% and 10% of contemporary radiocarbon peaks. They can be eliminated with an additional electric analyser.

However, the final detector will not eliminate any remaining E/Q interference, after the two magnetic analysers, and an additional magnet is needed. This is because, as mentioned earlier, the 8 MeV C^{+3} ions are near the peak in the dE/dx curve. For carbon enriched in ^{14}C the abundant molecule $^{13}CH^-$ is also injected. In addition this molecule can fragment so that about 10^{-6} of the molecules become $^{13}C^-$ before entering the tandem. Thus, for very old carbon studies it is desirable to optimise the number of magnetic analysers, as these can effectively attenuate this particularly dangerous interference. This will be discussed below.

Backgrounds at High Energy. The ME/Q^2 and E/Q Interferences: In AMS the MeV ions being analysed are characterised by the mass M in amu, by electric charge, Q, in units of the electronic charge, e, and by the kinetic energy, E, usually in MeV. The properties of the post-acceleration electric and magnetic analysers, through which these particles pass, have been discussed by many authors [17]. However, it is not usually emphasised that the two variables E/Q and M/Q form a set of constraints for the analyser arrangements that have been used in mass spectrometry:

Magnetic Field Deflection along a specific trajectory: $(M/Q \times E/Q) = ME/Q^2$

Electric Field Deflection along a specific trajectory: E/Q

Cyclotron Resonance at a specific frequency: M/Q

Velocity Analysis (E × B and time of flight): $MQ/E/Q = 1/V^2$

Any two of the above constraints define both E/Q and M/Q and these particles will be transmitted to the final detector without loss. In practice, however, all analysers also transmit other ions, due to scattering, molecular fragmentation and charge changing collisions, but usually at a greatly reduced efficiency. The cyclotron resonance is, to some extent, an exception to this rule. If one assumes a high energy analysis system made up of a series of pairs of electric and magnetic analysers, then each pair will attenuate ions of differing ME/Q^2 and E/Q from the

^{14}C ions of interest. This is an arrangement that is commonly used. If the attenuation by an electrostatic analyser of an ion having a particular ME/Q^2 value is α and the attenuation by a magnetic element of an ion having a particular E/Q value is β then the attenuation of the pair is $A = \alpha\beta$. Actual values for α and β are dependent upon details of the geometry of each analyser, including the angle of deflection, the vacuum pressure within each analyser and the location of any anti-scattering baffles. However, it is expected that the attenuation for unwanted particles by such a typical pair would be in the range 10^{-5} to 10^{-8}. It is possible to place 'p' pairs of these elements sequentially along the beam trajectory leading to a total attenuation of unwanted particles of A^p. Clearly, it is possible to attenuate unwanted ions, by an arbitrarily large factor, using a series of pairs of electric and magnetic analysers. This can be done without losing any of the ions of interest other than by rare scattering or charge changing. The sequence of analysers should not generate a continuum, in addition to the discrete values of ME/Q^2 and E/Q, unless a previous analyser has just rejected a strong interfering beam. The requirement for no continuum therefore decides the number of electric and magnetic analysers.

A decision on the resolution required for each of the analysers is more complicated and has been discussed in some detail [18]. This is because molecules with the same mass as the rare ion will be injected also. The fragments of these molecules after their dissociation in the tandem terminal can later interfere with the detection of the rare isotope when their masses and charges are such that $Mq = mQ$, where M is the mass of the rare isotope and Q is its charge. The numbers m and q are the corresponding quantities for the molecular fragment. Mass defects must be used for the heavier elements so that the equality is only a first approximation. Under these conditions the rare isotope can be swamped with a high counting rate of molecular fragments and this situation is made worse by the energy spread due to Coulomb explosion of the molecule. In some cases, pairs of molecular fragments can also arrive at the detector simultaneously to make matters worse. Consequently one must avoid the use of those charges and masses that satisfy the equation $Mq = mQ$.

The situation is even more complicated than this because those ions that satisfy the equation $Mq - mQ = n$, especially $n = \pm 1$, also require special attention. Note that for the heavier elements the actual masses must be used in this equation and consequently n can be a number <1, which exacerbates the problem. An example of the application of this equation, when n is nearly an integer, from the Woods Hole Laboratory, is instructive [19]. There the high-energy analysis system includes a 33° electric analyser coupled to the final magnetic analyser. This attenuates the ME/Q^2 interference quite effectively by about a further factor of 500 compared with the situation at IsoTrace. However, a strong higher energy peak is also observed.

This fortunately does not interfere with the analysis of normal radiocarbon samples. For the detection of low levels of ^{14}C from very old carbon, this peak must be understood and eliminated because of the presence of a tail. Even if the tail of the pulse spectrum were 10^{-6} of the peak that would imply a background at the radiocarbon position of 10^{-19}. The ions can be identified as coming from the contamination of the carbon target with traces of fluorine. Compared with the ^{12}C ions, the ^{19}F ions appear at the level of about 1 part in 10^{13}. In this case, as far as the high-energy mass spectrometry is concerned, the $^{14}C^{+3}$ and the $^{19}F^{+4}$ have a value of $n = -1$ and so the electric analyser requires a resolution greater than that needed to resolve the ME/Q^2 interference from the other carbon isotopes. Using the equations given in reference 15 we find that $E/\Delta E \sim 226$ is needed instead of 13. As a result the ^{19}F from the contamination of the target can if injected into the accelerator produce ions which have the same magnetic rigidity and an electric rigidity only -1.8% less than that of the $^{14}C^{+3}$. In this case the ^{19}F ions are injected into the accelerator because of the lack of an electric analyser at the ion source. As a result the sputter tail of lower energy ^{19}F ions is injected along with the other ions. About 1ppm of fluorine in the target could produce the counts observed. An electric analyser at the ion source would remove this peak by many orders of magnitude.

Values of $n = \pm 1$ can also create problems in the final detector because of the energy spread generated by the Coulomb explosion [18] of the molecule during the electron stripping. An interesting example of this from the analysis of heavy elements comes from the analysis of the rare long-lived radioactive isotope ^{129}I. The presence of the molecule $^{97}Zr^{16}O_2^-$, in the mass 129 negative ion beam, interferes with attempts to detect low levels of ^{129}I with charge +4. This is because the ^{97}Zr ion with charge +3 has a value of $n = -1$. The energy spread of the ^{97}Zr ions then ensures that the interference with the detection of rare ^{129}I is troublesome. The resolution of the electric analyser used to separate them then has to be larger than expected [18], being $E/\Delta E \sim 970$. If the intensity ratio of the interfering molecular ions is large then a much higher resolution is necessary. This situation is not practical and so charge state +4 is not used for ^{129}I analysis. Charge state +5 is much better because the interference from $^{103}Rh^{+4}$ is less likely because of the comparative rarity of rhodium.

A solution to the requirement of very high resolution is the use of a second charge change after acceleration so that the value of $Mq - mQ$ changes. For example in the case of $n = 0$, a charge change by only one unit can change the resolution needed dramatically. This is the situation in the case of for example $^{129}I^{+3}$, which is quite unusable due to molecular interference. If the ^{129}I ions are changed to charge state +4 after a first magnetic analysis the resolution requirement for a subsequent electric analyser is reduced. These considerations indicate the complexity of the molecular interference problem in relation to the magnetic and electric analyser resolution requirements.

Both $^{13}CH^-$ (0.74 eV binding energy) and $^{12}CH_2^-$ (0.21 eV binding energy) are injected into the accelerator as mass-14 ions. At the high voltage terminal these mass-14 molecules, are dissociated to $^{13}C^{+Q}$ and $^{12}C^{+Q}$ plus H^+ ions. These mass 13 and 14 carbon ions can be troublesome as they leave the acceleration section with a continuous energy spectrum because of charge changing collisions during both the negative and the positive-ion acceleration stages. As noted earlier, analysis in the terminal could reduce these ions to much lower levels.

For the IsoTrace AMS system some of the most troublesome events are those that start as C^{+4} in the terminal and charge change to C^{+3} at along the positive acceleration tube. Within this continuum of energies there are $^{13}C^{+3}$ and $^{14}C^{+3}$ ions that have the same magnetic rigidity as the wanted $^{14}C^{+3}$ ions. These ions are readily observed and form a principle background to the detection of ^{14}C by tandem accelerators. It should be emphasised that the intensity of these $^{12}C^{+3}$ and $^{13}C^{+3}$ beams bear no relationship to the true $^{12}C^-$ and $^{13}C^-$ beams. They are therefore useless for normalisation purposes. Also, their intensities fluctuate, relative to the wanted $^{14}C^{+3}$, as they are sensitive to the pressure in the accelerating tubes.

Low energy Accelerator Mass Spectrometers like those at the IsoTrace and Woods Hole Laboratories are susceptible to E/Q interference from the major isotopes. This is because the ions are, in the case of ^{14}C analysis, at the peak of the dE/dx or Bragg curves. Consequently it is difficult to distinguish between ^{12}C, ^{13}C and ^{14}C ions of the same energy. Processes that result in carbon ions of the same energy therefore produce a potential background, which cannot be distinguished from the ^{14}C ions of interest. In the case of the IsoTrace [20] machine and the Woods Hole Laboratory type of machine at Kiel [21] the total attenuation for ^{13}C ions of the same energy emerging from the machines has been measured to be 10^{12} and 3×10^{16} respectively. These are very satisfactory numbers showing that E/Q interference is small for normal radiocarbon dating. The ^{12}C and ^{13}C injected along with the ^{14}C is expected to be near the 10^{-6} and 10^{-8} level of the ^{12}C intensity. Consequently, the addition of one more magnetic analyser should suffice for a ratio of 10^{-22} to be studied with either machine due to the E/Q interference. For machines with only one high-energy magnet a larger E/Q interference is expected.

If complete stripping of the ions is possible, as it is for very large tandems such as the Holifield 25MV folded tandem [22] at Oak Ridge, then the ME/Q^2 ^{12}C and ^{13}C interference disappears from the $^{14}C^{+6}$ spectrum because no higher energy carbon ion can be created. However, if the 180° magnet in the terminal is used to select $^{14}C^{+5}$ carbon ions from the stripping foil, then higher energy ^{12}C and ^{13}C ions can be generated from charge changing during their further acceleration. These ions, which are from the mass 14 molecular ions, will be rare because they must undergo a scattering, in the residual gas of the magnet, to enter the second stage of acceleration. They can then as C^{+6} ions pick up extra energy and then charge change back to C^{+5} ions. As a result the ME/Q^2 problem can again exist, but at a very much lower level compared with a straight through tandem. This feature of the Holifield Tandem shows the value of terminal analysis after the first stripping.

Pilot Beams: Although scattering and charge changing backgrounds can be made as small as is necessary by increasing the number of high energy attenuating elements, p, there is a class of backgrounds that are not removed. Multiples and sub-multiples of each of the variables M, E and Q represent classes of particles that cannot be distinguished by typical analysers. An example could be the ion $^{28}Si^{+6}$ with twice the energy of a $^{14}C^{+3}$ ion. The $^{28}Si^{+6}$ ions can then be used as a pilot beam in the study of very low level $^{14}C^{+3}$ ions. This can, for example, be facilitated by the injected ion $^{28}Si^-$. If these negative ions change charge to Si^{+7} in the terminal and then change back to +6 after acceleration they will penetrate all the subsequent analysers. Such a beam can provide a continuous monitor of the operation of the system and confidence of stability when the counting rates are measured in events per day. If the molecular ion $^{28}Si_2^-$ decays spontaneously to $^{28}Si^- + ^{28}Si$, or is encouraged to break up in a low-pressure region of gas after electric analysis, it will be transmitted by the magnetic analyser and injected into the tandem along with the $^{14}C^-$. Then after tandem acceleration the $^{28}Si^{+7}$ can be encouraged to capture an electron in a region of slightly elevated gas pressure. These charge changes would have to be planned in advance but they could then be used to produce a variable intensity pilot beam.

Conclusions

To go further in measuring low levels of radiocarbon at the IsoTrace Laboratory is certainly possible but to do it successfully will require upgrading of many of the procedures and modification of the apparatus.

Isotope Enrichment: The CO used for the early experiments of Beukens [4] had a $^{13}C/^{12}C$ ratio of about 100 and was obtained [12, 13] from Isotec Inc. The $^{14}C/^{13}C$ ratio was estimated to be near 200, assuming [13] that the original $^{14}C/^{12}C$ was 10^{-18}. The vapour distilled liquid CO was derived from natural gas obtained commercially so that it is indeed remarkable that the ^{14}C level is so low. The actual level of the ^{14}C in the CO produced by Isotec Inc. must be studied in the future. It is clear from table II that isotope enrichment will be absolutely necessary to reach a ratio of $^{14}C/^{12}C$ of 10^{-20} if existing AMS ion source technology is used. However, with 7mA of C^- the level of 10^{-22} can be reached in a day or so with enriched carbon. Ion sources producing such currents have been produced [23] but they have not yet been used for AMS. This will be one of the prime challenges for the measurement of very old carbon.

At the ratio $^{14}C/C$ of 10^{-22} there are only 5 atoms of ^{14}C per gram, so that, for an enrichment of a factor of 100 for $^{13}C/^{12}C$, 100g of carbon would be needed to provide 1 gram of material enriched to the ratio $^{14}C/^{13}C = 200$. This material would then produce about 30 counts of ^{14}C in about one day of operation at 7mA [23], which is clearly not an impossible task. However, the construction of an ion source with such a low level of contamination, as discussed below, and the enrichment of such material without any contamination will be formidable tasks.

Sample Preparation: Sample preparation, without contamination from contemporary levels of ^{14}C, will be a critical problem. The use of CO_2 should ideally be avoided during the production of targets for the ion source, because of the possibility of contamination with atmospheric CO_2. A good starting point could be methane, which is less abundant in the atmosphere.

Ion Source Requirements: Reducing ion source backgrounds, to a level where the extended high-energy analysis makes sense, will involve research. The evidence from experience at IsoTrace suggests that the ion source region will be a contributor to the ultimate ^{14}C background. Research needs to be done to understand the contributions to backgrounds originating from the ion source walls and from the caesium beam itself. The contribution of C^+ ions within the incoming Cs^+ beam is presently unknown and must be measured. For example, 10ppb of contemporary C^+ ions travelling along with the wanted caesium ions could give an apparent $^{14}C/C$ ratio of 10^{-20}. It is possible that magnetic analysis will be needed for purifying the caesium beam. The contamination of the aluminium supports for the carbon targets will also have to be studied. Fortunately, the amount of carbon in aluminium measured using photon excitation is less than 100ppb. And it is expected that this will be mainly old carbon. Larger negative ion currents are certainly possible [23].

High Energy Analysis: An additional electric and magnetic analyser, each with an appropriate energy and momentum resolution, will be needed. Then with the addition of all the ingredients described above the study of very old carbon may be possible and very old carbon identified.

Acknowledgements

We wish to thank R. P. Beukens and R. S. Raghavan for most valuable discussions on this topic. In addition we are grateful to J. P. Doupe and H. W. Lee for providing material from their unpublished theses. We thank M-J. Nadeau for providing us with results from the Kiel Laboratory and A. Galindo-Uribarri of Oak Ridge provided us with a deeper insight into the remarkable Holifield folded tandem accelerator. We are also grateful to W. E. Kieser and X-L. Zhao of the IsoTrace laboratory for continuing discussions on these unfamiliar problems. A. E. Litherland was supported by an operating grant from the Natural Sciences and Engineering Research Council of Canada.

References

(1) Reines, F. and Cowan Jr., C.L. (1959) Phys. Rev., **113**, 273-279.
(2) Reines, F. and Sellschop, J.P.F. (1965) Phys. Rev. Lett.,**15**, 429-433.
(3) Reines, F. and Sellschop, J.P.F. (1966) Scientific American (Feb 1966), 40-48.
(4) Beukens R. P., *Nucl. Instr. and Meth.* **B79**, 620 - 623 (1993).
(5) Raghavan R. S. Solar Neutrinos - From Puzzle to Paradox, *Science* **267**, 45 - 51 (1995).
(6) Alimonti G. et al., *Physics Letters* **B422**, 349 - 358 (1998).
(7) Gratta G. KamLAND: Neutrinos from heaven and Earth, *CERN Courier*, 22 - 24, April 1999.
(8) Bahcall J. & Suzuki Y. Solar Neutrinos below 1MeV, Neutrino 2000, Pre-Conference Workshop, Sudbury June 16 - 21, 2000 (unpublished).
(9) Hortig, G. (1969), *IEEE Trans. Nucl. Sci.* **NS-16,** 38.
(10) Piangos N. A. and Nicolaides C. A (1998), J. Phys. B: At. Mol. Opt. Phys. **31,** L147-L154.
(11) Tuniz C., Bird J. R., Fink D. & Herzog G. F. Accelerator Mass Spectrometry, pp 371, 1998, CRC Press, Boca Raton, Boston, London, New York and Washington, D.C.
(12) The use of enriched isotopes was suggested by R. S. Raghavan, of Lucent Technologies. He also supplied the $\times 100$ $^{13}C/C$ as CO gas, obtained from Isotec Inc.
(13) T. Fahey, Isotec Inc. private communication. See also www.isotec.com
(14) Middleton, R., (1974), *Nucl. Instr. and Meth.* **B122,** 35-43, and references therein.

(15) Kilius, L. R., Rucklidge, J. C. and Litherland, A. E. (1988), Nucl. Instr. and Meth., **B31**, 433-441.
(16) Gnaser, H., (1999), *Nucl. Instr. and Meth.* **B149**, 38-52.
(17) Septier, A, *"Focusing of Charge particles"* (Academic Press, New York, 1967)
(18) Kilius, L. R., Zhao, X-L., Litherland, A. E. and Purser, K. H. (1997), Nucl. Instr. and Meth., **B123,** 10-17.
(19) von Reden, K. F., et al., (1999) Conference Proceedings **473**, 410-421, The American Institute of Physics, *Heavy Ion Accelerator technology: Eighth International Conference,* edited by K. W. Shepard. See figure 2.
(20) Lee, H. W. PhD Thesis (1988) University of Toronto, unpublished.
(21) Nadeau, M-J. Private communication.
(22) Galindo-Uribarri, A. Private communication.
(23) Mori, Y. *Rev. Sci. Instrum.* **63** (4) 1992, 2357-2362.

7. Science Policy and Anticipations

Choosing good science in a developing country

R M Adam

*Department of Arts, Culture, Science and Technology,
Private Bag X894, Pretoria, 0001, South Africa.*

Models for national systems of innovation generally involve the interaction of a strong science base with local industry. However, in developing countries the term "science base" may be over-optimistic. Instead, at best we have isolated peaks of scientific excellence. Is it possible to optimise this configuration of peaks to support national development? Or is the useful coupling of scientific endeavour to developmental goals inevitably scale-dependent, with science in low GDP per capita countries linked only to global knowledge paradigms somewhat remote from the local concerns? This paper outlines a system for generating a research portfolio which uses local advantages to generate world class science.

1. Introduction

There are two basic principles underpinning Government support for scientific programmes:

1. The programme could potentially contribute towards addressing social or economic goals.
2. The programme is potentially world class and could contribute to leading edge global knowledge.

Increasingly, particularly in developed countries, these principles tend to overlap. Fiercer competition due to globalisation and the rapid growth of information technology and of biotechnology have led to the shortening of the innovation cycle. The new global economy is more dependent on knowledge than ever before and it is obviously *new* knowledge that delivers competitive advantage. How should developing countries position themselves with respect to the knowledge economy, which depends critically on the national research portfolio and on linkages between this portfolio and the national system of innovation?

I would like to introduce the central question addressed in this paper by quoting from a lecture given by Sir William Stewart, former Scientific Advisor to the British Prime Minister. According to Stewart "Britain does 5% of the world's research and tries to maintain sufficient competence in the other 95% to be able to move into it if necessary." [1] Is this a viable strategy for South Africa, which performs about 0,5% of the world's research, and for other catch-up nations? It

---

[1] W. Stewart, Invited Talk, FRD Policy Lunch Club, 1 March 1995.

seems unlikely that we will be able to cover all possibilities within the remaining 99,5% from our much narrower base. This realisation has stimulated a discussion on whether Government has the responsibility to optimise the national research portfolio and if so, what criteria should drive this optimisation. Should we choose, and if so, how?

I begin by discussing international trends in science-industry linkages and move on to depict the most serious constraint on South Africa's entry into the knowledge economy, namely human resources. Finally, I attempt a framework for determining a national science portfolio and show how it could be applied to South Africa.

2. Science-industry linkages

It is widely held among both scientists and economists that public science is a driving force behind technological and economic growth. Recent research[2] has shown that 73% of the papers cited by U.S. industry patents are public science, authored at academic, governmental and other public institutions; only 27% are authored by industrial scientists. A strong national component of this linkage has also been found[3], with each country's inventors preferentially citing papers authored in their own country, by a factor of between two and four. The pronounced diagonal in Figure 1 depicts this clearly. The growth in linkage has been rapidly increasing[4] - see Figure 2. In general, the cited papers are from the mainstream of modern science; quite basic, in influential journals, authored at top-flight research universities and laboratories, relatively recent and supported by

Figure 1: Percentage research article citations in US patent applications by country

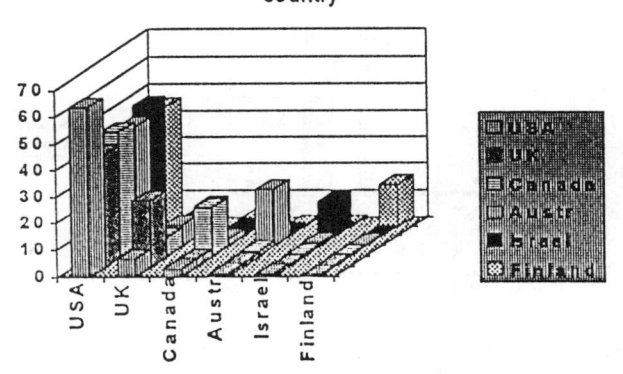

"Science and Industry Linkages", Australian Research Council Report, 2000.

---

[2] F. Narin, K.S. Hamilton and D. Olivastro, Research Policy 26 (1997) 317-330.
[3] "Science and Industry Linkages", Australian Research Council, 2000 (http://www.arc.gov.au)
[4] F. Narin, K.S. Hamilton and D. Olivastro, Research Policy 26 (1997) 317-330.

NIH, NSF and other public agencies. The linkages are much stronger in "new economy" areas than in "old economy" areas[5]. For example, the average number

Figure 2: Citations from patents to papers versus time (smoothed)

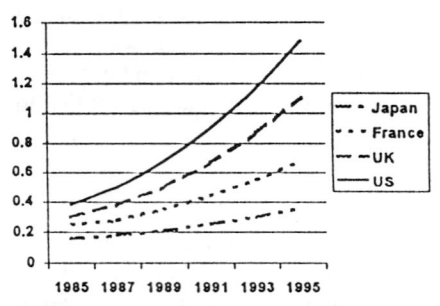

F. Narin, K. Hamilton and D. Olivastro, "The Increased Linkage between US Technology and Public Science", Research Policy 26, p317

of citations of research articles per U.S. patent application differ by a factor of approximately 200 between the fields of biotechnology and machinery parts – see figure 3. This indicates that basic science will play an increasingly important role in global competition. Unfortunately the linkage between science and industry in what the OECD terms "catch-up" countries (e.g. South Africa, Brazil) is much

Figure 3: New Economy is Dependent on Basic Scientific Research

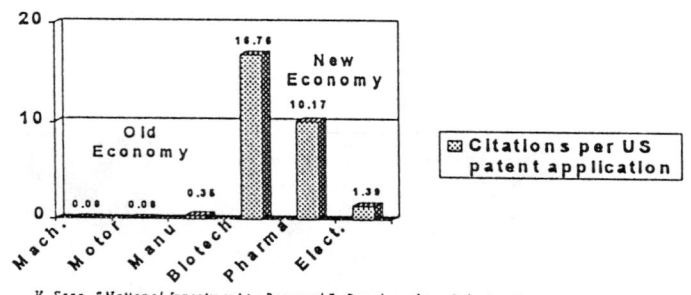

V. Sara, "National Investment in Research", Benchmarking Industry-Science Relationships, Berlin 2000.

weaker. The rates of application for U.S. patents are much lower to begin with and the national components are less significant. Where good science exists it tends to be linked to programmes in developed countries. Local industries generally

---

[5] V. Sara, "National Investment in Research", Benchmarking Industry-Science Relationships, Berlin (2000) - in press. http://www.industry-science-berlin2000.de

purchase technology and related know-how from abroad rather than connect to local scientific thrusts. Towards the lower end of the development spectrum there is an almost total disjuncture between science and industry. The only way that scientific programmes contribute is via human resource development; scientists leave science and join industry and the thinking processes they have acquired are of assistance in the new environment.

The "scale-dependence" of industry-science relationships is summarised pictorially in Figure 4. Given the increasing science-dependence of innovation and

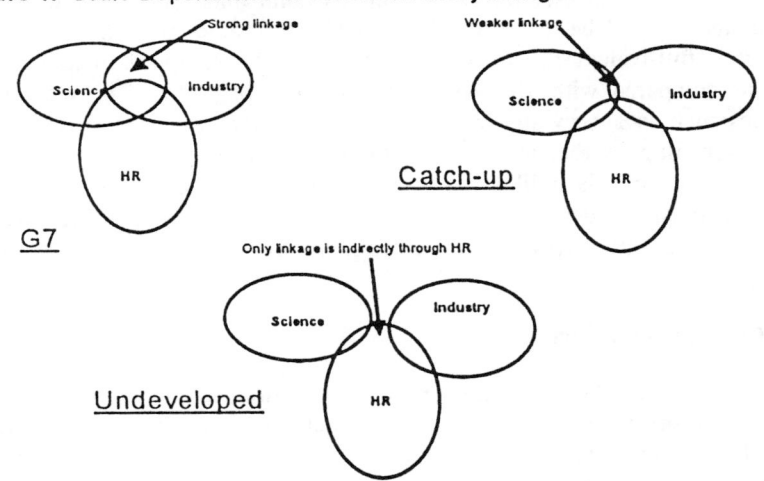

Figure 4: Scale Dependence of Science/industry linkages

consequently of economic development, three key issues facing developing nations are: (i) How to increase the scale of scientific activity; (ii) How to optimise the portfolio of scientific activities and (iii) How to generate linkages be between science and industry.

3. Human resources constraints

Science is a highly globalised activity. Even in advanced economies (e.g. Germany, Canada) there is a worrying trend of the best scientists being drawn towards the highly dynamic United States system. To counteract this trend the affected countries are attempting a range of interventions. For example, Canada has set aside funds for the creation of two thousand university chairs in science and engineering over the next five years. Both France and Germany are in the process of radically overhauling their legislation for the purpose of promoting science-industry linkages in line with the United States' highly successful Bayh-Dole Act.

In South Africa recent studies[6] show attrition rates of approximately 11% per annum from government laboratories and 15% per annum from universities for researchers. Of those who leave, about 5% of the government laboratory scientists and about 22% of the academics emigrate. Given the current fiscal environment and skills scarcity, the vacant posts are not automatically filled.

South Africa spends approximately 0,69% of its GDP on research and development[7]. This figure does not compare well with the OECD average of 2,15%. Nevertheless, it is interesting to note that the average expenditure per full-time equivalent researcher in South Africa is about $100 000, more than the equivalent figures for Australia and New Zealand and about 70% of those for Canada and Japan. The reason for this apparent contradiction is the extremely low number of full-ivalent researchers in the labour force, approximately 0,72 per 1000, as compared with 10,1 for Japan and 4,8 for Korea. The indicators for countries such as Turkey and Mexico are very similar to the South African ones. This clearly implies that any strategy for improving research outputs from less developed countries is unlikely to succeed unless it incorporates a bold human resource acquisition, development and maintenance plan. Individual researchers in such countries suffer less from underfunding than they do from a dearth of colleagues.

4. Choosing niche areas

It is often said that Governments are bad at choosing research priorities. This is less a criticism of the acumen of government officials than a viewpoint that detailed planning can never capture the quicksilver nature of innovation. Nevertheless, technology foresight studies have become the norm rather than the exception over the past decade in developed countries. Japan has a regular five year cycle and Great Britain is beginning its second exercise. Increasing interest is being shown in non-OECD countries too, with South Africa, Thailand and the Commonwealth Science Council taking the lead. The methodology underpinning these studies involves creating a shared vision among decision makers and within key sectors regarding the threats and opportunities likely to be faced in the future and how to respond to these in terms of broad portfolio planning. It does not pretend to forecast the details of actual technologies.

In 1998 a review (somewhat similar to Britain's *Forward Look*[8]) was conducted of the South African national system of innovation[9]. Out of this review there emerged a set of criteria for recognising what were termed "core competence clusters". A

---

[6] Loss of Skills from the Public Sector, Department of Arts, Culture, Science and Technology (1999).
[7] Survey of Resources Allocated to Research Development, Department of Arts, Culture, Science and Technology (2000).
[8] Forward Look, Department of Trade and Industry (UK) (1998).
[9] System-wide Review, Department of Arts, Culture, Science and Technology (1998).

core competence cluster is an effectively functioning set of infrastructure, technology, capabilities and platforms which provide outputs to clients and stakeholders which they can appropriate and evaluate. A nation's technological competitiveness would rest on an appropriate portfolio of such clusters.

Criteria for recognizing core competence clusters were also developed. In general they should:

- Provide a global competitive edge;
- Be sustainable;
- Be hard to emulate;
- Have multiple applications;
- Attract international collaboration and investment.

What type of core competence clusters should South Africa try to establish? The information society concept lies at the heart of the knowledge economy. Increasingly, biotechnology is being seen as equally critical. The South African Foresight study proposes that we invest in establishing platforms in R&D, training and technology transfer in base information and communication technologies such as (i) access technologies (e.g. connectivity, low-cost satellites and stratospheric communication), (ii) spatial numeric environments, (iii) Human language technologies and (iv) security technologies. The Foresight also recognizes that biotechnology, the term given to the wide range of agriculture, industrial and medical technologies that make use of genetic manipulation of living organisms to create new products, will be a key factor in improving the quality and quantity of the world's supplies.

Clearly we cannot afford to be on the wrong side of information divide or the bio-divide. The generic technologies underpinning rapid development in information technology and biotechnology are nonnegotiable in the modern world. Are there other areas in which it is possible to achieve a world class standard? If so, how do we recognise them? In general we can define two additional broad features which are likely to underpin the more specific criteria for core competence clusters listed above:

1. **Scientific areas where there is an obvious geographic advantage.**

In the case of South Africa there are several of these which stand out. Examples are:

    i   Astronomy (we have good access to the Southern skies and the engineering capability to build telescopes locally);

> ii   Human palaeontology (we have excellent sites in the Krugersdorp region dating from shortly after the bifurcation between apes and humans occurred);
>
> iii   Biodiversity (the Cape Floral Kingdom is the most diverse of all the seven floral kingdoms).

Other possible phenomena or systems on which it has been or would be possible to base good science include the Kaapvaal Kraton (geology) and the South Atlantic Magnetic Anomaly (geomagnetism and space science).

2. Scientific areas where there is an obvious knowledge advantage.

Important South African examples are:

> i.   Indigenous knowledge (clearly the collective inherited and evolving knowledge systems of indigenous peoples constitute a national competitive advantage);
>
> ii.   Technology for deep mining (geological conditions and economic imperatives have pushed South Africa to the forefront);
>
> iii.   Microsatellite engineering (although large multinational consortia now dominate the communication satellite market, South Africa has retained a niche competence in microsatellites deriving from a fusion of defence spin-offs and university research);
>
> iv.   Encryption technology (spin-offs from State investment in the defence sector have generated significant foreign exchange recently);
>
> v.   Fluorine technology (high entry barriers mean that the competence developed in the uranium enrichment programme could be turned to advantage).

The lists above are intended to be illustrative rather than exhaustive. They illustrate a way of looking at science which attempts to prioritize in terms of likely outcomes, moderated by what is deemed necessary in terms of national competitiveness in the broadest sense of that term. Despite the fact that the motivations of individual scientists are generally fired by intellectual curiosity rather than by the weighing of potential outcomes, it is necessary for decisions to be made unsentimentally. Not to prioritize in a way which attempts to optimize impact is irresponsible and potentially wasteful. The public and their elected decision makers respond positively to success. It is a sound strategy for scientists, science administrators and policy specialists to develop a common approach towards maximising the chances for success.

5. Conclusion

In this paper I have attempted to show three things:

i. The evolution and the scale-dependence of industry-science relationships;
ii. The critical importance of human resources to the national research and development effort;
iii. An approach towards optimizing the national science and technology portfolio.

The space available has dictated suggestive rather than rigorous arguments. Nevertheless, I believe that there are important lessons here for the nurturing of a healthy science base in a developing country.

SCIENCE PARTNERSHIPS FOR AN AFRICAN RENAISSANCE: A FRAMEWORK FOR *NGUMZO*

A.M. KINYUA

Institute of Nuclear Science, University of Nairobi, P.O. Box 30197, Nairobi, Kenya

E-mail: antonykinyua@insightkenya.com

Presented is a conceptual framework for African Renaissance that has to involve scientific, social and technological partnerships between universities in Africa together with those in developed countries and industry emphasizing the need for *Ngumzo* (i.e. dialogue). This framework is based on the model of a microprocessor. Inputs for this partnership when processed do lead to different outputs. It is being proposed that national policy and delivery of inputs to our national institutions are insufficient to change these relationships or partnerships for an effective African Renaissance. The alternative is that it will be the central role of each individual institution in Africa, in all its complexity, to pay more attention in the planning, management and development of its mission and vision strategies, enhancement of staff performance and evaluation of its technical education among other inherent activities. The role that science can play in nurturing a culture of learning and capacity building in Africa will be discussed and presented by highlighting the future challenges and justification of an Africa Renaissance Group (ARG).

1 Introduction

During this century there will be a great need for African intellectuals, universities and industry to work together as partners in research, science and technology for the development of the African continent. His Excellency, President Thabo Mbeki of Republic of South Africa has already given the African intelligentsia a challenge to come up with a renaissance program offering African solutions to solve the many problems facing the continent. This is in terms of health and diseases, scarcity of food, energy, environmental degradation, recovery of African pride, the confidence in ourselves that we can succeed as well any other in building a humane and prosperous society [1].

To create intelligentsia universities rely on industries, sound economy and good governance to educate its students while industries and governments rely on an educated workforce to remain effective in policy implementations, competitive and profitable [2]. These institutions have therefore to have a symbiotic relationship that is at the same time interdependent.

Satisfactory education-industry relationships have long been recognized and given very high priority in the industrialized countries [3,4]. UNESCO has been organizing workshops where the main topic of discussion was education–industry cooperation. This started in Paris, July 1970 within the framework of Resolution 2.221, which was adopted by the General Conference at its fifteenth session.

UNESCO organized a meeting of internationally recognized experts on engineering education and training. The meeting urged that *"future activities in education-industry co-operation should be on a larger scale."* Since that time, UNESCO sponsored many regional meetings on the future activities on education-industry co-operation: Nairobi, Kenya (December 1972); Cordoba, Argentina (May 1973); and Manila, Philippines (October 1973). This led to the formation of the International Working Group on Education–Industry Co-operation which met in Paris (May 1974, September 1976) and then in Cairo (September 1978).

Other meetings by UNESCO on this partnership theme included: (i) the Kilimanjaro declaration on the Application of Science and Technology Development in Africa in Arusha, Tanzania (July 1987); and (ii) the Workshop on Development of the University-Industry-Science Partnership Programme in Africa in Nairobi (February 1994). These two meetings called upon the African Scientific communities to organize themselves to be able to contribute more effectively to African development and to improve their co-operation with one another, as well as, with scientists from other countries and regions of the world. Invitations were also extended to African industrialists and other persons playing an active role in the economy. This was with the hope that they were to make better use of science and technology for the improved production of goods and services, and also to contribute financially to the funds specially set up in order to stimulate, develop and undertake research and development.

Other organizations that have organized these partnership programmes include UNDP and UNIDO [5]. In April 1997, UNDP launched the Otto Essien Young Professionals Programme (OEYP) in Nairobi, Kenya. UNDP has also continued with OEYP activities where in June 1998 an Industry/Student consultative meeting was held in Nairobi which was then followed in July 1999 by another University/Industry Links Symposium, *"Partnership for Effective Industrialization"* also held in Nairobi. The University of Nairobi in 1997 appointed an OEYP Liaison Officer. This was despite having established a University's Industry Link Committee in 1987 [6] whose activities and membership are not well known.

Barriers to effective communication between our universities and other institutions in Africa and the rest of the world are very real. Industrial support for university research in most African universities is non-existent while the total academic Research and Development expenditures is less than one per cent of GNP of most countries in the continent. Considering the effects of budgetary reductions by the exchequer and a strong desire within our universities to seek other sources of funding, the increasing pressure to access fundamental research within other universities in developed countries and industries, one wonders why this scenario has not been reversed.

This paper does not seek to provide a universal solution to this dilemma but to identify some of the factors that need to be worked out so that the parties concerned do not keep each other at arm's length. The issues pertaining to lack of African pride, confidence, science driven economies and industrialization in Africa has been addressed previously [1,6]. The issues raised ranged from lack of proper systematic policies and coherent institutional frameworks, to resource availability, expansion and integration of markets plus the development of appropriate cultural practices and attitudes.

African Renaissance cannot be effective without a common strategy and understanding of the respective inputs, objectives and willingness by the trained human resources and institutions involved. This would involve striking a common denominator so as to take advantage of their respective strengths in order to maximize the outcomes to all stakeholders. The role that African scientists can play in nurturing a culture of learning and development of human resources in Africa is of paramount importance. This is in the light of the pan-African interests of Profs. Jesse Mugambi, Micere Mugo, Ngugi wa Thiongo, Dialo Diop, Namane Magau, Ivan Sertima, Karima Bounemra, Mongane Serote, Friedel Sellschop among many others. They have all been relevant to scientific and social research, capacity building in the continent and do continue to act as a catalyst for the formation of an African Renaissance Group within the continent and the Diaspora.

The issues discussed in this paper are not from a formal research exercise but are the issues that have found their way into various discussions and/or fora of academics, industrialists, and government representatives. Such groups now exist in most countries as National Academy of Sciences, National Councils' for Science and Technology, Industrial Research and Development Institutes, Institution of Engineers, the Medical Societies, Meteorological Societies, Physicist Societies, Students Associations and Chemical Societies among other professional bodies. The author hopes that the comments on *Ngumzo* framework will provide a useful resource to these groups and especially to those who seek to benefit and improve the collaboration of universities in the continent and Diaspora, industry and government linkages that will enhance mutual research and development endeavours of Africa.

2 Conceptual Framework for *Ngumzo*

For Ngumzo to be effective the hypothesis being put forward is that national policies and the delivery of inputs to our institutions are not sufficient to change partnerships. Instead, the alternative hypothesis being put forward and argued is that it is the central role of the individual institution, in all its complexity, that needs more attention in the planning, management and evaluation of its research, technical education and extra-curricular activities. Figure 1 is utilized to demonstrate this hypothesis pictorially. This framework is presented from the perspective of computer architecture, consisting of the input, central processing unit (CPU) and the output. It is adapted from the conceptual framework of the facts that determine school effectiveness [7].

The power of CPU has been increasing in leaps and bounds such that a Pentium CPU in 1998 had 7.5 million transistors enhancing its capability ten thousand-fold [8]. Such large number of transistors is necessary if various inputs have to be

Factors of ARG Effectiveness

Institutional Environment
1. Student Expectations
2. Staff Attitude
3. Order and Discipline
4. Curriculum Organization

Enabling Conditions
1. Effective Leadership
2. Capable Staff
3. Autonomy and Flexibility
4. Residence Time in Institution

Training and Learning
1. High Learning Time
2. Variety in Strategies
3. Regular Assignments
4. Regular Staff/Student Assessment Feedback

ARG Support Services
1. Institution and Community
2. Authorities
3. Adequate Materials
   a) Textbooks, Manuals
   b) Adequate Facilities
   c) Frequent and Appropriate Trainer Development
   d) Student Extra-curricular Activities

Student and Staff Characteristics

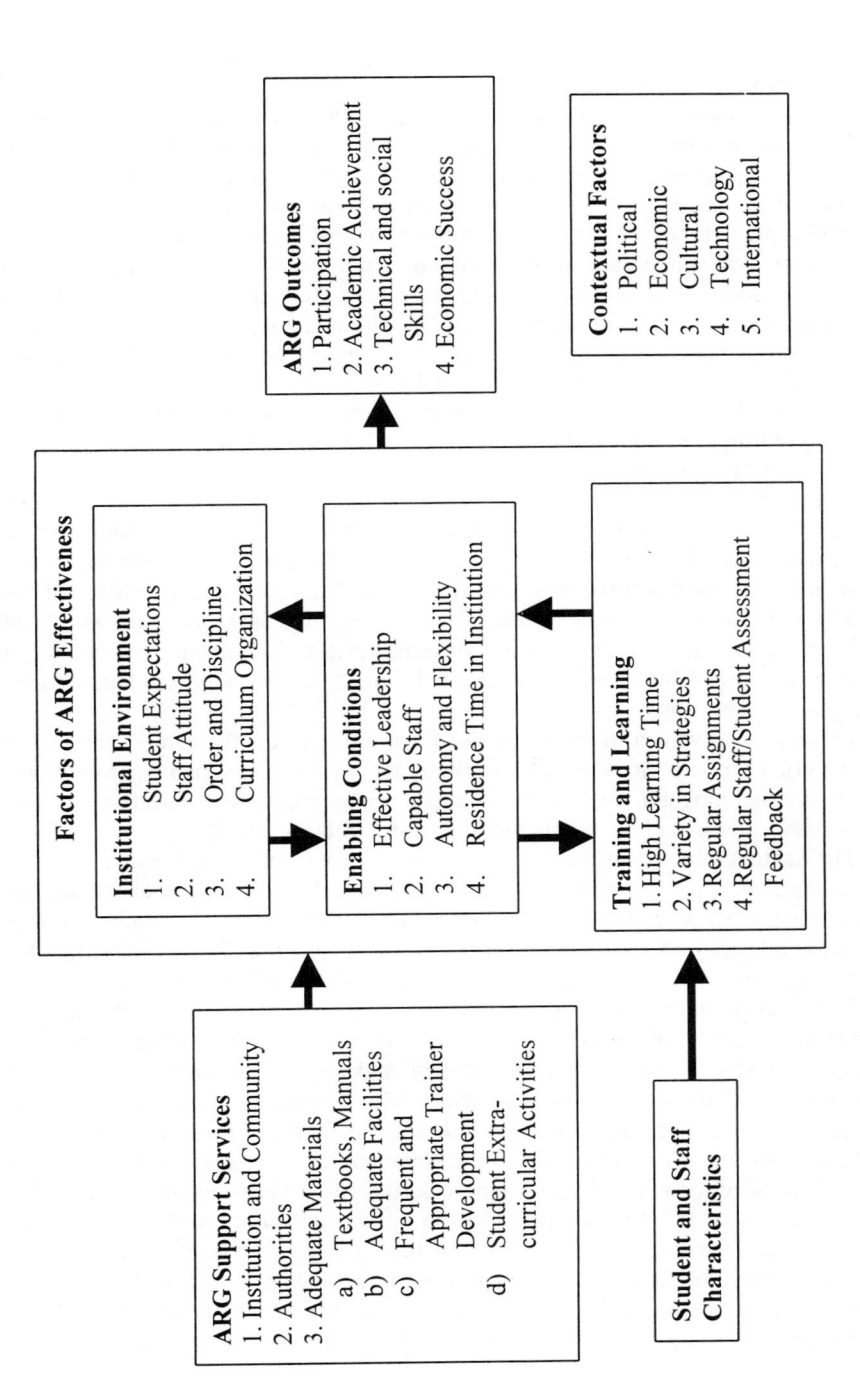

ARG Outcomes
1. Participation
2. Academic Achievement
3. Technical and social Skills
4. Economic Success

Contextual Factors
1. Political
2. Economic
3. Cultural
4. Technology
5. International

Figure 1. A Conceptual Framework for Science Partnerships for an African Renaissance

processed within a short time. The input to the CPU is very important. We are reminded that *"garbage in, garbage out."* It is therefore necessary to make sure that the data input is correct and of high quality. The inputs can be classified into two groups; i.e. support services of ARG plus the staff and student characteristics. The support services include the roles of the individual institutions and the community, authorities plus supply of adequate materials to staff and students. Development models show that where there is strong support from the community, authorities and institutions (both human and capital infrastructures), the activity being initiated will definitely succeed. It is therefore necessary that these factors are adequately thought out and agreed upon before any ARG project support commences. Of course the need for adequate materials for both staff and students whether at the industry or the university is also very important. This would improve on the technical expertise especially during the industrial attachment of university staff and students so as to enhance their professional development.

The characteristics of both staff and students are also very important to the CPU output. This entails the need for high social skills of students especially during their upbringing. Previous parental guidance and support in their education endeavors becomes therefore very important. In addition, the characteristics of the teachers such as their subject mastery, experience, time input to their work and confidence are some of the many parameters that will also determine the quality of the student output and their relevance to the needs of the industry and African society.

Factors of effectiveness by our African institution forms the central processing unit (CPU). There are three sections: institutional environment, enabling conditions, training and learning activities. These sections can be viewed to respectively mimic the memory, control and the arithmetic logic unit (ALU) in a CPU.

The factors encompassing the institution environment, though not restrictive and exhaustive, encompass the memory section of the CPU. This can either be viewed both in RAM and ROM categories. The institutional environment, whether it is in an industry or a university department in the continent, is very important for inculcating good performances, tradition, order and discipline where high ethical standards and productivity are maintained [9]. The need to remember and store all these requirements is of paramount importance to the well being of our institutions. This is especially if expectations of staff and students can be raised as a result of clear understanding of the mission and vision of their institution, plus the development of curriculum and programme of activities suited to each of the institutional needs. Curriculum within the university can therefore be developed so as to suit some of the industrial and society needs. The benefits accrued from such a curriculum would be very beneficial to the nation, university staff, students and industry. The incentives and rewards, whether monetary or otherwise, are also necessary. They are usually ingrained in peoples' memory and will always be remembered as shown previously by B.F. Skinner [10]. This was through his experiments that showed an animal rewarded for good behavior (positive reinforcement) did learn much more rapidly and retained what it learned more effectively than an animal punished for bad behaviour (negative reinforcement) [10]. The staff and students can then serve the purpose of remembering the need to excel both as teams or as an individual.

The enabling conditions serve the role of a control logic unit as in the CPU. Effective leadership is necessary and is of paramount importance in any given institution. Good leadership in any place of work will always enhance **harm**ony, **tea**mwork, **co**nsensus and **de**cision (HAM TEA CODE) [11]. This is very important in enhancing both university interactions within and outside of the African continent. The spin-offs could lead to capacity building, better industrial processes, patents and higher profit margins. Staff development and recognition of capable and potential individuals also emanate from an effective institutional leadership. The benefits of good leadership are always enormous as stated by Mr. John Whitmore, "... *leadership is a deft, discreet activity securing provision of whatever is missing in order that sparks fly and reactions take place*" [12].

Autonomy and flexibility of private sector operations is one area that public universities and institutions in Africa would benefit from the interactions with other universities in the developed countries plus the industrial sector. This is especially in the finance and day-to-day management practices. African public universities should therefore seek ways and means to get this autonomy from the central government. It would enhance their financial operations and fundraising activities especially through the industrial and collaborative linkages with educational institutions in the developed world.

Another aspect of this enabling condition would be the residence time of staff and students in both the industry and universities. In a typical CPU, the clock speeds are in the nano and pico-second ranges or less. This is important for fast computer operations and control. However, residence time of staff/students at any given institution is very critical to the level of expertise gained. It is therefore necessary that university and government authorities in Africa ensure that residence time of students especially when learning or when on attachment at industries is as long as possible. Unnecessary closures due to external factors should be avoided as much as possible.

The training and learning section can be viewed as the ALU in a typical CPU. Again, the factors depicted here are neither restrictive nor exhaustive. It can therefore be hoped that due to the high learning and residence time of students and staff at any given institution, there will good learning and development of necessary skills. This will accrue from the regular assignments especially those, which have direct product applications as per customer and market demands. The regular feedback systems from both the staff performance and student assessment are important tools to assess possible problems and solutions in an institution. However, this is only possible if the institution has developed strategies from its mission and vision statements such that it can be expected to respond to any challenge in some kind of a digital nervous system [8]. This can help the African universities to respond quickly to various society and market demands by industry especially new product development and management of competition. This aspect mimics the ALU where numbers are crunched very fast such that the CPU can provide an output quickly. The institutions would then be able to complete and perhaps improve on their products and services.

The output from the CPU can be sent to the monitor, saved onto the diskette or stored on the hard disk. For the university partnerships, the outputs are expected to grow like a tree (Fig. 2) growing from some of the rich African soils with plenty of natural minerals. This might develop into new curricula design, exchange of visits,

Ph.D. and M.Sc. Programs
Exchange Visits
Fellowships
Conference and Workshops
Distance Learning
Industrial Liaison Offices
Consultants
Industry Based Projects
Regional Co-operation
University Councils and Advisory Boards
Equipment and Spares Support

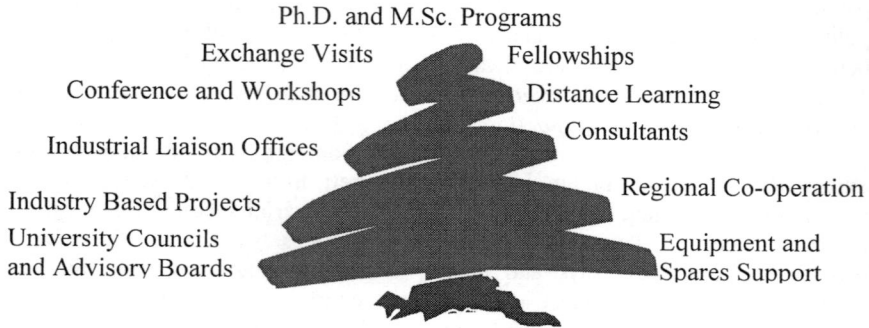

Figure 2. "Tree" of African Renaissance Group Collaboration

research collaboration, equipment and spares support, participation of industry in university councils and advisory boards, joint conferences, seminars and workshops, creation of industrial liaison offices at our universities, participation of staff and students in industry-based project consultancies, fellowships and distance learning among many other benefits [3,4,13]. All this would be expected to lead into economic success and a new renaissance of African public and private institutions.

This conceptual framework takes cognizance of the fact those contextual factors such as politics, culture, technological backwardness, globalization and economics do play a very crucial role to the outcomes of this model. It is the hope that the basic inputs to the "CPU", which governs our African institutional effectiveness, will be of high quality such that no "*garbage-in-garbage out*" scenario is perpetuated.

3 African Renaissance Group (ARG)

The ARG team is planned to be composed of all those group of scientists, sociologists and technocrats interested in the re-awakening of Africa, both scientifically and technologically. The performance of ARG will depend very much on the staff in our universities and those collaborating with them both within and outside of Africa. This is not an easy task as discussed in the conceptual framework model for *Ngumzo* [11]. It is a subject that will continue to attract much writing, seminars and workshops because human nature is very dynamic. In addition, the requirements and satisfaction of different individuals will always be different now and in the future.

3.1 Justification of ARG

- ☐ Proposed activities are supportive of our nations' human resource development endeavors;
- ☐ For sustenance of current and future programs at our institutions, it is necessary to enhance collaboration and support in the training of students and staff to Ph.D. and M.Sc. levels;
- ☐ ARG activities will lead to capacity building, introduction of new analytical techniques in our nations and the region by way of enhancing monitoring of environmental pollution, understanding the trans-boundary movement of pollutants and their health effects;
- ☐ Results of ARG projects would benefit tremendously the activities of planning and policy formulation of our Institutions, the government especially in the sectors of higher education, health, industry, science and technology. This is depicted in an ARG 'Tree of Knowledge and Activities' that can be enhanced for the mutual benefit of our peoples (Fig. 2).

Among many other issues that might not be addressed in this paper, the following will be addressed and should be rectified where possible. First, the changing nature of information as previously discussed by de Bono [14], is also important and ARG team needs to get involved in its acquisition, generation, organization, processing, dissemination and decision making especially during this information age era. Data or information has to be processed by the brain so that there can be some action by an individual or an organism. After receipt of the information or data the individual can then be involved in seven activities. These encompass those aspects of sitting and waiting so that other individuals are the ones to do something rather than you getting involved! Included also will be those aspects of managing the data by way of exposing, borrowing, organizing, processing, clarifying and finally, the need to make a decision. If Africa Renaissance has to be realized in this millennium, then there is a great need for the ARG team to get active especially in the last six categories shown above. It is also important also to remember: *"Knowledge is doubling every 18 months. It's impossible to know everything. Your best hope is to learn how to access information. Your biggest mistake may be your unwillingness to pay for information"* [15].

Secondly, all ARG needs to *THINK BIG*. This is as illustrated by Carson and Murphy [16] in their book with same title. They discuss issues pertaining to individual talent, being honest, having an insight for future progress, being nice to other people, having a knack for knowledge acquisition by reading good books, striving to have an in-depth knowledge of your profession and trusting in God. These are very pertinent issues to ARG team.

Thirdly, The S*even (7) "C's"* in management as discussed by Martin [17] in her book *Martin's Magic Motivation Book* also needs to be part of the ARG team. The team will need to perform their duties taking it as a challenge with a charming attitude and total commitment. This aspect will then require that the ARG team do become good communicators with confidence in themselves, being creative in tackling various problems encountered and also being current in their field of training. The latter will be as a result of the knowledge acquired from reading good books, participating in informative conferences such as the recent Lüderitz 2000 and collaborating with people of good will.

Fourth, the future challenges to the ARG team will also encompass those aspects of knowing that you are a knowledge worker. Drücker [18] discussed this in his recent book, *Management Challenges for the 21st Century*. The ARG team needs to know clearly *what is their task, what is their strength, what will be their contribution* to this African Renaissance, *how do they manage themselves, whether they treat themselves as an asset or liability* to their institutions and whether *quality is the essence of their output*.

Finally, a dynamic development of an African Renaissance Group must encompass perspectives in energy, appreciation and understanding our environment, insights into health and diseases, pathways for informatics, and the structure of matter. Therefore the future challenges to ARG will involve the knowing and participating in the accomplishment of an ARG Mission and Vision. The ARG Mission can then be proposed as: *to engage in activities where its members have competitive advantage including the commercialization of its resources (human, intellectual property, specialized services and facilities) for academic and non-academic benefits to its members and other stakeholders*. The ARG Vision can be proposed as: *to be viable and competitive contributing academically or otherwise to the achievement of the Mission of our institutions*. It is therefore necessary that all parties in our team need to be clear on the Mission and Vision of the ARG. Among other many factors not discussed in this paper, understanding of the Mission and Vision of ARG is hoped that it will enable staff to perform their duties with diligence, know where they belong, their expected contribution to society and whether they do regard themselves as assets or liabilities to the institution. It is therefore hoped that ARG teams will strife to acquire knowledge and build a humane and prosperous society and always remember that: *people are destroyed for lack of knowledge because they have rejected knowledge* [18].

References

1. Makgoba, M.W. (ed.), African Renaissance: The New Struggle (Mafube Publishing (Proprietary) Ltd., Sandton and Tafelberg Publishers Ltd., Cape Town, 1999) pp. xiii-xxi, 305-417.
2. Burlington, J.D. *Int. J. Technology Management* **8**, Nos. 6/7/8, (1993) pp. 440-446.
3. Fishwick, W., Strengthening Co-operation between Engineering Schools and Industry, Studies in Engineering Education 8, UNESCO (1983) pp. 4-49.
4. Konecny, E., Quinn, C.P., Sachs, K. and Thompson, D.T., Universities and Industrial Research (The Royal Society of Chemistry, Thomas Graham House, The Science Park, Cambridge CB4 4WF (1995) pp. 1-9.
5. UNIDO, Industry and Development: Global Report 1986 (UNIDO, Vienna (1986).
6. Anyang' Nyong'o, P. and Coughlin, P., Industrialization at Bay: African Experiences (Editors) (African Academy of Sciences, Academy of Science Publishers, Nairobi, Kenya (1991) pp.vi-27.
7. Heneveld, W., Planning and Monitoring the Quality of Education in Sub-Saharan Africa (AFTHR Technical Note No. 14, Human Resources and Poverty Division, Technical Department, Africa Region, Washington D.C., World Bank (1994).
8. Gates, B., "Business @ The Speed of Thought: Using a Digital Nervous System" (Warner Books Inc., New York, NY 10020 (1994) pp. 143.
9. Kinyua, A.M., University-Industry Partnership for effective Industrialization: A Framework for *Ngumzo* (M.J. Alport and E.C. Zigu (Eds.), Programme and Abstracts of 3rd International Conference on Physics and Industrial Development, Bridging the Gap, University of Natal (2000) pp. 20.
10. Carnegie, D., How to Enjoy Your Life and Your Job (Pocket Books, Simon & Schuster Inc., New York, NY 10020 (1986) pp. 68.
11. Kinyua, A.M., Enhancing into the Next Millennium Staff Performance (Unpublished Seminar Report on Technical Staff Training, College of Physical and Biological Sciences, University of Nairobi (1999).
12. Thompson, H., Can you Sing? -The Secrets of a Creative Group (The East African, October 25-31 (1999) pp. 30.
13. Antonio, J., Promoting University-Industry interaction through formal collaboration agreements: The case of the University of Science and Technology, Kumasi and the Volta River Authority (Meeting on Development of the University–Industry–Science Partnership Programme in Africa, UNESCO, Nairobi, Kenya, February 16-22 (1994).
14. De Bono, E., What can one do about Thinking? *In: Lateral Thinking for Management: A Handbook*, Peguin Books (1971) pp. 43.
15. Ogden, F. alias *Dr. Tomorrow*, *Sunday Nation*, January 2, 1994.

16. Carson, B. and Murphey, C.B. *Think Big*, Zondervan Publishing House (1996) pp. 278.
17. Martin, P., *Martin's Magic Motivation Book*, St. Martin's Press, New York, NY, (1984) pp.117.
18. Drücker, P., *Management Challenges for the 21st Century* (1999) pp.135-195.
19. Hosea, *God's love for his people*, **Chp. 4**: 6, Holy Bible, Revised Standard Version, British and Foreign Bible Society (1980) pp.794.

POLICY FRAMEWORKS IN SCIENCE AND TECHNOLOGY: THEN, NOW AND TOMORROW

ADI PATERSON
*CSIR and University of Pretoria, PO Box 395, Pretoria,
Gauteng, 0001, South Africa.*
apaterso@csir.co.za

Science and technology policy has become very dynamic since the first set of interventions in the early to mid-eighties. These interventions represented the end of the "Endless Frontier" era of very high S&T spending following the Second World War. The way S&T policy developed from the mid-eighties and, in particular, how these developments have impacted South African science and technology are analysed. The current situation of S&T in South Africa will be discussed from the point of view of the system of innovation policy that was introduced in 1996. The analysis will suggest that good progress has been made with this policy approach, but some risks, including the capacity of the system and its responsiveness will be made. This policy framework will be further examined from the perspective of a knowledge society and economy as contrasted with innovation policy. A framework for developing and analysing strategy is introduced and applied to the domain of knowledge and innovation. From this review and analysis current challenges in S&T policy, as a result of the genuine ambiguity in global responses, will be presented. In the light of this, an attempt is made to propose a future direction for S&T policy in South Africa.

1.1.1 The Recent History of Science Technology and Innovation

After the positing of the Endless Frontier by Vannevar Bush in the late 1940's science entered a golden era with very significant expenditures, complemented by massive military technology spending in the developed economies. This pattern was followed in developing economies including South Africa. Key responses to this spending were the massive increase in the pool of scientific human resources, the extension of the scientific paradigm into the social sciences and the creation of national research infrastructures in Universities and National Laboratories. In the governmental and private sectors "big was beautiful" with the creation of state corporations and the huge success of large corporate organisations (which themselves had large R&D capacities to rival those of National Laboratories).

Economic woes and the emergence of small firms as an important sector, together with the recognition that economies without a strong science base (like that of Japan) were posting massive successes, led to a re-evaluation of the performance of the science and technology system.

Globally, response to this re-evaluation varied. In the United States funding levels stabilised but direction-setting for the system was not invoked. Rather there was an attempt to make S&T investments more productive for small businesses.

This included the Bayh-Dole Act which placed an obligation (and an incentive) on institutions receiving public funds to seek to secure the intellectual property developed and to make it available to American firms. In addition government guaranteed venture funds were created (by the Small Business Investment Corporation). Both of these initiatives recognised that the large corporates were not truly the engines of new economic growth.

In Europe large research organisations were reoriented from discipline-based activities towards market-oriented strategies, which was also associated with reduced real funding levels in many cases. The market focussing of these national research laboratories converted them into what are now known as Research and Technology Organisations (if one looks at the main competences) or Contract Research Organisations (if the focus is on the primary mode of interacting with the market).

John Ziman wrote of "Science in the steady state", a metaphor which characterised this era very effectively, but perhaps underestimated the extent to which science was undergoing a change with respect to the public perception of how it provides (or does not provide) benefits to society at large.

The more careful measurement of science and technology systems grew during this period driven by the various manuals developed by the OECD and the NSF in the United States. Within science and technology there was a different tectonic shift taking place. This was the emergence of new interdisciplinary fields which broke down the disciplinary rigidity of the sciences and proved to be happy hunting grounds for application oriented scientists and engineers.

The combination of the collapse of these disciplinary barriers was also fuelled by mega science projects (such as CERN and less obviously NASA), which demanded the type of inter-disciplinary collaborations (and inter-institutional partnerships), which had, in the military sphere, been the characteristic of the large weapons programmes.

Observers of science and technology systems started to see inherent virtues in such collaborations, particularly where there were positive spin-offs and transfer and embodiment of knowledge in new products and services. This process, now termed innovation, sparked a new revolution in science policy: rather than simply orienting institutions towards the market in a utilitarian way the innovation system analysis suggested that the system as a whole could operate in a more virtuous way if the relationships and interconnections were optimised through careful design of policy instruments. One of the key modes of achieving this was collaborative research and the use of more competitive inter-institutional funding instruments.

In the United States the second generation of venture funds had stimulated entirely new industries in software, communications and computers, science and technology parks had mushroomed (also at some key locations in Europe (Cambridge and Eindhoven are well-known examples) and economic analysts began to consider the crucial importance on knowledge institutions in economic development. This shift brings us to the present.

The South African Responses

The South African science, technology and innovation systems were not isolated from these changes. The de facto system in South Africa in the mid-eighties was a military research and technology driven system (including massive expenditures on a nuclear weapons capability) that had an institutional tapestry similar to that of science intensive commonwealth countries (but with a more European flavour than most). In addition significant components of the system were embedded in the civil service (including the geological survey and agricultural research). The international changes were replicated in South Africa with a shift to contract research in the largest national laboratory, the CSIR, (mirroring the moves in large European public research institutions). In addition the civil service based research infrastructures were corporatised and turned into parastatals. The period from the late eighties to the present was also characterised by disengagement by government departments followed by re-engagement with the parastatal research and technology organisations (a process still taking place in the case of the most recently created institutions).

In contrast to the developed economies, however, government funded R&D expenditure has been falling relative to GDP. The quantum of spending is also low compared to successful emerging economies. In the context of constrained funding there is a focus on efficiencies and performance in the short-term.

Notwithstanding these pressures South African S&T policy has adopted a "system of innovation" design in order to redirect resources within the context of the post-cold war, post apartheid demands in a changing society. This has included the introduction of new funding instruments such as the Technology for Human Resources for Industry Programme and the Innovation Fund, which emphasise partnership and collaboration.

A national Foresight exercise, which covered key sectors such as Energy and Manufacturing as well as more innovative analyses on, for instance, Youth and Financial Services, was conducted in the late 1990s. All of these processes, and the policy interventions, have so far failed to achieve agreement and commitment that higher levels of investment in science, engineering and technology are crucial for the future of the country. The R&D "underpinnings" of science, technology and innovation remains a poorly articulated part of national policy for the future in South Africa, at a time where there is no doubt in the developed world that R&D spending is critical to national success. "[The] difference between private and social rates of return is the primary reason why governments must support R&D spending…If governments don't support R&D spending, much too little R&D will be done…The economic payoff from more social investment [government funding] in basic research is as clear as anything is ever going to be in economics." [1]

1.1.2 Enter the knowledge age

While the "systems" debates have dominated the policy stage (the science system, the science and technology system, the innovation system and so forth) a second less recognised strand has always been present, but has now reached centre stage: the knowledge required for the systems to work.

It is the major thesis of this analysis that both innovation (as a process) and knowledge (as required content) are necessary conditions in using S&T to improve the quality of life and competitiveness of a nation. It is critical to improve the relationships, linkages and flows within the system (process improvement and re-engineering) as well as paying attention to the essential content. The utility of Foresight, roadmapping and review processes, including the National Research and Technology Audit (in the South African context) are essentially attempting answers to the knowledge (current or future) question. However many of these initiatives are bedevilled by the intangibility of much important scientific knowledge (technology is somewhat easier, and product and service "knowledge" is apparent to every consumer). In addition, surveys like Foresight, in focussing on explicit aspects of knowledge pay insufficient attention to "capability" knowledge – the "things" that, for instance an industrial system can do repeatedly – such a deep-level mining in South Africa and "competences" - knowledge that is genuinely valuable as a platform for further industrial and organisational development in new markets. The Science Council review, mandated and executed by South Africa's Department of Arts Culture Science and Technology was unique in this respect, in seeking to identify genuine national and institutional competences (notwithstanding considerable resistance from some institutions).

Competences and capabilities represent the easiest and most available platform for the extension of national knowledge. Indeed if these platforms are not treated strategically and managed as assets they rapidly become "rigidities" and merely encode a memory of historical spending, that rapidly "hollows out" in relation to knowledge development globally.

At present South Africa has no explicit knowledge strategy within the policy space of science, technology and innovation – in common with many other nations. In very large economies that are well endowed with resources, knowledge strategies become less important as there exist an abundance of knowledge institutions, which effectively compete to "overproduce" new knowledge. In these sophisticated economies there are many willing receptors for new knowledge: venture capitalists, business school graduates, consultancies, large corporations, NGOs and activist groups. These players rapidly take up new knowledge and put it to use economically or in the greater interest of society (such as the case with the environmental movement which has become a very effective adopter of scientific knowledge). In less developed economies these modes of adoption and receptor capacity are low (relatively speaking) and in that sense the innovation system has a structural deficiency with respect to effective uptake of locally generated knowledge.

South Africa is not excluded from this deficiency. The industrial sector develops essentially as an evolution of current capabilities. New sectors (in silicon based materials and devices and biotechnology for example) are not represented in our industrial landscape to any significant degree. If the system of innovation alone is interrogated, it seems as if only incremental strategies are available to develop the economy.

To move forward both innovation (the process) and knowledge (the content), need to be treated strategically. Table 1 indicates some of the differences that result from innovation centred and knowledge centred strategies.

Table 1: Comparison of aspects of knowledge and innovation strategies

Policy Paradigm	Innovation	Knowledge
High level analysis	System of innovation	Competences/niches
Basis of effectiveness	Linkages to enhance innovation	Ability to develop and harness knowledge
Basis of Planning	Market led	Education and research led
"Getting it out the door"	Technology transfer and diffusion instruments and institutions	Clusters, communities of practice and knowledge networks
Basis of value	Contracts, licenses and ventures	Intellectual capital

What strategic approach is required?

In a seminal article on strategy [2] four different levels of uncertainty are identified which inform the type of strategies that have to be developed. In the case of each level of uncertainty different strategic analysis tools are available to infrom the strategist of the most effective way forward. Of course there is a fifth basis for strategy which could be termed the "wisdom of the current" which leads policy people to effectively avoid taking action. Table 2 captures the four levels of uncertainty and typical tools which can be used by the strategist. Using this framework it is possible to analyse the de facto strategies of different countries. In addition it is possible to look, at the level of national government and say the private sector, at the policy approaches being adopted. In the case of South Africa this presents a tapestry of different views of the importance or otherwise S&T to the future of the country.

Table 2: The different levels of uncertainty and some of the "tools" available given these different levels of uncertainty.

Level of Uncertainty	Description	Toolkit
I	A clear enough future	Competitor analysis, forecasting, five forces
II	Alternate futures	Discrete scenarios, value analysis, change models
III	A range of futures	Indicative scenarios, response sets
IV	True ambiguity*	Analogies, indicators, triggers

For instance the fiscal approach adopted has led to a decline in real R&D spending over the last few years. This is at best a level I strategy, which has "determined" that declines in S&T spending do not have a negative impact on the future viability of the system of innovation. Of course, if the level of spending was in the same range as competitor nations it would be a genuinely level I strategy, possibly level II if the competitor nations were "betting" on at least two future scenarios.

By contrast the Department of Arts, Culture, Science and Technology has four possible scenarios, explicitly defined and an extensive set of foresight documents which allow reflection on the future. This is strategically at level III (assuming it is backed up by adequate resources).

The Department of Education may have a level II strategy for tertiary institutions but it is not clear whether alternative futures inform this strategy. The analysis may be continued in this way. But what is the actual situation from the perspective of the levels of uncertainty? Innovation is subject to two major uncertainties: the market itself and the new products and services which future markets will require.

The predictive capacity of small economies has typically been poor in this respect. Knowledge is subject to at least two major uncertainties, can the knowledge be turned into useful products and services and does the economy have the inherent competence to capture the knowledge it requires to be successful? These four uncertainties are not the flip sides of a coin – rather they are essentially to a large degree orthogonal. Therefore the strategic analyst is placed in the position where, with respect to science and technology at present, there is genuine ambiguity.

In strategic terms this requires a policy/strategy environment which has two necessary responses:

1) Reserving the right to play
2) Agility

Reserving the right to play has (in the S&T space) some key features:

Investment must be at a level where genuine options are created. This level of R&D intensity seems from historical analysis to be a level of between 0.75 and 1% government funding of civilian R&D expenditure relative to GDP. Currently the South African government spends 0.29%. A higher level of investment has to be sustained for a significant period, for benefits to be realised given the low levels of spending currently (this is essentially a "capacitation" exercise).
The instruments of a modern economy must be in place for uptake of useful knowledge. This would include technology incubators, venture capital guarantee schemes from government, fiscal incentives for private sector research, strong protection of intellectual property rights and reward and recognition for tertiary education post-graduate outputs and the creation of Centres of Excellence in attractive knowledge areas.
It is not possible in this domain to "pick winners" but rather to seed a portfolio of "high potential domains".

Agility has some additional requirements:

Agility strategies have to make the assumption that a sound knowledge base is in place and move on from that point. Agility then requires continual investment in the consortia and networks that will effectively "link up" the players in the system of innovation.
Many science systems display the ability to respond to new needs rapidly. This is sometimes criticised as "window dressing" but in reality knowledge workers are inherently responsive (hopefully in a more than Pavlovian sense) to stimuli and rewards received for doing the "right" and relevant science.
Agility is also based on a sense of fairness and integrity in the system of innovation. If non-performers are rewarded or even tolerated this has a negative impact on the contributors in the system. It is necessary therefore that non-performers attract sanctions which are toxic to them but not toxic to the system. This is necessarily not just a mechanistic process.
Under genuine ambiguity non-performance does not automatically include contributions in fields that are not fashionable or attractive, since genuine ambiguity implies that foreknowledge of what "pays off" is difficult to *determine a*

priori. Performance is rather based on a negotiated set of outcomes that take into account the ambiguity rather than dismiss it from consideration.

Agility must include the ability to create economic and social value from science and technology using multiple mechanisms. Put differently multiple mechanisms are a precondition for agility. If particular mechanisms (such as those created in the United States by the Bayh-Dole Act) are not permitted in, say, the South African context, agility will be impaired. This is a serious consideration for policy-makers in science and technology. The key question is not: how can this be better controlled/stabilised or monitored. But, how do our policies contribute to the flexibility and responsiveness of the system of innovation as a whole?

Agility includes not doing everything yourself. International networks are crucial for sensing where change is coming from and what the new challenges are for the S&T system.

1.1.3 The way forward

Given these challenges what are the requirements for the South African system at present?

The first challenge is to recognise that the system of innovation approach is essentially process oriented and makes no particular demands on content (this is of course never totally the case – the Innovation Fund for instance has three "focusing themes"). However South Africa is not making sufficient investment in civilian R&D funded by government to be a player in a number of new knowledge areas.

The level of investment is simply too low. The best comparisons and benchmarks suggest that a "knowledge" strategy would require 2 to 3 times the current levels of government civilian R&D investment targeted on areas that had high potential for the country. Without this investment South Africa's long-term "right to play" will be severely compromised. How this funding ought to be deployed will be a matter for debate but a significant portion should be used to create centres of excellence at universities and research councils, preferably by insisting on partnerships.

Such areas would also have to have technology and industrial incentives linked to them to make the knowledge platforms attractive for foreign direct investment.

It will also be important to enhance the attractiveness of the system as a whole to the private sector. The most comprehensive policy studies show that increased government spending will not displace but rather leverage private sector spending. Private sector spending could be made even more attractive by the adoption of fiscal incentives, particularly tax credits which are used effectively in many countries. The absence of such credits makes South Africa a less attractive site for medium and long-term R&D investment.

In addition the University research infrastructure needs to be developed to be very attractive for post-graduate research that is not linked to up front commitments to large companies. This "freeing up" of intellectual resources combined with instruments such as venture capital will create the new channels (or at least one of them) that will permit greater flexibility and agility within our system. A necessary condition for success in this arena is better awareness of the intellectual property domain by institutions engaged in public research and development activities.

1.1.4 Conclusion

South Africa has made good use of the "system of innovation" approach to create more positive processes and value from its S&T base. However against the background of the global war for talent and the need to have effective knowledge bases for the future, current government investment is dangerously low and many new channels for innovation need to be built. This would require a policy shift as complex and profound as that engineered in South Africa in 1996/7 and the political will to carry it through.

1.1.5 Bibliography

1) Lester Thurow, *Creating Wealth*, pg 113, Nicolas Breasley Publishing, London, 2000.
2) Courtney, Kirkland and Viguerie, Strategy under uncertainty, Harvard Business Review, November/December, 1997

8. Posters

LOCALISED SOLUTIONS OF THE PARAMETRICALLY DRIVEN COMPLEX GINZBURG-LANDAU EQUATION

I.V. BARASHENKOV
Department of Mathematics, University of Cape Town, Private Bag Rondebosch, 7701, South Africa
E-mail: igor@maths.uct.ac.za

S.D. CROSS
Department of Mathematics, University of Cape Town, Private Bag Rondebosch, 7701, South Africa
E-mail: scross@maths.uct.ac.za

The complex Ginzburg-Landau and related equations describe many phenomena in non-linear optics including the self-focusing and braiding of light. In this paper we find qualitatively new numerical solutions to a particular form of the complex Ginzburg-Landau equation.

1 Introduction

The complex Ginzburg-Landau equation arises in the description of a variety of physical systems, including optical transmission lines [1], fluid dynamics [2] and convection in binary fluid mixtures [3]. This note is devoted to the parametrically driven cubic Ginzburg-Landau equation in one dimension:

$$i\psi_t + (1 - ic)\psi_{xx} + 2|\psi|^2 \psi - (1 - i\gamma)\psi = h\overline{\psi} , \qquad (1)$$

where $\gamma \geq 0$, $c \geq 0$ and h can also be chosen positive without loss of generality.

In the limit $c = 0$ this becomes the parametrically driven damped nonlinear Schrödinger equation, which has explicit time-independent pulse-like solutions [4] of the form

$$\psi_\pm(x,t) = A_\pm \, \text{sech}\,(A_\pm x) \, e^{-i\theta_\pm} , \qquad (2)$$

where A and θ are given by

$$A_\pm = \sqrt{1 \pm \sqrt{h^2 - \gamma^2}} \quad ; \quad \cos 2\theta_\pm = \pm\sqrt{1 - \frac{\gamma^2}{h^2}} . \qquad (3)$$

The stability analysis for these two solutions was performed in [4]. The soliton ψ_- was found to be unstable for all h and γ, while ψ_+ can be stable over a range of values of h and γ.

The aim of this work is to analyze the existence and stability of pulses beyond the nonlinear Schrödinger limit, i.e. for $c \neq 0$. In the first part of section 2 we derive a continuability condition which suggests that all stationary solutions to the associated nonlinear Schrödinger equation, when considered as solutions for the Ginzburg-Landau equation with $c = 0$, may be continued to non-zero c provided the solution and its derivatives decay to 0 as $x \to \pm\infty$. In the second part of section 2 we analyse the stability of the flat stationary solutions of the Ginzburg-Landau.

In section 3 we find new stationary solutions of the Ginzburg-Landau equation by numerically continuing the solitons (2) to nonzero c. Lastly, in section 4 we check the stability of the solutions found by examining the associated linearised eigenvalue problem.

2 Preliminaries

2.1 Continuability condition

In the limit $c \to 0$ we may expand any stationary solution ψ_s in powers of c:

$$\psi = \psi_0 + c\psi_1 + c^2\psi_2 + \ldots \qquad (4)$$

Having done so, we subsitute this expression into equation (1) with $\psi_t = 0$ and match like powers of c. Decomposing ψ_0 and ψ_1 as

$$\psi_0 = u + iv \, ; \quad \psi_1 = u_1 + iv_1 \, , \qquad (5)$$

the equation for the c^1 terms becomes

$$L \begin{pmatrix} u_1 \\ v_1 \end{pmatrix} = \begin{pmatrix} v_{xx} \\ -u_{xx} \end{pmatrix}, \qquad (6)$$

where $L = \mathcal{H}|_{c=0}$ and $\mathcal{H}$ is the operator obtained by linearising the Ginzburg-Landau equation about the soliton ψ:

$$\mathcal{H} = (1 - \partial_{xx})I + (c\partial_{xx} - \gamma)J \\ + \begin{pmatrix} h - 6u^2 - 2v^2 & -4uv \\ -4uv & -h - 6v^2 - 2u^2 \end{pmatrix}. \qquad (7)$$

Here I is the unit matrix and $J = \begin{pmatrix} 0 & -1 \\ 1 & 0 \end{pmatrix}$.

Fredholm's alternative states that equation (6) has bounded solutions if and only if the right-hand side is orthogonal to the kernel of $L^\dagger$. If L is the operator obtained by linearising the associate nonlinear Schrödinger equation with parameters h and γ about ψ_0, then $L^\dagger$ is the operator obtained by linearising the associate nonlinear Schrödinger equation with parameters h and $-\gamma$ about $\overline{\psi_0}$. Any continuous symmetry of the nonlinear Schrödinger equation will generate a subspace of the kernel of $L^\dagger$. Applying the infinitesimal generator of the symmetry ($\hat{G}$) to the conjugate of equation (1) with $\psi_t = 0$ and $c = 0$ gives $L^\dagger(\hat{G}\overline{\psi_0}) = 0$. The only symmetry of the associate nonlinear Schrödinger equation is a translation in x, with the generator $\hat{G} = \frac{\partial}{\partial x}$. The resulting continuability condition is

$$\int_{-\infty}^{\infty} (u_x v_{xx} + v_x u_{xx}) dx = 0, \tag{8}$$

which is satisfied as long as $\psi_0 \to 0$ as $x \to \pm\infty$.

2.2 Flat Backgrounds

The Ginzburg-Landau equation has 3 stationary flat solutions, namely $\psi = 0$ and $\psi = \phi_\pm$, where

$$\phi_\pm = \frac{1}{\sqrt{2}} A_\pm e^{-i\theta_\pm}, \tag{9}$$

with $A_\pm$ and $\theta_\pm$ as in equation (1.3). Note that the solutions $\phi_\pm$ are independent of c.

By linearising the stationary Ginzburg-Landau equation about $\psi_{\text{flat}} = \phi_\pm$ or 0, we can analyze the stability of the flat backgrounds and determine when decay to the flat backgrounds can occur. Letting $\psi = \psi_{\text{flat}} + \delta\psi$ gives

$$\mathcal{H}\delta\psi = 0, \tag{10}$$

where $\mathcal{H}$ is as in equation (7) with $u + iv = \phi_\pm$ or 0. Assuming a perturbation of the form $\delta\psi = (a + ib)e^{i(\omega t - kx)}$, gives a (complex) matrix equation which has non-trivial solutions if the determinant of the matrix is zero. At this stage we can either solve for ω in terms of k, or set $\omega = 0$ and solve for k. The first will allow us to determine the stability of the flat backgrounds while the second will allow us to determine when decay to the flat backgrounds can occur.

Stability of the flat backgrounds:

Solving for ω in terms of k results in the expression

$$i\omega = -(ck^2 + \gamma) \pm i\sqrt{Z}, \tag{11}$$

where

$$Z = \left(1 - 2A^2 + k^2\right)^2 + A^2\left(A^2 - 2\right) - h^2. \tag{12}$$

Here A is one of $A_\pm$ or 0 depending on which solution we have linearised about. The flat background is stable if $\text{Re}(i\omega) \leq 0$ for all k and unstable otherwise. In the case of the zero background ($A = 0$) equation (11) simplies considerably. In this case $i\omega$ will have the greatest real part when $k = 0$. Thus the stability of the zero background does not depend on c (as long as $c \geq 0$) and it will be stable when $Z|_{k=0} \geq -\gamma^2$, i.e.

$$h^2 \leq 1 + \gamma^2 . \tag{13}$$

Consider now the flat nonzero backgrounds. The $\phi_\pm$ solutions will be stable when either $Z \geq 0$, or when $-(ck^2 + \gamma)^2 \leq Z < 0$. These can be combined into a single condition $Z \geq -(ck^2 + \gamma)^2$ which must hold for all real k. This condition amounts to the inequality

$$\left(1 + c^2\right) s^2 + 2\left(1 - 2A^2 + \gamma c\right) s + 4A^2 \left(A^2 - 1\right) \geq 0 \tag{14}$$

where

$$s = k^2 , \tag{15}$$

which must be true for all real k.

Since $A^2 < 1$ for ϕ_-, the inequality (14) does not hold for $k = 0$ and hence ϕ_- is unstable for all γ, h and c.

For ϕ_+ we have $A^2 \geq 1$. Since the coefficient of s^2 in the quadratic on the left hand side of inequality (14) is positive, the quadratic will be concave up. In addition, the coefficient of s^0 is non-negative thus the real roots (if they exist) of the quadratic in (14) will either have the same sign or one of them will be zero. Inequality (14) will hold for all real k if it holds for all $s \geq 0$, i.e. when the roots of the quadratic in s are negative, zero, complex, or equal. Since the coefficient of s^2 is positive the sign of the roots (or of the nonzero zero root) depends only on the sign of the coefficient of s. The real roots (if they exist) will be negative or zero when

$$c \geq \frac{2A_+^2 - 1}{\gamma} . \tag{16}$$

If the roots are complex the solution ϕ_+ will be stable, so ϕ_+ is stable whenever c obeys inequality (16), regardless of whether or not the roots are real. We now turn to the case when the roots are equal or complex which occurs when the discriminant of quadratic (14) is less than or equal to zero. The discriminant will be less than or equal to zero when

$$\left[\gamma^2 - 4A_+^2 \left(A_+^2 - 1\right)\right] c^2 + 2\left(1 - 2A_+^2\right)\gamma c + 1 \leq 0 . \tag{17}$$

[We immediately note that when $c = 0$ the solution ϕ_+ is unstable for all h and γ since neither (16) nor (17) will hold. Setting $c = 0$ in (16) gives the inequality $0 \geq \frac{1}{\gamma}$ (after using $A_+^2 \geq 1$) while setting $c = 0$ in (17) gives the inequality $1 \leq 0$, both of which are false.]

The discriminant of the quadratic expression in c (17) is equal to

$$4A_+^2 \left(A_+^2 - 1\right) \left(1 + \gamma^2\right) \tag{18}$$

which is positive, thus the quadratic in (17) has real roots:

$$c_\pm = \frac{\left(2A_+^2 - 1\right)\gamma}{\gamma^2 - 4A_+^2 \left(A_+^2 - 1\right)} \mp \frac{\sqrt{4A_+^2 \left(A_+^2 - 1\right)\left(1 + \gamma^2\right)}}{\gamma^2 - 4A_+^2 \left(A_+^2 - 1\right)} \tag{19}$$

When $h^2 > \frac{1}{4}\left(\sqrt{1+\gamma^2} - 1\right)^2 + \gamma^2$, the coefficient of c^2 in the quadratic is negative. Thus the graph of the quadratic is concave down and the quadratic's roots have opposite sign (since the constant part of the quadratic is positive). The ϕ_+ solution is stable for $c \geq c_+$, the positive root of the quadratic, which is given in equation (19).

When $h^2 < \frac{1}{4}\left(\sqrt{1+\gamma^2} - 1\right)^2 + \gamma^2$, the graph of the quadratic is concave up and the quadratic's roots have the same sign (since the constant part of the quadratic is positive). Both the quadratic's roots are positive since the first term in the expression for the roots, equation (19), is positive. The ϕ_+ solution is stable only for $c_+ \leq c \leq c_-$.

Lastly, when $h^2 = \frac{1}{4}\left(\sqrt{1+\gamma^2} - 1\right)^2 + \gamma^2$, the coefficient of c^2 is zero and the graph of (17) is a straight line. The ϕ_+ solution is stable for $c \geq \frac{1}{2(2A_+^2-1)}$.

Decay to the flat backgrounds:

Setting $\omega = 0$ in equation (11) we can solve for k and determine when decay to the solutions $\phi_\pm$ can occur. Setting $\omega = 0$ results in a biquadratic

$$\left(1 + c^2\right)s^2 + 2\left(2A^2 - 1 - \gamma c\right)s + 4A^2\left(A^2 - 1\right) = 0 \tag{20}$$

where

$$s = (ik)^2 . \tag{21}$$

Decay to $\phi_\pm$ will be possible unless $\text{Re}(ik) = 0$ for all four roots which occurs when both roots, $s_{1,2}$, of (20) are real and non-positive. Thus decay to $\phi_\pm$ can occur either when the roots of the quadratic (20) are complex (the discriminant is less than zero) or when one of the roots is positive. The discriminant of (20) is given by

$$\left(2A^2 - 1 - \gamma c\right)^2 - 4A^2\left(A^2 - 1\right)\left(1 + c^2\right) . \tag{22}$$

For ϕ_- the discriminant is always greater than zero as $A^2 < 1$ making the discriminant a sum of two positive quantities. However, the coefficient of s^2 is positive so the graph of the quadratic is concave up and it then follows that the roots have opposite sign since the constant term in the quadratic is negative. Thus decay to ϕ_- may occur for all h, γ and c.

In the case of ϕ_+ the quadratic (20) will have a positive real root when the inequality (16) is satisfied but not saturated. This occurs since the biquadratics in equations (14) and (20) are related by the simple transformation $s \to -s$. Similarly, quadratic (20) will have complex roots when the inequality (17) is satisfied but not saturated. Thus decay to ϕ_+ can occur when either inequality (16) or (17) is satisfied but not saturated.

Setting both $A = 0$ and $\omega = 0$ in equation (11) we can determine when decay to the zero background can occur. Setting $A = 0$ and $\omega = 0$ gives

$$\left(1 + c^2\right) s^2 - 2\left(1 + \gamma c\right) s - (h^2 - \gamma^2) = 0 \tag{23}$$

where

$$s = (ik)^2 \tag{24}$$

As before decay to the flat background will be possible when the quadratic for s has a negative discriminant or a positive root. Since the biquadratic has a positive turning point it has a positive root if the discriminant is non-negative. Thus decay to the zero flat background is possible for all h, γ and c.

3 Continuation of $\psi_\pm$

The continuation of $\psi_\pm$ was performed using the AUTO94 software package [5]. The continuation of the soliton ψ_+ is depicted in the left-hand diagram of Figure 1. Here we fixed $\gamma = 0.5$ and $h = 0.8$. The ψ_+ solution of the parametrically driven damped nonlinear Schrödinger equation corresponds to the lower of the two points where the graph crosses the $c = 0$ axis. As we continue from ψ_+ into the region $c < 0$ the solution develops into a three hump state, with the three humps close together. Continuing past the turning point (i.e. proceeding to the top branch in the $c < 0$ region), the three humps move apart and reshape to give us $\psi_{(-+-)}$ once we return to $c = 0$. This complex of three solitons of the parametrically driven nonlinear Schrödinger equation was previously found in [6]. Crossing over into the $c > 0$ region we reach a series of turning points (saddle-node bifurcations). Each passing of a turning point results in the creation of another hump at the centre of the solution. These additional humps may merge if the turning points are close together. Eventually only a large single composite hump and the two lateral

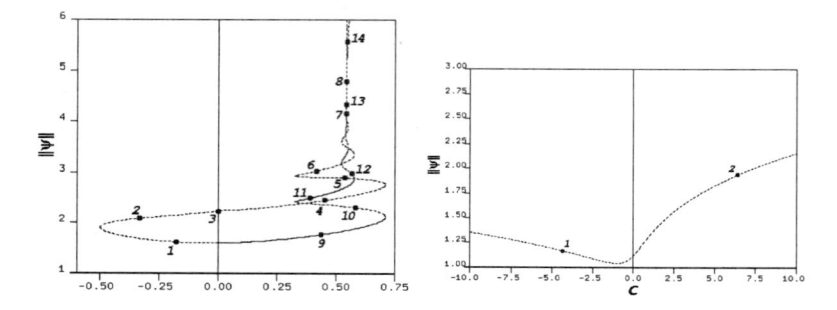

Figure 1. Bifurcation diagrams for ψ_+ (left diagram) and ψ_- (right diagram) when $h = 0.8$ and $\gamma = 0.5$. The Sobolev norm $||\psi||^2 = \int_{-\infty}^{\infty} \left(|\psi|^2 + |\psi_x|^2 \right) dx$ is plotted against c (the filtering parameter). Stable (unstable) branches are shown as solid (dashed) lines. Points 1-8 and 9-14 in the left diagram mark solutions on the branch found by initially continuing into the regions $c < 0$ and $c > 0$ repestively.

humps from the $\psi_{(-+-)}$ solution are clearly distinguishable. Near the end of the branch ($||\psi|| > 5$) the lateral humps encounter the boundaries of the numerical system causing the branch to bend away from vertical.

Moving away from ψ_+ (the lower intersection point on the c=0 axis) into the $c > 0$ region, we find that the solution gradually changes its shape, becoming two-humped after passing through the turning point near $c = 0.7$. It then enters a region where additional humps are added through a series of turning points. As before (on the branch found by continuing $\psi_{(-+-)}$ to $c > 0$), a stage is reached when the central humps merge. However, in this case, the two lateral humps are absent.

The continuation of the other nonlinear Schrödinger soliton, ψ_-, is depicted in the right-hand diagram of Figure 1. Again, $\gamma = 0.5$ and $h = 0.8$. As we move along the branch into the $c < 0$ region the single hump of the ψ_- soliton splits in two forming a double hump. The double humped solution then simply broadens as c is decreased. Continuing ψ_- into the $c > 0$ region results only in the broadening of the central hump.

4 Stability Analysis

Given a solution, $\psi_s = u + iv$, to the Ginzburg-Landau equation we test it for stability by adding a small perturbation of the form

$$\delta\psi(x,t) = [\delta u(x) + i\delta v(x)] e^{\lambda t}, \tag{25}$$

where δu and δv are real and λ may be complex, and linearising in $\delta\psi$. This gives an eigenvalue problem of the form

$$\mathcal{H}\begin{pmatrix}\delta u \\ \delta v\end{pmatrix} = \lambda J\begin{pmatrix}\delta u \\ \delta v\end{pmatrix}, \qquad (26)$$

where $\mathcal{H}$ and J are as in equation (7).

We solve the above eigenvalue problem by expanding δu, u, δv and v in the Fourier series on the interval $[-\frac{L}{2}, \frac{L}{2}]$. Truncating the series converts the problem to one of finding the eigenvalues (λ) of a complex square matrix. We truncated the series to 300 Fourier modes on the interval $[-28, 28]$.

The continuation of the nonlinear Schrödinger soliton ψ_+, for $h = 0.8$ and $\gamma = 0.5$, is shown in the left-hand diagram of Figure 1. Continuing ψ_+ into the region $c < 0$ results in an immediate loss of stability since the zero background is unstable for $c < 0$. The $\psi_{(-+-)}$ solution of the nonlinear Schrödinger equation (the top intersection with the $c = 0$ line in the left-hand diagram of Figure 1) is found to be unstable in agreement with [6]. Stability is only regained once the composite hump begins to form.

While continuing ψ_+ into the region $c > 0$ the branch remains stable until the first turning point. It then becomes unstable and alternates stability at each successive turning point. The branch gains stability and remains stable once the composite hump forms.

The bifurcation diagram for ψ_-, for $h = 0.8$ and $\gamma = 0.5$, is shown in the right-hand diagram of Figure 1. The solutions found by continuation in c from ψ_- are all found to be unstable.

References

[1] A. Meccozzi, J.D. Moores, H.A. Haus and Y. Lai, Opt. Letters 16 (1991) 1841
[2] K. Stewartson and J.T. Stuart, Journal of Fluid Mechanics 48 (1971) 529
[3] P. Kolodner, D. Bensimon and C.M. Surko, Phys. Rev. Lett. 60 (1988) 1723
[4] I.V. Barashenkov, M.M. Bogdan, V.I. Korobov, Europhys. Lett. 15 (1991) 113
[5] E.J. Doedel, X.J. Wang and T.F. Fairgrieve, *AUTO 94: Software for continuation and bifurcation in ordinary differential equations*, Applied Mathematics Reports, California Institute of Technology
[6] I.V. Barashenkov and E.V. Zemlyanaya, Phys. Rev. Lett. 83 (1999) 2568

π^0 AND η PHOTOPRODUCTION OFF THE PROTON AT GRAAL

O. BARTALINI, M. CAPOGNI, A. D'ANGELO, R. DI SALVO, D. MORICCIANI, C. SCHAERF
INFN, sezione di Roma II and Università di Roma "Tor Vergata", Roma, Italy

V. BELLINI, M.L. SPERDUTO
Laboratori Nazionali del Sud and Università di Catania, Catania, Italy

J.P. BOCQUET, A. LLERES, L.NICOLETTI, D. REBREYEND, F. RENARD
IN2P3, Institut des Sciences Nucléaires, Grenoble, France

M. CASTOLDI, A. ZUCCHIATTI
INFN, Sezione di Genova and Università di Genova, Genova, Italy

J.P. DIDELEZ, M. GUIDAL, E. HOURANY, R.KUNNE
IN2P3, Institut de Physique Nucléaire, Orsay, France

G. GERVINO
INFN, sezione di Torino and Università di Torino, Torino, Italy

F. GHIO, B. GIROLAMI
INFN, sezione di Roma I and Istituto Superiore di Sanità,, Roma, Italy

A. LAPIK, V. KOUZNETSOV, V. NEDOREZOV
Institute for Nuclear Research, Moscow, Russia

P. LEVI SANDRI
INFN, Laboratori Nazionali di Frascati, Frascati, Italy

N. RUDNEV
Institute of Theoretical and Experimental Physics, Moscow, Russia

A. TURINGE
RRC Kurchatov Institute of Atomic Energy, Moscow, Russia

Some recent results on π^0 and η photoproduction off hydrogen from the Graal experiment are presented and discussed.

1 Introduction

Several new experiments have been recently performed or are ongoing to obtain a more precise and complete knowledge of the baryon spectrum. This new generation of experiments exploits high intensity and high polarisation electron or photon beams associated to large acceptance detectors. Such large experimental effort is motivated by the still unsatisfactory knowledge of the nucleon excited states. Predicted states are not sufficiently well established, and many properties of the observed states (e.g. coupling constants, branching ratios, helicity amplitudes) are often poorly known. The present information [1] comes almost entirely from partial-wave analyses of pion-nucleon scattering. It is therefore urgent to improve the study of baryon resonances by exploiting the features of the electro-magnetic probe: small coupling constant and easy polarisability of beams. In the pseudo-scalar meson photoproduction reaction:

$$\gamma + p \to PS + nucleon$$

there are eight possible combination of particle spins. The scattering amplitude is thus described by eight matrix elements, but because of rotational invariance and parity only four of them are independent. With these four complex amplitudes, 16 bilinear products can be constructed, corresponding to 16 observables: the differential cross section, three single polarisation observables (beam, target, recoil) and twelve double polarisation observables. If we could measure the cross section, the three single polarisation observables and four appropriately chosen double polarisation observables the scattering amplitude would be completely determined [2].

The differential cross section ($d\sigma/d\Omega$), the beam polarisation asymmetry (Σ), the target polarisation asymmetry (T) and the recoil polarisation asymmetry (P), can be adequately expressed in terms of helicity amplitudes H_i. In that case, the following relations [3][4][5] hold:

$$\frac{d\sigma}{d\Omega} \approx H_1^2 + H_2^2 + H_3^2 + H_4^2$$

$$\Sigma \approx \text{Re}(H_1 H_4^* - H_2 H_3^*)$$

$$T \approx \text{Im}(H_1 H_2^* - H_3 H_4^*)$$

$$P \approx \text{Re}(H_1 H_3^* - H_2 H_4^*)$$

It is clear that the general structure of the scattering amplitude is contained in the differential cross section but its details can be explored more clearly in the study of polarisation observables, where the interference among the helicity amplitudes can play a fundamental role in revealing more subtle effects [6].

A completely polarised and versatile tagged photon beam, associated to a large acceptance detector is an excellent experimental tool to perform part of the ambitious program of a full determination of the transition amplitudes.

2 The Graal Beam and Detectors

The Graal facility provides a polarised and tagged photon beam by the backward Compton scattering of laser light on the high energy electrons circulating in the ESRF storage ring [7]. Using the UV line (350 nm) of an Ar-Ion laser we produce a gamma-ray beam tagged from 550 to 1470 MeV with typical intensity of $1 \cdot 10^6 s^{-1}$. With the laser green line of 514 nm the average intensity is $2 \cdot 10^6 s^{-1}$ and the energy is tagged between 550 and 1100 MeV. The polarisation is 0.98 at the maximum photon energy and the tagging energy resolution is 16 MeV (FWHM). The detector assembly is formed by a central part surrounding the target and a forward part. Particles leaving the target at angles from 25 to 155 degrees are detected in sequence by two cylindrical wire chambers, by a barrel made of 32 strips of plastic scintillator aligned with the beam axis, used to determine the $\Delta E/\Delta x$ of charged particles, and finally by the BGO calorimeter composed of 480 pyramidal crystals which are 21 radiation lengths long (24 cm). This calorimeter has an excellent energy resolution for photons [8], a good response to protons [9] and is very stable in time due to a continuous monitoring and calibration slow control system [10].

Particles moving at angles smaller than 25 degrees encounter first two plane wire chambers, (xy and uv), then, at 3 m from the target, two walls of plastic scintillator bars, 3 cm thick, to provide a measurement of the time-of-flight for charged particles (700 ps FWHM resolution). Finally we have a sampling electromagnetic calorimeter (a sandwich of 4 layers of lead and plastic scintillators 4 cm thick) that provides a full coverage of the solid angle for photon detection (with 95 percent efficiency) and a 20 percent efficiency for neutron detection.

In the backward direction two disks of plastic scintillator separated by a disk of lead complete the solid angle coverage . The beam intensity is continuously monitored by a flux monitor, composed of three thin plastic scintillators and by a lead/scintillating fibre detector that measures energy and flux [11]. The experiment trigger is given by the coincidence between the tagging counter and the detector trigger. The latter is formed when the total energy collected by the calorimeter is larger than 160 MeV or when a preset charged particles multiplicity is reached in the forward direction thus allowing events that are below threshold in the calorimeter to be recorded. The rate of the data acquisition, is typically between 100 and 200 s^{-1}. During data taking, the polarisation of the gamma-ray beam is very easily rotated just by rotating the laser beam polarisation. This is done approximately every twenty minutes. The contribution of the bremsstrahlung in the residual vacuum of the storage ring (typically two orders of magnitude lower, with respect to the Compton beam) is measured taking data without laser beam on.

3 Recent Results

3.1 η photoproduction

Since the isospin of η is I=0, the (γ,η) process offers selective access to N^* resonances, being insensitive to the propagation, in the intermediate state, of I=3/2 resonances, strongly coupled to the pion photoproduction channel. The new Graal data on the Σ beam asymmetry for η photoproduction [12] has stimulated a number of refinements in the existing theoretical approaches. Li and Saghai [13] investigate the process of η photoproduction within a quark model approach and find that significant contribution from $D_{13}(1520)$, $F_{15}(1680)$ and $P_{13}(1720)$ are required to riproduce the beam asymmetry data. Tiator and collaborators [14] have performed a combined analysis of η photoproduction cross section and asymmetry data. They have confirmed the role of $D_{13}(1520)$ and $F_{15}(1680)$ and have extracted their ηN branching ratios. N. Mukhopadhyay and N. Mathur [15] have combined cross section [16] and single polarisation observables from Bonn (target asymmetry) [17] and from Graal, and by making use of an effective Lagrangian approach have extracted the electro-strong parameters for the $N^*(1520)$ providing a critical test for many QCD inspired hadron models.

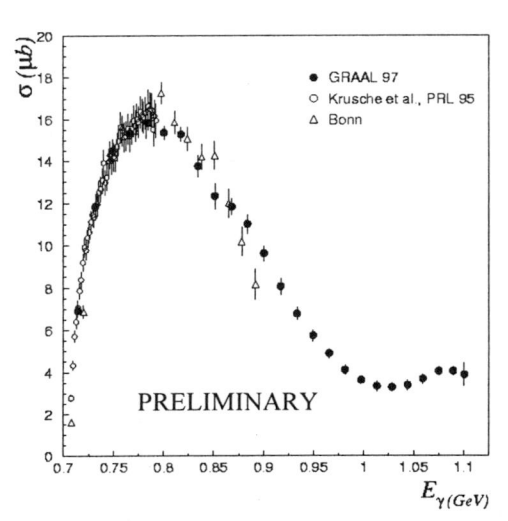

Figure 1. Preliminary total cross section for h photoproduction on the proton (close circles) compared with the existing Mainz (open circles) and Bonn (open triangles) data sets.

In figure 1 we show the preliminary total η photoproduction cross section, which we have recently analysed together with differential cross sections. There is an excellent agreement with the existing Mainz data [16] and the data set is now extended up to 1.1 GeV. Below 900 MeV the shape of the total cross section is dominated by the $S_{11}(1535)$ resonance, whose parameters can be determined by a Breit-Wigner fit, with energy dependent width as in [16], that covers now the entire resonance region. We have extracted a value of Γ_R between 150 and 170 MeV compatible with π–N analysis.

3.2 π^0 photoproduction

Pion photoproduction (π^0, π^+) is one of the most extensively studied photoreaction and the main source of information on the structure of nucleons and nuclei. However, many of the measurements included in the large data base on differential cross section, already existing, are not sufficiently accurate.

The importance of Graal in this field lies in the excellent statistical and systematic error that is achieved and in the large energy and angular range covered by asymmetry measurements thus providing a new, consistent data base from 500 to 1500 MeV.

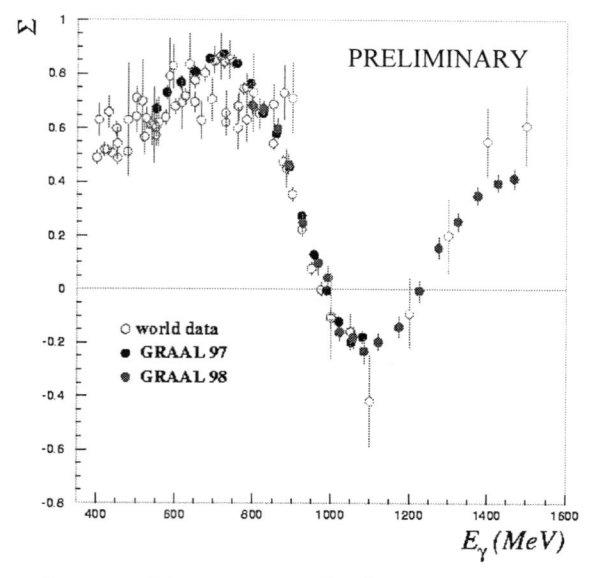

Figure 2. Preliminary 90 degrees Σ beam asymmetry for π^0 photoproduction on the proton. Light full dots are taken 1470 MeV, black dots at a maximum beam energy of 1100 MeV.

In Figure 2 the preliminary 90 degrees Σ beam asymmetry in π^0 photoproduction data are shown togheter with all existing old data. The improvement of data quality assured by Graal, is evident both in the reduced total (statistic + systematic) error bars and in the good control of beam polarisation. This shows up in the very good correspondence of asymmetry values taken at two different end point energies of the photon beam, corresponding to different polarisation values.

In Figure 3 is shown a part of the collected and analysed data for Σ beam asymmetry in π^0 photoproduction. Preliminary data from Graal are now available, with continuity from 500 to 1500 MeV. The curve shown is from ref [6] We can see that the fine details of the asymmetry are only qualitatively reproduced: the small statistical error is a really strong constraint for all models and analyses.

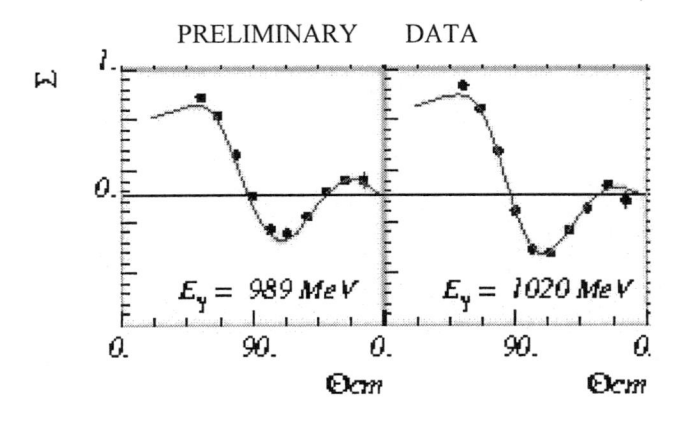

Figure 3. Preliminary Σ beam asymmetry for π^0 photoproduction on the proton.

4 Conclusions

The Graal experiment started data taking in 1997. It runs both with the green laser line, giving rise to a photon beam of maximum energy of 1100 MeV, and with a UV line corresponding to a gamma-ray maximum energy of 1470 MeV. Asymmetry data and cross sections have been produced for η, π^0 (and π^+) photoproduction channels providing, for these reactions, the most extended, precise and coherent data base available until now.

References

1. Particle Data Group, Eur. Phys. Journ. C3, (1998), 613
2. W.T. Chiang and F. Tabakin, Phys. Rev. C 55, (1997), 2054.
3. D. Drechsel, O Hanstein, S. Kamalov and L. Tiator, Nucl. Phys A645, (1999), 145.
4. T. Feuster and U. Mosel, Phys. Rev. C59, (1999), 460.
5. B. Saghai and F. Tabakin, Phys. Rev. C55, (1997), 917 and Phys. Rev. C53, (1996), 66 and C. Fasano, F. Tabakin and B. Saghai, Phys. Rev. C46, (1992), 2430.
6. R. Arndt et al., Phys. Rev., C42, (1990), 1853.
7. Graal Collaboration, Nucl. Phys. A622, (1997) 110c
8. P. Levi Sandri et al., Nucl. Inst. and Meth. in Phys. Research A370, (1996), 396.
9. A. Zucchiatti et al., Nucl. Inst. and Meth. in Phys. Research A321, (1992), 219.
10. F. Ghio et al., Nucl. Inst. and Meth. in Phys. Research A404, (1998), 71.
11. V. Bellini et al., Nucl. Inst. and Meth. in Phys. Research A386, (1997), 254.
12. J. Ajaka et al., Phys. Rev. Lett. 81, (1998), 1797.
13. Z. Li and B. Saghai, Nucl. Phys. A644 (1998) 345.
14. L. Tiator, D. Drechsel, G. Knochlein and C. Bennhold, Phys. Rev.n C60, (1999), 035210.
15. N. Mukhopadhyaya and N. Mathur, Phys. Lett. B444, (1998), 7.
16. B. Krusche et al., Phys. Rev. Lett. 75, (1995), 40.
17. A. Bock et al., Phys. Rev. Lett. 81, (1998), 534.

MUON(IUM) IN NITROGEN-RICH AND ^{13}C DIAMOND

IZ MACHI[1,2]*, SH CONNELL[1], K BHARUTH-RAM[3] AND JPF SELLSCHOP[1]

[1] *Schonland Research Center for Nuclear Sciences, University of the Witwatersrand, PO WITS, Johannesburg 2050, South Africa*

[2] *Physics Department, University of South Africa, Box 392, UNISA 0003, South Africa*

[3] *Physics Department, University of Durban-Westville, Private Bag X54001, DURBAN 4000, South Africa*

We report on investigations on muon(ium) interactions with H2/H3 defects induced in a nitrogen-rich type Ia diamond by photon-irradiation, as well as on preliminary Transverse Field muon Spin Rotation (TF-μSR) measurements on a pure ^{13}C diamond. The H2/H3 trapping centres in the type Ia diamond were produced by irradiating the sample with 2 MeV photons to a dose of 10^{19} cm^{-2}, and then annealing at 1270 K for 2 hrs. The prompt fraction (f) and spin relaxation rate (λ) of the diamagnetic (μ^+) and the paramagnetic (Mu$_T$) states, formed in the sample, were determined from TF-μSR measurements made at temperatures ranging from 5 K to 300 K, in applied magnetic fields of 1.0 mT and 7.5 mT. The spin relaxation rates of the μ^+ and the Mu$_T$ ($\approx$ 0.018 μs^{-1} and 6.5 μs^{-1}, respectively) show no change with temperature. The prompt fraction of the μ^+ state shows a value almost twice that in a "virgin" nitrogen-rich sample, and increases from 20% at 5 K to 30% at 300 K, while that of the Mu$_T$ state decreases from 40% at 5 K to about 20% at 300 K. These results suggest an interaction between the muon(ium) states and the irradiation induced H2/H3 defects. A surprising result in the ^{13}C diamond is that the prompt fractions of the μ^+ and the Mu$_T$ states at 200 K add to 100%, showing no formation of the bond-centred Mu$_{BC}$ state which is observed in all other pure diamond samples.

1 Introduction

Hydrogen is known to be a common impurity in diamond, with concentrations ranging from 100 to 1000 ppm[1-3]. The considerable effort made in recent years to investigate hydrogen behaviour in diamond reveal that special techniques are required to study its dynamic behaviour [4-11]. For example, electron paramagnetic resonance (EPR) studies of synthetic high pressure high temperature (HPHT) produced diamonds implanted with protons, failed to detect any signals with hydrogen-related hyperfine parameters, indicating that implanted hydrogen existed mostly in non-paramagnetic form[12]. On the other hand, substantial contributions to the understanding of the behaviour of hydrogen in diamond (and other semiconductors) has come from methods

---

*PHYSICS DEPARTMENT, BOX 392, UNISA, 0003, SOUTH AFRICA, FAX. ++27 12 429 3643, E-MAIL: MACHIIZ@UNISA.AC.ZA

such as Muon Spin Rotation/Relaxation/Resonance (μSR). This is made possible by the similar electronic properties of muonium and the hydrogen atom[13] (see Tab. 1). This chemical analogy has allowed hydrogen states in diamond

Table 1. Comparable properties of muonium and hydrogen.

	Muonium	Hydrogen
Reduced mass (m_e)	0.995	0.999
Radius (Å)	0.531	0.529
Ionisation energy (eV)	-13.54	-13.59
H.F. Frequency (GHz)	4.46	1.42

to be identified through three main muonium configurations, namely i) a diamagnetic state (μ^+), ii) a tetrahedral interstitial state (Mu_T) with isotropic hyperfine parameters, and iii) a bond-centred state (Mu_{BC}) with anisotropic hyperfine parameters. The Mu_{BC} state is known to be the most stable[14-16] of the three configurations. The Mu_T state is known to form soon after implantation, and to diffuse rapidly. It does not display any appreciable spin relaxation in pure samples, but exhibits a large spin relaxation in samples with impurities[13,17,18,19,20]. In this work we present the results of transverse field muon spin rotation[13] (TF-μSR) measurements on nitrogen-rich (type Ia) diamonds. The prompt fractions (f) and the spin relaxation rates (λ) of the μ^+ and the Mu_T states were determined with the type Ia diamond in the "virgin" state and after nitrogen-related defects, known as H2/H3[21,22], were induced by photon-irradiation. Preliminary results of measurements on a pure ^{13}C diamond are also presented.

2 Experimental

The TF-μSR measurements were preceded with characterization of the diamond samples with the aid of infra-red and optical spectroscopy. The type Ia diamonds contained about 600 ppm A-centres, which is a pair of two nearby substitutional nitrogen atoms. To produce the nitrogen-related H2/H3 centres, three pieces of the type Ia diamonds were irradiated with 2 MeV gamma photons to a dose of 10^{19} cm^{-2}, and then annealed at 1270 K for 2 hrs. The H2/H3 defects are thought to be formed by a vacancy located near the two adjacent nitrogen atoms in the A-centre. The ^{13}C diamond contained more than 90% of ^{13}C atoms.

The TF-μSR measurements were performed at the Paul Scherrer Institute, Switzerland. The prompt fractions (f) and the spin relaxation rates (λ) of the

μ^+ and the Mu$_T$ states were determined as functions of sample temperature in the range 5-300 K, and in applied magnetic fields of 1.0 mT and 7.5 mT.

3 Results and Discussion

3.1 Natural Type Ia Diamond (nitrogen-rich)

The prompt fractions and the spin relaxation rates determined for the μ^+ and the Mu$_T$ states in the type Ia diamond (after photon-irradiation and anneal) are displayed as functions of temperature in Fig. 1. The results reflect several significant features

1. The μ^+ state shows a considerable increase in its prompt fraction (f) after photon irradiation and anneal of the Ia diamond, its value almost doubling in magnitude from 10% in the virgin sample[18] (or as in other nitrogen-rich diamond samples [23-25]) to 20% (at 5 K) after irradiation. Evidently, the H2/H3 centres created by the photon-irradiation and subsequent annealing have produced an appreciable increase in trapping centres for the diamagnetic μ^+.

2. The prompt fraction (f) of the μ^+ state shows an increase from 20% at 5 K to about 30% at 300 K (in the applied magnetic field of 7.5 mT), while that of the Mu$_T$ state shows a decrease from 40% at 5 K, to about 20% at 55 K (in the applied magnetic field of 1.0 mT). This suggests either a Mu$_T$ to μ^+ transition as the sample temperature increases, or the formation probability of the two states is altered by effects related to electron temperature dependent dynamics at the trap site.

3. The μ^+ state shows a spin relaxation rate λ of about 0.02 μs^{-1} after photon irradiation; in virgin Ia diamond the spin relaxation rate $\lambda = 0\ \mu s^{-1}$.

4. The spin relaxation rate λ of the Mu$_T$ state is relatively constant at about 5 μs^{-1}, showing little dependence on temperature.

Our recent Longitudinal Field muon Spin Relaxation (LF-μSR) measurements on the same sample (before the creation of H2/H3 defects) had shown that Mu$_T$ was trapped at nitrogen related defects, with the the A-centres being strong candidates for the trap. In addition, a weak resonance at about 50 mT field in the repolarization curve for the "old muons" (see Fig. 2) indicated that a little μ^+ might also be trapped at/in A-centres. [For more detailed discussion on the repolarization curves for the "young" and "old" muons, see

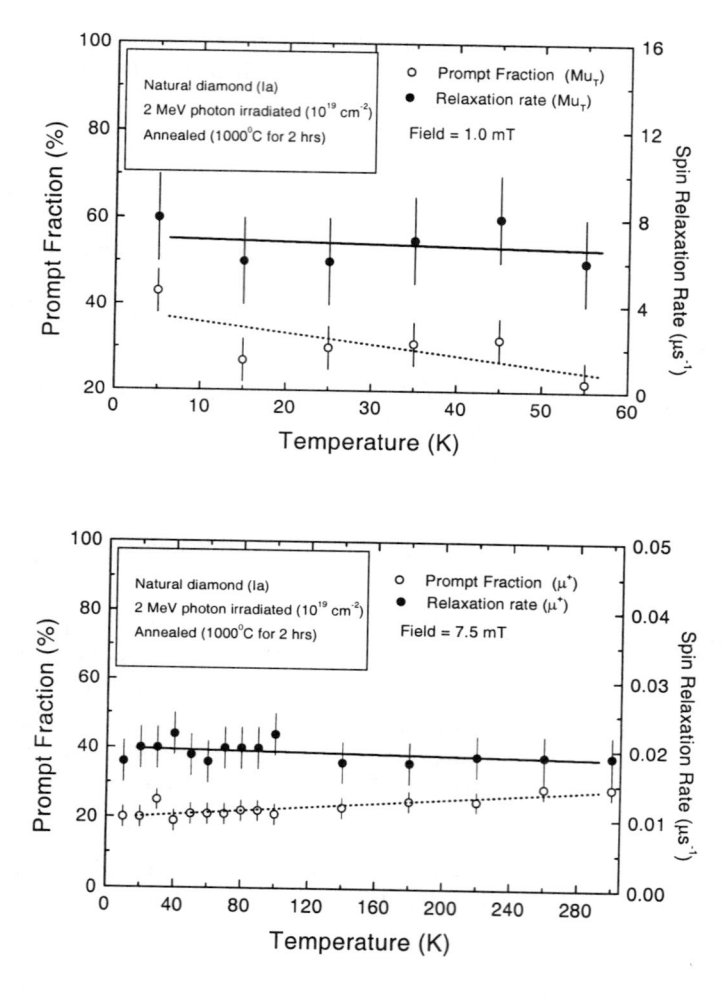

Figure 1. Prompt fractions and spin relaxation rates of the Mu_T and the μ^+ states in nitrogen-rich diamond with H2/H3 defects.

Machi et al [18]. However, the very weak resonance shown in Fig. 2 cannot account for the strong interaction experienced by the μ^+ state in the current results.

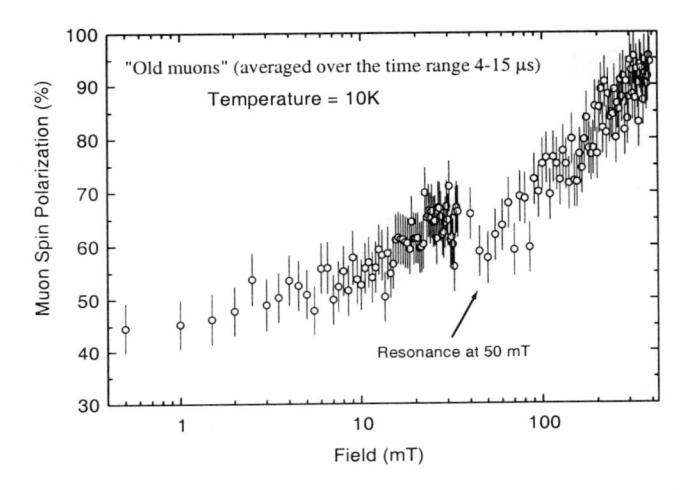

Figure 2. Repolarization curve of old" muons in virgin" state nitrogen-rich diamond.

It is possible that the A-centre is a shallower trap than the induced H2/H3 defects which are composed of vacancies near the A-centres. The vacancies are predicted to represent well depths of about 5 eV[26]. In any event, these results clearly show that the μ^+ state does interact with the nitrogen-vacancy-related defect in diamond.

3.2 Pure ^{13}C Diamond

The prompt fraction (f) and the spin relaxation rate (λ) for the μ^+ and the Mu$_T$ states in a ^{13}C diamond were determined from measurements in a transverse magnetic field of 7.5 mT, and sample temperatures of 10 K, 100 K and 200 K. The results are shown in Fig. 3.

These three data points were aimed at surveying the possibility of future investigations of quantum diffusion of the Mu$_T$ state in diamond. The presence of the spin-$\frac{1}{2}$ nuclei of the ^{13}C atoms in the sample allow the observation of diffusion via the motional narrowing mechanism. It was necessary to verify that any possible additional magnetic impurities (derived from the parental melt) did not obliterate the muonium signals. However as evident in Fig. 3, very strong signals for both Mu$_T$ and μ^+ state are observed.

The prompt fraction of the μ^+ state is relatively constant at about 20%.

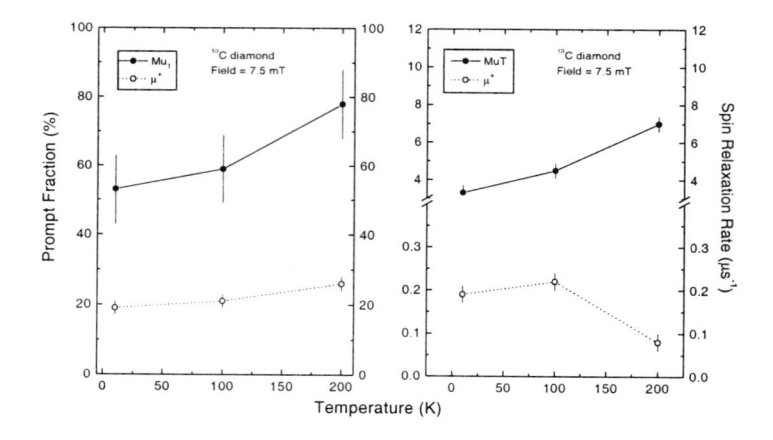

Figure 3. Prompt fractions and spin relaxation rates of the Mu_T and the μ^+ states in ultra-pure ^{13}C diamond sample.

This value is about a factor 2 larger than those generally observed in other diamonds [23] (except for the type Ia diamond with H2/H3 defects also reported here). For the Mu_T state, the prompt fraction shows a strong temperature dependence, increasing from about 50% at 5 K to about 80% at 200 K (see Fig. 3). The relatively large prompt fraction (f) for the Mu_T state is comparable, especially at low temperatures, with those obtained in relatively pure natural type II diamond samples[23].

A result of particular interest is that the prompt fractions for the μ^+ and Mu_T states add up to 100% at 200 K, suggesting non-population of the bond-centre Mu_{BC} state (which is observed in all other pure diamond samples). No proper explanation of this phenomenon can be provided until more detailed measurements at temperatures over the range 5 - 300 K, and in applied fields that could be expected to show the Mu_{BC} precession signal more clearly.

The μ^+ state has a spin relaxation rate λ of 0.15 μs^{-1}, which remains relatively constant with increasing temperature. This shows that the μ^+ state experiences an interaction in the sample. This interaction could be associated with the presence of ^{13}C atoms, as zero values of the spin relaxation rate for the μ^+ state have thus far been obtained in pure type II diamond samples[23]. Little has been known up to now about the μ^+ state in diamond, and these first observations of μ^+ interactions may provide the opportunity to study this species.

For the Mu$_T$ state, on the other hand, the spin relaxation rate λ increases from about 3 μs^{-1} at 10 K to about 7 μs^{-1} at 200 K. From our previous measurements on quantum diffusion of Mu$_T$ in natural diamonds[19], we would expect the spin relaxation rate to drop in the temperature range from 100 K to 300 K. However, as time did not permit measurement at higher temperatures, it was not possible to verify the expected drop in temperature here. Thus far, the use of this sample in investigations of quantum diffusion is not precluded.

4 Conclusions

The behaviour of the μ^+ state (its prompt fraction and spin relaxation rate) observed in our measurements suggest significant interaction between this state and nitrogen-vacancy-related H2/H3 defects produced in type Ia diamond by photon irradiation followed by high temperature annealing. This interaction is expected to be due to the μ^+ trapped at H2/H3 defects. Our results also suggest a Mu$_T$ to μ^+ transition, or the formation probability of the two states is altered by effects related to electron temperature dependent dynamics at the trap site.

In the case of the ^{13}C diamond sample, the data indicate a surprising result. The prompt fractions of the μ^+ states at 200 K account for the total muons implanted, suggesting non-population of the Mu$_{BC}$ state. Also, there is again evidence of μ^+ interactions.

The particular impurities in these samples are therefore capable of elucidating the behaviour of the μ^+ state in diamond. Clearly these results provide the basis for more detailed investigations of muonium dynamics.

References

1. G. Davis, Phys. Rev. B, **185** (1993) 1-15
2. S.J. Pearton, J.W. Corbett and M. Stavola, *Hydrogen in Crystalline Semiconductors* (Springler-Verlag), **16** (1992)
3. JPF Sellschop, CCP Madiba and HJ Annegarn, Diam. Conf. (Cambridge) (1979) p.43-44
4. S.K. Estreicher, Mater. Science and Engin. R14, **7-8** (1995) 320
5. H. Jia, J. Shinar, D.P. Lang and M. Pruski, Phys. Rev. B, **48** (1993) 17595
6. K.M. McNamara, D.H. Levy, K.K. Gleason and C.J. Robinson, Appl. Phys. Lett., **60** (1992) 580

7. G. Popovici, R.G. Wilson, T. Sung, M.A. Prelas and S. Khasawinah, J. Appl. Phys., **77** (1995) 5103
8. E. Vainonen, J. Likonen, T. Ahlgren, P. Haussalo, J. Keinonen and C.H. Wu, J. Appl. Phys., **82** (1997) 3791
9. C.G. Smallman, S.H. Connell, C.C.P. Madiba and J.P.F. Sellschop, Nucl. Intr. and Meth., **B118** (1996) 688
10. S.H. Connell, E. Sideras-Haddad, C.G. Smallman, J.P.F. Sellschop, I.Z. Machi and K. Bharuth-Ram, Nucl. Intr. and Meth., **B118** (1996) 332
11. I.Z. Machi, P. Schaaff, S.H. Connell, B.P. Doyle, R.D. Maclear, P. Formenti, K. Bharuth-Ram, and J.P.F. Sellchop, Nucl. Instr. and Meth., **127/128** (1997) 212
12. J. Isoya, S. Wakoh, M. Matsumoto and Y. Morita, JAERI TIARA Anual Report, **3** (1993) 21
13. B.D. Patterson, Rev. Mod. Phys., **60** (1988) 70
14. T.A. Claxton, A. Evans and M.C.R. Symonds, J. Chem. Soc. Faraday. Trans. II., **82** (1986) 2031
15. T.L. Estle, S. Estreicher and D.S. Marynick, Phys. Rev. Lett., **58** (1987) 1547
16. P. Briddon, R. Jones and G.M.S. Lister, J. Phys. C, **21** (1988) L1024
17. S.F.J. Cox and M.C.R. Symons, Chem. Phys. Lett., **126** (1986) 516
18. I.Z. Machi, S.H. Connell, M. Baker, J.P.F. Sellschop, K. Bharuth-Ram, C.G. Fischer, R.W. Nilen, S.F.J. Cox and J.E. Butler, Physica B, **289/290** (2000) 507
19. I.Z. Machi, S.H. Connell, J.P.F. Sellschop, K. Bharuth-Ram, B.P. Doyle, R.D. Maclear, J. Major and R. Scheuermann, Physica B, **289/290** (2000) 486
20. J.W. Schneider, R.F. Kiefl, K.H. Chow, S. Johnston, J. Soier, T.L. Estle, B. Hitti, R.L. Lichti, S.H. Connell, J.P.F. Sellschop, C.G. Smallman, T.R. Anthony and W.F. Banholzer, Phys. Rev. Lett., **71** (1993) 557
21. Y. Mita, Y. Nisida, K. Suito, A. Onodera, and S. Yazu, J. Phys.: Condens. Matter, **2** (1990) 8567
22. G. Davies, S.C. Lawson, A.T. Collins, A. Mainwood, and S.J. Sharp, Phys. Rev. B, **46** (1992) 13157
23. K. Bharuth-Ram, R. Scheuermann, I.Z. Machi, S.H. Connell, J. Major, J.P.F. Sellschop and A. Seeger, Hyp. Int., **105** (1997) 339
24. E. Holzschuh, W. Kündig, P.F. Meier, B.D. Patterson, J.P.F. Sellschop, M.C. Stemmet and H. Appel, Phys. Rev., **25** (1982) 1272
25. D.P. Spencer, D.G. Fleming and J.H. Brewer, Hyp. Int., **17-19** (1984) 567
26. Mehandru S.P., and A.B. Anderson, *J. Mat. Res.*, **9**(1994)383

POSITRONS IN DIAMOND

C.G. Fischer , R.W.N. Nilen , S.H. Connell, D.T. Britton*, J.P.F. Sellschop

Schonland Research Centre for Nuclear Sciences, University of the Witwatersrand, Private Bag 3, Johannesburg, WITS 2050, South Africa
*Department of Physics, University of Cape Town, Private Bag, Rondebosch 7700, South Africa

We present an overview of the results of investigations on the interaction of the positron with the diamond lattice. Various experimental configurations were employed in an effort to understand the interaction of positrons with the undefected diamond lattice and with intrinsic and extrinsic defects.

1 Introduction

Carbon in the form of coal or charcoal has been familiar since the discovery of fire. In the form of diamond, known as a gem stone for centuries, it has become the focus of much applied and fundamental scientific interest.

The greatest research activity into diamond is of an applied nature. Many of the properties relevant to diamond's technological significance, such as phonon limited transparency from the infrared to the ultraviolet region, and its wide band gap and high breakdown voltage, depend on the crystalline quality of the sample under investigation [1]. Defects, impurities, disorder and strain compromise the crystallinity of the sample and thus diamond's optical and electrical properties. However, defects are not always unwanted in the diamond crystal; the controlled doping, by ion implantation, of diamond with acceptors and donors is actively pursued as a research topic and hopefully will result in methods for the routine precision doping of diamond. Methods for the implantation and subsequent treatment of diamond which allow for little unwanted damage in the final product must be developed.

Defects in diamond thus need to be characterised and so is evident the need for spectroscopies sensitive to defects in diamond. Only recently has positron annihilation spectroscopy (PAS) been applied to the study of defects in diamond and thus it has not yet reached the maturity it exhibits in the study of materials such as silicon [2].

Positrons implanted into most solids rapidly thermalise and subsequently diffuse throughout the solid and annihilate with an electron either in an untrapped or a trapped (localised) state. Open volume defects are well known traps for positrons due to the absence of the repulsive ion core.

Two photon annihilation exhibits the dominant cross section with each emitted γ-ray Doppler broadened corresponding to the component of the C.M. momentum along the emission direction, $\Delta E_\gamma = cp_l/2$. Transverse momentum components are taken into account by deviations from collinearity, $\theta = p_l/mc$ where m is the rest mass of the electron/positron. Since positrons are thermalised or close to thermalised, their contribution to the C.M. momentum is negligible. The Doppler broadening forms the basis of Doppler broadening spectrosocopy where the momentum distributions of the annihilation electrons are probed by investigating the annihilation curve which is a plot of counts, centered at 511 keV, versus photon energy. In lifetime measurements, when using Na^{22} as the source, the 1.274 MeV decay γ-ray is used as start signal. The lifetime components conventionally are extracted according to a two-state trapping model. The lifetime of the positron is indicative of the electron density it samples, i.e., it increases/decreases with decreasing/increasing electron density.

When the positron traps, and thus localises, at an open volume defect such as a vacancy, it samples a smaller electron density and its overlap with the core electrons is reduced. This leads to a reduction in Doppler broadening and increase in lifetime.

The Doppler broadening curves are characterised by so-called S(hape)-parameters and W(ing)-parameters. The S-parameter is the ratio of counts in a region around 511 keV to the total counts in the photopeak, and thus is sensitive to annihilations with low momentum electrons. The W-parameter is defined as the ratio of counts in an interval towards the outer regions and the total counts in the photopeak and thus is sensitive to annihilations with high momentum electrons.

2   Investigations on diamond

Age Momentum Correlation (AMOC) Measurements, where the lifetime of the positron is correlated with the momentum distribution of the annihilating electron, were carried out on a suite of diamonds [3]. The results for a natural type IIa (virtually nitrogen free) are shown below in figure 1 and compared with quartz.

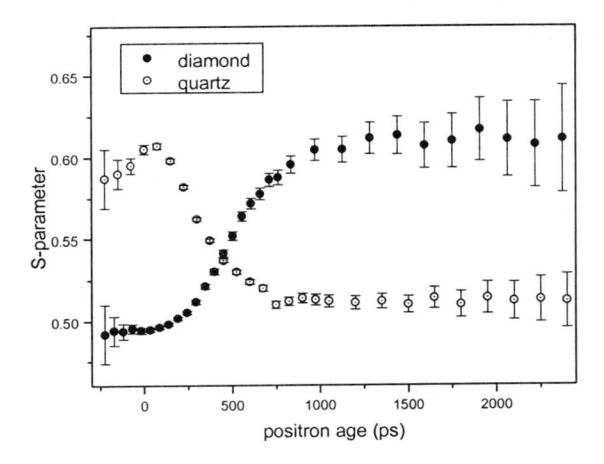

Figure 1. AMOC measurements on diamond and fused quartz. For the diamond, a very high Doppler broadening (low S-parameter) is correlated with a very short lifetime. There is no indication of the formation of Ps, unlike for quartz.

A giant Doppler broadening (~ 3.5 keV/c FWHM) is found to be correlated with a very short lifetime (~ 100 ps). This is rather surprising since one would have expected these values to be comparable with covalently bonded Si which also crystallises in the diamond structure, and exhibits a bulk lifetime of ~ 220 ps [4] and a Doppler broadening of ~ 2.7 keV/c FWHM [5].

In insulators, the thermalisation of the positron is hindered by the large discrepancy between the band gap (and thus threshold energy for electron-hole excitations) and energies of below 1 eV where phonon scattering becomes dominant. Besides exciton formation, so-called "sub-excitation" thresholds are available, where the positron, though of an energy less than the band gap, can cause an electron-hole excitation and subsequently form a bound state with the electron to form quasi Positronium (qPs) [6] as opposed to vacuum Positronium (Ps). Quartz is a classic qPs former with 2-γ annihilation of the shorter lived (~ 125 ps) singlet para state leading to a large S-parameter (as the electron is unbound) followed by the longer lived 2-γ pick-off annihilation (with a bound electron of opposite spin to that of the positron) of the triplet ortho state leading to a lower S-parameter. From the AMOC curve diamond clearly is not a classical Ps former.

The bulk annihilation configuration

How is the correlation of a giant Doppler broadening with a very short lifetime in diamond to be explained? Besides an improbable enhanced interaction of the positron with the carbon 1s electrons, one should consider a possible interaction of a thermalised positron with the high spatial density of the intrabond electrons. The small lattice constant means that the electrons exhibit a high momentum distribution. In figure 2 we show the results of a two-component density functional theory calculation of the electron-positron overlap density [7], which clearly peaks in the intrabond region, unlike in Si where it peaks in the interstitial region. For diamond, this results in a 1:1 (± 5%) ratio of annihilation intensities for the intrabond and interstitial regions.

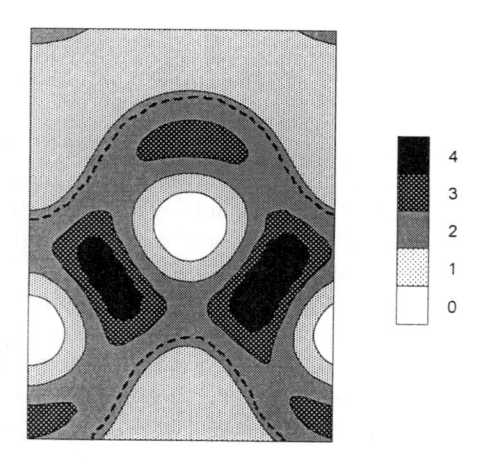

Figure 2. The overlap of the electron-positron ground state densities as calculated in the density functional theory calculation. The distribution represents a map of the annihilation probability (arbitrary units) and reveals a maximum in the intrabond region. The dashed line indicates the chosen interface between the intrabond and interstitial domains for the annihilation site calculation.

An experimental investigation of the electron-positron momentum distribution in diamond was undertaken [8]. Figure 3 illustrates a high resolution Doppler broadening study of the diamond electron-positron momentum density, showing the momentum anisotropy, that is, the difference between the <100> and <110> momentum projections.

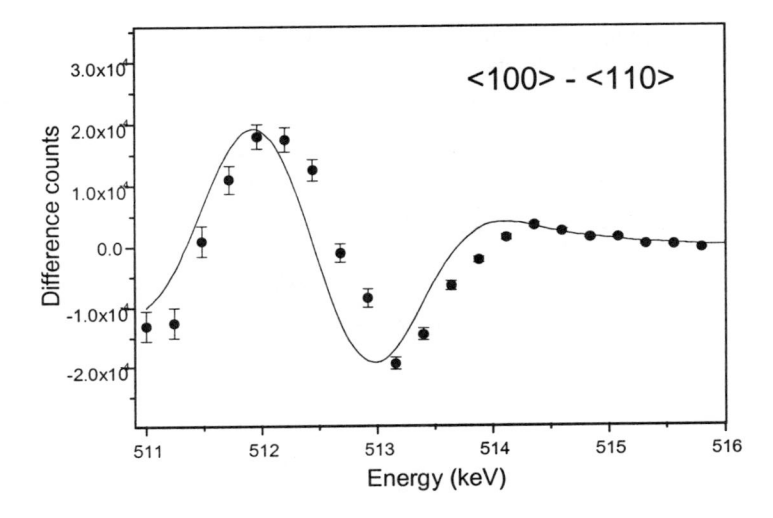

Figure 3. The <100> - <110> difference spectra for the momentum profiles of a natural type IIa diamond. The predictions of the density functional theory calculation (solid line) show good qualitative agreement with experiment.

Good qualitative agreement between experiment (dots) and theory is found. The positron can be understood to be thermalised and interacts as a delocalised Bloch wave with the valence electron density whereby the positron density has sufficient amplitude at the intrabond region so that the annihilation characteristics are dominated by the bond electron density and momentum distribution. This annihilation configuration explains the correlation, as observed by AMOC, between a short lifetime and a large Doppler broadening. The higher S-parameters with increasing positron age, as observed in the AMOC measurement (Figure 1), may then be interpreted in terms of the annihilation of trapped positrons.

Positron-defect interactions

The efficiency of positron trapping depends in part on the mechanism by which the binding energy gets transferred. This usually is done via electron-hole excitations which clearly is suppressed in insulators. Phonon excitation is rather inefficient; indeed, it has been suggested that trapping in insulators results in the formation of Ps [9]. From an analysis of lifetime spectra, the trapping rate into defects can be extracted, and is given by $\kappa = \mu C$ where C is the defect concentration and μ the specific trapping rate. Negative charging of defects leads to increases in the specific trapping rate. Figure 4 shows the trapping rate as a function of illumination and annealing temperature for diamond [10].

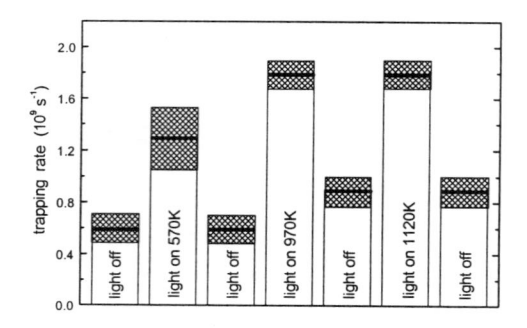

Figure 4. The trapping rate, extracted from the lifetime spectra, as a function of temperature and illumination. The 'light off' measurements are at room temperature. The hatched areas represent the error bars.

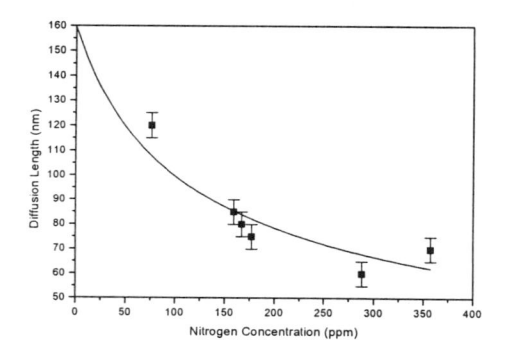

Figure 5. Best fit to the diffusion lengths for the synthetic Ib diamonds assuming the positron to trap at the substitutional nitrogen and/or defects whose concentrations are proportional to the substitutional nitrogen.

A clear temperature and illumination dependence can be observed and is best understood in terms of the photochromicity and thermochromicity of trapping defects – we thus see the trapping of positrons in the insulator diamond. The mechanism by which this proceeds is not clear however. A possibility is that, due to diamond's extremely high Debye temperature, phonon assisted trapping may be considerably more efficient than is normally assumed.

Diamond is classified into different types according to the nitrogen content and configuration. Type Ib diamond differs from type IIa by the presence of nitrogen, which in type Ib diamond is located substitutionally. Type Ib diamond is found to exhibit a greater Doppler broadening and this general observation has lead to much speculation on a possible positron-nitrogen interaction [11,12].

Efficiencies of positron moderators have improved considerably in the last two decades [13] and this has allowed for the routine application of collimated, monoenergetic slow positron beams. When using a slow positron beam the S-parameter is measured as a function of implantation energy, to give us a so-called S(E) curve. Measurements of positron backdiffusion to the surface as a function of implantation energy allows for an extraction of the positron diffusion length, whereby the diffusion length is given by the expression

$$L_+ = \sqrt{\frac{D_+}{\lambda_b + \kappa}}$$

where D_+ is the positron diffusion coefficient, λ_b the bulk annihilation rate and κ the total trapping rate.

Profiling, by backdiffusion measurements, a number of type Ib diamonds with varying nitrogen concentrations allows us to investigate whether the positron does trap at the substitutional nitrogen. Below in figure 5 we show the diffusion length as a function of substitutional nitrogen concentration, as well as the best fit to the diffusion lengths assuming the positron to trap at the substitutional nitrogen and/or defects whose concentrations are proportional to that of the substitutional nitrogen. It is found that the data are not consistent with the above assumption.

The data however are consistent with the positron trapping at the substitutional nitrogen and/or defects whose concentrations increase disproportionately with that of the substitutional impurity.

Diamond exhibits certain planes where the rate of material removal depends on the direction of polishing [14]. Here we investigated the (110) plane, which is known to exhibit wear anisotropies. Backdiffusion curves for diamond polished in the "hard" (lowest wear rate) and "soft" (highest wear rate) directions vary considerably and, on closer analysis, which will be described in more detail elsewhere, it can be seen that polishing in the "hard" direction induces subsurface damage.

The backdiffusion technique is applied extensively to the study of ion implantation induced damage [4]. Virtually no such measurements have been carried out on ion implanted diamond. Below are shown curves of the S-parameter versus implantation energy for a diamond sample self-implanted with 2 MeV C atoms both "off"-axis and in a roughly "on"-axis orientation.

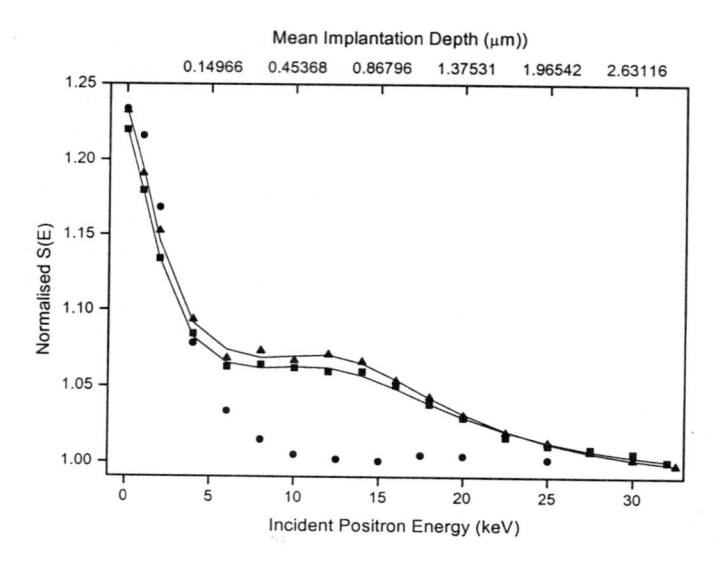

Figure 6. S(E) curves for a type Ib diamond polished along the "hard" and "soft" polishing directions in the (110) plane. Squares: "Soft" direction. Circles: "Hard" direction. The top x-axis denotes the mean implantation depth of positrons implanted at the corresponding energy.

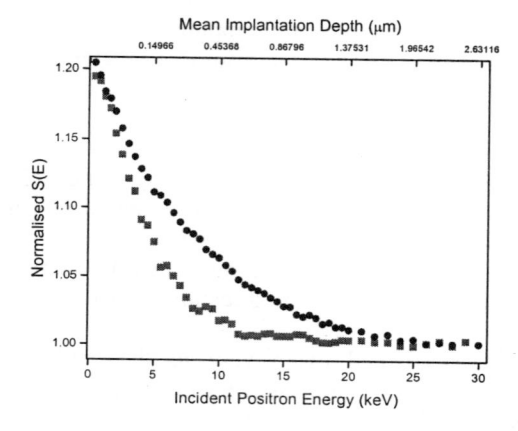

Figure 7. S(E) curves for a synthetic type Ib diamond implanted "on-" and "off-axis" with 2 MeV carbon ions. "On-axis": Triangles. "Off-axis": Squares. Unimplanted side: Dots. Lines are fits to the data. A clear effect can be observed over the positron mean implantation depth range of 0.2 µm – 1 µm.

A clear difference can be observed between the S(E) curves for "on-axis" and "off-axis" self-implanted diamond. This is in agreement with the expectation that greater damage is induced when implanting in the "off-axis" direction where crystal effects are suppressed. In the "off-axis" direction there is a greater probability for nuclear collisions, while in the "on"-axis direction the cross section for energy loss via electronic collisions is substantial. We find that the data are consistent with the positron trapping at implantation induced vacancy clusters.

3 Conclusions

An overview of some of the results of investigations on the positron-diamond interaction has been given. The research has allowed for a good understanding of the interaction of the positron with the undefected diamond lattice. The existence of trapping in the diamond lattice, as well as the photochromic and thermochromic behaviour of vacancy clusters has been demonstrated. The presence of substitutional nitrogen leads to a broader annihilation curve. It has been shown that an increase in the concentration of substitutional nitrogen correlates with a disproportional increase in the concentration of defects which trap the positron. Backdiffusion measurements have shown that polishing in the (110) plane in the direction with the lowest wear rate results in subsurface damage. Finally the sensitivity of the backdiffusion technique to ion implantation induced damage has been demonstrated.

References

1. The properties of natural and synthetic diamond, J. E. Field, Academic Press, 1992
2. Positron Spectroscopy of Solids, A. Dupasquier and A. P. Mills, Jr., IOS Press, 1995
3. M. Koch, K. Maier, J. Major, A. Seeger, J. P. F. Sellschop, E. Sideras-Haddad, H. Stoll, S. H. Connell, *Mat. Sci. Forum,* **105-110**, 671 (1992)
4. P. Asoka-Kumar, K.G. Lynn, D.O. Welch, *J. Appl. Phys.*, **76**, (1994), 4935
5. J.C. Erskine, J. D. McGervey, *Phys. Rev.*, **151**, (1966), 615
6. A. Dupasquier in *Positrons in Solids*, P. Hautojärvi, Springer, New York, 1979, 235
7. W.G. Schmidt, W.S. Verwoerd, *Phys. Lett. A*, **222**, (1996), 275
8. R.W.N. Nilen, S.H. Connell, D.T. Britton, C.G. Fischer, E.J. Sendezera, P. Schaaf, W.G. Schmidt, J.P.F. Sellschop, W.S. Verwoerd, *J. Phys. C*, **9**, (1997), 6323
9. M. Shi., W.B. Waeber, W. Triftshäuser, *Appl. Surf. Sci.*, **116**, (1997), 203

10. U. Lauff, R.W.N. Nilen, S.H. Connell, H. Stoll, K. Bharuth-Ram, A. Siegle, H. Schneider, P. Harmat, P. Wesolowski, J.P.F. Sellschop, A. Seeger, *Appl. Surf. Sci.*, **116,** (1997), 330
11. S. Fujii, Y. Nishibayashi, S. Shikata, A. Uedono, S. Tanigawa, *Appl. Phys. A*, **61**, (1995), 331
12. A. Uedono, S. Fujii, N. Morishita, H. Itoh, S. Tanigawa, S. Shikata, *J. Phys. C*, **11**, (1999), 4109
13. P.J. Schultz, K.G. Lynn, *Rev. Mod. Phys.*, **60**, (1988), 701
14. M. Tolkowsky, Ph.D. Thesis, University of London, 1992

STUDY OF THE MOMENTUM TRANSFER TO TARGET-LIKE RESIDUES IN HEAVY ION REACTIONS BY PROMPT GAMMA MEASUREMENTS

K. A. KORIR, S. H. CONNELL, J. P. F. SELLSCHOP, E. SIDERAS-HADDAD

Schonland Research Centre for Nuclear Sciences, University of the Witwatersrand, Johannesburg, South Africa

S. V. FÖRTSCH, J. J. LAWRIE, R. T. NEWMAN, G. K. MABALA, F. D. SMIT, G. F. STEYN

National Accelerator Centre, Faure, South Africa

R. BASSINI, C. BIRATTARI, M. CAVINATO, E. FABRICI, E. GADIOLI, E. GADIOLI ERBA

Dipartimento di Fisica, Università di Milano, Istituto Nazionale di Fisica Nucleare, Sezione di Milano, Milano, Italy

Z. VILAKAZI, B. BECKER

University of Cape Town, Cape Town, South Africa

Doppler shift and Doppler broadening of prompt γ lines have been measured for many reaction residues in the interaction of ^{12}C with ^{63}Cu at 33 A MeV incident ^{12}C energy using the AFRODITE detector array at NAC, Faure, Cape Town. A preliminary analysis of these data, which carry information of the momentum transferred in the reaction, shows that very useful information regarding reaction mechanisms can be obtained by this technique.

1 Introduction

Many studies of heavy ion reactions are based on the detection of the light particles, and/or the projectile-like (PLF) and the target-like (TLF) fragments. Different experimental techniques have been applied, depending on the reaction mechanism under study. One is based on the simultaneous measurement of the residue time-of-flight and kinetic energy with detector telescopes[1,2]. Another is based on the measurement of the recoil ranges of the radioactive residues in catcher foils[3-13]. A third, limited to fission fragments, is based on the measurement of linear momentum transferred to the intermediate fragments, by the observation of the folding angles of coincident fission fragments[14,15].

In our previous studies of the reactions induced by ^{12}C, made at the National Accelerator Centre at Faure, South Africa, we used the stacked foil technique for measuring the production cross-sections, the forward recoil range distributions and the angular distributions of many residues. We also measured the spectra of the emitted α-particles as well as of ^{8}Be[10−13,16, 17]. In agreement with the suggestions of other authors[18−26], we concluded that, for incident energies up to 400 MeV and for all the target nuclei considered, the incomplete fusion (or the transfer to the target) of α-particle ^{12}C fragments, and, at the higher energies, of single nucleons also are, together with the complete fusion, the dominant reaction mechanisms. The main mechanism leading to incomplete fusion is the break-up of the projectile. These mechanisms are incorporated in a comprehensive model which seems to reproduce quite reasonably many of these experimental results[11−13,16,17].

All the studies which are quoted above show that the transferred longitudinal momentum expressed as a fraction of the projectile momentum decreases with increasing projectile energy. An additional way to study the momentum transferred to TLFs is the in-beam identification of residues using a large solid angle gamma array of high granularity, measuring the intensity distributions of the Doppler broadened and shifted residue's γ lines as a function of the detector angle. For us a strong motivation for undertaking such a measurement is the possibility of observing also residues which cannot be studied with the stacked foil technique. Near target residues are for us of particular interest, as the comprehensive model mentioned above predicts that they are produced with very large yields and a very low linear momentum. The mechanism leading to this effect is related to the high probability of fast re-emission of the projectile's fragments.

Prompt γ spectroscopy has typically been used in the study of the properties and decay modes of highly excited nuclei. Its effectiveness in elucidating reaction mechanisms at intermediate energies is not yet completely explored. In fact, the technique has been mainly utilised in the study of the GDR and continuous high energy γ ray spectra. The main limitation of prompt γ spectroscopy in nuclear reaction mechanism studies is the extreme complexity of the in-beam γ ray spectra and the neutron damage of the detectors at the higher projectile energies. It must be mentioned that our work was also motivated by calculations made with the above mentioned model which predict that the neutron damage is tolerable at the energies provided by the NAC accelerator. The objective of this report is to describe some of the results obtained for the ^{63}Cu+^{12}C reaction at an incident ^{12}C energy of E=33 A MeV using the AFRODITE 4π gamma detector array, which indicate the potential of this technique for reaction mechanism studies.

2 Experimental Setup

A 5nA, 400 MeV ^{12}C beam was delivered by the cyclotron of the National Accelerator Centre, Faure, South Africa[27]. The ^{63}Cu targets used were 150-170 μgcm^{-2} thick, mounted in the centre of the target chamber of the AFRODITE gamma detector array[28]. The array consists of 8 segmented LEPS and 7 segmented escape-suppressed Clover detectors, making a total of 60 detector elements. The detection limits range from about 20 to 800 keV and 35 to 1800 keV for LEPs and Clovers respectively. Calibration of the detectors was performed using ^{133}Ba and ^{152}Eu point sources located at the target position of the chamber. The detectors were distributed as follows, 2 Clovers and 2 LEPS at both 45° and 135°, 3 Clovers and 4 LEPS at 90°. Aluminum discs of 0.9mm thickness for the Clover and 1.55mm thickness for the LEPs detectors were used to attenuate the low energy background. The trigger logic was set to record data when at least 2 out of the 15 detectors fired. A total of about 1 billion events was recorded on DLT tapes for off-line analysis.

3 Data Analysis

For the most part of the analysis (residue identification, peak fitting and also Doppler shift determination) the Radware computer software has been used. For identification of the reaction residues, at least 2γ coincidences of the 90° detectors were used in a $\gamma - \gamma$ coincidence matrix[29]. Extraction of Doppler shifted (and broadened) gamma spectra gated for particular residues was then achieved with the Xsys sorting code, operating on the calibrated and timing filtered event file.

For evaluating the average residue's velocity $\beta = v/c$, one exploits the Doppler effect according to which the energy of a gamma line emitted by a residue recoiling with velocity **v** is given by

$$E = E_0(1 + \beta \cos\alpha) \qquad (1)$$

where E_0 is the gamma transition energy for a residue at rest, $\beta = v/c$ and α is the angle of the detected gamma with respect to the residue velocity. Due to the residue's azimuthal symmetry around the beam direction, the average energy of the γ lines measured by a detector at 90° to the beam coincides with E_0, while that of the lines measured by non-perpendicular detectors (in the case of AFRODITE at 45° and 135°) are *shifted* to higher or lower energies. The observed residue's γ lines are broadened since the residue has a distribution of **v** which depends on the reaction path for its production. One must also take into account that in part the broadening also depends on the opening angle of the detector and in the analysis of the data one must take into account this geometrical effect. The residue average velocity β (in units of c) is given by

$$\beta = \frac{E_{\gamma 1} - E_{\gamma 2}}{\sqrt{2} E_\gamma} \qquad (2)$$

where $E_{\gamma 1}$, $E_{\gamma 2}$ and E_γ are the average energies of the γ lines measured by the 45°, 135° and 90° detectors, respectively.

4 Results and discussion

It is immediately apparent on considering the Doppler broadened and shifted γ spectra for the various residues that information on the residue recoil momentum distribution may be extracted. Even if the momentum transfer to the residues is in general rather low, they nonetheless evidence appreciable changes in their line shapes as a function of detection angle. The case of the 212 keV line of ^{54}Mn (figure 1) and the 948 keV line of ^{61}Ni (figure 2) are selected to illustrate this point.

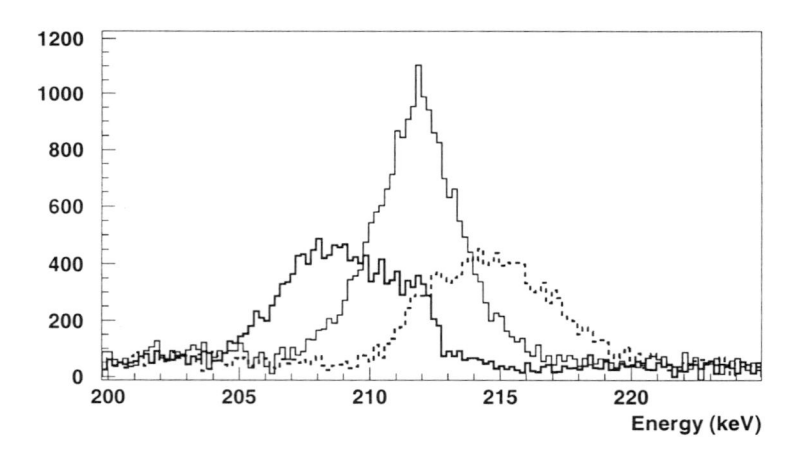

Figure 1. ^{54}Mn Doppler Shift of 212 keV line (LEPs spectra)

Altogether, we have been able to analyse the Doppler broadened and shifted γ lines of the following residues, for which we give also in parenthesis the value of the ratio β/β_{cn} of the residue's average velocity (in units of c) to the compound nucleus velocity : ^{64}Cu(0.033), ^{62}Cu(0.047),

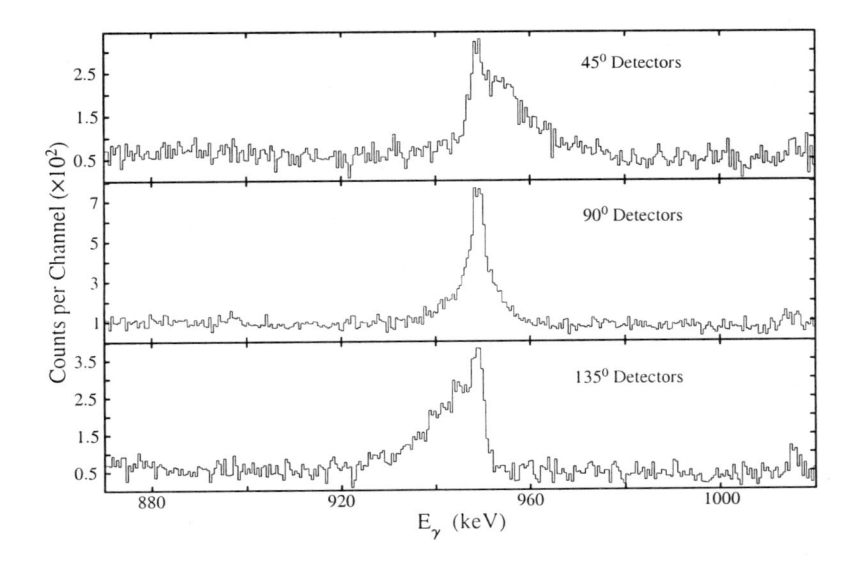

Figure 2. ^{61}Ni Doppler Shift of 948 keV line (Clover spectra).

^{61}Ni(0.13), ^{60}Cu(0.22), ^{59}Ni(0.31), ^{59}Co(0.34), ^{57}Fe(0.34), ^{55}Mn(0.41), ^{55}Fe(0.45), ^{54}Mn(0.45), ^{47}V(0.47), ^{42}K(0.49). These values of β/β_{cn} and their estimated uncertainties are given by the black dots in figure 3 which shows that most of the residues are produced with quite small velocity with respect to the compound nucleus, that is in processes with quite a small momentum and energy transfer from the projectile to the target.

Even the residues with mass considerably smaller than that of the target nucleus, which are expected to be produced in comparatively larger energy and momentum transfer processes, have an average velocity which is only about one half of the compound nucleus velocity. Thus, these measurements provide strong evidence that, with increasing energy, the break-up of ^{12}C followed by the incomplete fusion of only one of its fragments becomes the dominant reaction mechanism.

The shift and broadening of the observed γ lines are estimated with the model which has been mentioned above. The preliminary results of the analysis, only part of which are shown, show that the calculation reproduces quite

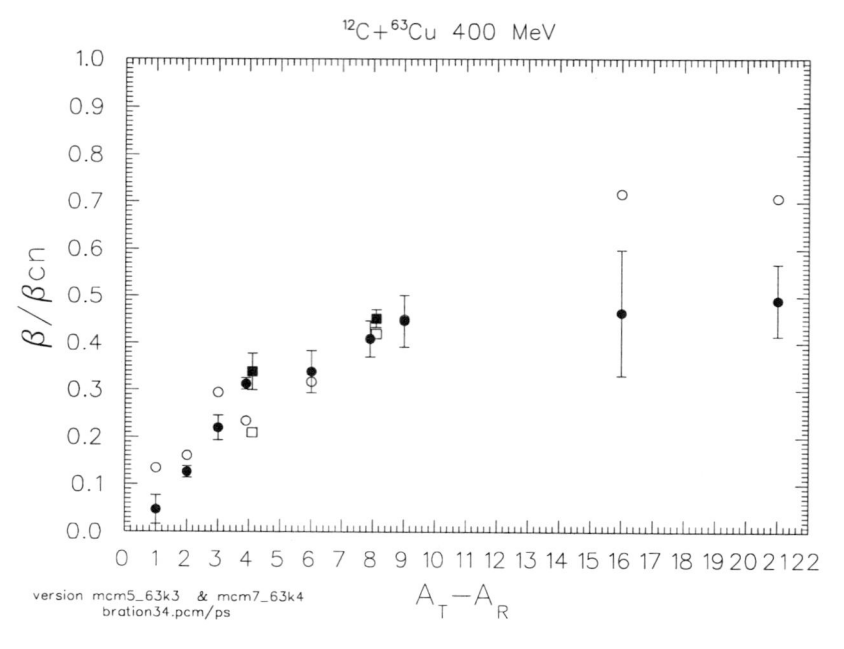

Figure 3. Ratio between the average residue velocity in units of c and the compound nucleus velocity β_{cn} for residues with mass smaller than the target nucleus mass.

reasonably the Doppler shift and broadening of most of the observed lines and their general trend. The estimated average velocities are shown in figure 3 by the open symbols.

However, there are a few discrepancies which call for an improvement of the model. These essentially concern the overestimation of the average velocities of the target isotopes (^{62}Cu, and ^{64}Cu) which are most presumably produced in direct interactions which we do not consider explicitly in our present approach, but which may be of great relevance in the production of these residues in spite of the small contribution to the total reaction cross-section. In the case of ^{64}Cu, which is not shown in the figure, and for which the experimental value of β/β_{cn} is 0.033 while the theoretical estimate is 0.161, this direct reaction could be the capture by the target of a very low energy neutron from ^{12}C. This process is dynamically unfavoured at high ^{12}C energies, but it might be spectroscopically favoured in comparison to the transfer of a higher energy neutron. The very low shift of the ^{62}Cu γ lines could be due to the inelastic scattering of ^{12}C followed by the emission of one neutron by the excited target nucleus. In principle, the processes which we consider include these reaction paths, but their importance might be significantly underestimated. We are looking for an improvement of our calculations to take more realistically into account such contributions. Also the momentum transfer to the two residues with smallest mass seems to be overestimated by our calculations. This might imply that either we overestimate the contribution of complete fusion to their production and/or we underestimate the amount of pre-equilibrium emissions.

5 Conclusion

A significant number of the reaction products in the interaction of ^{12}C with ^{63}Cu at an incident energy of 33 A MeV has been observed and the Doppler broadening and shift of their lines studied. These data indicate that a large number of the observed residues are produced with velocities considerably smaller than the CN velocity, suggesting that in most of the interactions there is a reduced transfer of linear momentum and energy between the projectile and the target. A preliminary analysis of these the data indicates that they are sensitive to more than one reaction mechanism and of sufficient quality to test and improve the theoretical reaction models. A more detailed comparison of the data to predictions of our comprehensive model, as well as other models, is in progress.

6 Acknowledgements

We wish to thank the NAC accelerator staff who worked tirelessly to provide us with a beam of excellent quality.

References

1. S. Leray, Journ. Phys. (Paris) C 4 (1986) 275.
2. G. A. Souliotis et al., Phys. Rev. C **57** (1998) 3129.
3. S. Y. Cho, N. T. Porile and D. J. Morrissey, Phys. Rev. C **39** (1989) 2227.
4. S. Y. Cho, Y. H. Chung, N. T. Porile and D. J. Morrissey, Phy. Rev. C **36** (1987) 2349.
5. J. P. Whitfield and N. T. Porile, Nuclear Physics A **550** (1992) 553.
6. J. P. Whitfield and N. T. Porile, Phys. Rev. C **39** (1993) 1636.

7. L. Kowalski, P. E. Haustein and J. B. Cummings, Phys. Rev. Lett. **51** (1983) 642.
8. T. Lund et al., Phys. Lett. **102** B (1981) 239.
9. Y. K. Kim et al., Nucl. Phys. A **578** (1994) 621.
10. C Birattari et al., Phys Rev C **54** (1996) 3051 - 3055
11. E Gadioli et al., Phys. Lett. B. **394** (1997) 29-36
12. E Gadioli et al., Nucl Phys A **641** (1998) 271
13. E Gadioli et al., Heavy Ion Physics **7** (1998) 275 - 287
14. V. E. Viola et al., Phys. Rev. C **36** (1982) 178.
15. M. Fatyga et al., Phys. Rev. Lett. **55** (1985) 1376.
16. E. Gadioli et al., Nucl. Phys. A**654** (1999) 523
17. E. Gadioli et al., Eur. Phys. J. A **8** (2000) 373-376
18. D. J. Parker et al., Phys. Rev. C**44** (1991) 1528
19. R. Bimbot, D. Gardès and M.F.Rivet, Nucl. Phys. A**189** (1972) 193
20. Y. Le Beyec, M. Lefort and M. Sarda, Nucl. Phys. A**192** (1972) 405
21. K. Siwek-Wilczyńska et al., Nucl. Phys. A**330** (1979) 150
22. D. J. Parker et al., Phys. Rev. C**30** (1984) 143
23. B. S. Tomar et al., Z. Phys. A**343** (1992) 223
24. P. Vergani et al., Phys. Rev. C**48** (1993) 1815
25. M. Crippa et al., Z. Phys. A**350** (1994) 121
26. K. Siwek-Wilczyńska et al., Phys. Rev. Lett. **42** (1979) 1599 Phys. Rev. C **48** (1993) 633.
27. J. V. Pilcher, A. A. Cowley, D. M. Whittal and J. J. Lawrie, Phys. Rev. C **40** (1989) 1937.
28. R. T. Newman et al. NAC Report, *High-Spin Studies with the AFRODITE Array*, Nov. (1998).
29. D. C. Radford, Nucl. Instr. Meth. A **361** (1995) 297.
    D. C. Radford, Nucl. Instr. and Meth. A **361** (1995) 306. (1998) 811. **24** (1981) 2162. (1979) 150.

THE SCHONLAND NUCLEAR MICROPROBE – AN IMPORTANT TOOL IN GEOSCIENCES.

R.K. DUTTA[1], E. SIDERAS-HADDAD[1], S.H. CONNELL[1], J.P.F. SELLSCHOP[1], R.J. HART[1], M.A.G. ANDREOLI[2], M. TREDOUX[3], P. FORMENTI[1] AND A. WITTENBERG[4]

[1]*Schonland Research Centre for Nuclear Sciences, WITS University, Johannesburg*

[2]*NECSA, PO BOX 582, Pretoria,*

[3]*Department of Geological Sciences, University of Cape Town, Rondebosch 7700, SA.*

[4]*Institut fur Mineralogie, University of Hannover, Hannover Germany.*

The Schonland Scanning Proton Microprobe has been used over several years to investigate various types of geological materials in order to understand the evolution of these materials in the earth. This paper represents an overview of some selected studies.

1 Introduction

The Schonland nuclear microprobe, a state-of-the-art facility, was commissioned in 1992 and since then has been used extensively in various research fronts, namely, solid state physics, materials science, archaeology and mineralogy. One of the most important features of this microprobe is that it is coupled to two accelerators, a 6 MV tandem and a 2.5 MV single ended Van de Graaff. The details have been discussed earlier [1]. A beam-line is dedicated for nuclear microprobe studies and this facility makes use of OXFORD quadrupole triplet [2]. The Van de Graaff is connected to the microprobe beam-line by means of two 90° magnets. The applications in geosciences mostly use the proton as a probe and thus it is widely referred to as proton microprobe. These studies comprise trace element mapping of the mineral phases, which is an important tool in understanding the geochemical features of the deposits. Most such studies reflect the correlations of trace elements with the matrix. In addition, the spatial distribution of the matrix and the trace elements in the geological materials are commonly studied. Further, the depth profiling of any desired trace elements is studied. The mapping of trace and matrix elements and their quantification usually refer to the geochemical evolution of the minerals, and these studies are carried out by a technique called micro-PIXE (Proton Induced X-ray Emission). The capabilities of nuclear microprobe are : a) non-destructive analysis; b) detection limits to parts per million; c) depth profiling; d) elemental mapping of both trace and major elements and e) micrometer spatial resolution.

This paper is a review of the nature of the various proton microprobe applications in geosciences, studied using Schonland microprobe facility.

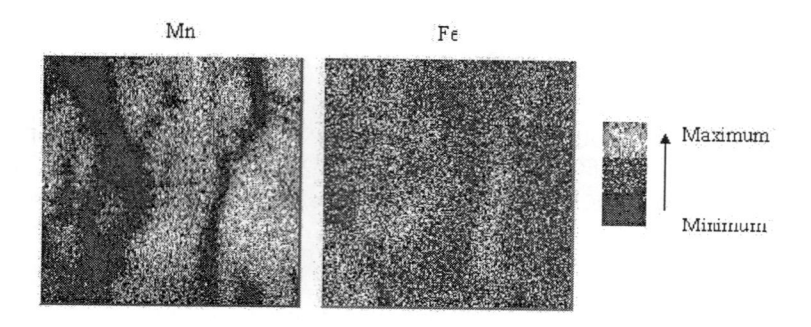

Figure 1: Micro-PIXE elemental map of typical layer formation of alternate Mn rich and Fe rich phases in a hydrogenous ferromanganese nodule from the Indian Ocean (scan size: 750μm x 750μm).

2 Analytical method

The microanalysis of the major and trace elements and their spatial distribution was performed by scanning micro-PIXE using a 3.2 MeV proton beam from the EN Tandem accelerator coupled with the OXFORD nuclear microprobe set–up. The spatial resolution of the proton beam at about ~ 2 - 4 nA beam current was chosen to be 10 μm to 30 μm as demanded for these kind of studies. The proton beam is usually scanned over an area ranging from 250 μm x 250 μm to 2.5 mm x 2.5 mm as required. Using the OMDAQ software the scan area can be matched to the region of interest (ROI) as was done for example, where the platinum speck was found in the meteorite sample shown in fig.6. The proton beam resolution was adjusted to 5 μm for obtaining the spatial distribution of platinum. The data were collected in the LISTMODE (event-by-event) mode, which facilitates off line micro-analysis after masking the ROI's, i.e., the various mineral phases. This enabled the characterization, at least in some cases, of the corresponding minerals. A LEICA microscope (x200 magnification), coupled to the microprobe chamber, was used to monitor the focused beam on the sample. In order to generate good statistics for the trace element distributions the integrated charge was often kept at 40 μC. The counting rates in the Si(Li) detector placed at 135° angle with respect to the beam direction, were maintained at less than 1 KHz. The quantification of the PIXE data obtained from the off-line mapping was carried out by GUPIX software [3]

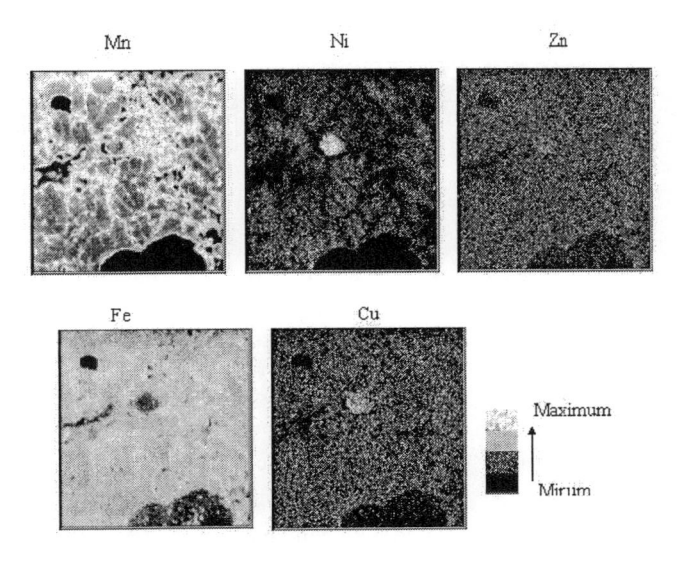

Figure 2: Micro-PIXE elemental map of scan size 25mm x 25mm showing strong spatial correlation of Ni, Cu, Zn and Mn and their anti-correlation with Fe in the deep sea ferromanganese nodule (Ni hot spot diameter ~ 75μm).

3 Results and Discussion

3.1 Deep sea ferromanganese nodule and crust from the Indian Ocean

Iron and manganese are the two major components in these oxide deposits that are found to occur as irregular layers as shown in the Figure 1. The opposite charge of the surface of oxides of Mn and oxyhydroxides of Fe in marine condition leads to the adsorption of these phases one over the other. The geochemical studies of these deposits concentrate on the understanding of the uptake mechanisms of the trace elements from the seawater by the matrix elements. Figure 2 exhibits a positive Ni correlation with Mn while Ni shows an anti correlation with Fe. Nickel in seawater is mostly present as Ni^{2+}, and thus is adsorbed on the negatively charged surface of d-MnO2, while iron oxyhydroxide bears a positive charge. It is further noticed that Cu and to some extent Zn too exhibit a positive correlation with Mn and anti correlation with Fe. Similar to Ni, Cu and Zn too exist as cations in seawater and are thus adsorbed on the Mn oxide phase of the nodule and the crust.

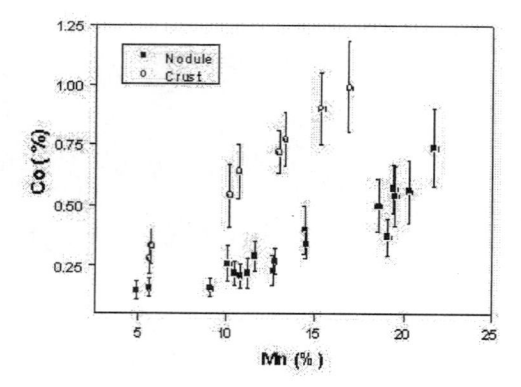

Figure 3: Linear positive correlation of Co with Mn.

Cobalt is found to be associated with Mn phase as shown in the Figure 3. At the depositional environment, which is highly oxidized, cobalt exists as Co^{3+} and has nearly the same ionic radii with Mn^{4+}. Conversely, arsenic, yttrium, niobium, zirconium and molybdenum exhibit a positive correlation with Fe. Except for yttrium, others occur as anions in the seawater and are thus adsorbed by the positive charge surface of iron oxyhydroxide. Yttrium has a strong chemical coherence with rare earth elements (REE) and since REEs are known to be carried by the Fe phase, the positive correlation of Y with Fe is understandable. Further details of this study may be obtained elsewhere [4]. Many other trace elements could be identified, namely, Ga, Se, Rb, Sr, Pb, Tl, Bi and Th. However their correlations with either the Fe or Mn phases were not significantly observed.

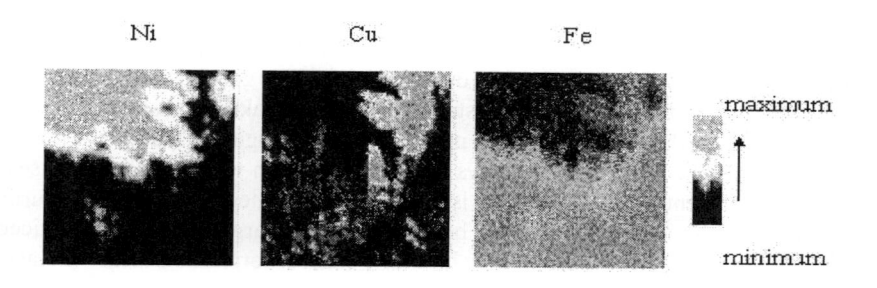

Figure 4: Spatial elemental distribution showing Cu enriched phase separating the Ni rich (millerite) and Fe rich phases in mineral assemblages from the Morokweng impact melt sheet. Scan: 750μm x 750μm.

3.2 Mineral assemblages in the Morokweng impact melt sheet

The samples comprise two major mineral phases, one rich in Ni (about 60 - 64 % concentration by weight) and the other rich in Fe (60 - 68 %). The Ni rich phase was found to be a sulphide bearing mineral phase with a close resemblance to millerite, as shown by complementary SEM studies [6], while the Fe-rich phase was quite complex and could not be identified conclusively. Previous studies suggested Fe to be present as trevorite (Ni-magnetite), however, the SEM analysis reflected the presence of Si in the Fe rich phase. This suggests that the Fe-rich rim is possibly an intergrowth of an oxide such as trevorite [5] and a silicate phase with a grain size too small to be resolved with the present probe. More notably, the millerite and the Fe mineral phases are typically separated by an interface region (width 30 μm – 60 μm) that is rich in Cu-bearing minerals (as shown in Fig 4). The elemental composition of one of these interstitial domains agrees well with talnakhite [Cu_9 (Fe, $Ni)_8S_{16}$], while other spots closely resemble pentlandite [(Fe, $Ni)_9 S_8$] and violarite ($Fe^{+2}Ni_2^{+3}S_4$) [6]. Most other phases remain unidentified because they seem to comprise complex Ni-Cu-Fe sulphides. Selenium, a trace element with chalcophile and volatile characteristics, is significantly enriched (range: 300 ppm - 550 ppm) in the Ni-rich (millerite) phase, and even so more in the Cu - Ni - Fe mineral interface (~ 800 ppm) relative to the Fe bearing (oxide) mineral phase (20 - 50 ppm). The distribution of Se in the Ni, Fe and the interface of Ni-Fe-Cu rich phases is revealed from the Figure 5. It is worth mentioning that the presence of the Cu-Se phases had previously been detected by proton probe in a millerite-rich veinlet in another borehole drilled in the Morokweng impact sheet [5].

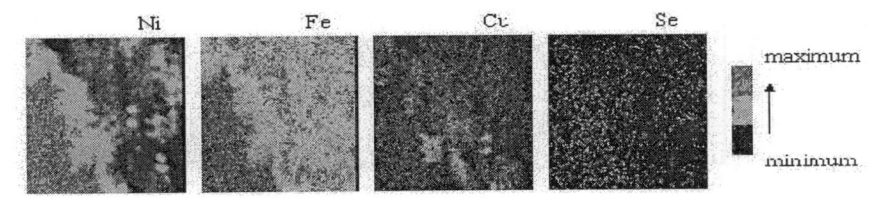

Figure 5: Selenium distribution in the Ni rich (millerite), Fe rich and Ni-Cu-Fe interface in mineral assemblages from the Morokweng impact melt sheet. Scan size: 750μm x 750μm.

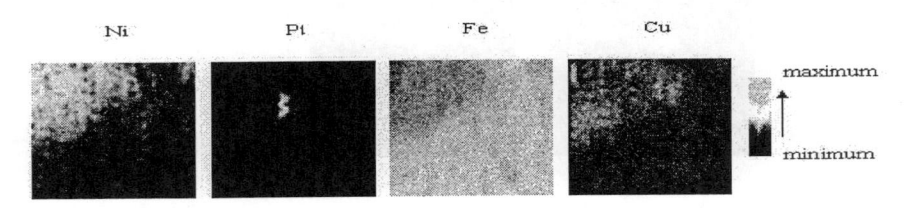

Figure 6: Spatial distribution of Pt in the Ni rich (millerite) and Fe rich interface in mineral assemblages from the Morokweng impact melt sheet. Scan size: 750μm x 750μm.

One of the important findings in this study is the identification of at least two new platinum-bearing minerals within the interstitial domain that carries the Ni - Fe - Cu intergrowths (see Fig. 6). A smaller scan (250 μm x 150 μm) more focused on the platinum speck revealed the nature of its distribution (see Figure 7) and the presence of platinum is reflected from the PIXE spectrum recorded from that scan (see Figure 8). In this map, platinum is found distributed in an area of 125 μm x 50 μm. The highest concentration of platinum (± 14.3 wt. %) is localized in smaller domains of ± 20 μm x 20 μm. We also observed that the spatial distribution of platinum is closely mirrored, though by far more weakly, by that of Se (see Figure7). We interpret this observation to be indicative of the platinum-bearing phases which are probably sulphides, although there is no record in the available literature for minerals with this composition [6]. Cobalt, another siderophile metal with chalcophile affinities is measured in a domain with intermediate Ni, Fe (Fe = 31.8%, Ni = 39.3 %, Co = 1.03 %, Cu = 1.62 %, Se = 337 ppm) rather than in phases such as millerite or trevorite. Finally, several other trace elements have been occasionally observed and quantified in various regions of the sample, e.g. Mo (690 ppm), Br (410 ppm) and Zn (147 ppm). The recognition of these elements is of great geochemical significance as Mo (like Se) is often associated with hydrothermal environments, while Br is a halogen associated with brines and Zn is an element also concentrated in lower temperature (e.g.volcanogenic) environments.

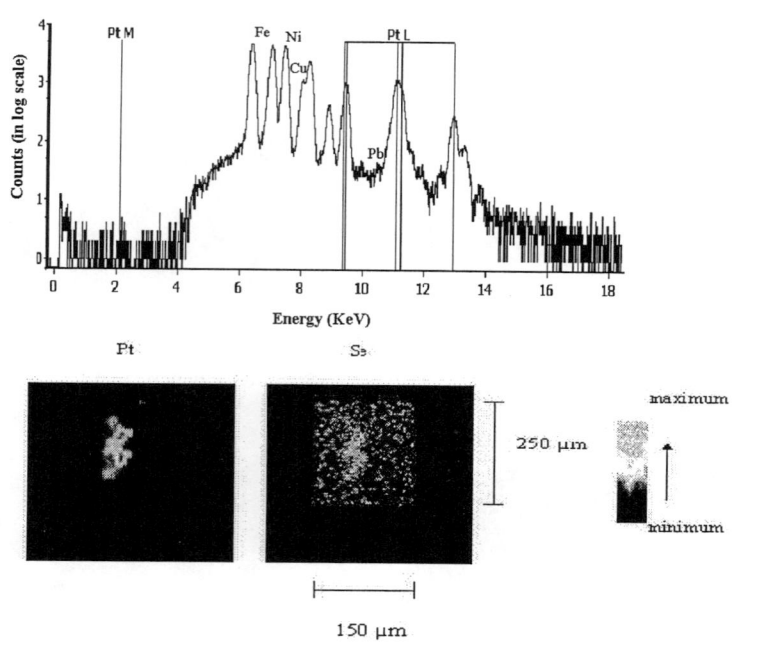

Figure 7: PIXE spectrum and spatial distribution of platinum and selenium in a scan area of 250μm x 150μm. in mineral assemblages from the Morokweng impact melt sheet

3.3 Microscopy of Polymineralic Eclogytic (clinopyroxene and garnet) samples.

Bi-polymineralic eclogytes commonly occur as xenoliths in Kimberlites and are formed due to subduction of oceanic crust. These are polymineralic phases separated by alteration material of micron scale (in the Figure 8, map obtained by proton microscopy). The kyanite bearing eclogites, in particular, grossular, pyroxene (omphacites) and kyanite disthen are less studied. The eclogites found in the South Africa Kimberlites known as Robert's Victor while those found in the Siberia are the Zagadochnaya type. These two types of eclogites differ in their mineralogy and mineral chemistry. In the present study the spatial distribution of Fe, Rb and Sr are measured to distinguish the two aforesaid types of eclogites. Details of this study have been reported earlier [7]

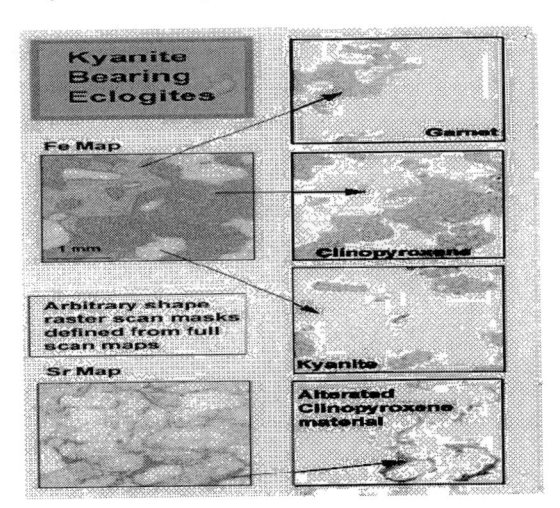

Figure 8: Elemental distribution maps of polymineralic eclogytes.

3.4 Microscopic mineral inclusion in diamond – an anomalous Fe-Mn heterogeneity.

Mineral inclusions in diamond provide direct information on the geological environment of the diamond and the type of geological processes that took place at the continental lithosphere at the time of genesis (~2-3 Gyrs ago). Once trapped the inclusions are thought to be isolated from further geological evolution and thus can act as a fingerprint of the conditions pertaining at the time of encapsulation. The elemental and isotopic composition of the inclusion is of particular importance to geosciences in terms of mantle evolution and geochronological studies.

Most diamond inclusions indicate homogeneous distribution of major and trace elements. This is in accordance with the belief that diamonds, in particular the peridotitic ones are much older than the associated kimberlites and resided for extended period of time in the sub-continental mantle where diffusion kinetics rapidly homogenized the elemental distribution. The significance of any chemical heterogeneity or zoning in diamond is that it should not exist given the high temperature and long time of encapsulation. An anomalous Fe and Mn heterogeneity effect was observed in a microscopic garnet inclusion in diamond using micro-PIXE (see fig.9). It would appear that the non-symmetric nature of the chemical zonation is indicative of formation of the inclusion prior to incorporation in the diamond.

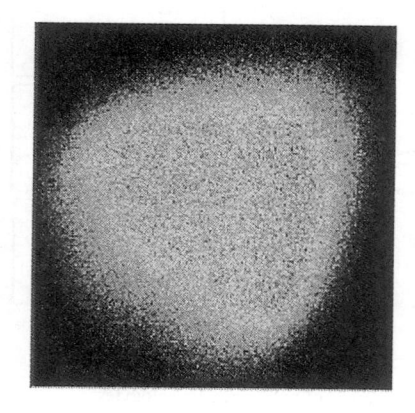

Figure 9: Elemental map of Fe showing its distribution in the microscopic garnet inclusion in diamond.

Conclusions

The selection of studies presented in this paper demonstrates vividly the capabilities of nuclear microprobe facilities in geosciences and earth sciences. The several virtues of the nuclear microprobe namely, being relatively non-destructive, high throughput, rapid scans for trace element distributions, reliability of theoretical modeling for accurate and fast quantification, and ability to combine a range of ion beam analysis techniques simultaneously (PIXE, RBS, STIM, ERDA), provide geosciences with certain unique benefits with respect to minor and trace element analysis.

References

1. J.I.W Watterson, R.W. Fearick, S.H. Connell, H.J. Annegarn, W.J. Przybylowicz, A.H. Andeweg, I. McQueen and J.P.F. Sellschop, Nuclear Instruments and Methods B77 (1993) 79.
2. G.W. Grime and F. Watt, Nuclear Instruments and Methods B30 (1988) 227.
3. J.A. Maxwell, W.J. Teesdale and J.L. Campbell, Nucl. Instr. Meth., B95 (1995), 407.
4. RK Dutta, E.Sideras Haddad and S.H. Connell, Nuclear Instruments and Methods B. (Proceedings of ICNMTA 2000) In Press.
5. M.A.G. Andreoli, L.D. Ashwal, R.J. Hart, J.M. Huizenga, In: B.O. Dressler, V.L. Sharpton, (eds)., Large Meteorite Impacts and Planetary Evolution II: Boulder, Colorado, Geological Society of America Special paper 339 (1999) p. 91-108.
6. J.W. Anthony, R.A. Bideaux, K.W. Bladh, M.C. Nichols, Handbook of Mineralogy, Mineral Data Publishing, Tucson (1990) 588.
7. S.H. Connell, R.J. Hart, P. Formenti, J.P.F. Sellschop and A. Wittenburg, Nuclear Instruments and Methods B130 (1997) 641.

ULTRA-THIN SINGLE CRYSTAL DIAMOND

D.B. REBULI, J.E. BUTLER*, J.P.F. SELLSCHOP, E. SIDERAS-HADDAD, S.H. CONNELL AND T.E. DERRY

Schonland Research Centre for Nuclear Sciences, University of the Witwatersrand, Private Bag 3, Johannesburg 2050, South Africa

Gas/Surface Dynamics Section, Code 6174, Naval Research Laboratory, Washington DC 20375, USA

The polishing of diamonds down to some microns in thickness has always been difficult to achieve. A process of damage, anneal and electrochemical etching has been used to produce single crystal ultra-thin diamonds of the order of 2 – 4 µm thick. Ion channeling experiments were performed using these crystals as a proof of the integrity of the crystal structure.

1 Introduction

Ultra-thin diamonds have been sought after in the physics community for many years. Some examples of experiments where such crystals are needed, is in the search for Coherent Resonance Excitation; impurity studies using ion channeling; and electron capture and electron loss in ion energy loss studies. A few samples of about 5µm thickness have been made previously by mechanical polishing but the difficulty (plastic deformation) in polishing down to the micron level has precluded this technique for routine production. New areas of diamond physics research could be developed with the introduction of ultra-thin diamonds and the need for numerous samples has led us on a search for an easier method of producing these ultra-thin films.

2 Method

Ion implantation leads to damage of the material up to the range of the implanted ions. The ion slows down in the material, losing energy by multiple collisions. It is at the end of the ion's range that it deposits most of its energy and creates the highest density of vacancies. This damage follows the Bragg peak whose depth is determined by the energy of the implanted ions as shown in figure 1.

Diamond displays an amorphisation threshold [1, 2] such that when the number of vacancies in the lattice is above a critical dose, then subsequent thermal annealing leads to amorphisation rather than restoration of the diamond lattice. This critical damage density has been found to be 10^{22} vacancies/cm^3. By implanting natural diamonds with C$^+$ ions to a dose of 10^{17} ions/cm^2, a Bragg peak is obtained

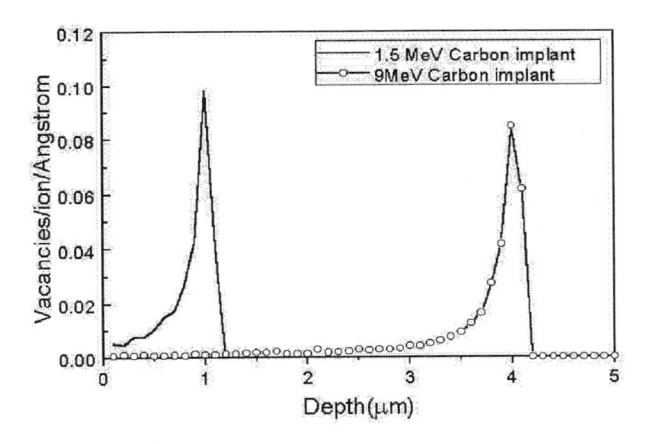

Figure 1: Trim [7] simulations of the depth distribution of vacancies created by C^+ implantation.

where the amorphisation threshold is only reached in a small region of the damaged area. After annealing, a thin and clearly defined amorphised subsurface layer of graphite is produced under a superficial layer of diamond. The damaged region nearer the surface (where the damage is below the amorphisation threshold) is essentially annealed back to the diamond structure. Channeling Rutherford Backscattering Spectrometry experiments were performed, in a micro beam mode, on a damaged diamond pre- and post-anneal to investigate the effect on the crystal structure. Rocking curves of the backscattered yields for these experiments can be seen in figure 2 showing the crystal structure is rather well maintained. The effect of the annealing cycle on the crystal structure is also seen by the reduction in the backscattered yield in the <100> axial

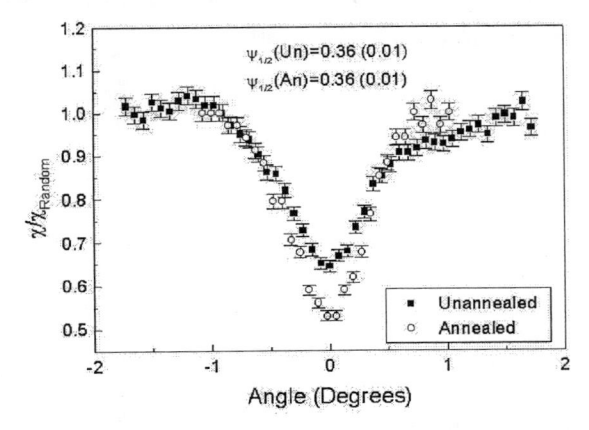

Figure 2: Rocking curve spectra showing the backscattered yields for a pre- and post-annealed diamond, after implantation, <100> direction.

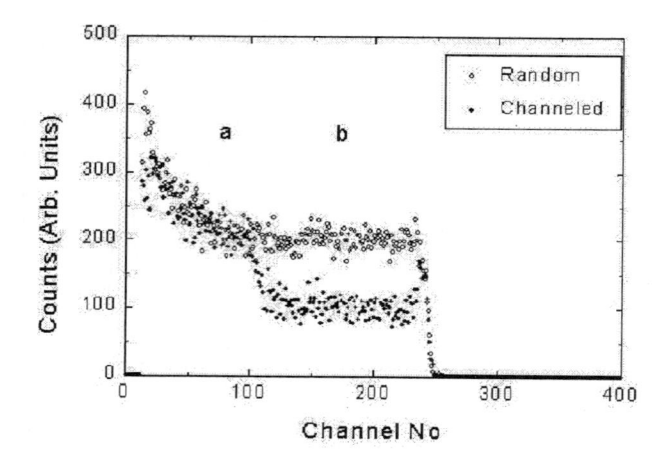

Figure 3: Rutherford backscattered spectra of the post-annealed diamond, showing (a) the graphitic region and (b) the superficial diamond layer.

The next step in the process is to etch away the graphitic layer. This etching is done in de-ionised water in an electric field [4-6]. Since diamond is non-conducting but the graphitic layer is, the electric field is focused on the damaged layer and the water etches the graphite away. Etching is believed to proceed via the OH ions as oxygen has an affinity for sp^2 hybridised carbon, which is released as a gas.

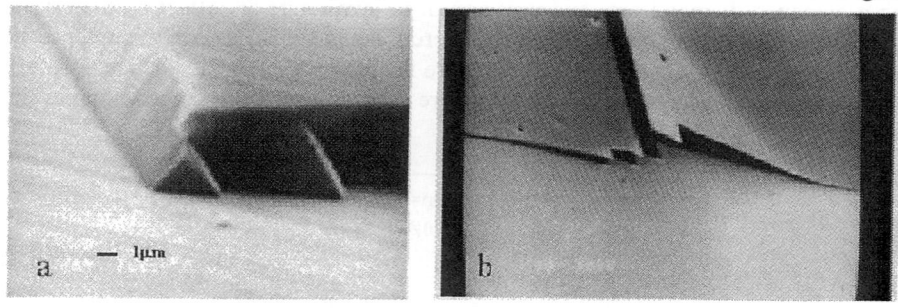

Figure 4: Scanning Electron microscope pictures showing (a) the original and etched surfaces of a diamond and (b) ultra-thin films about to lift off.

A new surface is left in a state mirroring the original surface as can be seen in figure 4. A slow rate of etching is needed as stresses in the superficial layer can cause small pieces to break off if the gas bubbles are large. The thickness of the ultra-thin diamond sample (figure 4b) can be determined by looking at the energy loss of ions through the sample. In this case an ^{241}Am source was used, which produces alpha particles of differing energies but with the majority of the particles having 5.486 MeV. The thickness of one of the samples was determined to be 4μm as seen in figure 5.

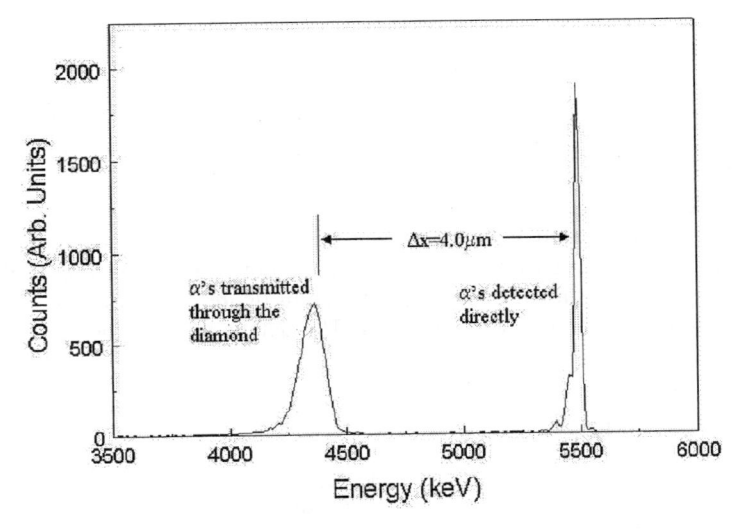

Figure 5: α spectrum from an ^{241}Am source where the broad low energy peak is of transmitted particles. The high-energy peaks are the naturally occurring energy distribution of ^{241}Am.

3 Experiments

The first experiment performed using the newly obtained ultra-thin diamond was to find a channeling pattern as a means of checking the crystal structure. This was necessary as the stresses in the ultra-thin diamonds cause the samples to curl and break. A collimated 4 MeV α beam was transmitted through the diamond onto a fluorescent screen placed approximately 1m behind the sample. The diamond was mounted on a goniometer and rotated, with respect to the incident beam, until a

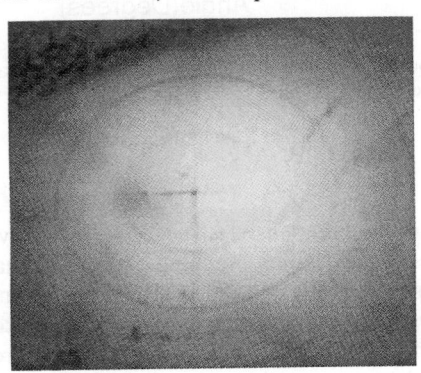

Figure 6: A fluorescent screen showing the blocking pattern observed from ions channeling down the <111> axial channel.

blocking pattern was seen on the fluorescent screen. The blocking pattern revealed a 3-fold symmetry corresponding to the <111> axial channel in the diamond lattice (figure 6).

A second experiment was performed where the crystal structure of the ultra-thin diamond was exploited to find the lattice location of a surface contaminant. The 4 MeV α beam was used in a channeling experiment down the <100> axial channel. The backscattered carbon yield, as per standard channeling experiments, was obtained and compared with the forward scattered yields of the transmitted and channeled beam, scattering off oxygen on the downstream surface of the crystal. These yields had a similar behaviour, as seen in figure 7, showing the oxygen to be aligned with the string of carbon atoms. The details of this work can be found in reference 8.

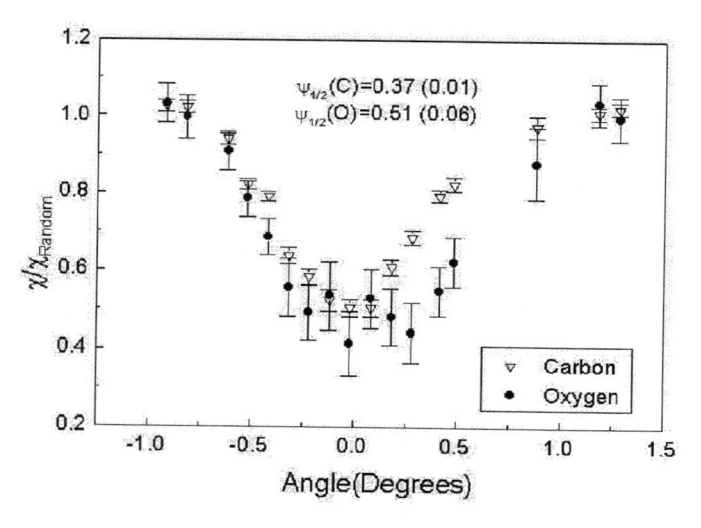

Figure 7: Rocking curve spectra of an ultra-thin diamond showing the backscattered yields from carbon and the forward scattered yields from oxygen.

4 Conclusions

The production of ultra-thin diamonds was successfully achieved. The quality of the crystal structure has been shown using blocking patterns of channeled ions showing the crystal structure is substantially preserved. These ultra-thin samples have been used in ion beam analysis experiments successfully. Work is continuing to improve the quality of these ultra-thin single crystal diamonds. New areas of study in diamond research have become possible using these thin diamond crystals.

References

1. R.A. Spits, J.F. Prins, T.E. Derry, Nucl. Instr. and Meth. in Physics Research B 85 (1995) 347.
2. C. Uzan-Saguy, C. Cytermann, R. Brener, V. Richter, M. Shaanan, R. Kalish, Appl. Phys. Lett. 67 (9) (1995) 1194.
3. J.C. McCallum, Ph.D Thesis, School of Physics, University of Melbourne (1987).
4. M.J. Marchywka, United States Patent, (1993), Patent number 5,269,890.
5. M.J. Marchywka, P.E. Pehrsson, D.J. Vestyck Jr., D. Moses, Appl. Phys. Lett. 63 (1993) 3521.
6. M.J. Marchywka, P.E. Pehrsson, United States Patent, (1996), Patent number 5,587,210.
7. J.F. Ziegler and J.P. Biersack, The Stopping and Range of Ions in Solids, Pergamon Press, New York (1985) ISBN/0/08/021603/X.
8. D.B. Rebuli, P. Aggerholm, J.E. Butler, S.H. Connell, T.E. Derry, B.P. Doyle, R.D. Maclear, J.P.F. Sellschop, E. Sideras-Haddad, Nucl. Instr. and Meth. in Physics Research B 158 (1999) 701.

Roasting Speeches at Dinner Party

"Roasting" speech at the dinner party

THE LIFE & TIMES OF PROF. JPF "FRIEDEL" SELLSCHOP FROM THE CHILDREN'S PERSPECTIVE.

RICHARD (1971), JACK (1969), INGRID (1961) & CELIA SELLSCHOP (1959).

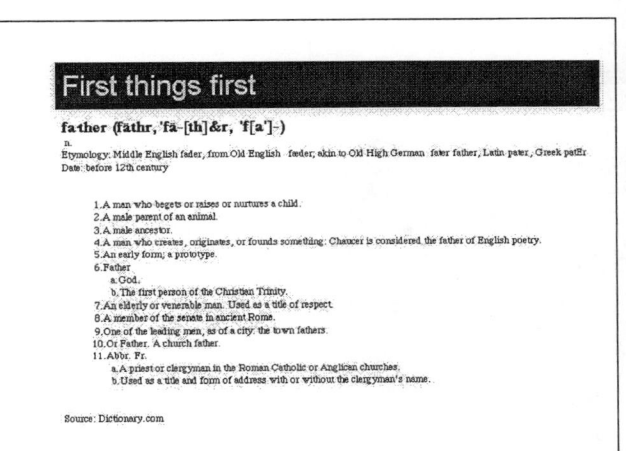

First things first

father (fäthr, 'fä-[th]&r, 'f[a']-)
n.
Etymology: Middle English fader, from Old English fæder; akin to Old High German fater father, Latin pater, Greek patēr
Date: before 12th century

1. A man who begets or raises or nurtures a child.
2. A male parent of an animal.
3. A male ancestor.
4. A man who treates, originates, or founds something: Chaucer is considered the father of English poetry.
5. An early form; a prototype.
6. Father
   a. God.
   b. The first person of the Christian Trinity.
7. An elderly or venerable man. Used as a title of respect.
8. A member of the senate in ancient Rome.
9. One of the leading men, as of a city: the town fathers.
10. Or Father. A church father.
11. Abbr. Fr.
    a. A priest or clergyman in the Roman Catholic or Anglican churches.
    b. Used as a title and form of address with or without the clergyman's name.

Source: Dictionary.com

Replicating our father's preference for starting speeches with the dictionary definition of the topic, we think this definition of "father" was apt. Ignore if you will the "parent of an animal" definition, and focus on "one of the leading men".

Friedel Sellschop, Rocket Scientist

Friedel supporting his son's endeavours to challenge NASA in the race to put Diamond experiments in space.
This was taken in Tennessee in 1980, when the family accompanied Friedel in his sabbatical to Oak Ridge.

...this is how WE see it

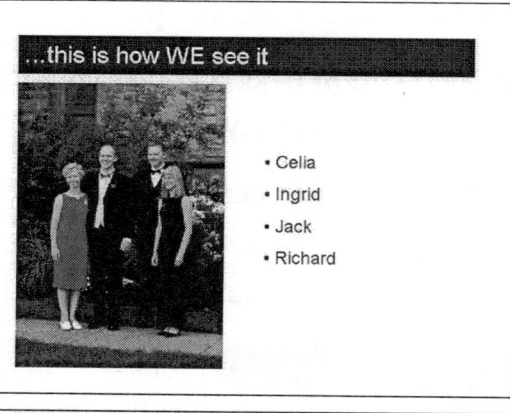

- Celia
- Ingrid
- Jack
- Richard

From left to right, Ingrid, Richard, Jack and Celia. Taken at Richard's wedding in 1999.
Richard and Jack both live in Johannesburg with their respectives wives. Ingrid and Celia live in the UK, Ingrid with the first two grandchildren in the family.

Less hair, different generation, same great effect on kids

Friedel with Jack on the left, and with Jessica (Ingrid's firstborn).

Beach bum

Friedel and the growing family on a beach in America.

Even he was a kid once

The father as a young man, showing off his scholarly attributes at an early point. It is often difficult for children to imagine their parents at a young age – so we all really enjoyed this photo!

This is as close as any of us got to following in his footsteps

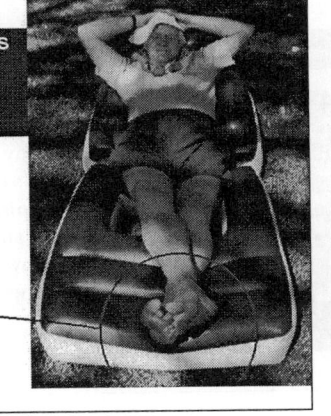

Rare that it is, quality relaxation is an important aspect of Friedel's life. This was taken at the family weekend retreat.

Luckily lately we have managed to break Friedel's habit of taking a 1 day annual holiday. We have increased it to several days on several occasions now, something we are very proud of!

Still messing about in boats.

It is here that Friedel spends quality time with the family and many friends. Sundowner drinks up the creek are a special occasion enjoyed as often as possible.

Always a good teacher

tie the thingy-ma-bob to the, no, no, to the left. Yes, that's it. Good.

Friedel has always dispensed good and honest advice to all of us whenever it is called for, often giving a tremendous amount of time, even when his own deadlines were pressing. This photo was also taken at the Vaal River, with Richard and Jack's Laser sailing boat in the background and an eager crowd of young students on the jetty.

A life full of learning

From the youngest age, we have always been included as adults in conversations and meals with Friedel's many friends and colleagues. This has proven to be very inspirational and enlightening to us, and created a desire for knowledge and excellence in all of us.

Inspiring us to achieve our best

If any of us got, say 85% in a test, the comment was usually, "what happened to the other 15%", but followed up with assistance and compassion. We have all grown up to aim to be the best we can, following in our father's footsteps and example.

Always a great orator, Friedel has enlightened many an occasion with his fantastic and well researched talks.

Thank you dad for being a great father to all of us.

"Roasting" speech at the Dinner party

Joseph Hamilton

I am delighted to speak at this roast of our dear friend Friedel Sellschop. Roasts are designed to give you insight into a person that otherwise may go unnoticed or be unknown and I speak in that tradition. Years ago I put forward a fundamental scientific theory. Doing theory is always a dangerous thing for an experimentalist – it is hard enough for theorists to formulate a fundamental theory. Nevertheless, I did so. What I am pleased to report here is that Friedel has proven my fundamental theory true without even knowing the theory. My theory is this: The true measure of greatness of a person is not what they can accomplish if they work in one of the world's best laboratories with everything at their disposal, but what they accomplish starting from scratch to build a laboratory, and a new tradition in a place where such work was unknown.

Friedel, you have proven experimentally many times over my theory by accomplishing true greatness in developing a world class laboratory in basic and applied nuclear physics in South Africa when there was nothing before you.

When I was a boy, my father would encourage me to be a stepping stone and not a stumbling block for others. R. L. Sharpe's poem expressed it well:

A BAG OF TOOLS

Isn't is strange
That princes and kings,
And clowns that caper
In sawdust rings,
And common people
Like you and me
Are builders for eternity?
Each is given a bag of tools,
A shapeless mass,
A book of rules;
And each must make–
Ere life is flown–
A stumbling block
Of a steppingstone.

R. L. Sharpe

Friedel through your remarkable work, you have taken a shapeless mass and built a stairway of stepping stones for physicists from South Africa and around the world. Warmest congratulations on your distinguished career which has put South Africa on the world map in nuclear physics.

Best wishes for continued success, good health and good life.

"Roasting" speech at the Dinner party

Paul Kienle

Dear Friedel,

At the official opening ceremony, I became absolutely fascinated by the proposition of honorable Fluchsman Samuel to create a Friedel Sellschop Center of Excellence under your leadership in Namibia!

I am sure that you will meet also this challenge, although it will become a tough task in the middle of a desert and on the shore of an Ocean which is dressed with many skeletons of brave discoverers. Also the number of meetings you must attend will rise and booth cruises at the Vaal river combined with bird watching will be replaced by sun downers in the desert, with snakes and scorpions.

In order to help you a bit, I decided to file an application as your assistant in this tough job. First I though, I might be useful with my German background to establish good relations with the landowners and richies of this desert land. But then I realized, that you may consider this completely useless, even hinderly, because science needs no farmers and members of the bourgeoisie, but just good know how.

This brought me the idea that I can offer you an unique experience which I acquired during our short visit of the famous Sossu Vlei-Dunes. It is not only that I learned to survive in a hot desert or climb Big Daddy and Big Mama, a real Sisiphos work. No, I got into the life sciences of the Namib Desert. To be more precise, I became an expert on the surviving strategies of an exceptional creature the Head, Standing, Black, Beatle (HSBB), yes you hear it right, the head standing black beatle which was also introduced by Dr. Henschel . It can run fast, but even more important by sticking its head in the sand and putting his asshole up, he collects water from up and down.

With this important technique, head down and bottom up, I think we both can survive if you should really become stuck in your desert home country. So please consider my application favourably.

"Roasting" speech at the Dinner party

Ettore Gadioli

Dear friends,
you all know the enormous energy of Friedel, his involvement in a barely numerable amount of collaborations and his ability to decide rapidly and well. I had first-hand experience of this when we met for the first time about ten years ago. We were in contact through exchange of letters. We arranged that he would meet me in Milano in order to speak about a possible collaboration (which in fact started at that time and continues until now) during one of his monthly travels in Europe. He arrived in the late afternoon, coming from CERN. In less than one hour the prospects of the collaboration were discussed and an agreement on all the points at issue was found. The morning of the next day he gave a marvelous talk on the Diamond Physics and in the early afternoon he left for Oxford to give a plenary talk to the annual Diamond Conference. After this he would go to Germany and other places in Europe. His ticket envelope was about 5 cm thick !
About two months later I visited Wits University in Johannesburg. There I had first-hand news of his tremendous working capacity. The first night I went to his home for dinner and he showed me his swimming pool. He told me incidentally that he would swim at night. I was extremely surprised. It was December and even if the days were good the nights were quite chilly. Then he explained to me that at night, after dinner, he would work a little. Late in the night he could get tired. In that case he would jump in the swimming pool and in a few minutes he would be completely restored and he could work a few hours more !
He is also able to infuse calm in his collaborators and get the best from them. Of this I had experience on occasion of my first visit to the iTT laboratory (at that time NAC). We were using for the first time the ^{12}C beam at the highest available energy. You can imagine how difficult it was to handle the beam properly for the first time. The control room was in a mess. No a single carbon ion was coming. Everybody was giving advice to the operators and they were very unhappy. Then the door opened and Friedel entered, greeting everybody. Everybody stopped and went to greet him too. It was like throwing a barrel of oil on a stormy sea. "Have we the beam ?" he asked and Willem (van Heerden) replied "Just a moment, Professor". You can imagine how skeptical I was, but in fact displaying a perfect anti-Pauli effect the Faraday cup ammeter, which at that time was even measuring a negative current, began to indicate a steady flow of carbon ions !
However, even a man of such qualities has his failures. That of Friedel concerns his name. Since the first time I visited Wits, everybody speaking of him pronounced his name as if **it was Friedel SELLSHOP**. So I too called him Friedel

SELLSHOP. Then, after several years, I was invited to a ceremony at the Stellenbosch University where he received an Honorary Degree. On that occasion the Chairman addressed him as Prof. SELLSKOP. I was then very surprised and I asked Friedel for the correct pronunciation. "Yes one should say SELLSKOP", and almost in tears he told me the numerous efforts he made to convince at least his closer collaborators (Simon, Elias...) to pronounce his name correctly. So I told Simon : "Simon, you know, the name of the Prof. is SELLSKOP !". "I know" he replied "but I prefer SELLSHOP !". However I did not realise the extent of the disaster until tonight when even Sue, his wife, said "SELLSHOP" !

Friedel, it was a great pleasure and an high honour for me to attend this unforgettable conference. I count you one of my best friends and a luminous example of devotion to Science.

Go well !

"Roasting" speech at the Dinner party

Helmut Appel

It is a great pleasure for me to join the long queue of friends and colleagues congratulating you, Friedel, on your reaching the biblical age of three score and ten years. I wish you health, the successful continuation of your many activities and joy all the time with your wonderful and admirable family.

My justification for speaking to this distinguished audience is the fact that I have known Friedel probably longer than anybody else in the audience: Our friendship dates back to eighteen hundred ..., I beg your pardon, 1959 and it has been extended mutually also to the younger generation of our families and cultivated for more than 40 years up to the present time.

It needed a conference in Japan in the middle of the seventies to decide about taking hyperfine-interaction to the then Nuclear Physics Research Unit at the University of the Witwatersrand in Johannesburg. I joined your working group for the first time in 1978. Since then more than 40 papers on diamonds and its allotropes, and on the other two modifications of carbon, graphite and fullerenes, have been published. This is a small number in comparison to all the papers about diamond that carry Friedel´s name in the top line which goes into the hundreds, many resulting from collaborations with colleagues all around the world. I recall the story, that when the second book on the properties of diamond was issued in 1992 our friend John Field - and this is a true story - apologized for the large volume of the book. He stated that the number of pages could have been cut down considerably if Friedel would have left out his .... acknowledgements.

It was always a great pleasure to meet you, Sue and Friedel and your children. We are so grateful for the warm hospitality which you always offered us.

Friedel, do you remember when you literally saved us from a catastrophe when a lion attacked our car in the Kruger Park. It needed your quick reaction time and your full body weight on the accelerator pedal to leave the lion no chance although the last picture on the film of my moving camera did not show much more than the two eyes of the lion.

It is not exaggerated, Friedel, to say that you changed the style of my life considerably. To pick out only one point - besides the fact that you introduced me to the culture of drinking sherry with enthusiasm: It was you who led me into temptation to touch a golf club with the effect that I am now a passionate golfer while you decided not to spend so much time on that sort of applied physics.

Friedel, I know you are fully aware to judge the situation what it means to step into the eighth decade of one`s life. In fact you gave me helpful advice in the form of a brilliant after dinner talk when I passed **this threshold** some time ago.

Ladies and gentlemen, you might have the impression that everything that should be said on such an occasion to a celebrated scientist and highly appreciated colleague and friend has already been said. Rightly so, but allow me to say, not by everybody. You, Friedel, deserve it that you are addressed with the kindest and warmest words that your language offer, so often.

Thank you.

"Roasting" speech at the Dinner party

Christian Toepffer

As many of you know, I stayed between 1973 and 1980 at Wits first as a guest, then on the chair of Theoretical Physics. This is not the occasion to dwell on the pleasant and fruitful scientific collaboration with Friedel, but rather to praise the extraordinary warmth of his personality and his and Sue's kindness, which contributed so much to make this period pleasurable for my family and myself.

The Sellschops made us really feel welcome even over and above the customary South African hospitality. Friedel taught me woodworking and helped to construct an elaborate huge shelf for my books and my Hifi. In the course of this work I learned much about the measurement problem in physics, about warped geodesics and non-euclidian geometry. In this connection I became acquainted and learned to love the African, non-cartesian approach to geometry: Rondavels instead of rectangular houses, irregular shaped flowerbeds and garden walls which meander into some general direction rather than running straight from one corner to the next.

Of course I tried my best to keep up with Friedel, but I could not always meet his standards. I particularly remember a sunny winter day on the river when we fitted doors and windows for the new house. I felt somewhat outprussianed when Friedel insisted that the slots of all screws point into exactly the same direction. Friedel, it would make me immensely happy, if I could come back one day to correct this.

Closing Ceremony — The Three Devils

CLOSING CEREMONY – THE THREE DEVILS

DEVIL 2 - FOR THE DELEGATES

J.A. DAVIES

As our conference chairman has indicated, the title for this closing session comes from the Swiss legend "Les Diableret" in which three devils on the surrounding mountain peaks hurl down either water or rocks on the village below. My role, as Devil # 2, is to speak for the 43 overseas delegates to Lüderitz-2000 and to make appropriate commentary (both positive and negative) on its success. Dr. F. Seitz had originally been invited to give this closing talk but unfortunately he was unable to come to Lüderitz, and so I have been given this privilege of representing the overseas delegates in paying tribute to Friedel Sellschop and also to our conference organizers - Simon Connell, Sharon Stoneley, and their colleagues. This is my first visit to South Africa and Namibia - and hence Lüderitz-2000 has given me a long-overdue opportunity to visit Friedel (whom I have known for almost 30 years) in his own country and indeed in the small town where he was born.

One possible "rock" - why choose Lüderitz? For most overseas delegates, this has involved several thousand kilometers and at least two extra days of travel - and of course such an isolated location has probably been much harder for Simon and his organizing group, since almost all conference supplies had to be brought here by car from either Johannesburg or Windhoek! However, thanks to Concorde Travel and Sharon Stoneley, the inconvenience and even the extra cost of coming to Lüderitz has been minimized - and we have all been given this unique opportunity of learning something special about Friedel's earliest roots, his birthplace here on Shark Island. We have also been given the opportunity to visit beautiful Namibia, which not only is the most sparsely populated country in Africa, but undoubtedly is also the friendliest. So, has the choice of Lüderitz been worthwhile? The answer is a resounding YES !

Many overseas delegates, myself included, set aside extra time in order to explore some of the Namibian landscape. For example, Jens Ulrik Andersen, Erik Uggerhoj and I arranged to meet in Windhoek early last week and we have spent 4 days exploring the Kalahari desert and its red sand dunes while driving from Windhoek to Lüderitz. En route, we saw hundreds of large mammals - antelope, giraffe, impala, oryx, wildebeest, zebra, lions and leopards. At Intu Afrika lodge, we even went on a 2-hour tracking expedition with 3 native bushmen. And finally we made a 1-day drive to Lüderitz across a daunting sea of red sand, the Namib desert. At one stage, just 15 km from Lüderitz, we had to stop while a SAND-PLOW cleared a rapidly drifting sand dune from the paved highway. On Canadian roads, a snow-plow is a very familiar sight - but I have never before needed a sand-plow! So, thank you, Simon and of course Friedel, for bringing us here to Lüderitz. The Namibian landscape is truly magnificent - and surprising.

The Lüderitz-2000 program has provided a very nice balance between pure science (reviewed by Dr. Greiner) and various discussion sessions on the science policy and aspirations of both Namibia and South Africa. In both countries, very impressive progress has clearly been made during barely ten years of democratic government, and an excellent rapport is being established between the scientists and their political leaders . This rapport is certainly a fitting tribute to Friedel's pioneering role in such matters. We overseas delegates have been deeply impressed by the Lüderitz-2000 presentations of the Namibian and South African politicians, such as the Hon. Fluksman Samuehl and Dr. Ben Ngubane. We were particularly excited by the Hon. Fluksman Samuehl's proposal for creating a new institute here in Lüderitz, dedicated to furthering Friedel Sellshop's initiative in science and the African renaissance. Following his suggestion, we have also all tasted and felt the atmosphere of the Lüderitz waterfront - and especially the incessant wind!

In organizing Lüderitz-2000, Simon Connell and his committee have spared no effort on our behalf. The opening talk on the Namib Desert by Dr. Johannes Henschel was an example of marvelous timing, since many of us had just driven across it and were puzzled as to how the trees and animals could survive in such an arid climate. The mentor program for young African delegates is a great idea, which hopefully we can emulate at other international meetings. A craft market from all over Namibia and even from Botswana, Zimbabwe and South Africa was assembled and brought to Lüderitz just for our benefit. Even the tour buses had to be driven here from Windhoek, more than 800 km away.

In addition to the technical talks, each day had its own distinctive 'extra-curricular activity': for example, the Seafood Braai at Lüderitz yacht club - the best crayfish I have ever tasted; the Wines-of-the-World tasting at the evening poster session; the conference outing to the diamond-processing plants (both new and old) at Elizabeth Bay; the conference banquet and its 'roasting' session. All were extraordinary and truly memorable events - and, from a delegate's perspective, were organized flawlessly. We have even been introduced to some new dress codes, such as 'desert casual' and 'Namib night'.

A major highlight of the week was our concert evening in Kolmanskoop ghost town, the site of Namibia's first diamond mine - and I still recall the haunting rhythm of those marvelous Namibian dancers and drummer! Kolmanskoop was also a rather fitting tribute to Friedel's life-long interest in diamond research. Many overseas delegates have collaborated for years with Friedel in his diamond-based studies and almost all of us (myself included) have benefitted from the many crystals he has so generously provided to various overseas laboratories. As one delegate remarked, Friedel is the one who continually reminds us that there is indeed an element lighter than silicon in Group IVA!

Now for a small 'rock', or rather a few small 'pebbles': The Delegate Contact Information pages in the Abstract Book contain several surprising errors. For example, my home town of Deep River has miraculously been transported 500 km to Hamilton; Peter Rose seems to have changed his affiliation from Orion Equipment to Orient Equipment; and Ken Purser has moved the University of Toronto to Massachusetts, thereby accomplishing undoubtedly the largest Canadian 'brain drain' of the century!

It is now time for a few thoughts about Friedel Sellschop himself. Like most overseas delegates, my friendship with Friedel extends back at least 30 years. We have met roughly once per year at various ion-beam conferences - IBA, IBMM, ICACS, and the Gordon Conference on Particle / Solid Interactions - but until Lüderitz-2000 we had never met in his own country. This must be blamed mainly on the apartheid policies of the previous South African regime but, in our mutual relationship, it has created a rather significant 'rock', which I shall comment on shortly.

Friedel's breadth of interest in science is truly phenomenal, as borne out by the scope of the Lüderitz-2000 scientific program. Indeed, whenever we would meet at scientific meetings around the world, Friedel was usually at the centre of some vigorous scientific discussion. But, equally often, we would seek out Friedel also for his latest input on the political situation in South Africa, knowing that he would always provide a balanced, knowledgeable and objective view of the apartheid problem.

Now for a few words on behalf of those colleagues who were unable to come to Lüderitz. A few weeks ago, Len Feldman phoned me to say that unfortunately he had to cancel his plans to attend Lüderitz-2000, and would I please say a few words on his behalf. So, for Len and all those other delegates who had to cancel out, let me quote briefly from his phone call. In 1973, Friedel joined Walter Brown's group at Bell Laboratories on a 6-month visit. Friedel arrived, full of gusto and enthusiasm, and immediately went around to each member of this well-known group (Len, Walt Gibson, Jack Macdonald, Laurie Miller, Lanzerotti, and many others) and asked each one to discuss their current research topics. From the resulting list of 24 topics, he eventually chose 12 - and during the next 6 months he managed to contribute significantly to each one! At that time, Len was working with a young German student on the atomic / nuclear polarization of a 100-keV deuterium beam - attempting to verify the so-called Kaminsky effect - but until Friedel's arrival their progress had been slow. However, Friedel's enthusiasm and leadership succeeded in motivating the student, and some rapid progress was made. This is perhaps one more example (as we heard at last night's banquet) where Friedel managed to out-Prussian even a Prussian. Twenty five years later, in 1999, he again visited Len Feldman - now at Vanderbilt University - in order to join Norman Tolk and others on the free-electron laser facility. Despite his increased age and jet lag, Friedel's energy and enthusiasm were undiminished and he insisted on working the night shift for the entire week. As Len put it, "throughout the 25 years that we have known him, Friedel seems to have been in a continuously excited energy level".

We return now to the apartheid problem, which for years was a major 'rock' for many overseas delegates. On several occasions, Friedel had invited me to make an extended visit to Wits, which I would have dearly loved to accept. Unfortunately, however, these invitations always had to be declined because, until my retirement a few years ago, I was working at Chalk River Nuclear Research Laboratories, a Canadian government research institute. Hence, as a government employee, I was subject to the federal boycott of South Africa.

Friedel himself also had difficulties joining us on certain occasions, despite his well-known opposition to the apartheid policies of the previous regime. One particularly painful memory for me (and probably even more painful for Friedel) occurred at the inaugural meeting of IBMM in Budapest in 1978. At that time, everyone required a visa to enter Hungary, and of course the communist-bloc countries would not issue visas to South African passport holders. Jozef Gyulai, the conference chair, had obtained a promise from his government that they would avoid the problem by quietly admitting such delegates without issuing a visa. Unfortunately, when Friedel arrived at Budapest airport, an overzealous official detained him overnight at the airport and then unceremoniously put him on the next flight back to Frankfurt - despite all the efforts of Jozef Gyulai and our international conference committee. Several of us visited Friedel while he was in custody, and it was even suggested that perhaps we should postpone the conference and move it elsewhere. However, Friedel urged us not to jeopardize the future of the new IBMM conference series.

This story eventually had a happy ending because, 15 years later, we all returned to Hungary for another ion-beam conference (IBA-11) at Balatonfurad - and again Jozef Gyulai was the conference chair. Despite his earlier 1978 rejection, Friedel graciously accepted Jozef's invitation to participate in IBA-11 and, together with Simon Connell, he presented five contributions, including the opening plenary talk. It was a very successful meeting. At the conference banquet, as a special tribute, Jozef presented Friedel with a symbolic and well-earned golden key to his country!

This brings me back to Lüderitz-2000, whose significance for the overseas delegates has certainly been enhanced by all the difficulties which previously hampered us from visiting Friedel in this part of the world. To have the opportunity (at long last) of coming not only to South Africa and Witwatersrand, but also to Friedel's birthplace here in Lüderitz, has been fantastic! Consequently, on behalf of the overseas delegates, I thank our organizing committee - and especially Simon Connell and Sharon Stoneley - for making Lüderitz-2000 happen.

Finally, as a small token of affection, I too would like to present Friedel and Sue with a bottle - in this case, neither wine nor whiskey; instead, it is home-made maple syrup from our family's maple bush at our cottage property on Lyell Lake. Each spring, the Davies family spend several weekends together at our cottage, collecting about two thousand litres of maple sap from the trees, boiling it for 10-12 hours on an outdoor evaporator in order to concentrate it 40-fold, and ending up with about 50 litres of home-made maple syrup. Making it has become a real family tradition and we hope you'll enjoy it too. Even better, if you are ever in North America in March or April, why not come up and join us for a weekend expedition.

CLOSING CEREMONY – THE THREE DEVILS

DEVIL 3 - FOR FRIEDEL

ANDREAS K. FREUND

This last contribution to our unique conference here in Lüderitz closes the circle that was opened when it was decided to organize this extraordinary event on behalf of Professor Sellschop's seventieth birthday. We are now about to leave this so special and wonderful place after having enjoyed several days together and having shared a wealth of scientific knowledge, old wisdom and new experience. Professor Jacques Pierre (Friedel) Sellschop has been at the centre of this conference during the whole week and we have heard many things about him, not only about his scientifically and politically important achievements, but also anecdotes and stories telling us details about his character. But do we therefore really know who he is? Depending on our personal preferences and experiences, we might see him from different angles. The aim of this brief presentation is to add some more aspects to a both complex and simple picture of the *"Prof"* and to share the last minutes of our presence in Lüderitz with him.

A good way to define a person is by describing what he/she likes or dislikes. Then many things come to my mind meaning that Friedel has always expressed himself in a very clear manner.

He likes:
- First of all: his family life.
- Good food, good wine, good company. And always more of this.
- To have and to be with friends: not only to love, but also to be loved.
- To be there for somebody, to care for and to take care of.

Moreover he likes:
- To understand and also to be understood.
- To know always more about things and people.
- To watch the stars at night while keeping his feet solidly on the ground.
- To dream, to imagine, to create, but also to take responsibilities.

He further likes:
- The arts: music, poetry, painting, etc.
- Beautiful cars and powerful engines.
- Philosophy: from mind games to the origin, principles and meaning of life.
- To discover and to be excited when discovering: there is no direct translation of the French word *"s'émerveiller"* that describes so well the deep feeling we experience when discovering the miracles of nature.

And of course he likes physics and science in general. His motivation is not just curiosity, but it has to do with the *"émerveillement"* being face-to-face with the mystery and miracle of creation and re-creation, like a musician re-creating the creation of a composer.

I still remember so well his expression when he first discovered the beautiful pictures of X-ray excited optical luminescence (XEOL) that made diamond glow in the dark:
"...it blows up my mind..." he repeated again and again.

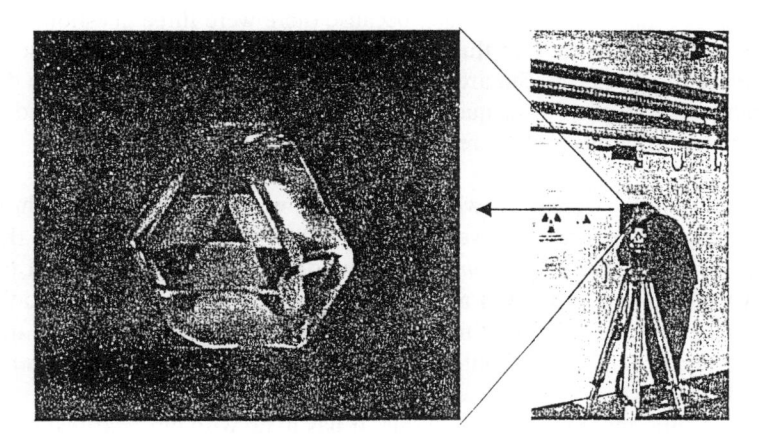

Figure 1: Friedel Sellschop looking through the lead glass window of the beamline BM5 at the ESRF (right) observing optical luminescence excited by synchrotron X-rays in a type Ib diamond single crystal (left).

And his face looked happy like that of a child discovering a beautiful seashell on the beach.

I was able to give a though incomplete list of what Friedel likes, but when I try to find out what (or whom...) he dislikes, nothing comes to my mind, maybe because he doesn't like to speak about these things or maybe not with me, or maybe he simply prefers to focus on the positive. Whatever the reason for this is, I then just suppose that he dislikes the absence of what he likes. Now I haven't yet said what we both like very much, and this is very important for me, because it is one of the main reasons why I like Friedel: he likes to *wonder*. I read somewhere that the Wise Old Man said: *"Beware of those who think and seek for those who wonder"* and I was very lucky to find Friedel. So let us wonder together for a little while.

First I would like to invite you to wonder about the three *"i"-s*. When the Dalai Lama visited Grenoble a few years ago, about 40 scientists from the local institutes were invited to a session with him. We were allowed to ask him questions, and these questions had of course been screened before the session by a small "question-review-committee" so that the stupid ones had been rejected saving our reputation so some extent. I have forgotten many details of this conversation with the Holy Man, but I still remember the three questions that one of us had asked and they were about the three *"i"-s*:

1. – what are the *"eyes"*?   – the observing tools…?
2. – what is $"i" = (-1)^{1/2}$?   – our imaginary part or feelings…?
3. – who am *"I"*?   – the one that observes…?

In fact, you can see that I made a mistake, because there were three questions, but four *"i"-s*. Now I am not sure that this is correct either if you consider that some believe that there is a third eye. But are you allowed to fully count the third eye if it is not independent from the *"i"* of question #2? This becomes too complicated as always when I start thinking, so let's return to wondering.

The French philosopher[1] Descartes (who, by the way, was mentioned last night in the roasting session) coined the well-known word: *"cogito ergo sum"* that translates as *"I think – therefore I am"*. This sounds excellent, but it doesn't tell me who or what I am and, anyway, I am wary of those who think according to the Wise Old Man. Following my own thinking experience (see above) I believe that Descartes' word is incomplete. It should read: *"I think – therefore I am confused"*.

Let me explain a little more why I believe this. It has to do with our feelings: when you separate feeling and thinking, then you become confused. Both must be in harmony and in permanent interaction, therefore feeling is at least as important as is thinking as emphasized many years ago by Albert Schweitzer. This has been widely recognized today and has even got a scientific base from brain research. Several books have been written recently on *"emotional intelligence"* or on *"relational intelligence"* as opposed to the traditional IQ that has become almost obsolete as a concept and as a criterion for employment. We can see man as a complex being consisting of a *real* and an *imaginary* part, the first corresponds to *thinking* or *rationalizing*, the second to *feeling* or *emotions*. There are no instruments other than we ourselves to measure feelings such as love or fear. We are dead if we stop thinking and/or feeling. Therefore I would invert Descartes' theorem and say: *"I am, therefore I (…am able to…) think"*. And the term *"wondering"* contains both thinking and feeling giving rise to inspiration and creation, to the happiness of being both an individual and one with the universe at every instant. So let us continue to wonder.

---

[1] A philosopher is a person that can help you a lot: once you hear the answer to a question you ask him, you forget what your question was – so no more problem! A French philosopher is then even much better for those whose knowledge of the French language is limited.

"To be or not to be", this is the famous existential question. For us scientists it becomes: "to know or not to know", because we firmly believe "knowing is better than not knowing". Consequently, we are what we know – but do we therefore know what we are? And are we able to know who Friedel is if we even don't know who we are? What do we really know and what is *new*? Often good hints for anwers to complicated questions can be found in language roots. I like the French word for knowledge that is "connaissance" and that can be translated directly as "to be born with". Then everything that exists is already known to us. When we discover we *re*-discover what we already have inside us and we thus *re*-create the world. When we incarnate ourselves on earth, we lose the global picture and during our lives we try to glue the pieces together driven by the home-sickness to return to the original one-ness. And when we perceive a glimpse of the original beauty and completeness we are happy – forthcoming instants of happiness might be the experimental proof of the Higgs Boson or watching the sunrise above snow-covered mountains or listening to the first notes of Mozart's requiem.

What to say about the "eyes"? At first glance, they stand for our instruments to observe the world around us. But this is not all, they also act on the world surrounding us when we look at something or somebody. Thus they are our tools to interact and to communicate and therefore they are essential for our existence. An oriental proverb says: "When you look at a flower, never forget that it looks at you too". Moreover, by looking at the world we make it exist: *does the rainbow exist without the observer*? The ultimate goal of a scientific researcher is to transcend his proper subjectivity to access the objective reality, the truth, if there is such an objective world. The German word for science is "Wissenschaft". It can be separated into "wissen" that has its root in the Latin word "videre" that is "to see", and in "schaft" that has to do with the word "schaffen" meaning "to create". Therefore the word "science" in German is equivalent to: "what our vision creates". We say: "seeing is believing", but also the inverse makes sense: "believing is seeing". Another proverb goes as follows : "if you believe in the invisible you will see the unbelievable". My feelings with respect to this are: limitations facing unlimited possibilities, growing awareness, respect, humility, confidence, vigilance.

Friedel likes wondering, he likes the stars and the desert and this inspires poetry. It was a slight surprise to me that he didn't know the magic story of "The Little Prince" written by the French airplane pilot, poet and writer Antoine de Saint-Exupéry [1]. This book has it all: stars, desert, silence, planets, and much more. This year we celebrate also Saint-Exupéry's 100th birthday and on this occasion the Lyon airport was recently renamed after him. The little prince is a person from another planet, a very, very small one. He represents the child inside us and children speak the truth. He is a messenger from the outer and inner space. During his many travels he meets various people each living alone on a different planet.

Figure 2: The little prince on his little planet with his rose and his volcanoes that he carefully rakes, even the one that is extinct, because you are never sure...

Finally, the little prince arrives on earth and meets the author of the book in the Sahara desert. Saint-Exupéry had been forced to land there because of a mechanical problem with his airplane. They had a long conversation and became friends. During his visit to the planet Earth the little prince met also the fox. They spent some time together and the fox became tamed. When they had to separate, they exchanged the following words:

"Good-bye," he said (the little prince).
"Good-bye," said the fox. *"Here is my secret. It's quite simple: one sees clearly only with the heart. Anything essential is invisible to the eyes."*
"Anything essential is invisible to the eyes," the little prince repeated, in order to remember.
"It's the time you spent on your rose that makes your rose so important."
"It's the time I spent on my rose..." the little prince repeated, in order to remember.
"People have forgotten this truth," the fox said. *"But you mustn't forget it. You become responsible forever for what you've tamed. You're responsible for your rose..."*
"I'm responsible for my rose..." the little prince repeated, in order to remember.

There is no need for further words. I see that Friedel has the same expression *"émerveillé"* that appeared on his face when he discovered the diamond glowing in the dark and I am happy.

Friedel, we all thank you for seeing with your heart and for feeling responsible for your many roses: science, Africa, colleagues, friends, students, diamonds.... We love you and wish to see you discovering many more things and to tame them. You have already succeeded to a great extent with diamond despite the fact that its name comes from a Greek word meaning *"the untamable"* as I learnt from you.

Reference:
1. Antoine de Saint-Exupéry, *"Le Petit Prince"*, Editions Gallimard, Paris, 1946, nouvelle édition 1999. See also the English version, 2000.

Conference Program

Sunday, 12 November 2000

Official Opening Ceremony
Master of Ceremonies — Conference Chair : Dr. SH Connell

16:45		- Delegates and guests take their seats.
17:00-17:05	Anthems	-The National Anthem of Namibia and the Anthem of the Organisation of African Unity.
17:05-17:15	Welcome	-The Mayor of Lüderitz : Her Worship, Mrs E. Amupewa
17:15-17:25	The setting	-Regional Councillor - Lüderitz Constituency, Karas : The Hon. Fluksman Samuehl (MP)
17:25-17:55	Opening Lecture	-The Deputy Minister : Higher Education, Training and Employment Creation : The Hon. B. Wentworth
17:55-18:15	Response	-The Minister : Arts Culture Science and Technology, Dr Ben Ngubane
18:15-19:00	Opening Presentation	-The Namib : the living desert by Dr Johannes Henschel
19:00-19:15	Closing	-The Governor - Karas region : The Hon Governor Stefaanus Goliath
19:15-	Welcome cocktails	Champagne and oysters on the terrace of The Nest

Monday, 13 November 2000

08:30-10:30 Welcome
Chairman : SH Connell
B Ngubane —Opening

Elementary Particle Physics
Chairman : Z Vilakazi

J Ellis	-Challenges and opportunities in particle physics (30')
J Gates	-Superstrings: Why Einstein would have loved spaghetti in fundamental physics (30')
J Cleymans	-Large features of particle multiplicities and strangeness production in central heavy-ion collisions between 1.7A and 158A GeV/c (30')

10:30-11:00 Tea

11:00-12:30 Nuclear (Astro)physics
Chairman : R Tegen

W Greiner	-Perspectives in Nuclear Physics : From Superheavies via Hypermatter to Antimatter and the Structure of a Highly Correlated Vacuum (30')
R Steenkamp	-H.E.S.S. - an array of stereoscopic Cherenkov detectors currently under construction in Namibia (30')
O de Jager	-Particle accelerators in the universe - leptonic or hadronic (30')

12:30-14:00 Lunch

14:00-15:30 **Science Policy and Anticipations**
Chairman : K Reed
Input and Panel Discussion (provisional format)

R Adam	-Choosing good science in a developing country (Director General - Dept Arts, Culture Science and Technology) (15)
A Paterson	-Policy Frameworks in Science and Technology: Then, Now and Tomorrow (Vice President - CSIR) (15')
K Reed	-American Physical Society Initiative for Collaboration between American and African Physics Programs (15')
R Uthui	-Science at the University of Maputo
Panel Discussion	(45')

15:30-16:00 Tea

16:00-18:30 **Nuclear Physics and Applied Nuclear Physics**
Chairman : E Sideras-Haddad

P Kienle	-Observation of deeply bound pionic states and the renormalised order parameter of spontaneous chiral symmetry breaking (30')
P Hodgson	-Nuclear Pre-Equilibrium Reactions (30')
S Krewald	-Meson-production at thresholds: physics with a cooler-synchrotron COSY-Jülich (30')
H Machner	-Meson production in p+d reactions (30')
J Fiase	-Semi-empirical effective interactions for inelastic scattering derived from the Reid potential (30')

Evening   Seafood Braai (barbecue) at the Yacht Club

Tuesday, 14 November 2000

08:30-10:30 **Atomic Physics and Applied Atomic Physics**
Chairman : J Davies

J-U Andersen	-Channeling revisited (30')
A Richter	-Radiation physics with diamonds (30')
A Solov'yov	-Channeling of Charged Particles Through Periodically Bent Crystals. On the Possibility of a Gamma Laser (30')
H Backe	-Novel interferometer in the X-ray region (30')

11:00-12:30 **Nuclear Physics and Applied Nuclear Physics**
        Chairman : P Hodgson
E Gadioli	-On the need for comprehensive studies of heavy-ion reactions (30')
J Hamilton	-New nuclear vistas from spontaneous fission (30')
D Ackermann	-The Synthesis of Superheavy Elements : The state of the Art (30')

12:30-14:00 Lunch

14:00-15:30 **Atomic and Applied Atomic Physics / Technology**
        Chairman : K Bharuth-Ram
P Sigmund	-Non-Perturbative Theory of Stopping for Swift Heavy Ions (30')
D Schardt	-Tumour therapy with high-energy heavy ion beams (20')
P Rose	-Applications of ion-implantation in advanced semi-conductors (20')
B Doyle	-Radiation effects microscopy (20')

15:30-16:00 Tea

16:00-18:30 **Specials**
        Chairman : M Tredoux
U Rosengard	-Application of Nuclear techniques to Humanitarian Demining (30')
RJ Hart	-Messengers from the deep: the application of INAA to the geochemical analysis of single diamonds. (30')
A Zucchiatti	-IBA techniques to study renaissance techniques (30')
A Kinyua	-Applications of the energy dispersive X-ray fluorescence technique in Kenya (30')
JA de Wet	-Nuclear geometry (30')

Evening   Musical Evening in the Ghost Town's Cabaret Hall

Wednesday, 15 November 2000

08:30-10:30 **Elementary Particle Physics**
        Chairman : H Appel
E Uggerhøj	-The influence of strong crystalline fields on QED-processes investigated using diamond crystals (30')
M Velasco	-Using crystals to solve the nucleon's spin crisis' TODAY, ... and look for physics beyond the Standard Model TOMORROW (30')
JP Coffin	-The search for exotic strange matter at new ultra-relativistic heavy-ion colliders. (30')
B Kämpfer	-Confinement in the Big bang and Deconfinement in the Little bangs at the CERN-SPS (30')

11:00-12:30 Atomic Physics and Applied Atomic Physics
             Chairman : B Doyle
  A Freund   -Scientific opportunities at 3rd and 4th generation X-ray sources (30')
  C Toepffer   -Hydrogen under extreme conditions (30')
  G Zwicknagel   -The interaction of charged particles with plasmas (30')

12:30-14:00 Lunch

14:00-15:30 Atomic and Nuclear Physics in the study of Diamond (and diamond-like materials)
             Chairman : J Hansen
  J Biersack   -Proton stopping and straggling diamond (30')
  J Butler   -Atomic hydrogen and the chemical vapour deposition (20')
  S Kalbitzer   -Hydrogen mobility in diamond (20')
  E Friedland   -Critical damage energies in diamond (20')

15:30-16:00 Tea

16:00-18:30 Nuclear Physics
             Chairman : J Cleymans
  C Greiner   -Signatures of Quark-gluon plasma: an overview (30')
  J Sharpey-Schafer   -The place of proton therapy and other sophisticated radiation cancer treatment modalities in Southern Africa (30')
  N Manyala   Extra-ordinarily large Hall effect in mono-silicides (30')

             Atomic and Nuclear Physics in the study of Diamond (and diamond-like materials)
             Chairman : A Kinyua
  JB Malherbe   -Bombardment induced topography on semiconductor surfaces (30')
  J Oyedele   -The bias in thickness calibration employing penetrating radiation (30')
  F Nabarro   -The activation volume for shear (30')
  K Purser   -How old is old carbon ? (30')

Evening   Poster Session and the "Wines of the World"
             Chairman : R Tegen (Diamond Room)

  B Becker   -Investigation of final state interactions by means of the inclusive ($^{12}C,^{7}Be$) and ($^{12}C,^{9}Be$) reactions on ^{59}Co and ^{93}Nb.
  S Cross   -Localised Solutions to the Parametrically Driven Complex Ginzburg-Landau Equation
  AM da Costa   -The technology of electronic contacts for diamond radiation detectors.
  RK Dutta   -The Schonland Nuclear Microprobe: an important tool for Geosciences.

CG Fischer	-Positrons in Diamond.
H Genz	-Investigation of channeling in plasma acceleration regime
R Groess	-Producing and Studying Polarization with 100 GeV gamma rays.
T Jili	- Characterisation of Ar+ - Implanted LiF by a Variable Energy Slow Positron Beam.
KA Korir	-Prompt gamma Doppler shift measurements of momentum transfer to residues after heavy ion collisions.
L Lekala	-Few Body Problem in Nuclear Physics Models.
IZ Machi	-Muon(ium) in Ia and ^{13}C Diamonds.
N Mhlahlo	-Studies of High Spin States with the AFRODITE
AU Naran	-Nuclear microprobe techniques for hydrogen diffusion studies in diamond.
DB Rebuli	-Resonance RBS and Channeling-FRS in diamond thin crystals.
M Roberts	-Vacuum Energy.
M Tredoux	-Nuclear methods and the development of platinum-group element geochemistry.
S Yacoob	-Determination of the Interaction Volume in Relativistic Heavy Ion Collisions from Particle Multiplicities
A Zucchiatti	-π^0 and η photo-production on the proton at GRAAL.

Thursday, 16 November 2000

09:00-17:30 Conference excursion: Elizabeth Bay, Agate beach and Atlas Bay, Peninsula and Kolmanskop

Evening Banquet (Diamond Room)

Friday, 17 November 2000

08:30-10:40 **Elementary Particle Physics**
Chairman : A Richter

G 't Hooft	-A confrontation with infinity (40')
H Stöcker	-Critical review of the Quark-gluon plasma (30')
S Schramm	-Chiral Model Calculations of Nuclear Matter and Finite Nuclei (30')
G Soff	-Parton showers and multijet events (30')

11:00-12:30 **Closing Ceremony - The Three Devils**

W Greiner	-For the scientific summary (30')
J Davies	-For the overseas delegates (30')
A Freund	-For Friedel (30')

LIST OF PARTICIPANTS

Ackermann, D	GSI Darmstadt, Darmstadt, Germany, d.ackermann@gsi.de
Adam, RM	Dept of Arts, Culture, Science & Technology, Pretoria South Africa adam@DACST5.pwv.gov.za
Adams, RM	University of Cape Town, Dept of Mathematics and Applied Mathematics, Rondebosch, South Africa, radams@maths.uct.ac.za
Andersen, J-U	University of Aarhus, Aarhus C, Denmark, jua@ifa.au.dk
Appel, H	Universitaet Karlsruhe, Karlsruhe, Germany, appel@itox.fzk.de
Backe, H	Institut für Kernphysik, Johann-Gutenberg-Universität, Mainz, Mainz, Germany, backe@kph.uni-mainz.de
Becker, B	University of Cape Town, Department of Physics, Rondebosch, South Africa, becker@physci.uct.ac.za
Bharuth-Ram, K	University of Durban-Westville, Department of Physics, Durban, South Africa, kbr@pixie.udw.ac.za
Biersack, J	Hahn Meitner Institut, Berlin, Germany, biersack@hmi.de
Brandt, T	University of Cape Town, Dept of Maths and Applied Maths, Rondebosch, South Africa,
Butler, JE	James Naval Research Laboratory, Washington DC, USA, butler@ccf.nrl.navy.mil

Chatu, A	University of Namibia, Department of Physics, Windhoek, Namibia, oyedelej@unam.na
Cleymans, J	University of Cape Town, Physics Department, Rondebosch, South Africa, cleymans@physci.uct.ac.za
Coffin, JP	Institut de Recherches Subatomiques (IReS), Strasbourg, France, coffin@in2p3.fr
Comins, JD	University of Witswaterand, Department of Physics, Johannesburg, South Africa, comins@physnet.phys.wits.ac.za
Connell, SH	Schonland Research Centre for Nuclear Sciences, University of the Witwatersrand, Johannesburg, South Africa, connell@inkosi.src.wits.ac.za
Cross, S	University of Cape Town, Department of Applied Mathematics, Rondebosch, South Africa, scross@maths.uct.ac.za
Cumbane, J	Eduardo Mondlane University, Maputo, Mozambique,
Da Costa, A	Schonland Research Centre, University of the Witwatersrand, Johannesburg, South Africa, dacosta@schonlan.src.wits.ac.za
Dabrowski A	University of Cape Town, Dept of Physics, Rondebosch, South Africa, dabrowski@physci.uct.ac.za
Davies, JA	Mc Master University, Department of Physics, Hamilton, Ontario, Canada, davies@magma.ca
de Jager, O	Potchefstroom University, Space Research Potchefstroom, South Africa, okkie@fskocdj.puk.ac.za

de Wet, JA	Plettenberg Bay, South Africa, jadew@global.co.za
Doyle, B	Sandia National Laboratories, Radiation Solid Interactions and Processing Department, Albuquerque, USA, bldoyle@sandia.gov
Dutta, R	Schonland Research Centre, University of the Witwatersrand, Johannesburg, South Africa, raja@inkosi.src.wits.ac.za
Ellis, JR	CERN, Theory Division, Geneva 23, Switzerland, John.Ellis@cern.ch
Erba, E	INFN and Milan University, Milano, Italy, Erba@mi.infn.it
Fiase, JO	University of Botswana, Department of Physics, Gaborone, Botswana, fiasejo@mopipi.up.bw
Fischer, C	Schonland Research Centre, for Nuclear Sciences, University of the Witwatersrand, Johannesburg, South Africa, Fischer@src.wits.ac.za
Freund, A	European Synchrotron Radiation Facility, GRENOBLE, CEDEX, France, freund@esrf.fr
Friedland, E	University of Pretoria, Pretoria, South Africa, fried@scientia.up.ac.za
Gadioli, E	U Milano (INFN Milano), Cattedra di Fisica Nucleare, Milano, Italy, Gadioli@mi.infn.it
Gates, SJ	University of Maryland at College Park, Dept of Physics, Maryland, USA, gates@umdgrb.umd.edu

Genz, H	Instut fur Kernphysik, TU Darmstadt, Germany, genz@ikp.tu-darmstadt.de
Goheer, N	University of Cape Town, Department of Physics, Rondebosch, South Africa, goheer@physci.uct.ac.za
Greiner, C	Department of Physics, Justus-Liebig-Universitaet, Giessen, Giessen, Germany, carsten.greiner@theo.physik.uni-giessen.de
Greiner, W	University of Frankfurt AM Main, Frankfurt, Germany, Greiner@th.physik.uni-frankfurt.de
Groess, R	Schonland Research Centre, University of the Witwatersrand, Johannesburg, South Africa, groess@src.wits.ac.za
Hamilton, J	Vanderbilt University, Dept. of Physics, Nashville, Tennessee, USA, hamilton@nucax1.phy.vanderbilt.edu
Hansen, JO	DeBeers Industrial Diamonds, De Beers Diamond Research Laboratory, Synthesis Divisions, Johannesburg, South Africa, lmaharaj@debid.db.za
Hart, R	Schonland Research Centre, University of the Witwatersrand, Johannesburg, South Africa, hart@schonlan.src.wits.ac.za
Hodgson, P	Oxford University, Nuclear Physics Laboratory, Oxford, UK, p.hodgson1@physics.ox.ac.uk
Jili, T	Schonland Research Centre University of the Witwatersrand, Johannesburg, South Africa, tjili@pan.uzulu.ac.za
Kämpfer, B	FZ Rossendorf, Dresden, Germany, kaempfer@fz-rossendorf.de

Kalbitzer, S	Max Planck Institute for Nuclear Physics, Heidelberg, Germany, SKalbitzer@aol.com
Kienle, P	Technical Universitat Munich, Garching, Germany, paul.kienle@physik.tu-muenchen.de
Kinyua, A	University of Nairobi, Institute of Nuclear Science, Kenya, Nairobi, antonykinyua@insightkenya.com
Korir, I	Schonland Research Centre for Nuclear Sciences, University of Witwatersrand, Johannesburg, South Africa, korir@inkosi.src.wits.ac.za
Krewald, S	IKP, Forschungszentrum Julich, Juelich, Germany, s.krewald@fz-juelich.de
Lekala, L	UNISA, Dept of Physics, Pretoria, South Africa, lekala@harry.unisa.ac.za
Levitt, C	Schonland Research Centre, University of the Witwatersrand, Johannesburg, South Africa, levitt@src.wits.ac.za
Machi, IZ	UNISA, Physics Department, Pretoria, South Africa, machiiz@unisa.ac.za
Machner, H	Forschungszentrum Julich, Inst fur Kernphysik, Juelich, Germany, h.machner@fz-juelich.de
Malherbe, J	University of Pretoria, Pretoria, South Africa, malherbe@scientia.up.ac.za
Manyala, N	University of Lesotho, Department of Physics, Maseru, Lesotho, ni.manyala@nul.ls
Maure, G	Dept of Physics, Universidade Eduardo Mondlane, Maputo, Mozambique, gmaure@hotmail.com

Mhlahlo, N	University of Cape Town, Physics Department, Rondebosch, South Africa, mhlahlo@physci.uct.ac.za
Murugan, J	University of Cape Town, Department of Mathematics, Rondebosch, South Africa, jeff@cosmology.mth.uct.ac.za
Muundjua, M	University of Namibia, Department of Physics, Windhoek, Namibia, oyedelej@unam.na
Nabarro, FRN	University of the Witwatersrand, Department of Physics, P O WITS, Johannesburg, South Africa,
Naran, A	Schonland Research Centre for Nuclear Sciences, University of the Witwatersrand, Johannesburg, South Africa, amino@inkosi.src.wits.ac.za
Ngubane, B	Department of Arts, Culture, Science & Technology, Pretoria, South Africa, min01@dacst3.wcape.gov.za
Oyedele, J	University of Namibia, Windhoek, Namibia, oyedelej@unam.na
Paterson, A	Council for Scientific and Industrial Research, Brummeria, Pretoria, South Africa, apaterso@csir.co.za
Purser, KH	Kenneth, University of Toronto, Peabody, USA, kpurser@shore.net
Rebak, M	De Beers Diamond Research Laboratory, Johannesburg, South Africa, rebak@schonlan.src.wits.ac.za

Rebuli, D	Schonland Research Centre, for Nuclear Sciences, University of the Witwatersrand, Johannesburg, South Africa, rebuli@src.wits.ac.za
Reed, K	Lawrence Livermore National Laboratory, Livermore, California, USA, reed5@LLnL.gov
Richter, A	Darmstadt University of Technology, TU Darmstadt, Darmstadt, Germany, richter@ikp.tu-darmstadt.de
Roberts, M	University of Cape Town,
Rose, P	Orient Equipment Inc., Massachussetts, USA, Peter Rose@orionequip.com
Rosengard, U	International Atomic Energy Agency, Vienna, Austria, U.Rosengard@iaea.org
Schardt, D	GSI Darmstadt, Darmstadt, Germany, D.Schardt@gsi.de
Schramm, S	University of Frankfurt, Inst fur Theoretische Physik, Frankfurt, Germany, schramm@th.physik.uni-frankfurt.de
Sellschop, JPF	Schonland Research Centre for Nuclear Sciences, University of the Witwatersrand, Johannesburg, South Africa, sellschop@src.wits.ac.za
Sharpey-Schafer, J	National Accelerator Centre, Faure, South Africa, director@nac.ac.za
Sideras-Haddad, E	Schonland Research Centre, University of the Witwatersrand, Johannesburg, South Africa, haddad@schonlan.src.wits.ac.za

Sigmund, P	Odense University, Physics Department, Odense M, Denmark, psi@dou.dk
Soff, G	T U Dresden, Technische Universität, Dresden, Germany, soff@physik.tu-dresden.de
Solov'yov, A	Russian Academy of Science, St. Petersburg, Russia, solovyov@th.physik.uni-frankfurt.de
Steenkamp, R	University of Namibia, Department of Physics, Windhoek, Namibia, rsteenkamp@unam.na
Stöcker, H	Goethe University, Frankfurt (Main), Germany, stoecker@th.physik.uni-frankfurt.de
't Hooft, G	Spinnoza Institute, Utrecht University, Utrecht, Netherlands, G.tHooft@phys.uu.nl
Tegen, R	University of the Witwatersrand, Johannesburg, South Africa, tegen@physnet.phys.wits.ac.za
Toepffer, C	Erlangen Universitaet, Physik, Erlangen, Germany, toepffer@theorie2.physik.uni-erlangen.de
Tredoux, M	University of Cape Town, Geology, Rondebosch, South Africa, mtd@geology.uct.ac.za
Uggerhoj, E	Institute of Synchrotron Radiation, University of Aarhus, Aarhus C, Denmark, ugh@ifa.au.dk
Uthui, R	HOD Physics Dept, Universidade Eduardo Mondlane, Maputo, Mocambique, ruthui@rei.uem.mz

Velasco, M	Department of Physics and Astronomy, Northwestern University, Illinois, USA, mayda.velasco@cern.ch
Vilakazi, Z	University of Cape Town, Cape Town, South Africa, vilakazi@physci.uct.ac.za
Weltman, A	University of Cape Town, Rondebosch, South Africa, Weltman@physci.uct.ac.za
Yacoob, S	University of Cape Town, Dept of Physics, Rondebosch, South Africa, sahal@hostess.phy.uct.ac.za
Zucchiatti, A	Istituto Nazionale di Fisica Nucleare, Genova, Italy, zucc@ge.infn.it
Zwicknagel, G	Erlangen Universität, Physik, Erlangen, Germany, zwicknagel@theorie2.physik.uni-erlangen.de

Author Index

Ackermann D, 18
Adam R M, 474
Alice Collaboration, 301
Andreoli M A G, 543
Appel H, 571
Andersen J U, 78
Backe H, 123
Bartalini O, 510
Barashenkov I V, 502
Bass S, 332
Bassini R, 535
Becattini F, 248
Becker B, 535
Beckmann C, 346
Bellini V, 510
Betigeri M, 62
Beyer C J, 11
Bharuth-Ram K, 517
Birattari C, 1, 535
Bleicher M, 332
Bocquet J P, 510
Bojowald J, 62
Bouquillon A, 441
Brachmann J, 332
Britton D T, 525
Budzanowski A, 62
Butler J E, 552
Capogni M, 510
Castaing J, 441
Castoldi M, 510
Cavinato M, 1, 535
Chatterjee A, 62
Chubarian G, 11
Clawiter N, 123
Cleymans J, 248
Coffin J P, 301
Cole J D, 11

Connell S H, 1, 517, 525, 535, 543, 552
Cowley A A, 1
Cross S D, 502
D'Angelo A, 510
Dambach S, 123
Daniel A V, 11
Davies J A, 577
de Jager O C, 411
Derry T E, 552
Didelez J P, 510
Di Salvo D, 510
Dodd P E, 188
Donangelo R, 11
Doyle B L, 188
Drigert M, 11
Dutta R K, 543
Ellis J, 219
Ernst J, 62
Euteneuer H, 123
Fabrici E, 1, 535
Fiase J O, 70
Fischer C G, 525
Formichev A S, 11
Förtsch V, 1, 535
Formenti P, 543
Freindl L, 62
Frekers D, 62
Freund A, 135, 582
Gadioli E, 1, 535, 569
Gadioli Erba E, 1, 535
Gallmeister K, 309
Garske L, 62
Gates S J, 235
Gerland L, 332
Gervino G, 510
Ghio F, 510

Ginter T N, 11
Girolami B, 510
Gore P M, 11
Greiner C, 363
Greiner W, xii, 11, 115, 332, 346, 373
Grewer K, 62
Guidal M, 510
Hagenbuck F, 123
Hamilton J H, 11, 566
Hamachner A, 62
Hart R J, 543
Hodgson P E, 42
Horn K M, 188
Hosaka A, 70
Hourany E, 510
Hwang J K, 11
Ilieva J, 62
Jandel M, 11
Janssens R V F, 11
Jarczyk L, 62
Jones E F, 11
Kaiser K-H, 123
Kalbitzer S, 422
Kämpfer B, 309
Keränen A, 248
Kettig O, 123
Kienle P, 28, 568
Kilian K, 62
Kinyua A M, 482
Kliczewski S, 62
Klimala W, 62
Kliman J, 11
Knaup M, 160
Kolev D, 62
Kormicki J, 11
Korir K A, 535
Korol A V, 115
Kouznetsov V, 510
Krause W, 115
Krauss F, 354
Krehl O, 54

Krewald S, 54
Krupa L, 11
Kube G, 123
Kuhn R, 354
Kunne R, 510
Kutsarova T, 62
Lanterna G, 441
Lapik A, 510
Lauth W, 123
Lawrie J J, 1, 535
Lee I Y, 11
Levi Sandri P, 510
Lieb J, 62
Litherland A E, 457
Lleres A, 453
Lucarelli F, 441
Ma W C, 11
Mabala G K, 535
Machi I Z, 517
Machner H, 62
Magiera A, 62
Malherbe J B, 201
Mando' P A, 441
Moricciani D, 510
Nabarro F R N, 211
Nann H, 62
Nedorezov V, 510
Newman R T, 535
Nicoletti L, 510
Nilen R W N, 525
Nortier F M, 1
Odendaal Q, 201
Oganessian Yu Ts, 11
Oyedele J A, 449
Paech K, 332
Paterson A, 493
Pavlenko O P, 309
Pentchev L, 62
Piercey R B, 11
Plendl H S, 62
Poenariu D, 11
Prati P, 441

Protić D, 62
Purser K H, 457
Ramayya A V, 11
Rasmussen J O, 11
Razen B, 62
Rebreyend D, 510
Rebuli D B, 552
Redlich K, 248
Reinhard P-G, 160
Renard F, 510
Richter A, 87
Rodin A M, 11
Roy B J, 62
Rudnev N, 510
Salomon J, 441
Schaerf C, 510
Schälicke A, 354
Schardt D, 433
Scherer S, 332
Schinner A, 178
Schramm S, 346
Sellschop J P F, 1, 517, 525, 535, 543, 552
Seweryniak D, 11
Sharma L K, 70
Sideras-Haddad E., 1, 535, 543, 552
Sigmund P, 178
Siudak R, 62
Smit F D, 535
Soff S, 332, 354
Solov'yov A V, 115
Sperduto M L, 510

Speth J, 54
Spieles C, 332
Star Collaborations, 301
Steenkamp R, 403
Steyn G.F., 1, 535
Stöcker H, 332, 346
Strzaikowski A, 62
Suhonen E, 248
Ter-Akopopian G M, 11
't Hooft G, 317
Toepffer C, 160, 573
Tredoux M, 543
Tsenov R, 62
Turinge A, 510
Uggerhøj E, 258
Vaccari M G, 441
Velasco M M, 269
Vilakazi Z, 535
Von Rossen P, 62
Walcher TH, 123
Walsh D S, 188
Wang Z S, 54
Weber H, 332
Winkoun P, 70
Wittenberg A, 543
Wu S-G, 11
Zhang X Q, 11
Zhu S J, 11
Zschiesche D, 332, 346
Zucchiatti A, 441, 510
Zwicknagel G, 168
Zwoll K, 62